T0346041

ORCHIDS
of MADAGASCAR SECOND EDITION

ORCHIDS
of MADAGASCAR SECOND EDITION

ANNOTATED CHECKLIST

Johan & Clare Hermans, David Du Puy, Phillip Cribb & Jean Bosser

ANNOTATED BIBLIOGRAPHY

Johan & Clare Hermans

Kew Publishing

Royal Botanic Gardens, Kew

PLANTS PEOPLE POSSIBILITIES

First published in 2007 by
Royal Botanic Gardens, Kew
Richmond, Surrey, TW9 3AB, UK
www.kew.org

ISBN 978-1-84246-133-4

British Library Cataloguing in Publication Data
A catalogue record for this book is available from the British Library

Typesetting and page layout: Christine Beard, Kew Publishing

Cover design by Jeff Eden

FSC
Mixed Sources
Product group from well-managed
forests and other controlled sources
Cert no. TT-COC-002313
www.fsc.org
© 1996 Forest Stewardship Council

This book is printed on paper that has been manufactured using wood from well managed forests certified in accordance with the rules of the Forest Stewardship Council.

Printed in the United Kingdom by B.A.S. Printers

For information or to purchase all Kew titles please visit www.kewbooks.com or email publishing@kew.org

All proceeds go to support Kew's work in saving the world's plants for life

CONTENTS

LIST OF COLOUR PLATES

Key to photographers: J. Hermans (JH), D. Du Puy (DD), P. Cribb (PC), M. Grübenmann (MG).

FOREWORD

Madagascar has one of the richest and most distinctive orchid floras anywhere in the world. The island boasts almost 1000 species in 60 genera. Of these, 17% of the genera and nearly 90% of the species are endemic. Visitors to the island are almost certain to see orchids on their visit. The well-known nature reserves and parks are rich in species. Some orchids are spectacular and well-known in cultivation but most are cryptic and difficult to identify and name.

The first edition of *The Orchids of Madagascar* appeared in 1999 and was sponsored by the Weston Foundation. The preparation of this edition was undertaken as part of the Madagascan Threatened Plants Project between the Parc Tzimbazaza in Antananarivo and the Royal Botanic Gardens, Kew, funded by the Friends and Foundation of Kew. Orchids were one of the three target groups, the others being palms and succulents. Detailed field and propagation studies were undertaken on a select number of threatened orchid species, with a view to their eventual re-introduction. Re-introducing orchids is not simple, especially when the number of plants in the wild has been reduced to less than 50, as was the case in several of the target species. With a narrow genetic base, it is important to breed with the right plants from the appropriate populations and to find appropriate means of raising the seedlings to maturity. It is also necessary to understand the ecology of the orchid well before re-introduction can occur. Finally, re-introduction must be followed by protection and monitoring to assess the success or otherwise of the project.

Since the publication of the first edition of *The Orchids of Madagascar* (Du Puy *et al.*, 1999), considerable progress has been made in understanding the island's rich orchid flora. Notable additions to the flora include the recognition of two new genera *Bathiorchis* and *Paralophia*. The genus *Disperis* in Madagascar, the Comoros, Seychelles and Mascarenes has been revised, six of the 22 species being newly described. New taxa have also been established in *Aeranthes*, *Angraecum*, *Bulbophyllum*, *Cynorkis*, *Neobathiea* and *Satyrium*, whereas new names have been coined in *Bulbophyllum*, *Cynorkis*, *Habenaria* and *Liparis*. Many of these novelties and changes are validated by Johan Hermans in the Appendix to this volume. Included are those taxa, named by Perrier de la Bâthie but never validly published, that were listed in the first edition as *nomina nuda*, lacking Latin descriptions. New distributional records for many species have also been obtained from a detailed examination of Madagascan herbarium collections in Brussels, Geneva, the Natural History Museum, Kew, Missouri, Paris, Tzimbazaza and Vienna. Consequently the authors have been able to provide more detailed range and habitat data for many species in the checklist.

The literature on Madagascan orchids has continued to grow apace. In the Bibliography by Johan and Clare Hermans, all recent publications have been added, together with a brief summary of their contents. For the literature cited and summarized in the first edition of this work, they have only included the authors, titles and bibliographic references. The Annotated Bibliography would not have been possible without the help of the staff of the libraries at Kew, Paris, Tzimbazaza and Vienna, whose librarians we would particularly like to thank.

In the course of the Threatened Plants Project, the team visited many protected areas. The majority of local park rangers and guides are keen to learn more about orchids and to be able to name them, but the lack of guide books with good photographs hampers their work. The two parts of Henri Perrier de la Bâthie's seminal orchid account in the *Flore de Madagascar* were published in 1939 and 1941 and have long been out-of-print. Second-hand copies, when available, fetch high prices and are well out of the range of nearly everyone. The first edition of *The Orchids of Madagascar* was aimed mainly at orchid scientists and growers. This second edition should appeal to a broader audience and be more useful in Madagascar. It has twice the number of photographs of species than appeared in the earlier edition. The authors hope that this will assist identification of a broader range of species and genera than was possible using the first edition. It is, however, an important step on the way to producing a field guide, which is in preparation, that will satisfy a broader group, including forest rangers, guides and tourists interested in Madagascar's unique plant life of which orchids make up nearly 8%.

Professor Stephen Hopper
Director,
Royal Botanic Gardens, Kew

April 2007

PREFACE

The first edition of this Checklist and Bibliography, published eight years ago, has now sold out. The work sparked a renewed interest in the orchids of Madagascar and surrounding islands and a considerable amount of additional information has become available since its publication. Progress has also been made in the conservation of the orchids and their habitats; a well-publicised appeal by the Royal Botanic Gardens, Kew has provided a new awareness of the challenges and brought funding for a number of valuable projects. Considering this interest it was felt important to produce a new edition of the Checklist and Bibliography as soon as possible. This edition is another step towards a full revision of the orchid flora of the region.

A number of oversights and errors have been amended in this edition. The Checklist has been updated, and idiosyncrasies in nomenclature have been resolved. Much additional information on distribution, habitat and flowering time has been extracted from herbarium collections and incorporated. The long-standing problems with Schlechter's and Perrier's holotypes and isotypes between the Berlin and Paris Herbaria have been researched and resolved; it was found that most herbarium sheets that had been sent to Berlin were later returned to Paris. Lectotypes have been assigned where necessary and, where the authors felt qualified to do so, neotypes have also been chosen. Short diagnostic notes have been added to each species: these should not been seen as a definitive description, but a summary of a few obvious comparative characteristics. The number of photographs has also been greatly increased. Entries in the bibliography that were included in the first edition have been reduced to a reference only, new entries are accompanied by a concise annotation. We hope that this new edition will encourage more interest in the flora of Madagascar, especially the orchids, and encourage further conservation efforts.

The phytogeographical regions cited refer to those on Plate 59. Bracketed numbers are references to the bibliography.

Acknowledgements

We would like to express our sincerest gratitude to the authorities and staff of the many herbaria and libraries that were consulted during our work. We specifically acknowledge Jean-Noel Labat, Philippe Morat and Marc Pignal of the Laboratoire de Phanérogamie at the Muséum National d'Histoire Naturelle and Simon Owens, Keeper of the Herbarium at the Royal Botanic Gardens, Kew for allowing access to the collections, and to all their staff for their friendly assistance.

We are also most grateful to Mr. Paul Ormerod of Queensland, Australia for bringing a number of nomenclatural problems to our attention and Melanie Thomas for kindly checking the Latin. We also gratefully acknowledge Mrs. Joyce Stewart who provided valuable observations, especially on the genus *Aerangis*.

We would also like to thank the Weston Foundation and the Kew Friends and Foundation for their financial support for this work.

DISCOVERING THE ORCHIDS OF MADAGASCAR

A brief history of orchid exploration in Madgascar is outlined here, featuring those who have made the most significant contributions to knowledge. A more detailed account of those involved can be found in the following publications: {193}, {272}, {529}, {568}.

The Early Explorers

The Frenchman Etienne de Flacourt (1607 – 1660) was one of the first to write about the plants of Madagascar. Flacourt studied chemistry, medicine and botany before he travelled to the island in 1648 where he joined a small French settlement in Fort Dauphin (Taolanaro) in the South. He collected geological and biological specimens, including plants. In 1655 he returned to France to argue for an extended interest in the developing colony and this was eventually agreed. However, returning to Madagascar, his ship was attacked by pirates and Flacourt was killed. As a legacy he left an expansive tome on the history of the island {297}, which includes a listing of a number of medicinal and other useful plants. It is difficult to ascertain which are orchids and which are not but Flacourt describes various roots of flowering plants which resemble *Cynorkis* species. Identifiable with more certainty is the plant known locally as '*Singofau*' that Flacourt described as a fan-shaped plant attached to a tree with large leaves that resemble fingers pointing upwards. An extract from the leaves was rubbed around the eyes and was said to clear the sight. A variation on this name '*Tsingolo*' is still used in Madagascar as a general name for orchids {240}. Flacourt's description is likely to refer to an *Angraecum*.

Aubert Aubert du Petit Thouars (1758 – 1832).

The first person to explore the island systematically for orchids and other plants was the Frenchman Aubert Aubert du Petit Thouars (1758 – 1832). In the 1790s he and his brother Aristide sold their properties and decided to equip a ship and to go in search of foreign lands. However, the French revolution had started and before Aubert could reach the ship he was arrested. By the time he was released, Aristide had left on the voyage. Aubert tried to meet up with his ship at the island of Mauritius. As it turned out they never met again; Aristide was killed by Nelson in a sea battle at Alexandria and became a French national hero. Aubert on the other hand, remained on Mauritius for almost 10 years, six-months of which was spent on neighbouring Madagascar where he collected and studied a great number of plants. He also devised his own, idiosyncratic system of nomenclature. On his eventual return to France in 1802 he developed this system of naming further and published several books, including a finely illustrated work on the orchids of Mauritius, Réunion and Madagascar {1398}. Some of the plants that Thouars first described include: *Aerangis citrata* (Thouars) Schltr., *Angraecum sesquipedale* Thouars, *Bulbophyllum occultum* Thouars, *Cymbidiella flabellata* (Thouars) Lindl., *Cynorkis fastigiata* Thouars, *C. purpurascens* Thouars, *Gastrorchis tuberculosa* (Thouars) Schltr., *Graphorkis concolor* var. *alphabetica* F.N.Rasm., *Liparis purpurascens* (Thouars) Lindl., *Oeonia volucris* (Thouars) Spreng., *Oeoniella polystachys* (Thouars) Schltr. and *Solenangis aphylla* (Thouars) Summerh.

The Nineteenth Century

Robert Lyall, a botanist and diplomat born in Scotland in 1790, was appointed British agent to Madagascar in 1827. He arrived with his family in Mauritius and proceeded to Madagascar to see King Radama I and negotiate over trade and slavery. Unfortunately, before he could start talks the king died. Radama's successor, the fearsome Queen Ranavalona I, refused to recognise Lyall. Whilst waiting for a change of heart he spent time collecting plants in the Eastern Highlands and along the East coast. Lyall was finally accused of sorcery and expelled in 1829, one of the reasons quoted for this being that he "… transgressed by sending his servant to catch butterflies, reptiles and botanical specimens …" {861}. Lyall died two years later from malaria. A few of his introductions, all described by John Lindley (1799 – 1865) are *Cynorkis uniflora* Lindl., *Disa incarnata* Lindl. and *Satyrium trinerve* Lindl.

Around the same time the German Carl Hilsenberg (1802 – 1824) and the Bohemian Wenceslas Bojer (1795 – 1856) undertook a short expedition from Mauritius to Madagascar. Some of the orchids they discovered were later described by Henry Ridley (1855 – 1956), including *Cynorkis angustipetala* Ridl. and *Habenaria hilsenbergii* Ridl.

The heyday of orchid exploration in Madagascar started in the middle of the 19th century. It not only saw an influx of botanising missionaries but also of commercial collectors who were in search of novelties to satisfy a growing demand for exotic plants in Europe. The most eminent of the missionaries, the Reverend William Ellis (1794 – 1872), worked for the London Missionary Society and was sent to Madagascar to reintroduce Christian teaching. After several attempts he was finally given permission to proceed inland. During the long waits by the coast of Madagascar he used his time collecting plants to bring back to England. Amongst the treasures introduced by him were *Aerangis articulata* (Rchb.f.) Schltr., *A. ellisii* (B. S. Williams) Schltr. and *Grammangis ellisii* (Lindl.) Rchb.f. The latter he refers to in his book: " my greatest treasure was a large bulbed plant of quite a new species, it had a large flower stalk, a seed pod the size of an orange and the natives said the flower was scarlet and purple" {287}.

The Reverend Richard Baron (1847 – 1907), also of the London Missionary Society, spent 30 years on Madagascar, travelling widely and collecting a large number of plants. Many were later described by Henry Ridley. Baron also compiled a catalogue of the orchids of Madagascar {42}. His finds include *Angraecum madagascariensis* (Finet) Schltr., *Bulbophyllum baronii* Ridl., *Cynorkis lilacina* Ridl. and *C. nutans* (Ridl.) H.Perrier. Baron's main base on the island was the town of Fianarantsoa, where he buried his second wife, who died of typhoid aged 28 after only a few months of marriage. Baron himself died of a fever in 1907.

Another missionary, the Reverend William Deans-Cowan (1844 – 1923), worked on the island at about the same time as Baron. Again, Ridley described the new species he found. Deans-Cowan's sketchbook at the Natural History Museum in London features some of the orchids first collected by him, including *Cynorkis gibbosa* Ridl., *Eulophia pileata* Ridl., *Liparis ornithorrhynchos* Ridl., *Oeonia rosea* Ridl. and *Polystachya rosea* Ridl.

Johann Hildebrandt (1847 – 1881), a German botanist and explorer, first arrived in Madagascar

Reverend William Ellis (1794 – 1872).

Reverend Richard Baron (1847 – 1907).

in 1879 with the intention of finding the remains of Diedrich Rutenberg (1851 – 1878), another collector, who had been killed by his porters the previous year. Hildebrandt failed to locate him but he did find his herbarium material and notes. From this material emerged, for example, *Angraecum rutenbergianum* described shortly afterwards by the German Fritz Kraenzlin (1847 – 1934). Hildebrandt ruined his health during these Madagascan explorations. H. G. Reichenbach (1823 – 1889) in his description of *Microterangis hildebrandtii* (Rchb.f.) Senghas declared that Hildebrandt "... is just now home, but only for the improvement of his health, so that he may once more brave all the dangers and perils of mysterious grand tropical Africa. Our best wishes will follow the keen traveller wherever he may wander" {982}. Three years later Hildebrandt died of malaria in Madagascar. He is remembered for first discovering orchids like *Bulbophyllum hildebrandtii* Rchb.f., *B. occlusum* Ridl., *Cynorkis ridleyi* T. Durand & Schinz and *Lemurorchis madagascariensis* Kraenzl.

Johann Hildebrandt (1847–1881).

One of the first explorers and naturalists to systematically send orchids into the European trade was the Frenchman Léon Humblot (1852 – 1914). For some time he worked for the firm of Frederick Sander of St. Albans. Letters from him to Sander (1847 – 1920) survive in the Archives at the Royal Botanic Gardens Kew {571} and contain valuable information on early importations and endless packing lists. They are also rather sad in their unrelenting demand for payment from Sander who was obviously stalling attempts at parting with money. In a letter of 1882 Humblot recorded sending 9 crates with 1000 *Aerangis citrata*, and also that he was leaving soon for the interior to find *Gastrorchis humblotii* (Rchb. f.) Schltr. which had been first found by him and described by Reichenbach a few years earlier. Other discoveries of his were *Aerangis fastuosa* (Rchb.f.) Schltr., *Angraecum germinyanum* Hook. f., *Cymbidiella falcigera* (Rchb.f.) Garay, *C. pardalina* (Rchb.f.) Garay and *Neobathiea grandidieriana* (Rchb.f.) Garay, named by Reichenbach for a contemporary of Humblot, the naturalist Alfred Grandidier (1836 – 1921). Humblot's system of numbering his collections is somewhat confusing but his catalogues kept in the Library of the Herbarium of the Museum d'Histoire Naturelle in Paris tabulate his different systems {572}.

A less attractive side of humanity comes with the French collector L. Hamelin (fl.1890 – 1900). His career centred around the wholesale importation of *Eulophiella elisabethae* Linden & Rolfe, an attractive species from Eastern Madagascar. Hamelin had contacted the Belgian firm of Linden to tell them of the death of another collector named Sallerin (fl. 1890s) who according to Hamelin "had drowned when his pirogue had been upset". Sallerin had discovered *Eulophiella elisabethae* and Hamelin offered his exclusive services to find more plants for Linden. After some months a few inferior pieces of the *Eulophiella* were offered to Linden. But in the meantime strange stories started to appear in the horticultural press how poor Hamelin had almost been killed by ferocious wild beasts and unfriendly locals, how he had to marry a chief's daughter and so on. News also spread that he had destroyed all plants of the *Eulophiella* in its native habitat. Suddenly and not by coincidence, quantities of *Eulophiella* started to turn up at Sander's nursery in England where they fetched high prices. Following this episode there follows a vitriolic exchange in the press involving Lucien Linden (1851 – 1940), Robert Allen Rolfe (1855 – 1921), then editor of the *Orchid Review*, Baron and Hamelin. This debacle is fully described in {467}.

Another little-known 19th century collector is W. R. Warpur (fl. 1900); even his nationality is uncertain, he was either Dutch or Flemish. A number of his plants were introduced into cultivation and described by Rolfe at Kew. Warpur's most important contribution is that he was one of the few explorers to observe and note a plant's habitat and growing conditions, information that he shared in the Orchid Review {1443}. Warpur's discoveries include *Calanthe madagascariensis* Watson ex Rolfe and *Cynorkis villosa* Rolfe ex Hook f.

The Twentieth Century

The twentieth century in Madagascar is largely dominated by the French; from 1896 until the Second World War they colonised Madagascar. By far the most important person collecting and describing Madagascan orchids was Henri Perrier de la Bâthie (1873 – 1958). Perrier came from a family of naturalists. He arrived in Madagascar in 1896 where he initially worked as a prospector for a gold-mining company, but from 1906 he started work for the Colonial Service as a geologist. He was also sent on a number of expeditions for the Museum National d'Histoire Naturelle in Paris. Initially, the orchids that Perrier discovered were described by Rudolf Schlechter (1872 – 1925). Later, and certainly after Schlechter's death in 1925, Perrier described his own discoveries. This culminated in his two-volume orchid account for the *Flore de Madagascar* {896}. During his travels Perrier reached the furthest corners of the island, bringing back numerous new species. He was responsible for hundreds of new additions to the Malagasy flora, including the orchids *Angraecum obesum* H.Perrier, *A. praestans* Schltr., *A. rhynchoglossum* Schltr., *Bulbophyllum alexandrae* Schltr., which was named after Schlechter's wife Alexandra Sobennikoff, *B. leandreanum* H.Perrier, *Gastrorchis françoisii* Schltr., *G. pulchra* Humbert & H.Perrier, *Microcoelia macrantha* (H.Perrier) Summerh., *Neobathiea perrieri* (Schltr.) Schltr., *Oeonia brauniana* var. *sarcanthoides* (Schltr.) Bosser, *Polystachya rhodochila* Schltr., *P. tsinjoarivensis* H.Perrier and *Physoceras violaceum* Schltr.

Several other colonial botanists and explorers collaborated with Perrier de la Bâthie and supplied plant material, they are often remembered in orchids named for them: Karl Afzelius (1887 – 1971) with *Goodyera afzelii* Schltr.; Charles Alleizette (1884 – 1967) with *Angraecum alleizettei* Schltr.; Edmond François (1882 – 1962) with *Bulbophyllum françoisii* H.Perrier; the infatigable Henri Humbert (1887 – 1967) with *Sobennikoffia humbertiana* H.Perrier; Henri Poisson (1877 – 1963) with *Sobennikoffia poissoniana* H.Perrier; René Viguier (1880 – 1931) with *Angraecum viguieri* Schltr.; Georges Waterlot (1877 – 1939) with *A. waterlotii* H.Perrier. Henri Jumelle (1866 – 1935), Professor of Botany and Director of the Botanic Garden in Marseilles, was slightly different. Although there is no evidence that he visited Madagascar, he collaborated extensively with Perrier on a number of publications. The genus *Jumellea* Schltr. was named for him. Raymond Decary (1891 – 1973) served as a military officer and administrator on the island in the 1930s. On his travels he reached some of the most inaccessible areas of the island and discovered a number of new orchids including *Aerangis decaryana* H.Perrier, *Cryptopus paniculatus* H.Perrier and *Oeceoclades decaryana* (H.Perrier) Garay & Taylor.

After Madagascar's full independence in 1960 French influence and research continued at a

Henri Perrier de la Bâthie (1873 – 1958).

Rudolf Schlechter (1872 – 1925).

Henri Humbert (1887 – 1967).

Henri Jumelle (1866 – 1935).

slower pace. A most respected taxonomist to work on Madagascar's orchids is Jean Bosser of the Laboratoire de Phanérogamie in the Paris Natural History Museum. He undertook extensive field work on the island but is best known for his revisions of a number of genera. He described several new entities like *Angraecum rubellum* Bosser, *Grammangis spectabilis* Bosser and *Phaius pulchellus* var. *sandragatensis* Bosser. One of the people who supplied Bosser with plants was an orchid enthusiast Jean-Pierre Peyrot, honoured with *Aeranthes peyrotii* Bosser, *Cynorkis peyrotii* Bosser and *Gastrorchis peyrotii* (Bosser) Senghas.

Another Madagascan orchid devotee is Marcel Lecoufle who, from his nursery near Paris, has brought Madagascan orchids to the enthusiasts' attention via his lectures and articles.

The last few decades have also seen some non-French people venturing into the Madagascan orchid field. Fred Hillerman, an American orchid-grower, published a work on the Madagascan angraecoids {529}. The German taxonomist Karlheinz Senghas described several new species from the island including *Angraecum dollii* Senghas and *Jumellea densefoliata* Senghas.

Since the publication of the first edition of this work much research and fieldwork has been done on the orchid flora of Madagascar and the Comoro Islands, not only on taxonomy and genetics but also on conservation, pollination and distribution patterns. There is no doubt that work by scientists at the Laboratoire de Phanérogamie of the Muséum d'Histoire Naturelle, the University and Parc Tsimbabaza in Antananarivo, the Missouri Botanic Garden and the Royal Botanic Gardens, Kew, are contributing greatly to our knowledge of the fascinating orchid flora of Madagascar.

Jean Bosser.

SUMMARY OF CHANGES IN NOMENCLATURE

Aerangis ellisii (*B.S.Williams*) *Schltr.*, Die Orchideen: 598 (1914).
Aerangis ellisii var. *grandiflora* J.Stewart in Amer. Orch. Soc. Bull. 55 (8): 798 (1986), **syn. nov.**

Aerangis mooreana (*Rolfe*) *R.J.Cribb & J.Stewart* in Orch. Rev. 91: 218 (1983)
Aerangis karthalensis R.Neirynck & M.Herremans in Caesiana., 26(2): 3 (2006), **syn. nov.**

Aeranthes aemula *Schltr.* in Repert. Sp. Nov. Regni Veg. Beih. 33: 274 (1925).
Aeranthes biauriculata H.Perrier in Notul. Syst. (Paris) 14: 161 (1951), **syn. nov.**

Aeranthes henricii *Schltr.* var. **isaloensis** *H.Perrier ex Hermans*, **var. nov.**

Agrostophyllum occidentale *Schltr.* in Beih. Bot. Centralbl. 33 (2): 413 (1915).
Agrostophyllum seychellarum Rolfe in Kew Bull. (Bull. Misc. Info.), 23 (1): 23 – 24 (1922), **syn. nov.**

Angraecum acutipetalum *Schltr.* var. **analabeensis** *H.Perrier ex Hermans* **var. nov.**

Angraecum acutipetalum *Schltr.* var. **ankeranae** *H.Perrier ex Hermans* **var. nov.**

Angraecum huntleyoides *Schltr.* in Bot. Jahrb. Syst. 38: 160 (1906).
Angraecum chloranthum Schltr. in Ann. Mus. Col. Marseille, sér. 3, 1: 189, t.23 (1913), **syn. nov.**

Angraecum calceolus *Thouars*, Hist. Orch.: t.78 (1822).
Angraecum laggiarae Schltr. in Repert. Sp. Nov. Regni Veg. Beih. 15: 337 (1918), **syn. nov.**

Angraecum pingue *Frapp.*, Orch. Réunion, Cat. Esp. Indig.: 13 (1880).
Angraecum nasutum Schltr. in Repert. Sp. Nov. Regni Veg. Beih. 33: 315 (1925), **syn. nov.**

Benthamia herminioides *Schltr.* in Repert. Sp. Nov. Regni Veg. Beih. 33: 27 (1924).
Benthamia herminioides forma *sulfurea* H.Perrier in Mem. Inst. Sc. Mad. 6: 253 (1955), *nom. nud.*

Bulbophyllaria pentastichum (*Pfitzer ex Kraenzl.*) *Schltr.* in Beih. Bot. Centralbl. 33 (2): 419 (1915).
Bulbophyllaria pentasticha Pfitzer ex Kraenzl. in Orchis 2: 135 (1908), *Bulbophyllum pentastichum* (Pfitz. ex Kraenzl.) Rolfe in Orch. Rev. 23: 181 (1915) & *B. matitanense* H.Perrier in Notul. Syst. (Paris) 6 (2): 85 (1937), **syn. nov.**

Bulbophyllum bryophilum Hermans **nom. nov.**
Bulbophyllum muscicola Schltr. in Ann. Mus. Col. Marseille, sér. 3, 1: 179, t.15 (1913).

Bulbophyllum henrici *Schltr.* var. **rectangulare** *H.Perrier ex Hermans* **var. nov.**

Bulbophyllum pentastichum (*Pfitzer ex Kraenzl.*) Schltr.
Bulbophyllum matitanense H.Perrier in Notul. Syst. (Paris) 6 (2): 85 (1937), **syn. nov.**

Bulbophyllum nitens *Jum. & H.Perrier* in Ann. Fac. Sci. Marseille 21 (2): 209 (1912).
Bulbophyllum nitens var. *minus* H.Perrier in Notul. Syst. (Paris) 6 (2): 112 (1937), *nom. nud.*

Bulbophyllum ophiuchus *Ridl.* var. **baronianum** *H.Perrier ex Hermans* **var. nov.**

Bulbophyllum ormerodianum *Hermans* **nom. nov.**
Bulbophyllum abbreviatum Schltr. in Repert. Sp. Nov. Regni Veg. Beih. 33: 198 (1924).

Bulbophyllum pantoblepharon *Schltr.* var. **vestitum** *H.Perrier ex Hermans* **var. nov.**

Bulbophyllum pentastichum (*Pfitzer ex Kraenzl.*) *Schltr.* subsp. **rostratum** *H.Perrier ex Hermans* **subsp. nov.**

Bulbophyllum perseverans *Hermans* **nom. nov.**
Bulbophyllum graciliscapum H.Perrier in Notul. Syst. (Paris) 6 (2): 107 (1937).

Bulbophyllum rubrigemmum *Hermans* **nom. nov.**
Bulbophyllum simulacrum Schltr. in Repert. Sp. Nov. Regni Veg. Beih. 33: 243 (1925).

Bulbophyllum sambiranense *Jum. & H.Perrier* var. **latibracteatum** *H.Perrier ex Hermans* **var. nov.**

Bulbophyllum sambiranense *Jum. & H.Perrier* in Ann. Fac. Sci. Marseille 21 (2): 214 (1912).
Bulbophyllum sambiranense var. *ankeranense* H.Perrier in Notul. Syst. (Paris) 6 (2): 86 (1937), **syn. nov.**

Bulbophyllum sarcorhachis *Schltr.* var. **flavomarginatum** *H.Perrier ex Hermans* **var. nov.**

Bulbophyllum vakonae *Hermans* **nom. nov.**
Bulbophyllum ochrochlamys Schltr. in Repert. Sp. Nov. Regni Veg. Beih. 33: 190 (1924).

Bulbophyllum zaratananae *Schltr.* subsp. **disjunctum** *H.Perrier ex Hermans* **subsp. nov.**

Calanthe sylvatica (*Thouars*) *Lindl.*, Gen. Sp. Orch. Pl.: 250 (1833).
Calanthe sylvatica var. *alba* Cordem., Fl. Réunion: 225 (1895), *C. sylvatica* var. *purpurea* Cordem., Fl. Réunion: 225 (1895), *C. sylvatica* var. *lilacina* Cordem., Fl. Réunion: 225 (1895), *C. sylvatica* var. *iodes* Cordem., Fl. Réunion: 225 (1895), *C. sylvatica* forma *imerina* Ursch & Genoud in Nat. Malg. 3 (2): 108 (1951): *nom nud.*, *C. sylvatica* forma *humberti* Ursch & Genoud in Nat. Malg. 3 (2): 108 (1951): *nom nud*: all **syn. nov.**

Cynorkis section **Gibbosorchis** *H.Perrier ex Hermans* **sect. nov.**

Cynorkis section **Imerinorchis** *H.Perrier ex Hermans* **sect. nov.**

Cynorkis section **Lowiorchis** *H.Perrier ex Hermans* **sect. nov.**

Cynorkis section **Monadeniorchis** *H.Perrier ex Hermans* **sect. nov.**

Cynorkis ampullacea *H.Perrier ex Hermans* **sp. nov.**

Cynorkis angustipetala *Ridl.* var. **moramangensis** *H.Perrier ex Hermans* **var. nov.**

Cynorkis angustipetala *Ridl.* var. **tananarivensis** *H.Perrier ex Hermans* **var. nov.**

Cynorkis decaryana *H.Perrier ex Hermans* **sp. nov.**

Cynorkis ericophila *H.Perrier ex Hermans* **sp. nov.**

Cynorkis flexuosa *Lindl.*, Gen. Sp. Orch. Pl.: 331 (1835).
Cynorkis fallax Schltr. in Beih. Bot. Centralbl. 34 (2): 308 (1916) **syn. nov.**

Cynorkis fastigiata *Thouars.*
Cynorchis fastigiata var. *decolorata* (Schltr.) H.Perrier in Humbert, Fl. Mad. Orch. 1: 141 (1939), *C. fastigiata* var. *diplorhyncha* (Schltr.) H.Perrier, in Humbert Fl. Mad. Orch. 1: 141 (1939), *C. fastigiata* var. *hygrophila* (Schltr.) H.Perrier in *loc. cit.*: 140 (1939), *C. fastigiata* var. *laggiarae* (Schltr.) H.Perrier in *loc. cit.*: 141 (1939), *Cynosorchis laggiarae* Schltr. in Repert. Spec. Nov. Regni Veg. Beih. 15: 326 (1918) & *C. laggiarae* var. *ecalcarata* Schltr. in *loc. cit.*: 326 (1918), **syn. nov.**

Cynorkis fastigiata *Thouars.* var. **ambatensis** *Hermans* **var. nov.**

Cynorkis fimbriata *H.Perrier ex Hermans* **sp. nov.**

Cynorkis hologlossa *Schltr.* var. **angustilabia** *H.Perrier ex Hermans* **var. nov.**

Cynorkis laeta *Schltr.* var. **angavoensis** *H.Perrier ex Hermans* **var. nov.**

Cynorkis lilacina *Ridl.* var. **comorensis** *H.Perrier ex Hermans* **var. nov.**

Cynorkis lilacina *Ridl.* var. **tereticalcar** *H.Perrier ex Hermans* **var. nov.**

Cynorkis lindleyana *Hermans* **nom. nov.**
Bicornella gracilis Lindl., Gen. Sp. Orch. Pl.: 334 (1835) & *Cynorkis gracilis* (Lindl.) Schltr. in Repert. Spec. Nov. Regni Veg. Beih. 33: 51 (1924).

Cynorkis × madagascarica (*Schltr.*) *Hermans* **comb. nov.**

Cynorkis pinguicularioides *H.Perrier ex Hermans* **sp. nov.**

Cynorkis quinqueloba *H.Perrier ex Hermans* **sp. nov.**

Cynorkis quinquepartita *H.Perrier ex Hermans* **sp. nov.**

Cynorkis raymondiana *H.Perrier ex Hermans* **sp. nov.**

Cynorkis ridleyi *T.Durand & Schinz,* Consp. Fl. Afric.: 92 (1895).
Cynorkis rhomboglossa Schltr. in Repert. Spec. Nov. Regni Veg. Beih. 33: 68 (1924), **syn. nov.**

Cynorkis spatulata *H.Perrier ex Hermans* **sp. nov.**

Cynorkis stenoglossa *H.Perrier* var. **pallens** *H.Perrier ex Hermans* **var. nov.**

Cynorkis tenuicalar *Schltr.* subsp. **andasibeensis** *H.Perrier ex Hermans* **subsp. nov.**

Cynorkis tenuicalar *Schltr.* subsp. **onivensis** *H.Perrier ex Hermans* **subsp. nov.**

Cynorkis uncinata *H.Perrier ex Hermans* **sp. nov.**

Cynorkis × madagascarica (*Schltr*) *Hermans.*
Cynosorchis lilacina × ridleyi H.Perrier in Arch. Bot. Bull. Mens. 5: 52 (1931) & in Humbert, Fl. Mad. Orch. 1: 92 (1939), **syn. nov.**

Cynorkis nutans (*Ridl.*) *H.Perrier* in Bull. Soc. Bot. France 83: 582 (1936).
Cynosorchis nutans var. *campenoni* H.Perrier in Humbert, Fl. Mad. Orch. 1: 156 (1939), *nom. nud.*

Eulophia angornensis (*Rchb.f.*) *Hermans* **comb. nov.** TYPE: Comoros, Anjouan (Angorna), *Peters* s.n. (holo. W!). Basionym *Galeandra angornensis* Rchb.f. in Linnaea 20: 680 (1847).

Oeceoclades gracillima (*Schltr.*) *Garay & Taylor* in Bot. Mus. Leafl. 24: 262 (1976).
Eulophidium roseovariegatum Senghas in Adansonia, sér. 2, 6: 561 (1967) & *Oeceoclades roseovariegata* (Senghas) Garay & Taylor in Bot. Mus. Leafl. 24: 270 (1976), **syn. nov.**

Solenangis aphylla (*Thouars*) *Summerh.* in Bot. Mus. Leafl. 11, 5: 159 (1943).
Gussonea aphylla var. *defoliata* (Schltr.) H.Perrier in Humbert, Fl. Mad. Orch. 2: 78 (1941), **syn. nov.** & *G. aphylla* var. *orientalis* H.Perrier in Humbert, Fl. Mad. Orch. 2: 78 (1941), *nom. nud.*

Habenaria clareae *Hermans* **nom. nov.**
Habenaria elliotii Rolfe in J. Linn. Soc., Bot. 29: 57 (1891)

Habenaria deanscowaniana *Hermans* **nom. nov.**
Habenaria stricta Ridl. in J. Linn. Soc., Bot. 21: 510 (1885).

Aerangis monantha *Schltr.* in Repert. Sp. Nov. Regni Veg. Beih. 33: 386 (1925).
Jumellea curnowiana (Finet) Schltr. in Beih. Bot. Centralbl. 33 (2): 429 (1915) *pro parte, non* Rchb.f. (1883) **syn. nov.**

Jumellea lignosa (*Schltr.*) *Schltr.* subsp. **acutissima** *H.Perrier ex Hermans* **subsp. nov.**

Jumellea lignosa (*Schltr.*) *Schltr.* subsp. **latilabia** *H.Perrier ex Hermans* **subsp. nov.**

Jumellea lignosa (*Schltr.*) *Schltr.* subsp. **tenuibracteata** *H.Perrier ex Hermans* **subsp. nov.**

Jumellea lignosa (*Schltr.*) *Schltr.* in Beih. Bot. Centralbl. 33 (2): 429 (1915).
Jumellea lignosa subsp. *ferkoana* (Schltr.) H.Perrier in Notul. Syst. (Paris) 7: 60 (1938), **syn. nov.**

Jumellea longivaginans *H.Perrier* in Notul. Syst. (Paris) 7: 56 (1938).
Jumellea longivaginans var. *grandis* H.Perrier in Notul. Syst. (Paris) 7: 57 (1938), **syn. nov.**

Liparis ambohimangana *Hermans* **nom. nov.**
Liparis monophylla H.Perrier in Notul. Syst. (Paris) 5: 248 (1936).

Liparis clareae *Hermans* **nom. nov.**
Liparis cardiophylla H.Perrier in Notul. Syst. (Paris) 5: 244 (1936).

Liparis clareae *Hermans* var. **angustifolia** *H.Perrier ex Hermans* **var. nov.**

Liparis perrieri *Schltr.* var. **trinervia** *H.Perrier ex Hermans* **var. nov.**

Neobathiea hirtula *H.Perrier* var. **floribunda** *H.Perrier ex Hermans* **var. nov.**

Oeceoclades maculata (*Lindl.*) *Lindl.,* Gen. Sp. Orch. Pl.:237 (1833).
Oeceoclades monophylla (A.Rich.) Garay & Taylor in Bot. Mus. Leafl. 24: 267 (1976), **syn. nov.**

Satyrium amoenum (*Thouars*) *A.Rich.* var. **tsaratananae** *H.Perrier ex Hermans* **var. nov.**

CHECKLIST

Acampe pachyglossa *Rchb.f.*, Otia Bot. Hamburg. 2: 76 (1881). TYPE: Kenya, nr. Mombasa, *Hildebrandt* 1991 (holo. W).

Acampe renschiana Rchb.f., Otia Bot. Hamburg.: 77 (1881); & in Bot. Zeit. 39: 449 (1881). TYPE: Madagascar, Nosi-be, *Hildebrandt* 3392 (holo. W).

Acampe madagascariensis Kraenzl. in Gard. Chron. sér. 3, 10: 608 (1891). TYPE: Madagascar, without locality, cult. Sander & Co (type not located).

Acampe pachyglossa subsp. *renschiana* (Rchb.f.) Senghas in Orchidee 15: 165 (1964).

Acampe rigida (Buch.-Ham. ex Sm.) P.F. Hunt; *sensu non* Seidenf. in Bot. Mus. Leafl. Harv. Univ. 25: 60 (1977).

DISTRIBUTION: Madagascar: Antsiranana, Mahajanga. Comoros: Anjouan, Grande Comore, Mayotte. Also in eastern Africa; Seychelles; southern Africa.

HABITAT: deciduous, seasonally dry forest, woodland and scrub, on tamarind trees in tropical, western forests, on mangroves, coastal cliffs on rock, secondary evergreen forest, limestone formations; on tree trunks, large branches or rocks.

ALTITUDE: sea level – 1000 m.

FLOWERING TIME: throughout the year.

LIFE FORM: epiphyte or lithophyte.

PHYTOGEOGRAPHICAL REGION: II, V.

DESCRIPTION: {221}, {896}, {1262}.

ILLUSTRATION: {34}, {50}, {84}, {221}, {292}, {281}, {459}, {482}, {768}, {896}, {906}, {914}, {941}, {1223}, {1241}, {1262}, {1270}, {1312}, {1319}, {1429}.

VERNACULAR NAME: Kisatrasatra.

NOTES: leaves distichous, coriaceous; inflorescence densely flowered, racemose; sepals and petals greenish cream, with transverse crimson or reddish brown stripes and spots; lip white spotted with red.

Aerangis articulata *(Rchb.f.) Schltr.* in Die Orchideen: 597 (1914). TYPE: Madagascar, *Ellis* s.n. (holo. W, Reichenbach Herbarium 39523).

Angraecum articulatum Rchb.f. in Gard. Chron.: 73 (1872).

Angraecum descendens Rchb.f. in Gard. Chron. 17: 558 (1882). TYPE: Madagascar, cultivated by Low (holo. W, Reichenbach Herbarium 39517).

Angraecum calligerum Rchb.f. in Gard. Chron. ser. 3, 2: 552 (1887). TYPE: cult. Bull (holo. W, Reichenbach Herbarium 39521).

Angorchis articulata (Rchb.f.) Kuntze, Rev. Gen. Pl. 2: 651 (1891).

Rhaphidorhynchus articulatus (Rchb.f.) Poiss., Recherch. Fl. Merid. Madag.: 185 (1912).

Aerangis venusta Schltr. in Repert. Sp. Nov. Regni Veg. 15: 334 (1918) & Repert. Sp. Nov. Regni Veg. Beih. 33: 389 (1925). TYPE: Madagascar, Moramanga, *Afzelius* s.n. (holo. S).

Aerangis stylosa sensu H.Perrier in H. Humbert ed., Fl. Mad. Orch. 2: 109 (1941), *non* Schltr. *nec* Rolfe.

Aerangis calligera (Rchb.f.) Garay in Bot. Mus. Leafl. 23 (10): 373 (1974).

DISTRIBUTION: Madagascar: Antananarivo, Antsiranana, Fianarantsoa, Toamasina, Toliara. Comoros: Anjouan.

HABITAT: humid, evergreen forest from coast to plateau, on trunks and major branches of large forest trees.

ALTITUDE: sea level – 2000m.

FLOWERING TIME: September – February.

LIFE FORM: epiphyte.

PHYTOGEOGRAPHICAL REGION: I, III, IV.

DESCRIPTION: {48}, {896}, {1335}.

ILLUSTRATION: {3}, {48}, {50}, {85}, {226}, {244}, {281}, {434}, {459}, {470}, {529}, {533}, {588}, {614}, {739}, {906}, {1026}, {1067}, {1342}, {1335}.

HISTORY: {226}

CULTIVATION: {48}, {529}, {799}.

NOTES: recognised by the medium- to large-sized flowers in which all the tepals are similar, the sepals winged on the outer surface, and its short column with a prominently beaked anther-cap. The only other species with the latter feature is *A. modesta* which has much smaller and usually many more flowers in each inflorescence.

Aerangis citrata *(Thouars) Schltr.*, Die Orchideen: 598 (1914), TYPE: without locality, *Thouars s.n.* (holo. P).
Angraecum citratum Thouars, Hist. Orch.: t.61 (1822).
Aerobion citratum (Thouars) Spreng., Syst. Veg. 3: 718 (1826).
Angorchis citrata (Thouars) Kuntze, Rev. Gen. Pl. 2: 651 (1891).
Rhaphidorhynchus citratus (Thouars) Finet in Bull. Soc. Bot. France 54, Mém. 9: 35 (1907).

DISTRIBUTION: Madagascar: throughout.
HABITAT: humid, evergreen forest from coast to plateau, usually on twigs or small trees.
ALTITUDE: sea level – 1500 m.
FLOWERING TIME: August – May.
LIFE FORM: epiphyte.
PHYTOGEOGRAPHICAL REGION: I, III.
DESCRIPTION: {47}, {896}, {1336}, {1319}.
ILLUSTRATION: {47}, {62}, {70}, {135}, {208}, {226}, {244}, {246}, {268}, {281}, {416}, {462}, {529}, {541}, {614}, {706}, {716}, {719}, {724}, {729}, {848}, {850}, {906}, {944}, {957}, {1025}, {1026}, {1039}, {1053}, {1062}, {1259}, {1281}, {1298}, {1319}, {1344}, {1398}, {1427}, {1442}, {1480}, {1495}.
HISTORY: {226}.
CULTIVATION: {3}, {27}, {47}, {85}, {529}, {601}, {860}, {1056}, {1319}, {1474}.
VERNACULAR NAME: Manta.
NOTES: recognised by the long raceme of closely spaced flowers which have a very small dorsal sepal and broad petals and lip. Flowers are usually a very pale yellow or cream-coloured.

Aerangis concavipetala *H.Perrier* in Notul. Syst. (Paris) 7: 35 (1938). TYPE: Madagascar, Bejofo, nr. Maromandia, *Decary* 2210 (holo. P).

DISTRIBUTION: Madagascar: Mahajanga.
FLOWERING TIME: June.
LIFE FORM: epiphyte on trunks.
PHYTOGEOGRAPHICAL REGION: II.
DESCRIPTION: {896}.
NOTES: only known from the type specimen but thought to be distinct.

Aerangis cryptodon *(Rchb.f.) Schltr.*, Die Orchideen: 598 (1914). TYPE: Madagascar, cult. Low (holo. W).
Angraecum cryptodon Rchb.f. in Gard. Chron. 19: 307 (1883).
Angorchis cryptodon (Rchb.f.) Kuntze, Rev. Gen. Pl. 2: 651 (1891).
Rhaphidorhynchus stylosus Finet in Bull. Soc. Bot. France 54, Mém. 9: 36, Pl. 9 – 13 (1907). TYPE: Madagascar, Imerina, *Lantz* s.n. (holo. P).
Aerangis malmquistiana Schltr. in Repert. Sp. Nov. Regni Veg. Beih. 33: 385 (1925). TYPE: Madagascar, Mananara river, *H.Perrier* 11347 (lecto. P).

DISTRIBUTION: Madagascar: Antananarivo, Fianarantsoa, Toamasina.
HABITAT: humid, evergreen forest and rocky outcrops on basalt slopes.
ALTITUDE: 200 – 1800 m.
FLOWERING TIME: April – August.
LIFE FORM: epiphyte.
PHYTOGEOGRAPHICAL REGION: I, III.
DESCRIPTION: {896}, {1337}.
ILLUSTRATION: {29}, {226}, {244}, {529}, {614}, {615}, {896}, {1223}, {1264}, {1271}, {1337}, {1470}.
HISTORY: {226}
CULTIVATION: {529}.
NOTES: the true *A. cryptodon* seems to have been collected rather rarely and is often confused with forms of *A. ellisii*. Different by two denticulate wings which stand upright above the anther at the apex of the column and the backward directed, teeth at the edge of the curved part of the lip where it joins the spur.

Aerangis decaryana *H.Perrier* in Notul. Syst. (Paris) 7: 34 (1938). TYPE: SW Madagascar, Ambovombe area, *Decary* 8585 (holo. P).

DISTRIBUTION: Madagascar: Toliara.

HABITAT: on *Alluaudia* spp., *Euphorbia* spp., tamarind trees and other spiny forest trees; in deciduous, dry, southern forest and scrubland, also in transition forest with evergreen, humid forest.

ALTITUDE: 60 – 900 m.

FLOWERING TIME: December – March.

LIFE FORM: epiphyte {896}.

PHYTOGEOGRAPHICAL REGION: VI.

DESCRIPTION: {896}, {1337}.

ILLUSTRATION: {3}, {281}, {480}, {484}, {529}, {896}, {1271}, {1337}.

CULTIVATION: {191}, {446}, {529}.

NOTES: one of the few epiphytic orchids that grow in the xerophytic forests of southern and south-western Madagascar. The leaves often have undulate margins, the spur is flattened at the tip and pale pink.

Aerangis ellisii *(B.S.Williams) Schltr.,* Die Orchideen: 598 (1914). TYPE: Madagascar, *Ellis* hort. J.Day (holo. W; iso. K).

Angraecum ellisii B.S. Williams, Orch. Grow. Man. ed. 4: 87 (1871); Rchb.f. in Flora: 278 (1872).

Angraecum dubuyssonii God.-Leb. in L'Orchidophile 7: 280 (1887) *loc. cit.* and L'Orchidophile 11: 283. 284 (1891) . TYPE: Photograph in L'Orchidophile 7: 281 (1887).

Angorchis ellisii (B.S. Williams) Kuntze, Rev. Gen. Pl. 2: 651 (1891).

Aerangis buyssonii God.-Leb. in L'Orchidophile: 282 – 286 (1891) TYPE: Madagascar, cult. Godefroy-Lebeuf (holo. P; iso. K).

Aerangis caulescens Schltr. in Beih. Bot. Centralbl. 34 (2): 334 (1916) and Repert. Spec. Nov. Regni Veg. Beih. 33: 383 (1925). TYPE: Madagascar, Central, Betafo, *H.Perrier* (II) 8093 (iso. P).

Aerangis platyphylla Schltr. in Repert. Sp. Nov. Regni Veg. Beih. 33: 387 (1925). TYPE: Madagascar, centre, massif du Manongarivo, *H.Perrier* 8038 (lecto. P, selected here).

Aerangis alata H.Perrier in Notul. Syst. (Paris) 7: 38 (1938) and in Fl. Madag. 49 (2): 107 (1941). TYPE: SE Madagascar, Manampanihy (Fitana), *Humbert* 6062bis (holo. P).

Aerangis cryptodon sensu H.Perrier in Not. Syst. 7: 31 (1938) and in Fl. Mad. Orch. 2: 101 (1941), *non* Schltr. 1914, 1918.

Aerangis ellisii var. *grandiflora* J.Stewart in Amer. Orch. Soc. Bull. 55 (8): 798 (1986), **syn. nov.** TYPE: Madagascar, Tampoketsa d'Ankazobe, *Malzy* s.n. (holo. P).

DISTRIBUTION: Madagascar: Antananarivo, Antsiranana, Fianarantsoa, Toliara.

HABITAT: humid, evergreen forest; plateau, rocky outcrops of basalt, quartz, granite and gneiss; amongst xerophytic vegetation.

ALTITUDE: 300 – 1800 m.

FLOWERING TIME: October, December – May.

LIFE FORM: epiphyte or lithophyte.

PHYTOGEOGRAPHICAL REGION: I, III, IV.

DESCRIPTION: {896}, {1335}, {1465}.

ILLUSTRATION: {3}, {28}, {191}, {208}, {232}, {244}, {246}, {281}, {343}, {406}, {470}, {478}, {486}, {504}, {529}, {672}, {700}, {724}, {746}, {819}, {842}, {896}, {906}, {944}, {1029}, {1170}, {1192}, {1294}, {1335}, {1448}, {1453}, {1466}, {1467}, {1468}, {1507}.

CULTIVATION: {28}, {232}, {529}, {680}, {746}, {819}, {1095}, {1192}.

NOTES: *A. ellisii* var. *occidentale* Kraenzl. is a synonym of *A. gravenreuthii* (Kraenzl.) Schltr. from mainland Africa. Flowers and plants are variable but the species generally has long stems with thick, succulent widely spaced leaves and long racemes of medium to large sized flowers. The perianth parts are strongly reflexed and there are two ridges on the lower part of the lip. Plants found in forest are different from those found in exposed rocky habitats but flower shape and size is variable within the two forms.

Aerangis fastuosa *(Rchb.f.) Schltr.,* Die Orchideen: 598 (1914). TYPE: Madagascar, SW coast, *Humblot* s.n. (holo. W, Reichenbach Herbarium 46259).

Angraecum fastuosum Rchb.f. in Gard. Chron. 15: 748 (1881) and 844 and 23: 533, fig. 96 (1885); Hooker in Bot. Mag. 117: t.7204 (1891).

Angorchis fastuosa (Rchb.f.) Kuntze, in Rev. Gen. Pl. 2: 651 (1891).

Rhaphidorhynchus fastuosus (Rchb.f.) Finet in Bull. Soc. Bot. France 54, Mém. 9: 38 (1907). TYPE: Madagascar, SW coast, *Grandidier* s.n. (holo. P).

Aerangis fastuosa var. *françoisii* H.Perrier in Humbert ed., Fl. Mad. Orch. 2: 93 (1941), *nom. nud.* Based upon: Madagascar, Mandraka valley, approx. 1400 m. *François* 12 in *H.Perrier* s.n. (P).

Aerangis fastuosa var. *grandidieri* H.Perrier in Humbert ed., Fl. Mad. Orch. 2: 93 (1941), *nom. nud.* Based upon: Madagascar, SW coast, *Grandidier* s.n. (P).

Aerangis fastuosa var. *maculata* H.Perrier in Humbert ed., Fl. Mad. Orch. 2: 95 (1941), *nom. nud.* Based upon: Madagascar, Mt. Kalambatitra, *Humbert* 11845 (P).

Aerangis fastuosa var. *rotundifolia* H.Perrier in Humbert ed., Fl. Mad. Orch. 2: 94 (1941), *nom. nud.* Based upon: Madagascar, Ivohibe, *Decary* 5700 (P).

Aerangis fastuosa var. *vondrozensis* H.Perrier in Humbert ed., Fl. Mad. Orch. 2: 94 (1941), *nom. nud.* Based upon: Madagascar, Vondrozo, *Decary* 5188 (P).

DISTRIBUTION: Madagascar: Antananarivo, Fianarantsoa, Toamasina, Toliara.

HABITAT: humid, evergreen forest; on twigs and small branches in sclerophyllous forest.

ALTITUDE: 900 – 1500 m.

FLOWERING TIME: September – October.

LIFE FORM: epiphyte.

PHYTOGEOGRAPHICAL REGION: I, III, VI.

DESCRIPTION: {50}, {896}, {1336}.

ILLUSTRATION: {3}, {50}, {239}, {244}, {253}, {352}, {529}, {546}, {604}, {614}, {700}, {896}, {906}, {944}, {1020}, {1259}, {1298}, {1336}, {1342}, {1389}, {1427}.

CULTIVATION: {191}, {239}, {376}, {529}, {533}.

NOTES: the varieties recognised by Perrier de la Bâthie were each represented by a single specimen, and the names were not validly published because they lack Latin descriptions. It is accepted here as a single and rather variable species. The species can be recognised by the large flowers on small plants with rounded succulent leaves and by the long rostellum.

Aerangis fuscata *(Rchb.f.) Schltr.*, Die Orchideen: 598 (1914). TYPE: Madagascar. cult. C.Marriott (syn. W, Reichenbach Herbarium 39415). Cult. Lawrence (syn. K).

Angraecum fuscatum Rchb.f. in Gard. Chron. 18: 488 (1882).

Rhaphidorhynchus umbonatus Finet in Bull. Soc. Bot. France 54, Mém. 9: 37 t.7: 21 – 27 (1907). TYPE: Madagascar, *Humblot* 383 (holo. P; iso. K).

Aerangis umbonata (Finet) Schltr. in Beih. Bot. Centralbl. 33 (2): 428 (1915) and 428 (1915) and 36 (2) (1918); H.Perrier in Fl. Madag. 49 (2): 110, fig. 53: 1 – 7 (1941); Stewart in Amer. Orch. Soc. Bull. 55 (9): 906 (1986).

DISTRIBUTION: Madagascar: Antananarivo, Antsiranana, Toamasina. Records from the Comoros are misidentifications {1336}, {1436}.

HABITAT: on twigs and branches in a range of habitats, on shrubs and small trees.

ALTITUDE: sea-level to 1500 m.

FLOWERING TIME: November.

LIFE FORM: epiphyte.

PHYTOGEOGRAPHICAL REGION: I, III.

DESCRIPTION: {896}, {1336}.

ILLUSTRATION: {20}, {177}, {244}, {440}, {501}, {506}, {529}, {615}, {716}, {838}, {896}, {906}, {944}, {1259}, {1336}, {1338}, {1346}.

HISTORY: {8}, {1336}.

CULTIVATION: {191}, {506}, {529}.

NOTES: small plants with 1- to 4-flowered racemes. The relatively large flowers are pale pink-brown except for the lip which is funnel-shaped at the base.

Aerangis hyaloides *(Rchb.f.) Schltr.*, Die Orchideen: 599 (1914). TYPE: Madagascar, cult. Veitch, (holo. W, Reichenabch Herbarium 46134).

Angraecum hyaloides Rchb.f. in Gard. Chron. 13: 264 (1880) and p.136 (1881), Xenia Orch., 73 t.238, 1 – 2 (1900); Godefroy-Lebeuf in Orchidophile, 347 (1889).

Angorchis hyaloides (Rchb.f.) Kuntze, Rev. Gen. Pl. 2: 651 (1891).

Aerangis pumilio Schltr. in Repert. Sp. Nov. Regni Veg. Beih. 15: 334 (1918); H.Perrier in Fl. Madag. 49 (2): 100 (1941). TYPE: Madagascar, East, without precise locality, *Laggiara* s.n. (holo. B†).

DISTRIBUTION: Madagascar: Antananarivo, Antsiranana, Toamasina.

HABITAT: humid, evergreen forest; forest, mainly in the shade on moss- and lichen-covered small trees. Often on twigs and small branches.

ALTITUDE: sea level – 1100 m.

FLOWERING TIME: September – January.

LIFE FORM: epiphyte

PHYTOGEOGRAPHICAL REGION: I, III.

DESCRIPTION: {896}, {1336}.

ILLUSTRATION: {3}, {244}, {529}, {655}, {802}, {849}, {906}, {1143}, {1163}, {1223}, {1336}, {1341}, {1471}.

CULTIVATION: {529}, {802}.

NOTES: a small plant carrying an abundance of small, hyaline flowers that do not open completely.

Aerangis macrocentra *(Schltr.) Schltr.* in Beih. Bot. Centralbl. 33 (2): 427 (1915). TYPE: Madagascar, Sambirano Mts., *H.Perrier* 1854 (93) (holo. B†; iso. P).

Angraecum macrocentrum Schltr. in Ann. Mus. Col. Marseille, sér. 3, 1: 195 (52), t.22 (1913).

Aerangis clavigera H.Perrier in Notul. Syst. (Paris) 7: 38 (1938) and in Fl. Madag. 49 (2): 103. fig. 54: 3 (1941). TYPE: Madagascar, Ankeramadinika forest, *François in H.Perrier* 18554 (holo. P).

DISTRIBUTION: Madagascar: Antananarivo, Antsiranana, Toamasina.

HABITAT: humid, evergreen forest on plateau; lichen-rich and mossy forest.

ALTITUDE: sea level – 2200 m.

FLOWERING TIME: March – June.

LIFE FORM: epiphyte.

PHYTOGEOGRAPHICAL REGION: I, III.

DESCRIPTION: {896}, {1336}.

ILLUSTRATION: {3}, {281}, {440}, {470}, {505} = *A. pallidiflora*, {1205}, {1223}, {1336}.

NOTES: flowers on a long inflorescence; perianth parts short, blunt, reflexed; spur inflated towards the tip.

Aerangis modesta *(Hook.f.) Schltr.,* Orchideen: 600 (1914). TYPE: cult. Lady Ashburton (holo. K).

Angraecum modestum Hook.f. in Curtis's Bot. Mag. 109: t.6693 (1883).

Angraecum sanderianum Rchb.f. in Gard. Chron. ser. 3, 1: 168 (1888); André in Rev. Hort.: 516 (1888); Sander, Orch. Guide, 11 (1901). TYPE: Comoro Is., without locality, *Humblot* s.n., hort. Sander (holo. W; iso. K).

Angorchis modesta (Hook.f.) Kuntze, Rev. Gen. Pl. 2: 651 (1891).

Rhaphidorhynchus modestus (Hook.f.) Finet in Bull. Soc. Bot. France 54, Mém. 9: 37 (1907). TYPE: Madagascar, SW coast, *Grandidier* (holo. P).

Rhaphidorhynchus modestus var. *sanderianus* Poiss., Recherch. Fl. Merid. Madagascar: 185 (1912).

Aerangis crassipes Schltr. in Repert. Sp. Nov. Regni Veg. 15: 333 (1918). TYPE: Madagascar, *Laggiara* s.n. (holo. B†)

Aerangis fastuosa var. *angustifolia* H.Perrier in Humbert ed., Fl. Mad. Orch. 2: 95 (1941), *nom. nud.* Based upon: Madagascar, between Moramanga and Anosibe, *H.Perrier* 18290 (P).

DISTRIBUTION: Madagascar: Antananarivo, Antsiranana, Fianarantsoa, Mahajanga, Toamasina. Comoros: Anjouan, Grande Comore.

HABITAT: humid, evergreen forest; often in forest remnants on trunks and small branches of trees.

ALTITUDE: 100 – 1500 m.

FLOWERING TIME: October – January.

LIFE FORM: epiphyte.

PHYTOGEOGRAPHICAL REGION: II, III.

DESCRIPTION: {529}, {896}, {1335}.

ILLUSTRATION: {9}, {199}, {251}, {253}, {529}, {544}, {605}, {614}, {615}, {746}, {870}, {1026}, {1097}, {1124}, {1191}, {1206}, {1222}, {1298}, {1335}, {1347}, {1502}.

CULTIVATION: {38}, {529}.

HISTORY: {9}, {199}, {392}, {571}, {1124}, {1346}.

NOTES: similar to *A. articulata* but smaller in all respects and flowers all a similar size along the inflorescence.

Aerangis monantha *Schltr.* in Repert. Sp. Nov. Regni Veg. Beih. 33: 386 (1925). TYPE: Madagascar, Manankazo, NE of Ankazobe, *H.Perrier* 11321 (holo B†; iso. P).

Rhaphidorhynchus curnowianus sensu Finet in Bull. Soc. Bot. France 54, Mém. 9: 37 (1907), *non* Rchb.f. in Gard. Chron. sér. 2, 19: 306 (1883). TYPE: Madagascar, *Campenon* s.n. (holo. P).

Jumellea curnowiana (Finet) Schltr. in Beih. Bot. Centralbl. 33 (2): 429 (1915) *pro parte, non* Rchb.f. (1883), **syn. nov.**

Aerangis curnowiana sensu H.Perrier in Notul. Syst. (Paris) 7: 34 (1938).

DISTRIBUTION: Madagascar: Antananarivo.
HABITAT: humid forest.
ALTITUDE: 800 – 1200 m.
FLOWERING TIME: October.
LIFE FORM: epiphyte.
PHYTOGEOGRAPHICAL REGION: III.
DESCRIPTION: {896}, {1221}.
ILLUSTRATION: {281} as *Aerangis curnowiana*, {529} as *Aerangis* 145.
NOTES: there has been some confusion over the identity of this species, Finet {292} made a mistake in assuming that *Rhaphidorhynchus curnowianus* was the same as *Aeranthus curnowianus* Rchb.f. of which the type is different and actually is an *Angraecum*, Perrier de la Bâthie {889} perpetuated this mistake. Similar to *Aerangis fuscata* but the plant is smaller, with a single flower, a differently shaped lip and a much longer and differently shaped rostellum.

Aerangis mooreana *(Rolfe) P.J.Cribb & J.Stewart* in Orch. Rev. 91 (1077): 218 (1983); TYPE: Comoros, hort. Sander (lecto. K).

Angraecum mooreanum Rolfe in Orch. Rev. 5: 126 (1897).

Aerangis ikopana Schltr. in Repert. Sp. Nov. Regni Veg. Beih. 33: 384 (1925); H.Perrier in Fl. Madag. 49 (2) (1941). TYPE: Madagascar, Firingalava (Ikopa), *H.Perrier* 681 (holo. B†; iso. P).

Aerangis anjoanensis H.Perrier in Humbert ed., Fl. Mad. Orch. 2: 104, (1941), *nom. nud.*

Aerangis karthalensis R.Neirynck & M.Herrmans in Caesiana., 26 (2): 3 (2006), **syn. nov.** TYPE: Comoro Is., Grand Comore (Ngazidja), 1989, *Herremans* s.n. deposited as Rik Neirynck 2002-01 (holo. GENT). The *Aerangis karthalensis* concept envelops the larger and salmon pink forms of this species, further research is needed to re-assess these different forms.

DISTRIBUTION: Madagascar: Antananarivo, Mahajanga. Comoros: Anjouan, Grande Comore.
HABITAT: lowland, humid, evergreen forest; on shrubs.
ALTITUDE: sea level – 600 m.
FLOWERING TIME: July – October.
LIFE FORM: epiphyte.
PHYTOGEOGRAPHICAL REGION: V.
DESCRIPTION: {218}, {281}, {896}, {1337}.
ILLUSTRATION: {1337}.
NOTES: recognised by the obovate lip and short, reflexed dorsal sepal which is smaller than the other perianth parts. Flowers often pale pink.

Aerangis pallidiflora *H.Perrier* in Notul. Syst. (Paris) 7: 36 (1938). TYPE: Madagascar, Central, Ankeramadinika Forest, *François* in *H.Perrier* 17207 (holo. P); Mt. Maromizaha, near Analamazaotra, Feb. 1924, *H.Perrier* 16048 (para. P).

Angraecum ramulicolum H.Perrier in Notul. Syst. (Paris) 7: 123 (1938). TYPE: Madagascar, Ankeramadinika, *E.François* in *H.Perrier* 18543 (holo. P).

DISTRIBUTION: Madagascar: Antananarivo, Fianarantsoa, Toamasina.
HABITAT: coastal and plateau, humid, mossy, evergreen forest; on branches and twigs.
ALTITUDE: 900 – 1500 m.
FLOWERING TIME: December – March, August.
LIFE FORM: epiphyte.
PHYTOGEOGRAPHICAL REGION: III.
DESCRIPTION: {896}.
ILLUSTRATION: {896}, {1256}.
NOTES: the leaves are distinctive, the remaining old inflorescences are typical. Flowers green, small; lip lanceolate-acute; spur c. 2.5 cm long. Similar to *A. seegeri* but the shape of the tepals and the spur are different.

Aerangis ×primulina *(Rolfe) H.Perrier* in Humbert ed., Fl. Mad. Orch. 2: 116 (1941). TYPE: Madagascar, without locality, cult., H. Low & Co. (holo. K).

Aerangis primulina (Rolfe) H.Perrier in Humbert ed., Fl. Mad. Orch. 2: 116 (1941).

Angraecum primulinum Rolfe in Gard. Chron. ser. 3, 7: 388 (1890).

> DISTRIBUTION: Madagascar.
> DESCRIPTION: {896}, {1338}.
> HISTORY: {1059}.
> NOTES: assumed natural hybrid of *Aerangis citratum* and *A. hyaloides*, known only from the type specimen.

Aerangis pulchella *(Schltr.) Schltr.* in Beih. Bot. Centralbl. 33 (2): 427 (1915). TYPE: Madagascar, Antanimena plateau (Boina), *H.Perrier* 1893 (59) (holo. B†; iso. P).

Angraecum pulchellum Schltr. in Ann. Mus. Col. Marseille, sér. 3, 1: 200, t.23 fig. 8 – 15 (1913).

> DISTRIBUTION: Madagascar: Mahajanga.
> HABITAT: seasonally dry, deciduous forest and woodland; low-elevation, humid, evergreen forest.
> ALTITUDE: sea level – 200 m.
> FLOWERING TIME: June.
> LIFE FORM: epiphyte.
> PHYTOGEOGRAPHICAL REGION: V.
> DESCRIPTION: {896}, {1338}.
> ILLUSTRATION: {1205}, {1223}.
> NOTES: Only known from the type specimen which is in bud. This may be conspecific with *A. mooreana.*

Aerangis punctata *J.Stewart* in Amer. Orch. Soc. Bull. 55 (11): 1120 (1986). TYPE: Madagascar, central, *H.Perrier* 16961 (holo. P).

> DISTRIBUTION: Madagascar: Antananarivo. Réunion.
> HABITAT: humid, evergreen forest on plateau.
> ALTITUDE: 1000 – 1500 m.
> FLOWERING TIME: February.
> LIFE FORM: epiphyte.
> PHYTOGEOGRAPHICAL REGION: III, IV.
> DESCRIPTION: {637}, {1005}, {1338}.
> ILLUSTRATION: {295} as *A. curnowiana,* {637}, {1338}, {1346}.
> CULTIVATION: {637}, {680}.
> NOTES: the species has flattened verrucose roots and leaves carrying tiny pale dots on the surface. Flowers similar to those of *A. monantha* but with a broader lip and straight spur.

Aerangis rostellaris *(Rchb.f.) H.Perrier* in Humbert ed., Fl. Mad. Orch. 49 (2): 116 (1941). TYPE: Grande Comore, *Humblot* s.n. (holo. W; iso. BM).

Angraecum rostellare Rchb.f. in Gard. Chron. n.s. 23: 726 (1885).

Angraecum avicularium Rchb.f. in Gard. Chron. ser. 3, 1: 40 (1887). TYPE: cult. T. Lawrence (holo. W).

Aerangis avicularia (Rchb.f.) Schltr. in Beih. Bot. Centralbl. 36 (2): 114 (1918); J.Stewart in Kew Bull. 34 (2): 314 (1979).

Aerangis buchlohii Senghas in Die Orchidee 20: 12 – 16 (1969), TYPE: Madagascar, Mt. Ambre, *Rauh & Buchloh* 8025 (holo. HEID).

> DISTRIBUTION: Madagascar: Antsiranana. Comoros: Grande Comore.
> HABITAT: humid, evergreen forest on plateau.
> LIFE FORM: epiphyte.
> PHYTOGEOGRAPHICAL REGION: III.
> DESCRIPTION: {896}, {1247}, {1337}.
> ILLUSTRATION: {1247}, {1259}, {1264}.
> NOTES: recognised by thick broad leaves, medium-sized flowers with distinctly incurved sepals and by the short column with an elongated rostellum.

Aerangis seegeri *Senghas* in Orchidee 34 (1): 23 (1983) . TYPE: Madagascar, Ile Sainte Marie, *Rauh & Senghas* 22723 (holo. HEID).

DISTRIBUTION: Madagascar: Antananarivo, Toamasina.
HABITAT: coastal forest; evergreen forest.
ALTITUDE: sea-level – 1500 m.
FLOWERING TIME: June.
LIFE FORM: epiphyte.
PHYTOGEOGRAPHICAL REGION: I, III.
DESCRIPTION: {1256}.
ILLUSTRATION: {1256}.
NOTES: similar to *A. pallidiflora* but flowers bronze-coloured, more star-shaped and the lower part of the spur thicked and flattened.

Aerangis spiculata *(Finet) Senghas* in Die Orchidee (Hamburg) 23 (6): 226 (1972). TYPE: Comoros, *Humblot* 414 (holo. P; iso. W).
Angraecum fuscatum sensu Carrière in Rev. Hort. 59: 235, fig. 49 (1887), *non* Rchb.f. (1882).
Leptocentrum spiculatum (Finet) Schltr. in Beih. Bot. Centralbl. 33 (2): 426 (1915).
Plectrelminthus spiculatus (Finet) Summerh. in Kew Bull.: 442 (1949) .
Rhaphidorhynchus spiculatus Finet in Bull. Soc. Bot. France 54, Mém. 9: 40, t.8, 1 – 12 (1907).

DISTRIBUTION: Madagascar: Antsiranana. Comoros: Grande Comore.
HABITAT: humid, lowland, evergreen forest.
ALTITUDE: 80 – 1000 m.
FLOWERING TIME: March.
LIFE FORM: epiphyte.
PHYTOGEOGRAPHICAL REGION: II.
DESCRIPTION: {1249}, {1337}.
ILLUSTRATION: {177}, {281}, {292}, {896}, {1249}, {1337}, {1340}.
CULTIVATION: {1340}.
NOTES: large plants with sometimes greyish green leaves which have a slightly undulate margin. Inflorescence long, flowers large.

Aerangis stylosa *(Rolfe) Schltr.* in Beih. Bot. Centralbl. 33 (2): 428 (1915). TYPE: Madagascar, cult. Sander (holo. K).
Angraecum stylosum Rolfe in Kew Bull.: 194 (1895).
Angraecum fournierae André in Rev. Hort. 1: 256 (1896). TYPE: Madagascar, cult. Fournier (holo.?).
Angraecum metallicum nomen in Proc., J. Roy. Hort. Soc. 20: CXIII (1896); Sander in Sander's Orch. Guide, 10 (1901). TYPE: Madagascar, cult. Glasnevin (lecto. K).
Rhaphidorynchus stylosus (Rolfe) Finet in Bull. Soc. Bot. France 54, Mém. 9: 36 (1907).
Aerangis fuscata sensu H.Perrier in Humbert ed., Fl. Mad. Orch. 2: 114 (1941).

DISTRIBUTION: Madagascar: Antananarivo, Antsiranana, Fianarantsoa, Toamasina. Comoros: Anjouan, Grande Comore.
HABITAT: humid, evergreen forest from coast to plateau; mossy forest.
ALTITUDE: sea level – 1400 m.
FLOWERING TIME: September – May.
LIFE FORM: epiphyte.
PHYTOGEOGRAPHICAL REGION: I, III.
DESCRIPTION: {50}, {896}, {1259}.
ILLUSTRATION: {3}, {11}, {50}, {292}, {529}? {629}? {710}? {1223}, {1259}, {1412}?, {1425}.
HISTORY: {805}.
CULTIVATION: {529}.
NOTES: close to *Aerangis spiculata* but the plants and flowers much smaller, spur shorter, margins of the lip rolled downwards.

Aeranthes adenopoda *H.Perrier* in Notul. Syst. (Paris) 7: 46 (1938). TYPE: Madagascar, Ambilo, S of Toamasina, *E.François* 6 (holo. P).

DISTRIBUTION: Madagascar: Antananarivo, Toamasina. There are also records from Réunion. {298a}.

HABITAT: humid, highland forest.
ALTITUDE: 1000 – 1300 m.
LIFE FORM: epiphyte on moss-covered branches.
PHYTOGEOGRAPHICAL REGION: I.
DESCRIPTION: {896}.
ILLUSTRATION: {896}.
NOTES: a small species similar to *A. pusilla* but with typical linear leaves and differently shaped flowers, ovary rough.

Aeranthes aemula *Schltr.* in Repert. Sp. Nov. Regni Veg. Beih. 33: 274 (1925). TYPE: Madagascar, Mt. Tsaratanana, *H.Perrier* 15968 (holo. B†; iso. P).

Aeranthes biauriculata H.Perrier in Notul. Syst. (Paris) 14: 161 (1951), **syn. nov.** TYPE: Madagascar, Andasibe, Onive, *H.Perrier* 17116 (lecto. P, selected here); Manankazo N of Ankazobe, *H.Perrier* 18460 (syn. P); & Andasibe, Onive, *H.Perrier* 17117 (syn. P).

DISTRIBUTION: Madagascar: Antananarivo, Antsiranana, Toamasina.
HABITAT: lichen-rich and mossy forest, on moss and lichen-rich covered trees.
ALTITUDE: 450 – 2000 m.
FLOWERING TIME: January – February.
LIFE FORM: epiphyte.
PHYTOGEOGRAPHICAL REGION: I, II, III, IV.
DESCRIPTION: {896}, {901}.
NOTES: related to *A. setipes* but the leaves are longer and wider, the flowers a little bigger and the flowers are more yellow, the lip and spur are also different. *A. biauriculata* H.Perrier is a form of this species with a slightly different lip and shorter spur.

Aeranthes albidiflora *Toill.-Gen., Ursch & Bosser* in Notul. Syst. (Paris) 16: 205 (1960). TYPE: Madagascar, Fatihita, Ifanadiana district, *J.Toilliez* 81 (holo. P).

DISTRIBUTION: Madagascar: Antananarivo, Fianarantsoa.
HABITAT: mossy forest and evergreen forest, on moss- and lichen-covered trees.
ALTITUDE: 30 – 1400 m.
FLOWERING TIME: November – April.
LIFE FORM: epiphyte.
PHYTOGEOGRAPHICAL REGION: I.
DESCRIPTION: {269}, {1404}.
ILLUSTRATION: {269}, {1404}.
VERNACULAR NAME: Velomihata.
NOTES: close to *A. sambiranoensis* but differs by the pendant single-flowered inflorescence and slender peduncle, broader lip of a different shape, and shorter strongly incurved spur which is strongly globular at the tip.

Aeranthes ambrensis *Toill.-Gen., Ursch & Bosser* in Notul. Syst. (Paris) 16: 209 (1960). TYPE: Madagascar, Ambre Forest near Diego-Suarez, *Begue* 246 (holo. P).

DISTRIBUTION: Madagascar: Antsiranana.
HABITAT: humid, evergreen forest.
FLOWERING TIME: July.
LIFE FORM: epiphyte.
PHYTOGEOGRAPHICAL REGION: III.
DESCRIPTION: {1404}.
ILLUSTRATION: {1404}.
NOTES: characterised by its ligulate, thin, flexible leaves, with slightly protruding dorsal veins. Only known from Mt. d'Ambre in N Madagascar.

Aeranthes angustidens *H.Perrier* in Notul. Syst. (Paris) 7: 44 (1938). TYPE: Madagascar, confl. of Onive & Mangoro, *H.Perrier* 17175 (holo. P).

DISTRIBUTION: Madagascar: Antananarivo?, Toamasina.

HABITAT: humid, evergreen forest.
ALTITUDE: c. 700 m.
FLOWERING TIME: January – February.
LIFE FORM: epiphyte.
PHYTOGEOGRAPHICAL REGION: I, III?
DESCRIPTION: {896}.
NOTES: the inflorescences can remain on the plant for several years, they extend and become more branched as they get older.

Aeranthes antennophora *H.Perrier* in Notul. Syst. (Paris) 7: 45 (1938). TYPE: Madagascar, Onive basin, Andasibe, *H.Perrier* 17154 (holo. P).

DISTRIBUTION: Madagascar: Toamasina.
HABITAT: mossy forest on moss- and lichen-covered trees.
ALTITUDE: c. 1000 m.
FLOWERING TIME: February.
LIFE FORM: epiphyte.
PHYTOGEOGRAPHICAL REGION: I.
DESCRIPTION: {896}.
ILLUSTRATION: {906}.
NOTES: distinct by its narrow sepals which can be up to 10 cm long, including the filiform tail. The inflorescence is normally shorter than the leaves and the sheaths are at most $^2/_3$ the length of the inter-nodes of the peduncle.

Aeranthes arachnitis *(Thouars) Lindl.* in Bot. Reg. t.817 (1824).
Dendrobium arachnites Thouars, Hist. Orch.: t.66 – 67 (1822).

NOTES: this species is endemic to the Mascarene Islands; records from Madagascar and the Comoros in the literature and herbaria seem to be mis-identifications.

Aeranthes bathieana *Schltr.* in Repert. Sp. Nov. Regni Veg. Beih. 33: 275 (1925). TYPE: Madagascar, Mt. Tsiafajavona, *H.Perrier* 13409 (holo. P).

DISTRIBUTION: Madagascar: Antananarivo.
HABITAT: mossy, montane forest on moss- and lichen-covered trees.
ALTITUDE: c. 2000 m.
FLOWERING TIME: January – February.
LIFE FORM: epiphyte.
PHYTOGEOGRAPHICAL REGION: III.
DESCRIPTION: {896}.
NOTES: only known from the Ankaratra Massif. Flowers whitish but pale yellow at the end of segments; lip with a bilobed callus at the base; spur conical-cylindrical, 5 mm long.

Aeranthes campbelliae *Hermans & Bosser* in Adansonia sér. 3, 25 (2): 215 (2003). TYPE: Comoros, Grande Comore, *Hermans* 3915 (holo. K).

DISTRIBUTION: Comoros: Grande Comore.
LIFE FORM: epiphyte {493}.
DESCRIPTION: {493}.
ILLUSTRATION: {493}.
NOTES: the small flowers are very distinct by the spur which is reduced to a small swelling at the base of the mentum. Differs from *Aeranthes ecalcarata* in its leaves and inflorescence.

Aeranthes carnosa *Toill.-Gen., Ursch & Bosser* in Notul. Syst. (Paris) 16: 209 (1960). TYPE: Madagascar, Maroantsetra, *Lambert* 485 (holo. P; iso. TAN).

DISTRIBUTION: Madagascar: Toamasina.
HABITAT: humid, mossy forest, on moss- and lichen-rich-covered trees.

ALTITUDE: low to medium elevations.
LIFE FORM: epiphyte.
PHYTOGEOGRAPHICAL REGION: I.
DESCRIPTION: {1404}.
ILLUSTRATION: {1404}.
NOTES: the lip is characterised by its short and narrow base and tip which is strongly dilated and globose, up to 7 – 8 mm long and 4 – 4.5 mm wide.

Aeranthes caudata *Rolfe* in Kew Bull.: 149 (1901). TYPE: Madagascar, cult. RBG Glasnevin (holo. K).
Aeranthes imerinensis H.Perrier in Notul. Syst. (Paris) 7: 44 (1938). TYPE: Madagascar, Ankeramadinika, *E.François* in *H.Perrier* 18404 (holo. P).

DISTRIBUTION: Madagascar: Antananarivo, Antsiranana, Toamasina, Fianarantsoa. Comoros: Grande Comore.
HABITAT: humid, mossy, evergreen forest on mossy branches and tree trunks.
ALTITUDE: 700 – 1500 m.
FLOWERING TIME: January – April.
LIFE FORM: epiphyte.
PHYTOGEOGRAPHICAL REGION: I, III.
DESCRIPTION: {211}, {896}.
ILLUSTRATION: {3}, {211}, {212}, {520}, {529}, {573}, {807}, {906}, {941}, {1259}, {1264}, {1410}.
CULTIVATION: {211}, {529}, {573}, {941}.
NOTES: distinguished by its short cylindrical spur, sepals with filiform acumens to 9 cm long, an obtrullate lip and ligulate leaves which have a marked unequally bilobed apex.

Aeranthes crassifolia *Schltr.* in Repert. Sp. Nov. Regni Veg. Beih. 33: 276 (1925). TYPE: Madagascar, Sambirano Valley, *H.Perrier* 15447 (holo. P).

DISTRIBUTION: Madagascar: Antsiranana.
HABITAT: on tamarind trees.
FLOWERING TIME: February.
LIFE FORM: epiphyte.
PHYTOGEOGRAPHICAL REGION: II.
DESCRIPTION: {896}.
NOTES: leaves thick and leathery; lip oblong-acuminate; spur almost straight, broadly cylindrical, obtuse.

Aeranthes denticulata *Toill.-Gen., Ursch & Bosser* in Notul. Syst. (Paris) 16: 207 (1960). TYPE: Madagascar, Perinet, Andasibe, *E.Ursch* 322 (holo. P).

DISTRIBUTION: Madagascar: Toamasina.
HABITAT: mid-elevation, humid, evergreen, eastern forests.
ALTITUDE: 1000 – 1500 m.
FLOWERING TIME: July – August.
LIFE FORM: epiphyte.
PHYTOGEOGRAPHICAL REGION: I.
DESCRIPTION: {1261}, {1404}.
ILLUSTRATION: {807}, {906}, {1259}, {1261}.
NOTES: lip acutely ovate, margins a little denticulate, carrying two characteristic diverging keels at the base; spur straight, clavate, 8 – 10 mm long.

Aeranthes dentiens *Rchb.f.* in Flora 68: 381 (1885), as *Aeranthus dentiens*. TYPE: Comoros, *Humblot* s.n. (holo. W).
Mystacidium dentiens (Rchb.f.) T.Durand & Schinz, Consp. Fl. Afric. 5: 52 (1892).

DISTRIBUTION: Madagascar?, Comoros.
FLOWERING TIME: March – May (in cultivation).
LIFE FORM: epiphyte.
DESCRIPTION: {896}, {1084}.

History: {1084}.

Notes: a dubious species said to differ from *A. grandiflora* in having much shorter less acuminate sepals and petals, and the spur simply clavate and not abruptly inflated.

Aeranthes ecalcarata *H.Perrier* in Notul. Syst. (Paris) 7: 40 (1938). TYPE: Madagascar, Mt. d'Ambre, *H.Perrier* 17715 (holo. P).

Distribution: Madagascar: Antananarivo, Antsiranana, Fianarantsoa, Toamasina.
Habitat: humid, evergreen, mossy forest, on tree trunks and branches of moss and lichen-covered trees.
Altitude: 1200 – 1400 m.
Flowering time: December – July.
Life form: epiphyte
Phytogeographical region: I, III.
Description: {896}.
Illustration: {896}, {1259}.
Notes: plant and flowers small; lip reduced to a slight swelling.

Aeranthes erectiflora *Senghas* in Orchidee 38 (1): 7 (1987). TYPE: Madagascar, S of Ambositra, *Rauh & Buchloh* 7692 (holo. HEID).

Distribution: Madagascar.
Habitat: humid, evergreen forest.
Altitude: 1500 – 2000 m.
Flowering time: September.
Life form: epiphyte.
Phytogeographical region: III.
Description: {1262}.
Illustration: {1262}, {1264}.
Notes: distinct by its erect inflorescence, differs from *A. longipes* by the less diaphanous flower, spur which is less curved, with the apex thickened and clavate, and the distinct long rostellar tooth.

Aeranthes filipes *Schltr.* in Ann. Mus. Col. Marseille, ser.3, 1: 185, t.19 (1913). TYPE: Madagascar, Manongarivo, *H.Perrier* (95) 1852 (holo. B†; iso. P.).

Distribution: Madagascar: Antsiranana, Toamasina, Toliara.
Habitat: riverine forest; humid, evergreen forest.
Altitude: 1000 – 1400 m.
Flowering time: March.
Life form: epiphyte.
Phytogeographical region: I, III.
Description: {522}, {896}.
Illustration: {3}, {529}, {1205}.
Cultivation: {529}.
Notes: inflorescence pendant, wiry; floral bracts acuminate, very small; lip broadly ovate-cordate at the base and acuminate at the apex; spur straight or a little inflated at the tip.

Aeranthes grandiflora *Lindl.* in Bot. Reg. 10: t.817 (1824). TYPE: Madagascar, Mt. d'Ambre, *Forbes* s.n. (holo. K).
Aeranthes brachycentron Regel, Gartenflora 40: 323 (1891). TYPE: Comoros, cult. Sander & Co. (holo. B†).

Distribution: Madagascar: Antsiranana, Toamasina, Toliara. Comoros.
Habitat: coastal and plateau, humid, evergreen forestand littoral forest, on trunks.
Altitude: sea level – 1200 m.
Flowering time: July – January.
Life form: epiphyte.
Phytogeographical region: I, III.

DESCRIPTION: {50}, {896}, {1085}, {1319}.

ILLUSTRATION: {3}, {50}, {51}, {74}, {199}, {226}, {244}, {292}, {308}, {370}, {388}, {440}, {470}, {504}, {514}, {529}, {615}, {747}, {771}, {806}, {896}, {906}, {944}, {957}, {1259}, {1285}, {1319}, {1425}, {1442} .

HISTORY: {226}.

CULTIVATION: {388}, {529}, {747}, {941}.

NOTES: *A. caudata* is distinguished by its smaller flowers and many-flowered inflorescences; and *A. ramosa* by its smaller flowers with shortly caudate segments and cylindrical spur which is not dilated above nor bent forward beneath the lip.

Aeranthes henrici *Schltr.* in Repert. Sp. Nov. Regni Veg. Beih. 33: 278 (1925). TYPE: Madagascar, Sambirano Mts, Manongarivo massif, *H.Perrier* 8128 (iso. P).

DISTRIBUTION: Madagascar: Antsiranana, Mahajanga.

HABITAT: humid, evergreen forest on plateau.

ALTITUDE: c. 1000 m.

FLOWERING TIME: April, August.

LIFE FORM: epiphyte.

PHYTOGEOGRAPHICAL REGION: II, III.

DESCRIPTION: {217}, {896}.

ILLUSTRATION: {77}, {135}, {217}, {281}, {479}, {501}, {507}, {529}, {535}, {724}, {896}, {1284}.

CULTIVATION: {529}.

NOTES: flowers white, very large; lip very broad, fimbriate; spur very long (up to12 cm). This species is to be re-classified in *Adansonia* as: **Erasanthe henrici** (*Schltr.*) *P.J.Cribb, Hermans & D.L.Roberts.*

Aeranthes henrici var. **isaloensis** *H.Perrier* ex *Hermans* **var. nov**. TYPE: Madagascar, Isalo, *H.Perrier* 16895 & Analavelona, N of Fiherenana, *Humbert* 14217 (holo. P). See Appendix.

Aeranthes henrici Schltr. var. *isaloensis* H.Perrier in Humbert ed., Fl. Mad. Orch. 2: 138 (1941), *nom. nud.*

DISTRIBUTION: Madagascar: Fianarantsoa, Toliara.

HABITAT: evergreen forest.

ALTITUDE: 500 – 1000 m.

FLOWERING TIME: March.

LIFE FORM: epiphyte.

PHYTOGEOGRAPHICAL REGION: III, V.

DESCRIPTION: {896}.

NOTES: this variety differs from the typical one in its triangular, acute floral bracts and smaller flowers which have a lip with an elliptic-transversal basal lamina extending into a square blade and with a shorter acuminate apex, a shorter spur and shorter pedicellate ovary, and a column that is longer and distinct in structure. It could be interpreted as a small, localised form. This subspecies is to be re-classified as: **Erasanthe henrici** (*Schltr.*) *P.J.Cribb & D.L.Roberts* subsp. **isaloensis** (*H. Perr. ex Hermans*) *P.J.Cribb & D.L.Roberts*.

Aeranthes laxiflora *Schltr.* in Repert. Sp. Nov. Regni Veg. Beih. 33: 279 (1925). TYPES: Madagascar, Andringitra massif, *H.Perrier* 14572 (lecto. P, selected here); Mt. Tsaratanana, *H.Perrier* 15722 (syn. P).

DISTRIBUTION: Madagascar: Antsiranana, Fianarantsoa.

HABITAT: montane forest.

ALTITUDE: 1600 – 1800 m.

FLOWERING TIME: January – February.

LIFE FORM: epiphyte.

PHYTOGEOGRAPHICAL REGION: III.

DESCRIPTION: {896}.

ILLUSTRATION: {896}.

NOTES: inflorescence pendant; flowers thin and medium-sized; lip broadly rounded-angular, then shortly acuminate; spur slender, cylindrical, 13 mm long.

Aeranthes leandriana *Bosser* in Adansonia, sér. 2, 11 (1): 83 (1971). TYPE: Madagascar, Tampoketsa d'Ankazobe, *Bosser* 16464 (holo. & iso. P).

DISTRIBUTION: Madagascar: Antananarivo.
HABITAT: humid, evergreen forest on plateau.
FLOWERING TIME: November – December.
ALTITUDE: between 1400 – 1500 m.
LIFE FORM: epiphyte.
PHYTOGEOGRAPHICAL REGION: III.
DESCRIPTION: {118}.
ILLUSTRATION: {118}.
NOTES: related to *A. albidiflora* but the sepals and petals are of a different shape and not folded backwards, the lip is not horizontal and cordate at the base, the column is shorter.

Aeranthes longipes *Schltr.* in Repert. Sp. Nov. Regni Veg. Beih. 33: 280 (1925). TYPE: Madagascar, Andringitra massif, *H.Perrier* 14581 (holo. B; iso. P).
Aeranthes rigidula Schltr. in Repert. Sp. Nov. Regni Veg. Beih. 33: 283 (1925). TYPE: Madagascar, Itremo, *H.Perrier* 12457 (holo. B†; iso. P).

DISTRIBUTION: Madagascar: Antananarivo, Fianarantsoa.
HABITAT: mossy, evergreen forest on moss- and lichen-covered trees.
ALTITUDE: c. 1400 m.
FLOWERING TIME: February – March, July.
LIFE FORM: epiphyte.
PHYTOGEOGRAPHICAL REGION: III.
DESCRIPTION: {896}.
ILLUSTRATION: {281}, {529}, {807}.
CULTIVATION: {529}.
NOTES: inflorescence long; tepals longly acuminate; lip rhomboid-suborbicular, truncate at the base and apiculate at the tip; spur short and almost cylindrical, obtuse.

Aeranthes moratii *Bosser* in Adansonia, sér. 2, 11 (1): 87 (1971). TYPE: Madagascar, Sambava Region, *P.Morat* 2835 (holo. P).

DISTRIBUTION: Madagascar: Antsiranana.
HABITAT: humid, lowland forest
LIFE FORM: epiphyte.
PHYTOGEOGRAPHICAL REGION: I.
DESCRIPTION: {118}.
ILLUSTRATION: {118}.
NOTES: recognisable by its thick, grey-green foliage, and flower with a compressed, short spur.

Aeranthes multinodis *Bosser* in Adansonia, sér. 2, 11 (1): 81 (1971).TYPE: Madagascar, E bank of Lake Tsiazompaniry, *Bosser & J.-P.Peyrot* 20009 (holo. P).

DISTRIBUTION: Madagascar: Antananarivo.
HABITAT: humid, evergreen, forest on plateau.
ALTITUDE: 1300 – 1400 m.
LIFE FORM: epiphyte.
PHYTOGEOGRAPHICAL REGION: III.
DESCRIPTION: {118}.
NOTES: this species is characterised by a peduncle with many nodes, and sheaths which are longer than the internodes. Close to *Aeranthes longipes*.

Aeranthes neoperrieri *Toill.-Gen., Ursch & Bosser* in Notul. Syst. (Paris) 16: 212 (1960). TYPE: Madagascar, Sandrangato, *Bosser* 13274 (holo. P).

DISTRIBUTION: Madagascar: Toamasina.
HABITAT: high-altitude, mossy, evergreen forest on moss- and lichen-covered trees.
FLOWERING TIME: October.

LIFE FORM: epiphyte.
PHYTOGEOGRAPHICAL REGION: I.
DESCRIPTION: {529}, {1404}.
ILLUSTRATION: {3}, {518}, {529}, {906}, {1404}.
CULTIVATION: {518}, {522}, {529}.
NOTES: inflorescence slender, long and pendent; lip subrectangular, obtuse acuminate, the base cordate with two rounded lobes, the blade with two lateral diverging ridges and one shorter central crest; spur cylindrical, a little broadened towards the tip, short.

Aeranthes nidus *Schltr.* in Repert. Sp. Nov. Regni Veg. Beih. 33: 281 (1925); Bosser in Adansonia 11 (1): 38 (1989). TYPE: Madagascar, Mt. Tsaratanana, *H.Perrier* 15293 (holo. B†; iso. P).
Aeranthes pseudonidus H.Perrier in Notul. Syst. (Paris) 7: 47 (1938). TYPE: Madagascar, Tsinjoarivo, *H.Perrier* 17118 (holo. P).

DISTRIBUTION: Madagascar: Antananarivo, Antsiranana.
HABITAT: mossy, evergreen forest, on moss and lichen-covered trees, encircling tree stems to catch debris, forming big clumps.
ALTITUDE: 1000 – 2000 m.
FLOWERING TIME: January – February, May.
LIFE FORM: epiphyte.
PHYTOGEOGRAPHICAL REGION: III.
DESCRIPTION: {529}, {896}.
ILLUSTRATION: {896}.
CULTIVATION: {529}.
NOTES: the dense plant with grass-like foliage is characteristic; flowers small; lip cordate at the base; spur straight.

Aeranthes orophila *Toill.-Gen.* in Nat. Malg. 10: 19 – 20 (1958). TYPE: Madagascar, Manjakatompo, Ankaratra, *Capuron* 996 (holo. P).

DISTRIBUTION: Madagascar: Antananarivo, Antsiranana.
HABITAT: mossy, evergreen, high-altitude forest.
ALTITUDE: 2000 – 2500 m.
FLOWERING TIME: November – January.
LIFE FORM: epiphyte.
PHYTOGEOGRAPHICAL REGION: IV.
DESCRIPTION: {1403}.
ILLUSTRATION: {1403}.
NOTES: small plant with copious roots and tiny flowers; lip rectangular, auriculate at the base; spur cylindrical, short.

Aeranthes orthopoda *Toill.-Gen., Ursch & Bosser* in Notul. Syst. (Paris) 16: 212 (1960). TYPE: Madagascar, Mandraka, *Rakotovazaha* in Jard. Bot. Tana. 810 (holo. P).

DISTRIBUTION: Madagascar: Antananarivo.
HABITAT: humid, evergreen, mossy forest on plateau, on moss- and lichen-covered trees.
LIFE FORM: epiphyte.
PHYTOGEOGRAPHICAL REGION: III.
DESCRIPTION: {1404}.
ILLUSTRATION: {514}, {518}, {529}.
CULTIVATION: {529}.
NOTES: lip rounded, orbicular with a narrow non-auriculate base; spur claviform, 10 – 12 mm. Northen {1412} refers to and illustrates it as *Aeranthes ramosus* var. *orthopoda* (*nom. nud.*).

Aeranthes parvula *Schltr.* in Ann. Mus. Col. Marseille, sér. 3, 1: 186, t.19 (1913). TYPE: Madagascar, Sambirano, *H.Perrier* 1853 (syn. B†; isosyn. P.); *H.Perrier* 15444 (syn. P); *H.Perrier* 16493 (syn. P).

DISTRIBUTION: Madagascar: Antsiranana.
HABITAT: on tamarind trees.

FLOWERING TIME: February – May.
LIFE FORM: epiphyte {308}.
PHYTOGEOGRAPHICAL REGION: II.
DESCRIPTION: {529}, {896}.
ILLUSTRATION: {896}, {1205}.
NOTES: plant and flowers small; lip sub rectangular and acuminate at the tip; column-foot big; spur broad at the opening, claviform, obtuse.

Aeranthes peyrotii *Bosser* in Adansonia, sér. 2, 11 (1): 89 (1971). TYPE: Madagascar, Lake Mantasoa, *Bosser & J.-P.Peyrot* 17570 (holo. P).

DISTRIBUTION: Madagascar: Antananarivo, Toamasina.
HABITAT: humid, evergreen forest.
ALTITUDE: 850 – 1200 m.
FLOWERING TIME: January – March.
LIFE FORM: epiphyte.
PHYTOGEOGRAPHICAL REGION: III.
DESCRIPTION: {118}.
ILLUSTRATION: {118}, {281}, {444}, {510}, {529}, {806}, {906}.
CULTIVATION: {529}.
NOTES: very distinct species, sometimes carrying plantlets at the internodes of the spike and sheaths. Leaves pendent, long and often glaucous; flowers large; lip oblong, sub-pandurate; spur club-shaped at the end, column wings long.

Aeranthes polyanthemus *Ridl.* in J. Linn. Soc., Bot. 22: 121 (1886). TYPE: Madagascar, Ambatovory (Imerina), *Fox* 32 (holo. BM).

DISTRIBUTION: Madagascar: Antananarivo, Fianarantsoa, Toamasina.
HABITAT: mossy, montane forest, on moss on trees and rocks.
ALTITUDE: 1100 m.
FLOWERING TIME: October, January.
LIFE FORM: epiphyte or lithophyte.
PHYTOGEOGRAPHICAL REGION: III.
DESCRIPTION: {281}, {896}.
NOTES: small plant with a long, pendent, many-flowered inflorescence; lip adnate to the base of the column, its base forming a somewhat deep concave depression; spur short, obtuse and incurved. Palacky {889} refers *Angraecum palmiferum* Thouars to this species.

Aeranthes ramosa *Rolfe* in Orch. Rev. 9 (107): 352 (1901). TYPE: Madagascar, Tanambe, *M.Warpur* s.n. (holo. K).
Aeranthes ramosus Cogn., Dict. Ic. Orch. Aer.: t.2 (1902), as *Aeranthus ramosus.*
Aeranthes vespertilio Cogn., Dict. Ic. Orch., Aer.: t.2 (1902), *nom. illeg.*
Aeranthes brevivaginans H.Perrier in Notul. Syst. (Paris) 7: 42 (1938), **syn. nov.** TYPE: Madagascar, Manerinerina, *H.Perrier* 16872 (lecto. P, selected here); Analamaitso, *H.Perrier* 1876 (syn. P).

DISTRIBUTION: Madagascar: Antananarivo, Mahajanga, Toamasina.
HABITAT: humid, evergreen, mossy forest, on the trunks of *Philippia*, moss- and lichen-covered trees.
ALTITUDE: 1000 – 1500 m.
FLOWERING TIME: May – August, October, December – February.
LIFE FORM: epiphyte.
PHYTOGEOGRAPHICAL REGION: I, III.
DESCRIPTION: {889}, {896}, {1137}.
ILLUSTRATION: {3}, {50}, {65}, {199}, {281}, {529}, {806}, {850}, {896}, {906}, {936}, {941}, {944}, {1026}, {1039}.
HISTORY: {199}, {1110}.
CULTIVATION: {65}, {527}, {529}, {615}, {1333}, {1440}.
NOTES: a large plant with a very long, slender, branching raceme; flowers large and variable in colour and shape; lip strongly auriculate at the base, then broadly obovate; spur club-

shaped and narrowed at the base. *A. ramosa* is a very variable species both in shape and size. *A. brevivaginans* H.Perrier is a small specimen of this species, with a slightly differently shaped mentum.

Aeranthes robusta *Senghas* in Orchidee 38 (1): 5 (1987). TYPE: Madagascar, Perinet, *Rauh & Buchloh* 7136 (holo. HEID); *W.Rauh & G.Buchloh* 7708 (para. HEID).

DISTRIBUTION: Madagascar: Antananarivo, Toamasina.
HABITAT: montane forest.
ALTITUDE: 1300 – 1600 m.
FLOWERING TIME: July, September.
LIFE FORM: epiphyte.
PHYTOGEOGRAPHICAL REGION: I.
DESCRIPTION: {1261}.
ILLUSTRATION: {1259}, {1261}.
NOTES: related to *Aeranthes crassifolia* and *A. carnosa*, but of a different texture and with bigger flowers, also differs by the pendent habit; lip very curved, mobile; spur linear, curved, slightly thickened at the tip.

Aeranthes sambiranoensis *Schltr.* in Repert. Sp. Nov. Regni Veg. Beih. 33: 284 (1925), TYPE: Madagascar, Sambirano Valley, *H.Perrier* 15721 (holo. B†; iso. P).

DISTRIBUTION: Madagascar: Antsiranana.
HABITAT: humid, evergreen forest at intermediate elevation; seasonally dry, deciduous forest and woodland.
ALTITUDE: c. 800 m.
FLOWERING TIME: January.
LIFE FORM: epiphyte.
PHYTOGEOGRAPHICAL REGION: II.
DESCRIPTION: {896}.
NOTES: inflorescence long, ascending; lip broadly ovate, 3-lobed at the front; spur broadly expanded at the opening, straight, thickly-globulose at the tip.

Aeranthes schlechteri *Bosser* in Adansonia, sér. 2, 9 (4): 547 (1969). TYPE: Madagascar, Sambirano Valley, *H.Perrier* 15446 (holo. B†; iso. P).
Neobathiea gracilis Schltr. in Repert. Sp. Nov. Regni Veg. Beih. 33: 370 (1925).
Neobathiea sambiranoensis Schltr. in Repert. Sp. Nov. Regni Veg. Beih. 33: 371 (1925). TYPE: Madagascar, Sambirano Valley, *H.Perrier* 1900 (holo. P).

DISTRIBUTION: Madagascar: Antsiranana, Mahajanga.
HABITAT: deciduous and semi-deciduous forest.
ALTITUDE: 90 – 1250 m.
FLOWERING TIME: February – March.
LIFE FORM: epiphyte on twigs, especially on *Erythroxylon*.
PHYTOGEOGRAPHICAL REGION: II.
DESCRIPTION: {114}, {896}.
ILLUSTRATION: {114}, {281}, {896}.
NOTES: inflorescence slender; sepals and lip very longly acuminate; spur long and filiform.

Aeranthes setiformis *Garay* in Bot. Mus. Leafl. 23 (4): 157 (1972), *nom. nov. pro A. pusilla* Schltr.
Aeranthes pusilla Schltr. in Repert. Sp. Nov. Regni Veg. Beih. 33: 282 (1925). TYPE: Madagascar, Mt. Tsiafajavona, *H.Perrier* 13429 (holo. B†; iso. P).

DISTRIBUTION: Madagascar: Antananarivo.
HABITAT: mossy, montane forest, on moss- and lichen-covered trees.
ALTITUDE: c. 2000 m.
FLOWERING TIME: October.
LIFE FORM: epiphyte.
PHYTOGEOGRAPHICAL REGION: IV.

DESCRIPTION: {896}.

ILLUSTRATION: {529}.

CULTIVATION: {529}.

NOTES: a small plant; lip suborbicular or broadly oval, very obtuse; spur 3 mm, curved at the front, cylindrical, obtuse.

Aeranthes setipes *Schltr.* in Repert. Sp. Nov. Regni Veg. Beih. 33: 285 (1925). TYPE: Madagascar, Mt. Tsaratanana, *H.Perrier* 15717 (holo. B†; iso. P).

DISTRIBUTION: Madagascar: Antsiranana, Toamasina.

HABITAT: mossy, montane forest on moss- and lichen-covered trees.

ALTITUDE: c. 1800 m.

FLOWERING TIME: January.

LIFE FORM: epiphyte.

PHYTOGEOGRAPHICAL REGION: III, IV.

DESCRIPTION: {896}.

NOTES: small, slender plant; plantlets can develop along the inflorescence; lip broad and then very acuminate, carrying 2 small calli near the base; spur straight, cylindrical-conical.

Aeranthes subramosa *(Schltr) Garay* in Bot. Mus. Leafl. Harvard Univ., 23 (4): 157 (1972). TYPE: Madagascar, Tsaratanana mountains, *H.Perrier* 15719 (holo. B†; iso. P).

Aeranthes gracilis Schltr. in Repert. Sp. Nov. Regni Veg. Beih. 33: 277 (1925), *non* Rchb.f. in Flora 1: 117 (1867) . TYPE: Madagascar, Tsaratanana mountains, *H.Perrier* 15719 (holo. B†; iso. P).

DISTRIBUTION: Madagascar: Antsiranana.

HABITAT: mossy and lichen-rich forest.

FLOWERING TIME: January.

LIFE FORM: epiphyte.

DESCRIPTION: {896}, {906}, {1221}.

NOTES: leaves fairly thick and short; inflorescences longer than the leaves; lip subrectangular, with 2 oblique ridges at the base.

Aeranthes tenella *Bosser* in Adansonia, ser.2, 11 (1): 85 (1971). TYPE: Madagascar, Marojejy massif, cult. Jard. Bot. Tananarive 1399 (holo. P).

DISTRIBUTION: Madagascar: Antsiranana.

HABITAT: humid, mossy, evergreen forest.

LIFE FORM: epiphyte.

PHYTOGEOGRAPHICAL REGION: IV.

DESCRIPTION: {118}.

ILLUSTRATION: {118}.

NOTES: related to *A. parvula*, but the leaves are wider, the sepals and petals are more longly acuminate in *A. parvula*, the lip is not denticulate and the spur clearly club-shaped at the tip.

Aeranthes tricalcarata *H.Perrier* in Notul. Syst. (Paris) 8: 43 (1939). TYPE: Madagascar, Isomono, Mandrare basin, *Humbert* 13164 (holo. P).

DISTRIBUTION: Madagascar: Toliara.

HABITAT: shaded, humid rocks.

ALTITUDE: c. 1200 m.

FLOWERING TIME: December.

LIFE FORM: epiphyte or lithophyte.

PHYTOGEOGRAPHICAL REGION: III.

DESCRIPTION: {896}.

NOTES: the base of the lip and column-foot are expanded into rounded lobes; the spur is nestled between these and narrowed in the middle.

Aeranthes tropophila *Bosser* in Adansonia, sér. 2, 11 (1): 91 (1971). TYPE: Madagascar, Tsingy d'Antsalova, *Bosser & P.Morat* 18113 (holo. P).

> DISTRIBUTION: Madagascar: Antsiranana, Mahajanga.
> HABITAT: dry, deciduous scrubland, and in tropical xerophilous forests.
> FLOWERING TIME: September – October.
> LIFE FORM: epiphyte.
> DESCRIPTION: {118}.
> ILLUSTRATION: {118}, {1259}.
> NOTES: the short inflorescences are similar to those of *A. laxiflora* but apart from coming from very different habitats, there also are differences in sepals, petals and lip.

Aeranthes virginalis *D.Roberts* in Kew Bull. (2005) 60: 139 – 141. TYPE: Comoros, Moheli, cultivated RBG Kew EN 1991 – 1049 (holo. K; iso. P, TAN)

> DISTRIBUTION: Comoros: Moheli.
> HABITAT: Mossy cloud forest.
> LIFE FORM: epiphyte.
> DESCRIPTION: {1049b}.
> ILLUSTRATION: {1049b}.
> NOTES: Somewhat similar to *A. arachnites* but the cross-legged appearance of the lateral sepals is typical.

Agrostophyllum occidentale *Schltr.* in Beih. Bot. Centralbl. 33 (2): 413 (1915). TYPE: Madagascar, Nosi-Faly (nr. Nosi-Ve), *Pervillé* s.n. (holo. B†; iso. P).
Agrostophyllum seychellarum Rolfe in Kew Bull. 23, 1: 23 – 24 (1922), **syn. nov.** TYPE: Seychelles, Morne Blanc, Mahé, *H.P.Thomasset* s.n. (holo. K).

> DISTRIBUTION: Madagascar: Antsiranana. Seychelles: Mahé, Silhouette.
> HABITAT: under large trees.
> ALTITUDE: 300 – 800 m.
> FLOWERING TIME: December – March.
> LIFE FORM: terrestrial.
> PHYTOGEOGRAPHICAL REGION: II.
> DESCRIPTION: {896}.
> NOTES: a tall plant forming dense clumps, with flattened stems completely covered in rigid sheaths and inflorescences in terminal heads of small flowers. *Agrostophyllum seychellarum* Rolfe differs only slightly in the shape of the flower.

Ambrella longituba *H.Perrier* in Bull. Soc. Bot. France 81: 655 (1934). TYPE: Madagascar, Mt. d'Ambre, *H.Perrier* 18811 (holo. P).

> DISTRIBUTION: Madagascar: Antsiranana.
> HABITAT: humid, evergreen forest on plateau, on branches of *Calliandra*.
> ALTITUDE: 500 – 1000 m.
> FLOWERING TIME: November.
> LIFE FORM: epiphyte.
> PHYTOGEOGRAPHICAL REGION: III.
> DESCRIPTION: {896}.
> ILLUSTRATION: {896}.
> NOTES: a peculiar species vegetatively resembling *Aerangis*. The lip completely envelops the column and is rolled into a narrow tube.

Angraecopsis parviflora *(Thouars) Schltr.*, Orchideen: 601 (1914). TYPE: Mauritius, *Thouars* 30 (holo. P).
Angraecum parviflorum Thouars, Hist. Orch.: t.60 (1822).
Aerobion parviflorum Spreng., Syst. Veg. 3: 718 (1826).
Oeceoclades parviflora (Thouars) Lindl., Gen. Sp. Orch. Pl.: 236 (1833).
Listrostachys parviflora (Thouars) S.Moore in Baker, Fl. Mauritius: 355 (1877).
Epidorchis parviflora (Thouars) Kuntze, Rev. Gen. Pl. 2: 660 (1891).

Saccolabium parviflorum (Thouars) Cordem. in Fl. Réunion: 197 (1895).

Mystacidium pedunculatum Rolfe in Fl. Trop. Africa 7: 175 (1897). TYPE: Mozambique, Namuli, *Last* s.n. (holo. K).

DISTRIBUTION: Madagascar: Antananarivo, Antsiranana, Fianarantsoa. Mascarenes: Réunion, Mauritius. Tropical Africa: Cameroon, Malawi, Mozambique, Tanzania, Zimbabwe .
HABITAT: humid, evergreen forest on plateau.
ALTITUDE: 700 – 1100 m.
FLOWERING TIME: March, July.
LIFE FORM: epiphyte.
PHYTOGEOGRAPHICAL REGION: II, III.
DESCRIPTION: {896}.
ILLUSTRATION: {34}, {221}, {292}, {676}, {1259}, {1398}.
NOTES: petals oblique, strongly expanded at the front; lobes of the lip broad.

Angraecopsis pobeguini *(Finet) H.Perrier* in *Humbert* ed., Fl. Mad. Orch. 2: 81 (1941). TYPE: Comoros, Grande Comore, *Pobéguin* s.n. (holo. P).

Rhaphidorhynchus pobeguini Finet in Bull. Soc. Bot. France 54, Mém. 9: 41 (1907).

Chamaeangis pobeguini (Finet) Schltr. in Beih. Bot. Centralbl. 33 (2): 426 (1915).

DISTRIBUTION: Comoros: Grande Comore.
FLOWERING TIME: June.
LIFE FORM: epiphyte.
DESCRIPTION: {896}.
ILLUSTRATION: {896}.
NOTES: a small plant; lip 3-lobed; spur curved and strongly expanded into a club at the tip, shorter than the ovary.

Angraecopsis trifurca *(Rchb.f.) Schltr.* in Beih. Bot. Centralbl. 33 (2): 428 (1915). TYPE: Comoros, *Humblot* 450 (holo. W, Reichenbach Herbarium 5344; iso. P).

Aeranthes trifurcus Rchb.f. in Flora 68: 540 (1885), as *Aeranthus trifurcus*.

Mystacidium trifurcum (Rchb.f.) T.Durand & Schinz, Consp. Fl. Afric. 5: 55 (1892).

Listrostachys trifurca (Rchb.f.) Finet in Bull. Soc. Bot. France 54, Mém. 9: 51 (1907).

Mystacidium thouarsii Finet in Bull. Soc. Bot. France 54, Mém. 9: 57 (1907), *pro parte*. TYPE: Mauritius, *Thouars* s.n. (holo. P).

Angraecopsis thouarsii (Finet) H.Perrier in Humbert ed., Fl. Mad. Orch. 2: 84 (1941), *pro parte*.

Angraecopsis comorensis Summerh. in Kew Bull. 4: 443 (1949), *non* Kraenzl. TYPE: Comoros, Grande Comore, *Pobéguin* s.n. (holo. P).

DISTRIBUTION: Comoros: Grande Comore. Africa: Ethiopia, Zambia, Zimbabwe.
HABITAT: evergreen forests.
ALTITUDE: 400 – 1800 m.
FLOWERING TIME: January, April, June, October.
LIFE FORM: epiphyte or lithophyte.
DESCRIPTION: {896}, {1366}.
ILLUSTRATION: {58}, {256b}, {292}, {896}, {955}.
NOTES: petals subequilateral; lip with 3 very narrow lobes; spur thread-like.

Angraecum acutipetalum *Schltr.* in Beih. Bot. Centralbl. 34 (2): 337 (1916). TYPE: Madagascar, MtTsiafajavona, *H.Perrier* XXXIX (holo. P).

DISTRIBUTION: Madagascar: Antananarivo, Antsiranana.
HABITAT: mossy and lichen-rich forest, on moss- and lichen-covered trees.
ALTITUDE: 1000 – 2000 m.
FLOWERING TIME: September – January.
LIFE FORM: epiphyte.
PHYTOGEOGRAPHICAL REGION: III, IV.
DESCRIPTION: {877}, {896}.
ILLUSTRATION: {505}, {896}, {1223}.
NOTES: inflorescence short, with 2 – 5 small flowers; spur short and saccate.

Angraecum acutipetalum var. **analabeensis** *H.Perrier* ex *Hermans* **var. nov.** TYPE: Analabe forest, N Imerina, *H.Perrier* 18516 (holo. P). See Appendix.
Angraecum acutipetalum var. *analabeensis* H.Perrier, Fl. Mad. Orch. 2: 229 (1941), *nom. nud.*

> DISTRIBUTION: Madagascar: Antananarivo.
> HABITAT: humid, highland forest
> LIFE FORM: epiphyte.
> PHYTOGEOGRAPHICAL REGION: III.
> DESCRIPTION: {896}.
> NOTES: differs from the typical variety in its shorter perianth parts, shorter spur and anther which is broadly indented at the front, the margins clearly acutely angular, and the 2 linear stipes being united at the base by a large viscous gland. Its flowers are white, with the sepals and petals slightly yellowish.

Angraecum acutipetalum var. **ankeranae** *H.Perrier ex Hermans* **var. nov.** TYPE: Madagascar, Ankeramadinika, *E.François* in *H.Perrier* 18517 (holo. P). See Appendix.
Angraecum acutipetalum var. *ankeranae* H.Perrier in Humbert ed., Fl. Mad. Orch. 2: 230 (1941), *nom. nud.*

> DISTRIBUTION: Madagascar: Antananarivo.
> HABITAT: humid, highland forest.
> LIFE FORM: epiphyte.
> PHYTOGEOGRAPHICAL REGION: III.
> DESCRIPTION: {896}.
> NOTES: differs from the typical variety in having a shorter spur, slightly downy disk of the lip and an anther with a rounded anterior indentation. Its flower is also generally smaller.

Angraecum alleizettei *Schltr.* in Repert. Sp. Nov. Regni Veg. Beih. 18: 325 (1922). TYPE: Madagascar, Ambatovary, Imerina, *Alleizette* 885 (holo. P).

> DISTRIBUTION: Madagascar: Antananarivo, Fianarantsoa.
> HABITAT: mossy forest on moss- and lichen-covered trees.
> ALTITUDE: 1400 – 1700 m.
> FLOWERING TIME: January, July, November.
> LIFE FORM: epiphyte.
> PHYTOGEOGRAPHICAL REGION: III.
> DESCRIPTION: {896}.
> NOTES: Stem long, branched with small oval leaves; flowers very small; lip deeply boat-shaped; spur straight.

Angraecum aloifolium *Hermans & P.J.Cribb* in Orch. Rev. 105, 1214: 108 (1997). TYPE: Madagascar, Andalaloka, *Hermans* 3423 (holo. K).

> DISTRIBUTION: Madagascar: Mahajanga.
> HABITAT: dry, deciduous forest.
> FLOWERING TIME: September (in cultivation).
> LIFE FORM: epiphyte.
> PHYTOGEOGRAPHICAL REGION: V.
> DESCRIPTION: {466}.
> ILLUSTRATION: {84}, {281}, {466}, {470}, {1270}.
> NOTES: distinct by its narrowly ensiform, thick, leathery leaves with a pitted surface and the shape of the spur and lip.

Angraecum ambrense *H.Perrier* in Notul. Syst. (Paris) 7: 118 (1938). TYPE: Madagascar, Mt. Ambre, *H.Perrier* 17719 (holo. P).

> DISTRIBUTION: Madagascar: Antsiranana.
> HABITAT: mossy forest on moss- and lichen-covered trees.
> ALTITUDE: c. 1200 m.
> FLOWERING TIME: September.

LIFE FORM: epiphyte.

PHYTOGEOGRAPHICAL REGION: III.

DESCRIPTION: {896}.

NOTES: known from the type only. The pandurate lip has a narrow base; spur short, its entrance expanded into a thick lobule.

Angraecum amplexicaule *Toill.-Gen. & Bosser* in Nat. Malg. 12: 11 (1960). TYPE: Madagascar, cult. Jard. Bot. Antananarivo, *Lambert* 1212 (holo. P).

DISTRIBUTION: Madagascar: Toamasina.

HABITAT: humid, lowland and coastal forest.

FLOWERING TIME: February, July.

LIFE FORM: epiphyte.

PHYTOGEOGRAPHICAL REGION: I.

DESCRIPTION: {1405}.

ILLUSTRATION: {1405}.

NOTES: stem pendent with thin, distichous leaves; lip uppermost, broadly acute-triangular, deeply concave with a broad base; spur long, thin, curved in the middle.

Angraecum ampullaceum *Bosser* in Adansonia, sér. 2, 10 (1): 109 (1970). TYPE: Madagascar, Mandrare, Marosohy, *Humbert* 6629 (holo. P).

Jumellea humbertii H.Perrier in Notul. Syst. (Paris) 7 (1): 60 (1938), *non Angraecum humbertii* H.Perrier.

DISTRIBUTION: Madagascar: Toamasina, Toliara.

HABITAT: humid, evergreen forest at intermediate elevation and on plateau, on trunks.

ALTITUDE: 500 – 1400 m.

FLOWERING TIME: March, November.

LIFE FORM: epiphyte.

PHYTOGEOGRAPHICAL REGION: I, III.

DESCRIPTION: {896}.

ILLUSTRATION: {896}.

NOTES: peculiar for its spur-shaped growths at the base of the peduncle surrounded by tissue that seems to contain a resin; lip triangular-acute; and anther clearly 3-toothed at the front.

Angraecum andasibeense *H.Perrier* in Notul. Syst. (Paris) 7: 124 (1938). TYPE: Madagascar, Andasibe forest, Onive, *H.Perrier* 17124 (holo. P).

DISTRIBUTION: Madagascar: Antananarivo, Antsiranana.

HABITAT: mossy and lichen-rich forest.

ALTITUDE: 1000 – 1700 m.

FLOWERING TIME: February.

LIFE FORM: epiphyte on moss- and lichen-covered trees.

PHYTOGEOGRAPHICAL REGION: III.

DESCRIPTION: {896}.

NOTES: known only from the type. Leaves narrowly linear; flowers small; lip c. 3.5 mm long, thickly apiculate; spur short and slender.

Angraecum andringitranum *Schltr.* in Repert. Sp. Nov. Regni Veg. Beih. 33: 344 (1925). TYPE: Madagascar, Andringitra massif, *H.Perrier* 14574 (holo. B†; iso. P).

DISTRIBUTION: Madagascar: Fianarantsoa.

HABITAT: mossy forest on moss- and lichen-covered trees.

ALTITUDE: c. 1600 m.

FLOWERING TIME: March.

LIFE FORM: epiphyte.

PHYTOGEOGRAPHICAL REGION: III, IV.

DESCRIPTION: {896}, {1221}.

NOTES: stem long; leaves ligulate; flowers small; lip uppermost; spur 22 mm long and filiform.

Angraecum ankeranense *H.Perrier* in Notul. Syst. (Paris) 7: 115 (1938). TYPE: Madagascar, Ankeramadinika, *H.Perrier* 18650 (holo. P).

DISTRIBUTION: Madagascar: Antananarivo.
HABITAT: mossy forest on moss- and lichen-covered trees.
ALTITUDE: 700 – 2000m.
FLOWERING TIME: February.
LIFE FORM: epiphyte.
PHYTOGEOGRAPHICAL REGION: III.
DESCRIPTION: {896}.
NOTES: sheaths with strong longitudinal ridges; pedicellate ovary very broadly winged; leaves fleshy; lip broadly oval; spur long.

Angraecum appendiculoides *Schltr.* in Repert. Sp. Nov. Regni Veg. Beih. 18: 325 (1922). TYPE: Madagascar, Ambatolampy, Onive, *Viguier & Humbert* 1892 (holo. B†; iso. P).

DISTRIBUTION: Madagascar: Antananarivo.
HABITAT: mossy, evergreen forest on moss- and lichen-covered trees.
ALTITUDE: 1700 – 2000 m.
FLOWERING TIME: September – December.
LIFE FORM: epiphyte.
PHYTOGEOGRAPHICAL REGION: III, IV.
DESCRIPTION: {896}.
ILLUSTRATION: {896}, {1223}.
NOTES: stem very long, pendent, densely leaved; floral bracts longer than the pedicellate ovary; flowers small; lip broadly oval; spur almost straight.

Angraecum arachnites *Schltr.* in Repert. Sp. Nov. Regni Veg. Beih. 33: 321 (1925). TYPE: Madagascar, Manankezo to the NE of Andazobe, *H.Perrier* 11318 (holo. P; iso. K).
Angraecum ramosum subsp. *typicum* var. *arachnites* (Schltr.) H.Perrier in Humbert ed., Fl. Mad. Orch. 2: 301 (1941).

DISTRIBUTION: Madagascar: Antananarivo, Antsiranana, Fianarantsoa, Toamasina. With the principal distribution in the centre of the island.
HABITAT: mainly in eastern evergreen forest.
ALTITUDE: 1380 – 1500 m.
FLOWERING TIME: November – March.
LIFE FORM: epiphyte.
PHYTOGEOGRAPHICAL REGION: III.
DESCRIPTION: {896}.
ILLUSTRATION: {20}, {528}, {529}, {838}.
CULTIVATION: {528}, {529}.
NOTES: plant small with a long stem carrying very small, somewhat thickened ovate leaves with the margins curved downwards; flowers long and thin; sepals and petals brownish-yellow; lip white rounded and narrow.

Angraecum aviceps *Schltr.* in Repert. Sp. Nov. Regni Veg. Beih. 33: 335 (1925). TYPE: Madagascar, Bemarivo, *H.Perrier* 8020 (holo. B†).

DISTRIBUTION: Madagascar: Antsiranana.
HABITAT: humid, evergreen forest on plateau.
ALTITUDE: c. 1000 m.
FLOWERING TIME: July.
LIFE FORM: epiphyte.
PHYTOGEOGRAPHICAL REGION: III.
DESCRIPTION: {896}.
NOTES: known only from the type which is missing. Plant small, stemless; flowers small; lip with a pointed end, resembling a bird's head, with the sepals and petals as wings.

Angraecum baronii *(Finet)* *Schltr.* in Beih. Bot. Centralbl. 33 (2): 431 (1915). TYPE: Madagascar, *Baron* 1999 (holo. P; iso. BM, K).

Macroplectrum baronii Finet in Bull. Soc. Bot. France 54, Mém. 9: 24, t.4 (1907).

Angraecum dichaeoides Schltr. in Beih. Bot. Centralbl. 34 (2): 336 (1916). TYPE: Madagascar, *H.Perrier* XL (holo. B†; iso. P).

DISTRIBUTION: Madagascar: Antananarivo, Antsiranana, Fianarantsoa.

HABITAT: moist, mossy, evergreen forest on moss- and lichen-covered trees.

ALTITUDE: 1400 – 2500m.

FLOWERING TIME: September – April, July.

LIFE FORM: epiphyte.

PHYTOGEOGRAPHICAL REGION: III, IV.

DESCRIPTION: {896}.

ILLUSTRATION: {1223}.

NOTES: plant pendent; stem long; leaves long and distichous; flowers small; lip cucullate, concave; spur curved and a little thickened towards the tip.

Angraecum bemarivoense *Schltr.* in Repert. Sp. Nov. Regni Veg. Beih. 33: 332 (1925). TYPE: Madagascar, Andraronga, Bemarivo, *H.Perrier* 11487 (holo. B†; iso. P).

DISTRIBUTION: Madagascar: Antananarivo, Antsiranana.

HABITAT: humid, evergreen forest on branches.

ALTITUDE: 800 – 1600 m.

FLOWERING TIME: March, September.

LIFE FORM: epiphyte.

PHYTOGEOGRAPHICAL REGION: I, III.

DESCRIPTION: {896}.

NOTES: plant small; leaves short; petals oblong, somewhat obtuse, a little narrower than the sepals; lip oval; spur a little longer than the ovary. Similar to *A. tenellum* and *A. microcharis*.

Angraecum bicallosum *H.Perrier* in Notul. Syst. (Paris) 7: 111 (1938). TYPE: Madagascar, Mt. Ambre, *H.Perrier* 18874 (holo. P).

DISTRIBUTION: Madagascar: Antsiranana.

HABITAT: mossy forest on moss and lichen-covered trees.

ALTITUDE: c. 1000 m.

FLOWERING TIME: November (in cultivation).

LIFE FORM: epiphyte.

PHYTOGEOGRAPHICAL REGION: III.

DESCRIPTION: {529}, {896}, {1344}.

ILLUSTRATION: {520}, {529}, {896}.

NOTES: illustrations cited above are unlikely to be this species which is only known from the type, only a flower being preserved. It is reportedly close to *A. obesum* and *A. didieri*, but differs by its glabrous roots and the shape of the lip with two pronounced calli at the base, the calli are not obvious in the type.

Angraecum brachyrhopalon *Schltr.* in Repert. Sp. Nov. Regni Veg. Beih. 33: 336 (1925). TYPE: Madagascar, Sambirano Valley, *H.Perrier* 15720 (holo. B†; iso. P).

DISTRIBUTION: Madagascar: Antsiranana.

HABITAT: seasonally dry, deciduous forest and woodland.

FLOWERING TIME: January.

LIFE FORM: epiphyte.

PHYTOGEOGRAPHICAL REGION: II.

DESCRIPTION: {896}.

NOTES: related to *A. setipes*, plant without a stem; flowers small; leaves narrow; lip narrow; spur short and thickened at the end.

Angraecum breve *Schltr.* in Repert. Sp. Nov. Regni Veg. Beih. 33: 324 (1925). TYPE: Madagascar, Mt. Tsaratanana, *H.Perrier* 15688 (holo. P).

DISTRIBUTION: Madagascar: Antsiranana.

HABITAT: lichen-rich, montane forest on moss- and lichen-covered trees.

ALTITUDE: c. 1800 m.

FLOWERING TIME: January.

LIFE FORM: epiphyte.

PHYTOGEOGRAPHICAL REGION: III.

DESCRIPTION: {896}.

NOTES: Related to *Angraecum rutenbergianum*, but has thicker leaves, smaller flowers with shorter petals and sepals, a shorter, wider lip and longer, thinner spur.

Angraecum calceolus *Thouars*, Hist. Orch.: t.78 (1822). TYPE: Mauritius, *Thouars* s.n. (holo. P).

Angraecum carpophorum Thouars, Hist. Orch.: t.76 (1822). TYPE: Réunion, *Thouars* s.n. (holo. P).

Aerobion calceolus Spreng., Syst. Veg. 3: 717 (1826).

Aeranthus calceolus (Thouars) S.Moore in Baker Fl. Maur.: 353 (1877).

Epidorchis calceolus (Thouars) Kuntze, Rev. Gen. Pl. 2: 660 (1891).

Epidorchis carpophora (Thouars) Kuntze, *loc cit.* (1891).

Angraecum paniculatum Frapp. ex Cordem., Fl. Réunion: 215 (1895). TYPE: Réunion, *Frappier* s.n. (MARS not found).

Angraecum anocentrum Schltr. in Bot. Jahrb. Syst. 26: 342 (1898). TYPE: Mozambique, (holo. B†).

Macroplectrum calceolus (Thouars) Finet in Bull. Soc. Bot. France 54, Mém. 9: 31 (1907).

Angraecum rhopaloceras Schltr. in Repert. Sp. Nov. Regni Veg. Beih. 33: 338 (1925). TYPE: Madagascar, Mt. Angavokely, E Imerina, *H.Perrier* 12940 (holo. P; iso. P).

Mystacidium calceolus (Thouars) Cordem., Fl. Réunion: 220 (1895).

Mystacidium carpophorum (Thouars) Cordem., *loc. cit.* (1895).

Angraecum patens Frapp. in. Cordem., Fl. Réunion: 206 (1895).

Angraecum laggiarae Schltr. in Repert. Sp. Nov. Regni Veg. Beih. 15: 337 (1918), **syn. nov.** TYPE: Madagascar, *Laggiara* s.n. (holo. B†); neo. the illustration in Repert. Spec. Nov. Reg. Veg., 68: t.90, 358 (1932), selected here.

Angraecum guillauminii H.Perrier in Bull. Mus. 22 (1): 114 (1959). TYPE: Madagascar, *Boiteau* F.72 (holo. P).

DISTRIBUTION: Madagascar: Antananarivo, Antsiranana, Fianarantsoa, Toliara, Tomasina. Comoros: Anjouan, Mayotte. Mascarenes: Réunion, Mauritius. Seychelles.

HABITAT: in humid, evergreen forest; at the base of ericaceous scrub; on rocks and bases of tree trunks amongst moss; dry, semi-deciduous, coastal forest; sclerophyllous woodland.

ALTITUDE: sea level – 2000 m.

FLOWERING TIME: October – May.

LIFE FORM: epiphyte, terrestrial, lithophyte.

PHYTOGEOGRAPHICAL REGION: I, II, III.

DESCRIPTION: {896}, {1260}.

ILLUSTRATION: {3}, {168}, {281}, {292}, {529}, {684}, {906}, {944}, {1026}, {1028}, {1223}, {1259}, {1260}, {1324}, {1392}, {1398}.

CULTIVATION: {529}, {932}, {1260}.

NOTES: a variable and widespread species. Inflorescence not normally branched; flowers small, greenish-yellow; lip cucullate, ovate, lanceolate; spur cylindrical sometimes a little thickened at the end, short. The type of *A. laggiarae* was destroyed and there is no other plant material available. From Schlechter's description and illustration cited above {1223} and the description by Perrier de la Bâthie {896} there is little doubt that this species is the same as *A. calceolus* Thouars.

Angraecum caricifolium *H.Perrier* in Notul. Syst. (Paris) 7: 128 (1938). TYPE: Madagascar, Ankeramadinika, *E.François* in *H.Perrier* 18520 (holo. P).

DISTRIBUTION: Madagascar: Antananarivo, Antsiranana, Toamasina.

HABITAT: mossy forest on moss- and lichen-covered trees.

ALTITUDE: 900 – 1400 m.

FLOWERING TIME: January – May, July, October.
LIFE FORM: epiphyte.
PHYTOGEOGRAPHICAL REGION: III.
DESCRIPTION: {891}, {896}.
ILLUSTRATION: {1324}.
VERNACULAR NAME: Velomitongoa.
NOTES: plant long, slender; leaves linear, thin-textured; flowers very small; lip with tiny hairs inside; spur thick, cylindrical, obtuse.

Angraecum caulescens *Thouars,* Hist. Orch.: t.75 (1822). TYPE: *Thouars* s.n. (holo. P).
Aerobion caulescens (Thouars) Spreng., Syst. Veg. 3: 717 (1826).
Epidorchis caulescens (Thouars) Kuntze, Rev. Gen. Pl. 2: 660 (1891).
Mystacidium caulescens (Thouars) Ridl. in J. Linn. Soc., Bot. 21: 488 (1885).

DISTRIBUTION: Madagascar: Antananarivo, Comoros. Mascarenes: Réunion, Mauritius. Rodrigues.
HABITAT: forest.
FLOWERING TIME: March.
LIFE FORM: epiphytic.
DESCRIPTION: {896}.
ILLUSTRATION: {167}, {246}, {1028}, {1398}.
NOTES: specimens cited from Madagascar may well be misidentifications. According to Perrier de la Bâthie {896} an enigmatic species, no type survives. The plant seems to differ from *A. multiflorum* by its inflorescence which is as long as the leaf, the stem shorter than the leaves and especially by the inflorescence with a simple and few-flowered raceme.

Angraecum chaetopodum *Schltr.* in Repert. Sp. Nov. Regni Veg. Beih. 33: 336 (1925). TYPE: Madagascar, Mt. Tsaratanana, *H.Perrier* 15686 (holo. P).

DISTRIBUTION: Madagascar: Antsiranana, Mahajanga ?
HABITAT: mossy, montane forest on moss- and lichen-covered trees.
ALTITUDE: 1200 – 1500 m.
FLOWERING TIME: January.
LIFE FORM: epiphyte.
PHYTOGEOGRAPHICAL REGION: IV, V?
DESCRIPTION: {896}.
ILLUSTRATION: {896}.
NOTES: close to *A. pergracile* from the Sambirano area, but differs by its narrower, more pointed and fleshier leaves, also related to *A. rhynchoglossum* but differs by the narrower sepals and petals and the more extended lip.

Angraecum chermezoni *H.Perrier* in Notul. Syst. (Paris) 7: 107 (1938). TYPE: Madagascar, Mt. Maromizaha, Analamazaotra, *H.Perrier* 16057 (holo. P).

DISTRIBUTION: Madagascar: Toamasina.
HABITAT: mossy, montane forest on trunks, moss- and lichen-covered trees.
ALTITUDE: 1000 – 1200 m.
FLOWERING TIME: February.
LIFE FORM: epiphyte.
PHYTOGEOGRAPHICAL REGION: III.
DESCRIPTION: {896}.
NOTES: similar to *A. obversifolium* from Réunion. Stems long, branched, with many short leaves; flowers produced in abundance, small; lip broadly triangular, acute, expanded abruptly above the narrow base, acuminate towards the thick apex; spur slightly expanded towards the hook-shaped tip.

Angraecum clareae *Hermans, la Croix & P.J.Cribb* in Orch. Rev. 109, 1237: 43 – 46 (2001). TYPE: Madagascar, Antananarivo province, Anjozorobe area, *Hermans* 3788 (holo. K).
DISTRIBUTION: Madagascar: Antananarivo.

HABITAT: mid-elevation, evergreen forest.

ALTITUDE: 1350 m.

FLOWERING TIME: September – October in cultivation.

LIFE FORM: epiphyte.

PHYTOGEOGRAPHICAL REGION: III.

DESCRIPTION: {474}.

ILLUSTRATION: {474}, {529}.

NOTES: close to *A. compactum* but differs in its glaucous, glabrous, succulent leaves which end in a short sheath placed at almost right angles to the blade, peduncle type, lip shape, and the more pronounced angle of the spur. Its closest ally is *A. kraenzlinianum* but the stem is shorter, the leaves longer and closer together on the stem, and the lip and spur are a different shape.

Angraecum clavigerum *Ridl.* in J. Linn. Soc., Bot. 21: 485 (1885). TYPE: Madagascar, Ankafina, *Deans Cowan* s.n. (holo. BM).

Angorchis clavigera (Ridl.) Kuntze, Rev. Gen. Pl. 2: 651 (1891).

Monixus clavigera (Ridl.) Finet in Bull. Soc. Bot. France 54, Mém. 9: 17 (1907).

DISTRIBUTION: Madagascar: Antananarivo, Antsiranana, Fianarantsoa, Toamasina.

HABITAT: mossy and lichen-rich forest on moss- and lichen-covered trees.

ALTITUDE: 1100 – 2000 m.

FLOWERING TIME: January – March.

LIFE FORM: epiphyte.

PHYTOGEOGRAPHICAL REGION: III.

DESCRIPTION: {896}.

ILLUSTRATION: {246}, {292}, {475}, {896}.

NOTES: plant small with relatively large flowers; lip similar in shape and size to the tepals, the area around the column bright green; the spur inflated at the end.

Angraecum compactum *Schltr.* in Beih. Bot. Centralbl. 34 (2): 339 (1916). TYPE: Madagascar, Tsaratanana, *H.Perrier* (141) 11358 (holo. B†; iso. P).

DISTRIBUTION: Madagascar: Antananarivo, Antsiranana, Fianarantosa, Toamasina, Toliara.

HABITAT: humid, evergreen forest on trees and shrubs including *Philippia*.

ALTITUDE: 700 – 2000 m.

FLOWERING TIME: August – December.

LIFE FORM: epiphyte.

PHYTOGEOGRAPHICAL REGION: I, III.

DESCRIPTION: {896}.

ILLUSTRATION: {20}, {67}, {75}, {135}, {285}, {513}, {519}, {529}, {615}, {866}, {896}, {906}, {944}, {1025}, {1223}, {1269}, {1412}, {1440}.

CULTIVATION: {513}, {519}, {527}, {529}, {866}, {932}, {941}.

NOTES: a widepread and variable species. A robust plant with leathery leaves; flowers large; tepals somewhat reflexed; lip oblong, concave, very rounded at the tip with a short terminal apicule; spur wide at the base, curved and then filiform and descending.

Angraecum compressicaule *H.Perrier* in Notul. Syst. (Paris) 7: 118 (1938). TYPE: Madagascar, Tsinjoarivo, *H.Perrier* 16964 (holo. P).

DISTRIBUTION: Madagascar: Antananarivo.

HABITAT: mossy, montane forest on moss- and lichen-covered trees & rocks.

ALTITUDE: c.1400m.

FLOWERING TIME: February.

LIFE FORM: epiphyte.

PHYTOGEOGRAPHICAL REGION: III.

DESCRIPTION: {891}, {896}.

ILLUSTRATION: {896}, {1255}.

NOTES: known from the type only. This is clearly a *Jumellea* in the *Spathulata* group. Plant tall with rigid, leathery leaves; lip broadly obovate-rhomboid, wider just above the middle, narrowed from this point to the base, with 2 ridges on the disk and with papillae on the veins; spur cylindric, 12 – 18 mm long.

Angraecum conchoglossum Schltr. in Repert. Sp. Nov. Regni Veg. Beih. 15: 336 (1918). TYPE: Madagascar, *Laggiara* s.n. (holo. B†). Central Madagascar, Mt. Tsaratanana, *H.Perrier* 16962bis (lecto. P).

Macroplectrum ramosum Finet in Bull. Soc. Bot. France 54, Mém. 9: 29, pl. 6 (1907), *non Angraecum ramosum* Thouars (1822).

Angraecum bathiei Schltr. in Repert. Sp. Nov. Regni Veg. Beih. 33: 322 (1925). TYPE: Madagascar, Mt. Tsaratanana, *H.Perrier* 15335 (in part) (holo. P).

Angraecum bathiei subsp. *peracuminatum* H.Perrier in Notul. Syst. (Paris) 7: 111 (1938), *nom. nud.*; Garay in Kew Bull. 28: 495 – 516 (1973).

Angraecum ramosum auct. non Thouars; H.Perrier in Humbert ed., Fl. Mad. Orch. 2: 299 (1941).

Angraecum ramosum subsp. *typicum* H.Perrier in Humbert ed., Fl. Mad. Orch. 2: 301 (1941), *nom. inval.*

Angraecum ramosum subsp. *typicum* var. *conchoglossum* (Schltr.) H.Perrier in Humbert ed., Fl. Mad. Orch. 2: 301 (1941).

Angraecum ramosum subsp. *typicum* var. *bathiei* (Schltr.) H.Perrier in Humbert ed., Fl. Mad. Orch. 2: 302 (1941).

Angraecum ramosum subsp. *typicum* var. *peracuminatum* H.Perrier in Humbert ed., Fl. Mad. Orch. 2: 302 (1941), *nom. nud.* Based upon: Madagascar, Mt. Ambre, *H.Perrier* 18876 & Mandraka, *E.François* in *H.Perrier* 16722 (both P).

DISTRIBUTION: Madagascar: Antananarivo, Antsiranana, Fianarantsoa, Toamasina. Mascarenes: rare on Réunion.

HABITAT: mainly in mossy, evergreen forest.

ALTITUDE: 1000 – 2000 m.

FLOWERING TIME: September – April.

LIFE FORM: epiphyte.

PHYTOGEOGRAPHICAL REGION: III, IV.

DESCRIPTION: {896}.

ILLUSTRATION: {3}, {167}, {168}, {281}, {292}, {295}, {501}, {528}, {529}, {545}, {850}, {896}, {906}, {944}, {1223}, {1319}.

CULTIVATION: {528}, {529}, {581}.

NOTES: this is a variable species. The leaves are always more than 2 cm long, more or less flat, thick or thin-textured, and placed closely together on the stem or not. The flowers are similar to *A. germinyanum* from the Comoros, but have thin and long tepals which are coloured; the lip is rounded and white; roots are often rough to verrucose.

Angraecum coriaceum *(Thunb. ex Sw.) Schltr.* in Beih. Bot. Centralbl. 33 (2): 432 (1915). TYPE: Madagascar, *Thunberg* s.n. (holo. S).

Limodorum coriaceum Thunb. ex Sw. in Schrad, J. Bot.: 234 (1799).

Aerides coriaceum Thunb. ex Sw. in Schrad, J. Bot.: 234 (1799).

Epidendrum coriaceum (Thunb. ex Sw.) Poir. in Lamarck, Tabl. Encycl. Suppl.: 385 (1823).

Saccolabium coriaceum (Thunb. ex Sw.) Lindl., Gen. Sp. Orch. Pl.: 224 (1833).

Gastrochilus coriaceus (Thunb. ex Sw.) Kuntze, Rev. Gen. Pl. 2: 661 (1891).

DISTRIBUTION: Madagascar.

DESCRIPTION: {896}.

NOTES: this is a dubious species: the flower is close to *Angraecum striatum* and *A. bracteosum* from the Mascarenes. Garay {313} places this species in *Monixus* Finet, as *M. coriaceum*. It is lectotypified by *Angraecum striatum* Thouars and was considered indeterminable by Ridley {1031}.

Angraecum cornucopiae *H.Perrier* in Humbert, Mém. Inst. Sci. Mad. sér. B, 6: 265 (1955). TYPE: Madagascar, *Humbert* 22536 (holo. P).

DISTRIBUTION: Madagascar.

HABITAT: lichen-rich forest; evergreen forest.

ALTITUDE: 1000 – 2000 m.

FLOWERING TIME: November – December.

LIFE FORM: epiphyte.

PHYTOGEOGRAPHICAL AREA: IV.

DESCRIPTION: {902}.

ILLUSTRATION: {292}, {902}.

NOTES: plant large, short-stemmed; leaves grass-like; inflorescences similar to *A. multiflorum*, arranged in a linear row; flowers small; lip widely boat-shaped; spur short and round.

Angraecum corynoceras *Schltr.* in Repert. Sp. Nov. Regni Veg. Beih. 33: 345 (1925). TYPE: Madagascar, Mt. Tsaratanana, *H.Perrier* 15698 (holo. B†; iso. P).

DISTRIBUTION: Madagascar: Antsiranana.

HABITAT: lichen-rich forest on moss- and lichen-covered trees.

ALTITUDE: c. 2000 m.

FLOWERING TIME: January.

LIFE FORM: epiphyte.

PHYTOGEOGRAPHICAL REGION: IV.

DESCRIPTION: {896}, {1221}.

NOTES: known from the type only. The species is characterised by its long slender leaves and short inflorescences, brown sheaths and 8 mm long spur, which is cylindrical then recurved and clavately inflated towards the tip.

Angraecum coutrixii *Bosser* in Adansonia, sér. 2, 10 (1): 107 (1970). TYPE: Madagascar, Itremo Massif, *Peyrot & Coutrix* 34 (holo. P).

DISTRIBUTION: Madagascar, Fianarantsoa.

HABITAT: rocky outcrops, confined to the quartzite mountain ranges of Itremo where it grows with *A. protensum*.

FLOWERING TIME: January, May.

LIFE FORM: lithophyte.

PHYTOGEOGRAPHICAL REGION:

DESCRIPTION: {115}.

ILLUSTRATION: {115}.

NOTES: plant tall with small leaves; lip ovate, apex subapiculate, slightly concave; spur filiform, 5 – 5.5 cm long.

Angraecum crassum *Thouars,* Hist. Orch.: t.70 – 71 (1822). TYPE: Madagascar, *Thouars* s.n. (holo. P).
Aerobion crassum (Thouars) Spreng., Syst. Veg. 3: 717 (1826).
Angorchis crassa (Thouars) Kuntze, Rev. Gen. Pl. 2: 651 (1891).
Angraecum crassiflorum H.Perrier in Notul. Syst. (Paris) 7: 132 (1938). TYPE: Madagascar, Mananara, *Decary* 58 (holo. P).
Angraecum sarcodanthum Schltr. in Repert. Sp. Nov. Regni Veg. Beih. 15: 339 (1918); Garay in Kew Bull. 28: 495 – 516 (1973). TYPE: Madagascar, without locality, *Laggiara* s.n. (holo. B†).

DISTRIBUTION: Madagascar: Toamasina.

HABITAT: humid, evergreen, lowland and coastal forest on trunks.

ALTITUDE: sea level – 400m.

FLOWERING TIME: September – March, July.

LIFE FORM: epiphyte.

PHYTOGEOGRAPHICAL REGION: I.

DESCRIPTION: {896}.

ILLUSTRATION: {896}, {906}, {1223}, {1398}.

NOTES: plant tall; stem thick with long leaves; lip ovate-cucullate, broadly ovate, very concave, very shortly acuminate-apiculate at the front, with 2 crests on the upper surface; spur short, 25 mm long, with a wide opening, thick, cylindrical and straight; anther with two acute teeth.

Angraecum curnowianum *(Rchb.f.) T.Durand & Schinz,* Consp. Fl. Afr. 5: 41 (1892). TYPE: Madagascar, cult. Hugh Low s.n. (holo. W).
Aeranthus curnowianus Rchb.f. in Gard. Chron. sér. 2, 19: 306 (1883).
Angorchis curnowiana (Rchb.f.) Kuntze, Rev. Gen. Pl. 2: 651 (1891).
Angraecum suarezense Toill.-Gen. & Bosser in Nat. Mal. 12: 9 (1960). TYPE: Madagascar, Antananarivo Botanic Garden, *Lamarque* 415 (holo. P).
?*Angraecum subcordatum* (H.Perrier) Bosser in Adansonia, sér. 2, 10 (1): 110 (1970).

Jumellea curnowiana (Rchb.f.) Schltr. in Beih. Bot. Centralbl. 33 (2): 429 (1915).

?*Jumellea subcordata* H.Perrier in Notul. Syst. (Paris) 7: 61 (1938). TYPE: Madagascar, Ankeramadinika, *H.Perrier* 18422 (holo. P).

Mystacidium curnowianus (Rchb.f.) Rolfe in Orch. Rev. 12: 47 (1904).

DISTRIBUTION: Madagascar: Antananarivo, Antsiranana, Toamasina.
HABITAT: mossy, evergreen forest on moss- and lichen-covered trees.
ALTITUDE: 800 – 1500 m.
FLOWERING TIME: February – May.
LIFE FORM: epiphyte.
PHYTOGEOGRAPHICAL REGION: III.
DESCRIPTION: {939}, {1405}.
ILLUSTRATION: {896}, {939}.
NOTES: plant robust; flowers large; lip widely ovate, apex acute-subapiculate; spur filiform, up to 9 cm long. *A. subcordatum* may be referable to this species, further work is needed. This species should not be confused with the name *Rhaphidorhynchus curnowianus* as interpreted by Finet {292} or *Aerangis curnowiana* in the sense that Perrier de la Bâthie used the name {889}.

Angraecum curvicalcar *Schltr.* in Repert. Sp. Nov. Regni Veg. Beih. 33: 325 (1925). TYPE: Madagascar, Mt. Tsaratanana, *H.Perrier* 15689 (holo. P).

DISTRIBUTION: Madagascar: Antsiranana.
HABITAT: lichen-rich forest on moss- and lichen-covered trees.
ALTITUDE: c. 2000 m.
FLOWERING TIME: January.
LIFE FORM: epiphyte.
PHYTOGEOGRAPHICAL REGION: IV.
DESCRIPTION: {896}.
NOTES: known from the type only. Plant small; flowers small; lip broadly rhombic-suborbicular, the central vein thickened into a keel at the base; spur 4 cm long, thread-like and gradually expanding towards the opening.

Angraecum curvicaule *Schltr.* in Repert. Sp. Nov. Regni Veg. Beih. 33: 346 (1925). TYPE: Madagascar, Analamazaotra, *H.Perrier* 11866 (holo. B†; iso. P).

DISTRIBUTION: Madagascar: Fianarantsoa, Toamasina.
HABITAT: humid, evergreen forest.
ALTITUDE: 900 – 1050 m.
FLOWERING TIME: January.
LIFE FORM: epiphyte.
PHYTOGEOGRAPHICAL REGION: III.
DESCRIPTION: {896}, {1060}.
ILLUSTRATION: {1223}.
NOTES: stem very long with many linear-ligulate leaves; inflorescence thin; lip broadly oval-apiculate, with a central ridge at the base; spur 7 mm long, cylindric, obtuse and as long as the lip.

Angraecum danguyanum *H.Perrier* in Notul. Syst. (Paris) 7: 107 (1938). TYPE: Madagascar, confl. of Mangoro & Onive, *H.Perrier* 17114 (holo. P).

DISTRIBUTION: Madagascar: Antananarivo, Fianarantosoa, Toamasina.
HABITAT: humid, evergreen forest.
ALTITUDE: 700 – 1000 m.
FLOWERING TIME: February.
LIFE FORM: epiphyte on trunks.
PHYTOGEOGRAPHICAL: REGION: I, III.
DESCRIPTION: {896}.
ILLUSTRATION: {896}.
NOTES: stem long and slender; leaves slightly rigid, semi-cylindric, linear; lip large, very broadly elliptic, obtuse; spur 6 – 7 cm long, becoming thinner towards the tip.

Angraecum dasycarpum *Schltr.* in Repert. Sp. Nov. Regni Veg. Beih. 15: 337 (1918). TYPE: Madagascar, *Laggiara* s.n. (holo. B†).

DISTRIBUTION: Madagascar: Toamasina.
HABITAT: evergreen forest; *Philippia* forest; coastal forest; on rocks covered in moss.
ALTITUDE: sea-level – 800 m.
FLOWERING TIME: January – April.
LIFE FORM: epiphyte or lithophyte.
PHYTOGEOGRAPHICAL REGION: I.
DESCRIPTION: {896}.
ILLUSTRATION: {50}, {1223}, {1267}.
NOTES: small semi-succulent plant with small leaves; lip oblong-ligulate, obtuse, slightly narrowed above the middle, concave, 6 mm long; spur oblong-cylindrical, obtuse, slightly narrowed below the middle, straight, pressed against the ovary, 4 mm long; pedicellate ovary and roots downy. A form without a spur has been recorded (*Hermans* 3427).

Angraecum dauphinense *(Rolfe) Schltr.* in Beih. Bot. Centralbl. 33 (2): 433 (1915). TYPE: Madagascar, Fort Dauphin, *Scott Elliot* 2499 (holo. K).
Mystacidium dauphinense Rolfe in J. Linn. Soc., Bot. 29: 54 (1891).

DISTRIBUTION: Madagascar, Toliara.
HABITAT: coastal forest.
LIFE FORM: epiphyte.
PHYTOGEOGRAPHICAL REGION: I.
DESCRIPTION: {896}.
NOTES: known only from the type. Allied to *A. caulescens*, but with much longer leaves and slightly different flowers.

Angraecum decaryanum *H.Perrier* in Notul. Syst. (Paris) 7: 129 (1938). TYPES: Madagascar, Andrahomana, *Decary* 4077 (lecto. P; isolecto. TAN, both selected here); Ampasimpolaka, Ambovombe, *Decary* 2817 (syn. P).

DISTRIBUTION: Madagascar: Toliara.
HABITAT: dry, deciduous scrubland on *Didiereaceae* and coastal forest.
FLOWERING TIME: June.
LIFE FORM: epiphyte.
PHYTOGEOGRAPHICAL REGION: I, V, IV.
DESCRIPTION: {896}.
ILLUSTRATION: {896}, {957}?, {1270}.
NOTES: stem long; leaves, small, acutely-linear, thick; flowers small; lip elliptic, apiculate-obtuse; spur short (c. 2 mm), tip globular.

Angraecum dendrobiopsis *Schltr.* in Repert. Sp. Nov. Regni Veg. Beih. 33: 356 (1925). TYPE: Madagascar, Mt. Tsaratanana, *H.Perrier* 15696 (holo. B†; iso. P).
DISTRIBUTION: Madagascar: Antsiranana, Toamasina.
HABITAT: mossy forest on moss- and lichen-covered trees.
ALTITUDE: 1000 – 2000 m.
FLOWERING TIME: January – April.
LIFE FORM: epiphyte.
PHYTOGEOGRAPHICAL REGION: III, IV.
DESCRIPTION: {896}, {1221}.
ILLUSTRATION: {3}, {529}, {896}.
NOTES: similar to *A. penzigianum* from the Comoros but differs in its more slender shape, narrower and longer leaves, the slightly smaller flowers with a shorter spur and the rhombic lip, as well as the much more slender stemmed inflorescences.

Angraecum didieri *(Baill. ex Finet) Schltr.* in Beih. Bot. Centralbl. 33 (2): 433 (1915); Finet in Bull. Soc. Bot. France 54, Mém. 9: 28, pl. 5 (1907). TYPE: Madagascar, *Grandidier* s.n. (holo. P).
Macroplectrum didieri Baill. ex Finet in Bull. Soc. Bot. France 54, Mém. 9: 28, pl. 5 (1907).

DISTRIBUTION: Madagascar: Antananarivo, Toamasina, Toliara.
HABITAT: humid, evergreen forest at intermediate elevation to plateau; on *Philippia*.
ALTITUDE: sea level – 1500m.
FLOWERING TIME: October – January.
LIFE FORM: epiphyte.
PHYTOGEOGRAPHICAL REGION: I, III.
DESCRIPTION: {896}, {1259}.
ILLUSTRATION: {3}, {292}, {295}, {529}, {1259}, {1444}.
CULTIVATION: {453}, {527}, {529}, {1444}.
NOTES: often misidentified in cultivation. Plants with a prominent stem; leaves ligulate, short; flowers large for the plant; lip elliptic or elliptic-oblong, shortly acute at the tip; spur narrowing to an acute tip, 8 – 15 cm long.

Angraecum dollii *Senghas* in Journ. Orchideenfreund 4, 1: 16 (1997). TYPE: Madagascar, nr. Antsirabe, *Senghas* 0 – 22177 (holo. HEID).

DISTRIBUTION: Madagascar; Antananarivo.
HABITAT: dry forest.
ALTITUDE: c. 1000m.
LIFE FORM: epiphyte.
PHYTOGEOGRAPHICAL REGION: III.
ILLUSTRATION: {1266}.
DESCRIPTION: {1266}.
NOTES: plant with long stem; leaves up to 14 cm long; flowers large, slightly glossy; lip 22 mm in diameter, apex acute; spur funnel-shaped first, curved, then thin, up to 14 cm long.

Angraecum drouhardii *H.Perrier* in Notul. Syst. (Paris) 7: 117 (1938). TYPE: Madagascar, Mt. Ambre, *H.Perrier* 19289 (holo. P).

DISTRIBUTION: Madagascar: Antsiranana.
HABITAT: mossy, montane forest on moss- and lichen-covered trees.
ALTITUDE: c. 1200m.
FLOWERING TIME: April, August.
LIFE FORM: epiphyte.
PHYTOGEOGRAPHICAL REGION: III, IV.
DESCRIPTION: {889}, {896}.
ILLUSTRATION: {896}.
NOTES: resembles *A. compressicaule*, but differs in its short and wide leaves, an inflorescence longer than the leaves, the ample and not contracted floral bracts, the wide lip folded around the column at the base, 15 × 8 mm, and the spur with a hook-shaped apex, which is slightly longer than the lip.

Angraecum dryadum *Schltr.* in Repert. Sp. Nov. Regni Veg. Beih. 33: 326 (1925). TYPES: Madagascar, Mt. Tsaratanana, *H.Perrier* 15316 (lecto. P, selected here); *H.Perrier* 15289 (syn. P).

DISTRIBUTION: Madagascar: Antananarivo, Antsiranana, Fianarantsoa, Toliara.
HABITAT: humid, lichen-rich, evergreen forest on plateau.
ALTITUDE: 1000 – 2100 m.
FLOWERING TIME: October – March.
LIFE FORM: epiphyte.
PHYTOGEOGRAPHICAL REGION: III, IV.
DESCRIPTION: {896}
ILLUSTRATION: {504}, {1269}.
NOTES: a widespread species. Leaves and flowers very variable in size. The closest relative is *A. rutenbergianum*, but it has a much longer stem and wider and thicker leaves, as well as narrower sepals and a differently shaped lip.

Angraecum eburneum *Bory,* Voy. Iles Afrique 1: 359, t.19 (1804). TYPE: lecto. Réunion, la Plaine des Chicots, illn. by Bory (t.19) cited above.

Limodorum eburneum (Bory) Willd., Sp. Pl. ed. 4: 125 (1805).

Angraecum virens Lindl. in Bot. Reg. 33: t.19 (1847). TYPE: cult. Serampore B. G., India (holo. K).

Angraecum eburneum var. *virens* Hook. in Curtis's Bot. Mag. 86: t.5170 (1860).

Angorchis eburnea (Bory) Kuntze, Rev. Gen. Pl. 2: 651 (1891).

Angraecum eburneum subsp. *typicum* H.Perrier in Humbert ed., Fl. Mad. Orch. 2: 314 (1941), *nom. superfl.*

DISTRIBUTION: Madagascar: Antsiranana, Toamasina. Comoros: Grande Comore. Mascarenes: Réunion, Mauritius. Seychelles.

HABITAT: on rocks or in littoral wet forest.

ALTITUDE: sea-level – 1500 m.

FLOWERING TIME: May – August {44}.

LIFE FORM: epiphyte or lithophyte/semi-terrestrial.

PHYTOGEOGRAPHICAL REGION: I, II.

DESCRIPTION: {50}, {896}, {1253}.

ILLUSTRATION: {3}, {45}, {50}, {106}, {149b}, {166}, {167}, {168}, {195}, {199}, {208}, {213}, {222}, {244}, {259}, {264}, {295}, {304}, {311}, {338}, {393}, {435}, {451}, {482}, {529}, {561}, {563}, {577}, {593}, {615}, {691}, {743}, {778}, {797}, {814}, {817}, {863}, {864}, {865}, {896}, {906}, {944}, {1025}, {1206}, {1222}, {1230}, {1253}, {1255}, {1264}, {1267}, {1285}, {1298}, {1343}, {1344}, {1386}, {1392}, {1398}, {1442}, {1466}, {1467}, {1468}.

HISTORY: {199}, {202}, {230}, {287}, {350}, {561}, {600}, {788}, {935}, {1080}, {1442}, {1486}.

CULTIVATION: {38}, {45}, {46}, {327}, {529}, {594}, {600}, {743}, {778}, {829}, {917}, {935}, {1386}.

VERNACULAR NAMES: Fontsylahyjamahavy, Ahaka.

NOTES: plant very large; flowers showy on a long inflorescence; lip uppermost, broadly ovate, concave; spur narrowing, 6 – 7 cm long.

Angraecum eburneum subsp. **superbum** *(Thouars) H.Perrier* in *Humbert* ed., Fl. Mad. Orch. 2: 315 (1941); Garay in Kew Bull. 28: 495 – 516 (1973). TYPE: lecto. Thouars, illustration in Hist. Orch., t.62 (1822).

Angraecum superbum Thouars, Hist. Orch.: t.62 – 64 (1822).

Aerobion superbum (Thouars) Spreng., Syst. Veg. 3: 718 (1826).

Angraecum brongniartianum Rchb.f. ex Linden, Pescatorea 1: t.16 (1854). TYPE: Ile de Bourbon, *Quesnel* s.n. (holo. W).

Angorchis brongniartiana (Rchb.f. ex Linden) Kuntze, Rev. Gen. Pl. 2: 651 (1891).

Angorchis superba (Thouars) Kuntze, *loc.cit.* 652.

Angraecum comorense Kraenzl. in Bot. Jahrb. Syst. 17: 60 (1893), *non* (Rchb.f.) Finet. TYPE: Comoros, *Schmidt* 154

Angraecum voeltzkowianum Kraenzl. in Bot. Jahrb. Syst. 36: 116 (1905). TYPE: Comoros, Grande Comore, *Voeltzkow* 193 (holo. not found).

Angraecum eburneum var. *brongniartianum* (Rchb.f. ex Linden) Finet in Bull. Soc. Bot. France 54, Mém. 9: 14 (1907).

DISTRIBUTION: Madagascar: Antananarivo, Antsiranana, Toamasina, Toliara. Comoros: Anjouan, Grande Comore, Mayotte. Seychelles: Mahé, Silhouette.

HABITAT: on rocky outcrops and in coastal forest.

ALTITUDE: sea level – 500 m.

FLOWERING TIME: September – June.

LIFE FORM: epiphyte or lithophyte.

PHYTOGEOGRAPHICAL REGION: I, II.

DESCRIPTION: {896}.

ILLUSTRATION: {3}, {135}, {199}, {287}, {414}, {470}, {535}, {724}, {736}, {749}, {817}, {896}, {904}, {924}, {1178}, {1259}, {1302}.

HISTORY: {199}, {202}, {1486}.

NOTES: this has much larger flowers than the typical subspecies with a subrectangular lip, reaching 4 cm in length and 4 – 7 cm in breath, and a spur up to 8 cm long.

Angraecum eburneum subsp. **xerophilum** *H.Perrier* in Notul. Syst. (Paris) 7: 133 (1938). TYPE: Madagascar, nr. Kotoala, *Decary* 9656 (holo. P).

DISTRIBUTION: Madagascar: Toliara.
HABITAT: dry, deciduous scrubland and xerophytic bush.
ALTITUDE: sea-level – 200 m.
FLOWERING TIME: February.
LIFE FORM: epiphyte or terrestrial.
PHYTOGEOGRAPHICAL REGION: VI.
DESCRIPTION: {891}, {896}.
ILLUSTRATION: {529}, {871}, {896}, {1239}, {1259}.
CULTIVATION: {529}, {871}.
NOTES: differs from the type in its smaller flowers, lip which is wider than long and the relatively longer 7 – 8 cm long spur.

Angraecum elephantinum *Schltr.* in Notizbl. Bot. Gart. Berlin-Dahlem 67 (7): 330 (1919). TYPE: Madagascar, *Ferko* s.n. (holo. B†).

DISTRIBUTION: Madagascar: Antananarivo.
LIFE FORM: epiphyte.
PHYTOGEOGRAPHICAL REGION: I.
DESCRIPTION: {896}.
ILLUSTRATION: {529}, {906}, {1355}.
CULTIVATION: {529}.
NOTES: the type of this species was destroyed and apart from Schlechter's description and Perrier's text {896} there is no other evidence about it. It is said to be related to *A. didieri* with a short stout plant, glabrous roots, leathery leaves, single-flower, an oval-obtuse lip and a filiform spur, c. 11 cm long. Many of the illustrations in the literature are of a different species.

Angraecum elliotii *Rolfe* in J. Linn. Soc., Bot. 29: 54 (1891). TYPE: Madagascar, Angalampena, *Scott Elliot* 2272 (holo. K).

DISTRIBUTION: Madagascar: Fianarantsoa, Toleara.
HABITAT: humid, lowland forest
LIFE FORM: epiphyte.
PHYTOGEOGRAPHICAL REGION: I {896}.
DESCRIPTION: {896}.
NOTES: allied to *A. expansum* from Mauritius, but with more acute and more distant leaves; flowers medium-sized; lip broadly oval, concave, longly acuminate; spur subclavate, 16 mm long.

Angraecum equitans *Schltr.* in Beih. Bot. Centralbl. 34 (2): 339 (1916). TYPE: Madagascar, Tsaratanana, *H.Perrier* 142 (holo. B†); Central, Mt. Tsaratanana, *H.Perrier* 11357 (*Schlechter* 142) (neo. P, selected here).

DISTRIBUTION: Madagascar: Antsiranana.
HABITAT: lichen-rich forest on moss- and lichen-covered trees.
ALTITUDE: c. 2000 m.
FLOWERING TIME: December.
LIFE FORM: epiphyte
PHYTOGEOGRAPHICAL REGION: III, IV.
DESCRIPTION: {896}.
ILLUSTRATION: {1223} all other illustrations are unlikely to be this species.
NOTES: an engimatic species often misinterpreted; the type was destroyed. It is an erect, compact plant with conduplicate overlapping leaves; the flower medium-sized and resembles *A. clareae*; lip concave, elliptic-acuminate; spur funnel-shaped at the base and then filiform, c. 8 cm long. A herbarium sheet in Paris with the same number as the type bears an elongate plant that does not correspond well with the original description.

Angraecum falcifolium *Bosser* in Adansonia, sér. 2, 10 (1): 104 (1970). TYPE: Madagascar, Forestry Station at Ampamaherana, *Bosser* 17738 (holo. P).

DISTRIBUTION: Madagascar: Fianarantsoa.
HABITAT: mossy forest on moss- and lichen-covered trees.
LIFE FORM: epiphyte.
FLOWERING TIME: February.
PHYTOGEOGRAPHICAL REGION: III.
DESCRIPTION: {115}.
ILLUSTRATION: {115}.
NOTES: known from the type only. Plant long and branching; leaves leathery; flowers small, fleshy; lip oval-lanceolate, acute, concave, downy-papillose on the inside; spur oblong at the back and thickened at the front, apex rounded, 3 mm long.

Angraecum ferkoanum *Schltr.* in Repert. Sp. Nov. Regni Veg. Beih. 15: 336 (1918). TYPE: Madagascar, *Laggiara* (holo. B†); neo. the illustration in Repert. Spec. Nov. Reg. Veg., 68 t.89, 355 (1932), selected here.

DISTRIBUTION: Madagascar.
LIFE FORM: epiphyte.
DESCRIPTION: {896}.
ILLUSTRATION: {1223}.
NOTES: the type was destroyed. It is thought to be related to *A. chloranthum*. Stemless plant with long leaves; flowers medium-sized; lip conical, very broadly oval, almost hollowed into a sac in the middle; spur curved, then straight, subfiliform, 24 mm long.

Angraecum filicornu *Thouars*, Hist. Orch.: t.52 (1822). TYPE: Madagascar, *Thouars* s.n. (holo. P).
Aeranthes thouarsii S.Moore in Baker, Fl. Mauritius: 351 (1877).
Aerobion filicornu (Thouars) Spreng., Syst. Veg. 3: 717 (1826).

DISTRIBUTION: Madagascar: Antananarivo, Toamasina, Toliara. Records from the Mascarenes refer to different species.
HABITAT: coastal forest; low elevation, humid, evergreen forest; on *Philippia*.
ALTITUDE: sea level – 800m.
FLOWERING TIME: September – July.
LIFE FORM: epiphyte.
PHYTOGEOGRAPHICAL REGION: I.
DESCRIPTION: {896}.
ILLUSTRATION: {199}, {246}, {292}, {372}, {896}, {1029}, {1398}.
HISTORY: {199}, {571}, {589}.
NOTES: stems long; roots verrucose; leaves linear, grass-like; flowers medium-sized: lip lanceolate, somewhat pandurate; spur filiform, 9 – 11 cm long.

Angraecum flavidum *Bosser* in Adansonia, sér. 2, 10 (1): 100 (1970)., TYPE: Madagascar, Andranonakoho Forest, *G.Cours* 5520 (holo. P).

DISTRIBUTION: Madagascar: Antsiranana.
HABITAT: low elevation, humid, evergreen forest; trees on limestone formations.
ALTITUDE: sea level – 500 m.
FLOWERING TIME: January.
LIFE FORM: epiphyte.
PHYTOGEOGRAPHICAL REGION: V.
DESCRIPTION: {115}.
ILLUSTRATION: {115}.
NOTES: small plant characterised by its flowers which are larger than any other species of the section *Boryangraecum*; its sepals and petals are narrow and tapering; spur 8 – 10 mm long.

Angraecum floribundum *Bosser* in Adansonia, sér. 2, 10 (1): 102 (1970). TYPE: Madagascar, nr. Anjozorobe, *Bosser* 19832 (holo. P).

DISTRIBUTION: Madagascar: Antananarivo.
HABITAT: mossy forest on moss- and lichen-covered trees.
FLOWERING TIME: August.
LIFE FORM: epiphyte.
PHYTOGEOGRAPHICAL REGION: III.
DESCRIPTION: {99}.
ILLUSTRATION: {99}.
NOTES: known only from the type. Stem and leaves long; inflorescence short; flowers abundantly; lip broad, elliptic, boat-shaped, papillose; spur 8 mm long, clavate, apex dilated into a sphere, laterally compressed.

Angraecum florulentum *Rchb.f.* in Gard. Chron. n.s. 23: 787 (1885) & in Flora 68: 380 (July 1885). TYPE: Comoros, Combani, *Humblot* 1417 (417) (holo. W; iso. BM).

DISTRIBUTION: Madagascar? Comoros: Grande Comore.
ALTITUDE: 900 – 1900 m.
FLOWERING TIME: November – December.
LIFE FORM: epiphyte on trunks.
DESCRIPTION: {896}.
ILLUSTRATION: {3}, {244}, {292}, {520}, {529}, {896}, {1317}.
HISTORY: {571}.
CULTIVATION: {529}.
NOTES: stem elongate and often pendent; flowers similar to *A. dendrobiopsis*; leaves short, 4.5 – 7 × 1 – 1.5 cm; spur 9 – 10 cm long, 3 times longer than the pedicellate ovary. Records from Madagascar are likely to be mis-identifications.

Angraecum germinyanum *Hook. f.* in Curtis Bot. Mag. 115, sér. 3, 45: t.7061 (1889). TYPE: Madagascar (in error), *Humblot* s.n. (holo. K).
Mystacidium germinyanum (Hook.f.) Rolfe in Orch. Rev. 12: 47 (1904).
Angraecum ramosum subsp. *bidentatum* H.Perrier in Humbert ed., Fl. Mad. Orch. 2: 301 (1941), *nom. nud.* Based upon: Comoros, Anjouan, *Lavanchie* 33 (P).

DISTRIBUTION: Comoros: Anjouan, Grande Comore, Mayotte.
HABITAT: humid forest
ALTITUDE: 800 and 1000 m.
FLOWERING TIME: September – December.
LIFE FORM: epiphyte.
DESCRIPTION: {545}.
ILLUSTRATION: {528}, {529}, {495}, {1319}.
CULTIVATION: {528}, {529}.
NOTES: possibly endemic to the Comoros and related to *A. conchoglossum* but has large white flowers with a broad, angular lip, the lateral sepals form a distinctive angle and the plant has smooth roots. The leaves are relatively wide but two distinct vegetative forms have been recorded; one with thin, shiny, ligulate leaves with a rounded, bilobed tip and the other with thicker, equally shiny, somewhat smaller leaves.

Angraecum humbertii *H.Perrier* in Notul. Syst. (Paris) 8: 46 (1939). TYPE: Madagascar, Mananara, *Humbert* 13734 (holo. P).

DISTRIBUTION: Madagascar: Toliara.
HABITAT: dry, deciduous scrub and rocky outcrops.
ALTITUDE: 800 – 900 m.
FLOWERING TIME: January – February.
LIFE FORM: lithophyte or epiphyte.
PHYTOGEOGRAPHICAL REGION: I, VI.
DESCRIPTION: {896}.
ILLUSTRATION: {529}.

CULTIVATION: {522}, {529}.

NOTES: stem short; leaves long and leathery; flower greenish white with a pure white lip; lip very concave, conch-shaped at the spur opening; spur at first broad, then acuminate-filiform, up to 14 cm long.

Angraecum humblotianum *(Finet) Schltr.* in Beih. Bot. Centralbl. 33 (2): 434 (1915); Summerh. in Kew Bull. 20 (2): 188 (1966). TYPE: Comoros, *Humblot* 670 (holo. P; iso. K).

Macroplectrum humblotii Finet in Bull. Soc. Bot. France 54, Mém. 9: 22, t.4 (1907).

Angraecum finetianum Schltr. in Beih. Bot. Centralbl. 36: 158 (1918). Based on same type as *A. humblotianum* (Finet) Schltr.

Angraecum humblotii (Finet) Summerh. in Kew Bull. 2: 133 (1947), *non* Rchb.f.

Angraecum abietianum Schltr. in Repert. Sp. Nov. Regni Veg. 33: 313 (1925). TYPE: Madagascar, *H.Perrier* 11861 (holo. P).

DISTRIBUTION: Comoros. Madagascar: Fianarantsoa, Toamasina.

HABITAT: humid, evergreen forest.

ALTITUDE: sea level – 900m.

FLOWERING TIME: January – March.

LIFE FORM: epiphyte.

PHYTOGEOGRAPHICAL REGION: I, III, IV.

DESCRIPTION: {896}.

ILLUSTRATION: {292}, {896}, {1221}.

NOTES: plant small; leaves long and narrow, almost cylindrical, short; lip ovate-oblong, obtuse; spur cylindric, concave, slightly inflated, partly hidden by the floral bracts; the ovary bears dark tufts of hair.

Angraecum huntleyoides *Schltr.* in Bot. Jahrb. Syst. 38: 160 (1906). TYPE: Madagascar, Toamasina, *Chr.Bang* s.n. (holo. B†); neo. the illustration in Repert. Spec. Nov. Reg. Veg., 68 t.89, 356 (1932), selected here.

Angraecum chloranthum Schltr. in Ann. Mus. Col. Marseille, sér. 3, 1: 189, t.23 (1913) **syn. nov.** TYPE: Madagascar, Analamazaotra forest, *H.Perrier* 1864 (= *Schlechter* 79) (holo. B†; iso. P).

DISTRIBUTION: Madagascar: Toamasina.

HABITAT: humid, evergreen forest on plateau; in the crown of *Dracaena*.

ALTITUDE: 580 – 1100 m.

FLOWERING TIME: February, July.

LIFE FORM: epiphyte.

PHYTOGEOGRAPHICAL REGION: I, III.

DESCRIPTION: {896}.

ILLUSTRATION: {281}, {1205}, {1223}.

NOTES: the holotype of this species has been destroyed and identification relies on Schlechter's illustration {1223} and his description. Plants large; flowers yellow, medium-sized, on short peduncles; lip very broadly elliptic, acuminate, concave, with numerous long hairs inside; spur subfiliform, arched-descendant, 20 mm long. There is little doubt that *A. chloranthum* described a few years later by Schlechter is the same. It differs by having several flowers on the inflorescence and by the slightly longer spur; it is not unusual to find *A. chloranthum* producing a mixture of single and multiple-flowered inflorescences especially on immature plants, the length of spur being variable in populations. The position of the spur shown in Schlechter's illustration may be a distortion in the dried specimen.

Angraecum imerinense *Schltr.* in Repert. Sp. Nov. Regni Veg. Beih. 33: 328 (1925). TYPE: Madagascar, between Betato & Soalazaina, *H.Perrier* 14937 (holo. B†; iso. P).

DISTRIBUTION: Madagascar: Antananarivo.

HABITAT: humid, evergreen forest on plateau.

ALTITUDE: c. 1700m.

FLOWERING TIME: September.

LIFE FORM: epiphyte.

PHYTOGEOGRAPHICAL REGION: III.

DESCRIPTION: {896}.

NOTES: known only from the type. Plant slender and long; leaves linear, thin-textured; flowers medium-sized; lip oval-elliptical, slightly obtuse and apiculate, concave at the base, the middle vein thickened into a ridge; spur pendent, filiform, slightly narrowed from the entrance to the apex, 9.5 cm long.

Angraecum implicatum *Thouars,* Hist. Orch.: t.58 (1822). TYPE: Madagascar, *Thouars* s.n. (holo. P).
Aerobion implicatum (Thouars) Spreng., Syst. Veg. 3: 716 (1826).
Angorchis implicata (Thouars) Kuntze, Rev. Gen. Pl. 2: 651 (1891).
Angraecum verruculosum Boivin ex Frapp. in Cordem., Fl. Réunion: 204 (1895).
Macroplectrum implicatum (Thouars) Finet in Bull. Soc. Bot. France 54, Mém. 9: 27, pl. 4 (1907).

DISTRIBUTION: Mascarenes: Réunion. Madagascar: East ?
LIFE FORM: epiphyte.
PHYTOGEOGRAPHICAL REGION: I.
DESCRIPTION: {896}, {1213}.
ILLUSTRATION: {292}, {1398}.
NOTES: species not well-known, further investigations on its identity and synonymy is necessary. The type seems to be missing.

Angraecum inapertum *Thouars,* Hist. Orch.: t.50 (1822). TYPE: Thouars, illustration cited here.
Aerobion inapertum (Thouars) Spreng., Syst. Veg. 3: 717 (1826).
Epidorchis inaperta (Thouars) Kuntze, Rev. Gen. Pl. 2: 660 (1891).
Mystacidium inapertum (Thouars) Ridl. in J. Linn. Soc., Bot. 21: 489 (1885).

DISTRIBUTION: Madagascar: Fianarantsoa. Mascarenes: Réunion, Mauritius.
ALTITUDE: 700 – 2200 m.
LIFE FORM: epiphyte.
DESCRIPTION: {896}.
ILLUSTRATION: {246}, {1398}.
NOTES: Thouars' illustration shows only a plant in fruit and there is no other reliable material available.

Angraecum kraenzlinianum *H.Perrier* in Humbert ed., Fl. Mad. Orch. 2: 276 (1941) TYPE: Madagascar, Ankafina, S Betsileo, *Hildebrandt* 4215 (holo. W; iso. K, P).
Aeranthes englerianus Kraenzl. in Bot. Jahrb. Syst. 17: 62 (1893). Based on the same type.
Angraecum englerianum (Kraenzl.) Schltr. in Beih. Bot. Centralbl. 33 (2): 433 (1915).
Angraecum robustum Kraenzl. in Ann. Naturh. Mus. Wien 44: 324 (1930), *non* Schlechter (1918).

DISTRIBUTION: Madagascar: Antsiranana, Fianarantsoa.
HABITAT: mossy and lichen-rich forest on moss- and lichen-covered trees.
ALTITUDE: 1501 – 2000 m.
FLOWERING TIME: January.
LIFE FORM: epiphyte.
PHYTOGEOGRAPHICAL REGION: III, IV.
DESCRIPTION: {896}.
NOTES: related to *Angraecum curnowianum*, but differs in the shape of the flower, the lip and the spur which is 3 times longer; lip broadly rounded-expanded at the edges, cuspidate-subacute at the front; spur 6 – 9 cm long.

Angraecum lecomtei *H.Perrier* in Notul. Syst. (Paris) 7: 119 (1938). TYPE: Madagascar, Mt. Maromizaha, Analamazaotra, *H.Perrier* 16021 (holo. P).

DISTRIBUTION: Madagascar: Toamasina.
HABITAT: mossy and evergreen forest on moss- and lichen-covered trees.
ALTITUDE: 1000 – 1200 m.
FLOWERING TIME: February – March.
LIFE FORM: epiphyte.
PHYTOGEOGRAPHICAL REGION: III.

DESCRIPTION: {896}.

NOTES: small plant with narrow linear leaves; flowers small; lip with a narrow base, the blade narrowed at the spur entrance, broadly oval; spur very slender, 10 – 12 cm, shortly narrowed below the wide opening.

Angraecum leonis *(Rchb.f.) André* in Rev. Hort.: 294 (1885) as *Angraecum leonii*. TYPE: Madagascar, *Humblot* s.n. (holo. W).

Aeranthes leonis Rchb.f. in Gard. Chron. n.s. 23: 726 (1885).

Angraecum humblotii Rchb.f. ex Rolfe in Sander, Reichenbachia 2 (2): 49 (1894), *non* (Finet) Summerh. TYPE: Comoros, cult. *Humblot* s.n. (holo. K).

Macroplectrum leonis (Rchb.f.) Finet in Bull. Soc. Bot. France 54, Mém. 9: 30 (1907).

Mystacidium leonis (Rchb.f.) Rolfe in Orch. Rev. 12: 47 (1904).

DISTRIBUTION: Madagascar: Antananarivo, Antsiranana, Toliara. Comoros: Anjouan, Grande Comore, Mayotte.

HABITAT: edge of streams, coastal forest, seasonally dry, deciduous forest and woodland, humid evergreen forest on plateau, on baobab on limestone, on trunks.

ALTITUDE: sea level – 1500 m.

FLOWERING TIME: November – March.

LIFE FORM: epiphyte, lithophyte.

PHYTOGEOGRAPHICAL REGION: II, III, V.

DESCRIPTION: {444}, {896}, {1018}, {1259}.

ILLUSTRATION: {3}, {50}, {199}, {222}, {244}, {281}, {292}, {295}, {353}, {387}, {470}, {508}, {529}, {532}, {533}, {590}, {602}, {745}, {896}, {906}, {924}, {944}, {1018}, {1206}, {1255}, {1259}, {1269}, {1312}, {1314}, {1319}, {1320}, {1412}, {1425}, {1427}, {1442}.

HISTORY: {8}, {51}, {150}, {174}, {273}, {346}, {380}, {571}, {590}, {935}, {1363}, {1377}, {1487}.

CULTIVATION: {529}, {533}, {602}, {745}, {935}, {1279}, {1314}, {1380}, {1442}.

NOTES: leaves narrowly ensiform, fleshy; flowers large; lip concave, broadly ovate-orbicular; disc with a short, central ridge; spur filiform below a funnel-shaped base, twisted-undulate, 7 – 9 cm long. The Comorean form generally has larger flowers.

Angraecum letouzeyi *Bosser* in Adansonia 11 (4): 374 (1989). TYPE: Madagascar, Sandrangato, *Bosser* 18425 (holo. P).

DISTRIBUTION: Madagascar: Antananarivo, Toamasina.

HABITAT: humid, mossy, evergreen forest.

ALTITUDE: 900 – 1200m.

FLOWERING TIME: February, October.

LIFE FORM: epiphyte.

PHYTOGEOGRAPHICAL REGION: III.

DESCRIPTION: {130}.

ILLUSTRATION: {130}.

NOTES: small plant, resembles *A. teretifolium* but the plant is almost stem-less; peduncle completely covered by the sheaths; lip spreading, oval, acutely attenuate at the tip, slightly concave and without median ridge at the base, spur filiform, 6 – 7 cm long.

Angraecum linearifolium *Garay* in Bot. Mus. Leafl. 23 (4): 160 (1972). Based on same type as *A. palmiforme* H.Perrier.

Angraecum palmiforme H.Perrier in Humbert, Mém. Inst. Sci. Mad. sér. B, 6: 268 (1955) *non A. palmiforme* Thouars (1822). TYPE: Madagascar, E summit of Marojejy, *Humbert & Cours* 23774 (holo. P).

DISTRIBUTION: Madagascar: Antananarivo, Antsiranana.

HABITAT: humid, mossy, evergreen forest.

ALTITUDE: 500 – 2500 m.

FLOWERING TIME: February, July, October – December.

LIFE FORM: epiphyte.

PHYTOGEOGRAPHICAL REGION: III, IV.

DESCRIPTION: {130}.

ILLUSTRATION: {130}, {902}, {906}.

NOTES: often confused in cultivation for *A. teretifolium.* Plant small, stocky, with filiform leaves; flowers spidery, with salmon pink tepals; lip very concave, semi-circular, edged rounded, with a strong central ridge at the base; spur conical at the base, then filiform, 9 – 10 cm long.

Angraecum litorale *Schltr.* in Repert. Sp. Nov. Regni Veg. Beih. 33: 327 (1925). TYPE: Madagascar, betw. Mahanoro & Nosy-Varika, *H.Perrier* 14225 (holo. B†; iso. P).

DISTRIBUTION: Madagascar: Fianarantsoa, Toamasina.
HABITAT: coastal forest.
ALTITUDE: sea level – 500 m.
FLOWERING TIME: October.
LIFE FORM: epiphyte.
PHYTOGEOGRAPHICAL REGION: I.
DESCRIPTION: {896}.
NOTES: the type is missing. Plant small; flower relatively large; lip elliptic, rather longly acuminate, the central vein thickened into a ridge at the base; spur filiform, 'S'-shaped, 9 cm long.

Angraecum longicalcar *(Bosser) Senghas* in Schltr., Orchideen, ed. 3: 1004 (1986). TYPE: Madagascar, nr. Analavory, Itasy, *Bosser & Peyrot* 17740 (holo. P).
Angraecum eburneum subsp. *superbum* var. *longicalcar* Bosser in Adansonia, sér. 2, 5: 408 (1965).

DISTRIBUTION: Madagascar: Antananarivo, Fianarantsoa.
HABITAT: lithophyte on trachyte rock; in xerophytic vegetation; in gallery forest.
ALTITUDE: 1000 – 2000m.
FLOWERING TIME: February.
LIFE FORM: epiphyte or lithophyte.
PHYTOGEOGRAPHICAL REGION: III.
ILLUSTRATION: {14}, {108}, {281}, {478}, {501}, {521}, {529}, {724}, {1254}.
CULTIVATION: {521}, {529}.
HISTORY: {223}.
NOTES: Close to *Angraecum eburneum* subsp. *superbum,* but has a much longer spur reaching up to 40 cm.

Angraecum longicaule *H.Perrier* in Humbert, Mém. Inst. Sci. Mad. sér. B, 6: 266 (1955). TYPES: Madagascar, NE, Lokoho valley, Beondroka Mount, north of Maroambihy, *Humbert* 23589 (lecto. P, selected here); E summit of Marojejy, *Humbert* 23646; summit of Marojejy, *Humbert* 22714; without exact loc., *Humbert & Cours* 22714 bis; Mt. Maimborondro, *Humbert* 23389 (all syn. P).

DISTRIBUTION: Madagascar: Antsiranana.
HABITAT: montane, ericaceous scrub and lichen-rich forest; terrestrial in moss.
ALTITUDE: 500 – 2100 m.
FLOWERING TIME: March April.
LIFE FORM: epiphyte or lithophyte.
PHYTOGEOGRAPHICAL AREA: IV.
DESCRIPTION: {902}.
ILLUSTRATION: {902}.
NOTES: plant slender; stem elongate; leaves small; lip suborbicular apiculate; spur 45 – 50 mm long, filiform.

Angraecum madagascariense *(Finet) Schltr.* in Beih. Bot. Centralbl. 36 (2): 162 (1918). TYPE: Madagascar, North, *Baron* 6365 (holo. K; iso. P).
Macroplectrum madagascariense Finet in Bull. Soc. Bot. France 54, Mém. 9: 25, t.5 (1907).

DISTRIBUTION: Madagascar: Antsiranana.
HABITAT: humid, highland forest.
ALTITUDE: 1500m.
LIFE FORM: epiphyte.
PHYTOGEOGRAPHICAL REGION: III.

DESCRIPTION: {896}.

ILLUSTRATION: {896}, {1267}.

NOTES: plant medium-sized; stem elongate; flowers very small; lip very concave, broadly oval, shortly acuminate-acute at the front; spur cylindrical, 1 mm, very obtuse at the apex.

Angraecum magdalenae *Schltr. & H.Perrier* in Repert. Sp. Nov. Regni Veg. Beih. 33: 354 (1925). TYPE: Madagascar, Mt. Ibity, *M.Drouhard* in *H.Perrier* 15456 (holo. P).

DISTRIBUTION: Madagascar: Antananarivo, Fianarantsoa.

HABITAT: rocky outcrops, *Uapaca* woodland.

ALTITUDE: 800 – 2000 m.

FLOWERING TIME: November – January.

LIFE FORM: lithophyte on quartz and granite.

PHYTOGEOGRAPHICAL REGION: III, IV.

DESCRIPTION: {50}, {576}, {896}.

ILLUSTRATION: {3}, {50}, {155}, {295}, {308}, {309}, {523}, {529}, {573}, {576}, {615}, {678}, {714}, {868}, {896}, {906}, {924}, {941}, {944}, {1232}, {1255}, {1259}, {1264}, {1269}, {1409}, {1425}.

CULTIVATION: {519}, {529}, {698}, {868}, {1232}.

NOTES: plant large; leaves leathery, finely veined across on both sides; flowers large; lip concave, suborbicular-ovate, shortly acuminate; spur with a wide mouth, attenuate-filiform, 10 – 11 cm long.

Angraecum magdalenae var. **latilabellum** *Bosser* in Adansonia, sér. 2, 5: 408 (1965).TYPE: Madagascar, Tsaratanana massif, *R.Paulian* 980 (holo. P; iso. TAN).

DISTRIBUTION: Madagascar: Antsiranana.

HABITAT: humid, high elevation forest.

FLOWERING TIME: November.

LIFE FORM: epiphyte.

PHYTOGEOGRAPHICAL REGION: III.

DESCRIPTION: {108}.

ILLUSTRATION: {108}.

NOTES: this is much bigger than the typical variety with larger wider leaves, reaching 30 cm. long by 7 cm. wide, and larger flowers, with the lip more spreading and a slightly different shape and blunter.

Angraecum mahavavense *H.Perrier* in Notul. Syst. (Paris) 7: 131 (1938). TYPE: Madagascar, Ambohipiraka, *H.Perrier* 18809 (holo. P).

DISTRIBUTION: Madagascar: Antsiranana.

HABITAT: humid, highland forest.

FLOWERING TIME: October – November.

LIFE FORM: epiphyte.

PHYTOGEOGRAPHICAL REGION: II, IV.

DESCRIPTION: {889}, {896}.

ILLUSTRATION: {529}.

CULTIVATION: {529}.

NOTES: plant stemless, medium-sized; leaves broadly linear; flowers medium-sized; lip acutely oval; spur with a narrow opening, linear or slightly narrowing towards the tip, 5 – 6 cm long.

Angraecum mauritianum *(Poir.) Frapp.*, Orch. Réunion, Cat. Especes Indig.: 13 (1880). TYPE: Mauritius, *Commerson* s.n. (holo. P).

Orchis mauritiana Poir. in Lamarck, Tabl. Encycl. 4: 601 (1794).

Angraecum gladiifolium Thouars, Hist. Orch.: t.53 (1822). TYPE: Mauritius, *Commerson* s.n. (holo. P).

Aerobion gladiifolium (Thouars) Spreng., Syst. Veg. 3: 716 (1826).

Aeranthes gladiifolius (Thouars) Rchb.f. in Walpers Ann. Bot. Syst. 6: 900 (1861), as *Aeranthus gladiifolius*.

Angorchis gladiifolia (Thouars) Kuntze, Rev. Gen. Pl. 2: 651 (1891).

Mystacidium mauritianum (Lam.) T.Durand & Schinz, Consp. Fl. Afric. 5: 53 (1892).
Mystacidium gladiifolium (Thouars) Rolfe in Orch. Rev. 12: 47 (1904).
Macroplectrum gladiifolium (Thouars) Pfitzer ex Finet in Bull. Soc. Bot. France 54, Mém. 9: 27 (1907).

DISTRIBUTION: Madagascar: Antananarivo, Fianarantsoa, Toamasina, Toliara. Comoros (unverified record). Mascarenes: Réunion, Mauritius.
HABITAT: humid, evergreen forest, edge of forests, in humus.
ALTITUDE: 200 – 1450 m.
FLOWERING TIME: February – March.
LIFE FORM: epiphyte or terrestrial.
PHYTOGEOGRAPHICAL REGION: I, III, IV.
DESCRIPTION: {896}.
ILLUSTRATION: {3}, {166}, {167}, {168}, {208}, {504}, {513}, {742}, {785}, {896}, {1259}, {1398}.
NOTES: stem elongate and slender; leaves oblong-lanceolate, evenly spaced; lip acutely lanceolate, widest in the lower fifth; spur filiform, c. 8 cm long.

Angraecum meirax *(Rchb.f.)* H.Perrier in Notul. Syst. (Paris) 7: 120 (1938). TYPE: Madagascar, Nossi-Ve, *Humblot* 384 (holo. W).
Aeranthus meirax Rchb.f. in Flora 68: 540 (1885).
Macroplectrum meirax (Rchb.f.) Finet in Bull. Soc. Bot. France 54, Mém. 9: 30 (1907).
Jumellea meirax (Rchb.f.) Schltr. in Beih. Bot. Centralbl. 33 (2): 429 (1915).

DISTRIBUTION: Madagascar: Toamasina. Comoros.
HABITAT: humid, lowland forest.
FLOWERING TIME: October.
LIFE FORM: epiphyte.
PHYTOGEOGRAPHICAL REGION: I.
DESCRIPTION: {896}.
ILLUSTRATION: {292}.
NOTES: small stemless plant; leaves linear; flowers relatively large; lip oval-lanceolate, very acutely narrowed above the middle, with a central ridge on top; spur filiform, 12 – 13 cm long.

Angraecum melanostictum *Schltr.* in Repert. Sp. Nov. Regni Veg. Beih. 15: 338 (1918). TYPE: Madagascar, *Laggiara* s.n. (holo. B†); NE, eastern slopes of Marojejy massif, *Humbert* 23640 (lecto. P, selected here).

DISTRIBUTION: Madagascar: Toamasina, Toliara.
HABITAT: low- to medium-elevation forest
ALTITUDE: 100 – 1450 m.
FLOWERING TIME: March.
LIFE FORM: epiphyte.
PHYTOGEOGRAPHICAL REGION: I, IV.
DESCRIPTION: {896}.
ILLUSTRATION: {1223}.
NOTES: stem long, the stem sheaths densely spotted black; leaves linear-ligulate; lip oval-lanceolate, obtuse, almost flat, 9 mm long; spur filiform, c. 7 cm long.

Angraecum microcharis *Schltr.* in Repert. Sp. Nov. Regni Veg. Beih. 33: 333 (1925). TYPE: Madagascar, Sambirano Valley, *H.Perrier* 15694 (holo. P).

DISTRIBUTION: Madagascar: Antsiranana.
HABITAT: seasonally dry, deciduous forest and woodland, on branches of shrubs.
ALTITUDE: c. 600m.
FLOWERING TIME: January.
LIFE FORM: epiphyte.
PHYTOGEOGRAPHICAL REGION: II.
DESCRIPTION: {896}.
NOTES: known from the type only. Very small plant; leaves obliquely ligulate; flowers tiny; lip very broadly rhombic-apiculate, obscurely 3-lobed, 1.1 mm long; spur oblong-obtuse, slightly dorsally compressed, 1.25 mm long.

Angraecum mirabile *Schltr.* in Repert. Sp. Nov. Regni Veg. Beih. 15: 338 (1918). TYPE: Madagascar, *Laggiara* s.n. (holo. B†).

> DISTRIBUTION: Madagascar: Toamasina.
> HABITAT: on *Philippia.*
> LIFE FORM: epiphyte.
> PHYTOGEOGRAPHICAL REGION: I.
> DESCRIPTION: {896}.
> ILLUSTRATION: {1223}, {1255}, {1266}.
> NOTES: an ambiguous species, the type is destroyed and further research on its identity is needed.

Angraecum moratii *Bosser* in Adansonia, sér. 2, 10 (1): 106 (1970). TYPE: Madagascar, S of Moramanga, Anosibe, *Bosser* 16422 (holo. P).

> DISTRIBUTION: Madagascar: Antsiranana, Toamasina.
> HABITAT: humid, evergreen and mossy forest.
> ALTITUDE: 501 – 1000 m.
> FLOWERING TIME: September.
> LIFE FORM: epiphyte.
> PHYTOGEOGRAPHICAL REGION: III.
> DESCRIPTION: {115}.
> ILLUSTRATION: {115}.
> NOTES: similar to *A. florulentum* from the Comoros, but the plant is more robust and also different in floral characteristics, especially the shape of the lip which is oblong, boat-shaped-concave with an apiculate apex; spur pendent, filiform, pale green, 12 – 13 cm long.

Angraecum multiflorum *Thouars,* Hist. Orch.: t.74 (1822). TYPE: Madagascar, *Thouars* s.n. (holo. P).
Aerobion multiflorum (Thouars) Spreng., Syst. Veg. 3: 717 (1826).
Angraecum caulescens var. *multiflorum* (Thouars) S.Moore in J. Bot. 14: 293 (1876).
Epidorchis multiflora (Thouars) Kuntze, Rev. Gen. Pl. 2: 660 (1891).
Mystacidium caulescens var. *multiflorum* (Thouars) T.Durand & Schinz, Consp. Fl. Afr. 5: 52 (1892).
Mystacidium multiflorum (Thouars) Cordem. in Rev. Gen. Pl. Bot.: 425 (1899).
Monixus multiflorus (Thouars) Finet in Bull. Soc. Bot. France 54, Mém. 9: 19 (1907).

> DISTRIBUTION: Madagascar: Antananarivo, Antsiranana. Mascarenes: Réunion, Mauritius. Seychelles.
> HABITAT: humid, evergreen forest on plateau.
> ALTITUDE: 1200 – 1500m.
> FLOWERING TIME: November – February.
> LIFE FORM: epiphyte.
> PHYTOGEOGRAPHICAL REGION: III, IV.
> DESCRIPTION: {896}.
> ILLUSTRATION: {167}, {292}.
> NOTES: plant medium-sized; inflorescence grouped in a unilateral linear series, some in flower, the others having flowered in previous years; lip 3 – 5 mm long, deeply bowl-shaped; spur with a broad mouth, 2 – 4 mm long, more or less expanded and curved at the apex.

Angraecum muscicolum *H.Perrier* in Notul. Syst. (Paris) 7: 126 (1938). TYPE: Madagascar, Tsinjoarivo, *H.Perrier* 16974 (holo. P).

> DISTRIBUTION: Madagascar: Antananarivo.
> HABITAT: mossy, montane forest on moss- and lichen-covered trees.
> ALTITUDE: c. 1400m.
> FLOWERING TIME: February.
> LIFE FORM: epiphyte.
> PHYTOGEOGRAPHICAL REGION: III.
> DESCRIPTION: {891}, {896}.
> NOTES: plant stemless; leaves linear; lip very concave, very broadly boat-shaped, 4 × 4 mm, rounded-subcordate at the base, shortly acuminate acute at the tip; spur 3 mm, narrowing from the wide entrance to the middle, then expanded bein an oval club-shape.

Angraecum musculiferum *H.Perrier* in Notul. Syst. (Paris) 7: 129 (1938). TYPES: Madagascar, Mt. Ambre, *H.Perrier* 17737 (lecto. P, selected here); Mt. Tsaratanana, *H.Perrier* 16085 (syn. P).

DISTRIBUTION: Madagascar: Antsiranana, Toamasina.
HABITAT: mossy, montane forest on moss- and lichen-covered trees.
ALTITUDE: 1200 – 1400m.
FLOWERING TIME: September – April.
LIFE FORM: epiphyte.
PHYTOGEOGRAPHICAL REGION: III.
DESCRIPTION: {891}, {896}.
ILLUSTRATION: {896}.
NOTES: plant medium-sized with small flowers; lip very concave, boat-shaped, 4 × 2.6 mm, acuminate-acute; spur 2 mm long.

Angraecum myrianthum *Schltr.* in Ann. Mus. Col. Marseille. sér. 3, 1: 197, t.21 (1913). TYPE: Madagascar, Menarandra, *H.Perrier* 22 (holo. B†; iso. P).

DISTRIBUTION: Madagascar: Toliara.
HABITAT: dry, deciduous scrub, in *Didieraceae* scrub on branches.
ALTITUDE: 250 – 1400 m.
FLOWERING TIME: March – June.
LIFE FORM: epiphyte or lithophyte.
PHYTOGEOGRAPHICAL REGION: VI.
DESCRIPTION: {896}.
ILLUSTRATION: {1205}, {1223}.
NOTES: plant almost stemless, very densely several-flowered; lip oval-lanceolate, acuminate; spur conical, subacute, a little contracted at the base, 0.75 mm long. One of the rare epiphytic orchids of southern Madagascar. A very distinct species, close to *A. pusillum* and *A. burchellii* from South Africa.

Angraecum obesum *H.Perrier* in Notul. Syst. (Paris) 7: 114 (1938). TYPES: Madagascar, Ankeramadinika, *E.François* in *H.Perrier* 18421 (lecto. P, selected here); Manjakandriana, *H.Perrier* 12975 (syn. P).

DISTRIBUTION: Madagascar: Antananarivo.
HABITAT: mossy forest on moss- and lichen-covered trees.
ALTITUDE: 1200 – 1500m.
FLOWERING TIME: January – March.
LIFE FORM: epiphyte, lithophyte.
PHYTOGEOGRAPHICAL REGION: III.
DESCRIPTION: {896}.
ILLUSTRATION: {281}, {529}, {896}, {906}.
CULTIVATION: {527}, {529}.
NOTES: stem long; leaves c. 12 cm long; flower thick-textured; lip with the base contracted but then immediately expanded into a very concave, broadly oval boat-shape, 3 × 2 cm; spur cylindrical, 5 – 10 cm long, gradually narrowing from the base to the apex.

Angraecum oblongifolium *Toill.-Gen. & Bosser* in Nat. Malg. 12: 13 (1960). TYPE: Madagascar, Bot. Gard. Antananarivo, *Jean de Dieu* 712 (holo. P).

DISTRIBUTION: Madagascar: Fianarantsoa.
HABITAT: humid, highland forest.
FLOWERING TIME: May.
LIFE FORM: epiphyte.
PHYTOGEOGRAPHICAL REGION: III, IV.
DESCRIPTION: {1405}.
ILLUSTRATION: {1405}.
NOTES: stem long and densely leaved; leaves short; flowers fleshy; lip concave, broadly triangular c. 13 mm long; spur narrowly cylindrical, narrowing towards the tip, 7 cm long.

Angraecum ochraceum *(Ridl.) Schltr.* in Beih. Bot. Centralbl. 33 (2): 436 (1915). TYPE: Madagascar, Ankafina, S Betsileo, *Deans Cowan* s.n. (holo. BM).
Mystacidium ochraceum Ridl. in J. Linn. Soc., Bot. 21: 488 (1885).
Macroplectrum ochraceum (Ridl.) Finet in Bull. Soc. Bot. France 54, Mém. 9: 26, t.4 (1907).

DISTRIBUTION: Madagascar: Fianarantsoa, Toamasina.
FLOWERING TIME: April.
LIFE FORM: epiphyte.
PHYTOGEOGRAPHICAL REGION: I, III.
DESCRIPTION: {896}.
ILLUSTRATION: {246}, {292}, {1267}.
NOTES: stem short; leaves narrowly linear; lip concave, acute at the tip, a little longer than the petals; spur slender, filiform, with the apex a little expanded.

Angraecum onivense *H.Perrier* in Notul. Syst. (Paris) 7: 122 (1938). TYPE: Madagascar, Andasibe forest, Onive, *H.Perrier* 17123 (holo. P).

DISTRIBUTION: Madagascar: Antananarivo.
HABITAT: mossy forest on moss- and lichen-covered trees.
ALTITUDE: c. 1000m.
FLOWERING TIME: February.
LIFE FORM: epiphyte.
PHYTOGEOGRAPHICAL REGION: III.
DESCRIPTION: {896}.
ILLUSTRATION: {896}.
NOTES: known from the type only. Small stemless plant; leaves lanceolate-linear c. 8 cm long; flowers small; lip very concave, oval, 1.8 × 1.3 mm; spur saccate, 1.4 mm, contracted in the middle.

Angraecum palmicolum *Bosser* in Adansonia 11 (4): 376 (1989). TYPE: Madagascar, Manankaza, Tampoketsa d'Ankazobe, *Bosser* 16240 (holo. P).

DISTRIBUTION: Madagascar: Antananarivo.
HABITAT: highland forest.
ALTITUDE: 1300 – 1400m.
FLOWERING TIME: September.
LIFE FORM: epiphyte on trunks of the palm *Dypsis decipiens*, amongst lichens.
PHYTOGEOGRAPHICAL REGION: III.
DESCRIPTION: {130}.
ILLUSTRATION: {130}.
NOTES: known from the type only. Resembles *A. letouzeyi*, but smaller and easily distinguished by the bluish-green, glaucous leaves and the smooth roots; lip broadly oval to suborbicular, acute and apiculate at the tip, 11 – 16 × 8 – 10 mm; spur filiform, 10 – 11 cm long.

Angraecum panicifolium *H.Perrier* in Notul. Syst. (Paris) 7: 105 (1938). TYPE: Madagascar, Mt. Maromizaha, Analamazaotra, *H.Perrier* 16052 (holo. P).

DISTRIBUTION: Madagascar: Toamasina.
HABITAT: evergreen and mossy forest on moss and lichen-covered trees.
ALTITUDE: 1000 – 1200 m.
FLOWERING TIME: February.
LIFE FORM: epiphyte.
PHYTOGEOGRAPHICAL REGION: III.
DESCRIPTION: {896}.
ILLUSTRATION: {896}.
NOTES: stem long and slender; leaves linear; flowers small; lip 7 – 8 mm, oval, concave, 4 mm wide above the base, a little contracted towards the middle; spur 5 – 6 mm, at first narrow, then cylindric.

Angraecum pauciramosum *Schltr.* in Repert. Sp. Nov. Regni Veg. Beih. 33: 350 (1925). TYPE: Madagascar, Analamazaotra, *H.Perrier* 11864 (iso. P).

Mystacidium graminifolium Ridl. in J. Linn. Soc., Bot. 21: 490 (1885). TYPE: Madagascar, Ankafana, *Hildebrandt* 3977 (holo. BM).

Epidorchis graminifolia (Ridl.) Kuntze, Rev. Gen. Pl. 2: 660 (1891).

Monixus graminifolius (Ridl.) Finet in Bull. Soc. Bot. France 54, Mém. 9: 19 (1907).

Angraecum graminifolium (Ridl.) Schltr. in Beih. Bot. Centralbl. 33 (2): 434 (1915), *non A. graminifolium* (Kraenzl.) Engl. (1908).

Angraecum poöphyllum Summerh. in Kew Bull. 8: 162 (1953), *nom. superfl.*

DISTRIBUTION: Madagascar: Antananarivo, Fianarantsoa, Toamasina.
HABITAT: humid, evergreen forest.
ALTITUDE: 800 – 1200 m.
FLOWERING TIME: September – July.
LIFE FORM: epiphyte.
PHYTOGEOGRAPHICAL REGION: III.
DESCRIPTION: {896}.
ILLUSTRATION: {292}, {896}, {1223}.
VERNACULAR NAME: Hazomiavona.
NOTES: long pendent plant; leaves ligulate or almost linear, 1 – 2 cm × 3 – 4 mm; flowers small, lip oval-acuminate, very concave; spur almost cylindrical, about 2 mm long. A variable species close to *A. caricifolium*.

Angraecum pectinatum *Thouars,* Hist. Orch.: t.51 (1822). TYPE: Réunion, *Thouars* s.n. (holo. P).

Aerobion pectinatum (Thouars) Spreng., Syst. Veg. 3: 717 (1826).

Aeranthes pectinatus (Thouars) Rchb.f. in Walpers Ann. Bot. Syst. 6: 900 (1864).

Mystacidium pectinatum (Thouars) Benth. in J. Linn. Soc., Bot. 18: 337 (1881).

Angorchis pectinata (Thouars) Kuntze, Rev. Gen. Pl. 2: 651 (1891).

Angorchis pectangis (Thouars) Kuntze in Bull. Herb. Boissier 2: 459 (1894).

Ctenorchis pectinata (Thouars) K.Schum. in Just's Bot. Jahresber.: 467 (1899).

Pectinaria thouarsii Cordem. in Rev. Gen. Pl. Bot. 11: 420 (1899).

Macroplectrum pectinatum (Thouars) Finet in Bull. Soc. Bot. France 54, Mém. 9: 25 (1907).

DISTRIBUTION: Madagascar: Antananarivo, Fianarantsoa, Toamasina. Mascarenes: Réunion, Mauritius. Comoros.
HABITAT: humid, low- to medium-elevation forest; in sclerophyllous woodland.
ALTITUDE: 200 – 1750 m.
FLOWERING TIME: December – May.
LIFE FORM: epiphyte.
PHYTOGEOGRAPHICAL REGION: III.
DESCRIPTION: {896}.
ILLUSTRATION: {168}, {292}, {1028}, {1398}.
CULTIVATION: {1243}, {1459}.
NOTES: medium-sized plant; leaves elongate-linear, thick, 12 – 16 mm long; flowers small, fleshy; lip similar to the sepals, almost flat; spur almost straight, a little inflated at the tip, just over 2 mm long, ovary a little rough.

Angraecum penzigianum *Schltr.* in Repert. Sp. Nov. Regni Veg. Beih. 33: 357 (1925). TYPE: Madagascar, Andringitra massif, *H.Perrier* 14585 (lecto. P, selected here); Ankafana, *Hildebrandt* 4214 (syn. P).

DISTRIBUTION: Madagascar: Fianarantsoa, Toamasina.
HABITAT: mossy forest on moss- and lichen-covered trees.
ALTITUDE: 1500 – 1800 m.
FLOWERING TIME: February – March.
LIFE FORM: epiphyte.
PHYTOGEOGRAPHICAL REGION: III.
DESCRIPTION: {896}.
ILLUSTRATION: {529}.
CULTIVATION: {529}.
NOTES: related to *A. florulentum,* but with narrower sepals and petals and a much longer spur.

Angraecum pergracile *Schltr.* in Repert. Sp. Nov. Regni Veg. Beih. 33: 337 (1925). TYPE: Madagascar, Sambirano Valley, *H.Perrier* 15687 (holo. P).

DISTRIBUTION: Madagascar: Antsiranana.
HABITAT: seasonally dry, deciduous forest and woodland, in dry woods on trunks.
ALTITUDE: c. 200 m.
FLOWERING TIME: January.
LIFE FORM: epiphyte.
PHYTOGEOGRAPHICAL REGION: II.
DESCRIPTION: {896}.
NOTES: related to *A. chaetopodum* but differs in its shorter and wider leaves and narrower flowers. It differs from *A. rhynchoglossum* which has narrower sepals and petals, a longer lip and a thinner spur.

Angraecum perhumile *H.Perrier* in Notul. Syst. (Paris) 7: 127 (1938). TYPE: Madagascar, Tsinjoarivo, *H.Perrier* 16976 (holo. P).

DISTRIBUTION: Madagascar: Antananarivo.
HABITAT: mossy forest on branches and lichen-covered trees.
ALTITUDE: c. 1400m.
FLOWERING TIME: February.
LIFE FORM: epiphyte.
PHYTOGEOGRAPHICAL REGION: III.
DESCRIPTION: {891}, {896}.
ILLUSTRATION: {1259}.
NOTES: known only from the type. Small plant; leaves narrow, c. 2 cm long; lip very concave, wider than high, 2.3 – 3 mm long; spur short, 1.5 mm long, cylindric.

Angraecum perparvulum *H.Perrier* in Notul. Syst. (Paris) 7: 123 (1938). TYPE: Madagascar, Mt. Tsaratanana, *H.Perrier* 16498 (holo. P).

DISTRIBUTION: Madagascar: Antananarivo, Antsiranana.
HABITAT: humid, evergreen forest on plateau on branches.
ALTITUDE: 1501 – 2000 m.
FLOWERING TIME: February – April.
LIFE FORM: epiphyte.
PHYTOGEOGRAPHICAL REGION: III.
DESCRIPTION: {891}, {896}.
NOTES: tiny plant; leaves less than 10 mm long; flowers small; lip almost pandurate with a small callus at the base; spur scrotiform, 1.4 × 1.2 mm.

Angraecum peyrotii *Bosser* in Adansonia 11 (1): 32 (1989). TYPE: Madagascar, nr. Anjozorobe, *Bosser & Peyrot* 16274 (holo. P).

DISTRIBUTION: Madagascar: Antananarivo.
HABITAT: Evergreen montane forest.
ALTITUDE: 1300 – 1500 m.
FLOWERING TIME: September.
LIFE FORM: epiphyte.
PHYTOGEOGRAPHICAL REGION: III.
DESCRIPTION: {128}.
ILLUSTRATION: {128}.
NOTES: related to *A. rutenbergianum* but differs essentially by its longer, very fleshy, semicylindrical leaves.

Angraecum pingue *Frapp.*, Orch. Réunion, Cat. Especes Indig.: 13 (1880) & *Frapp. ex Cordem.* Fl. Réunion: 214 (1895). TYPE: Réunion, *Frappier* s.n. (not found).
Mystacidium pingue (Frapp.) Cordem. in Rev. Gen. Pl. Bot.: 421 (1899).
Angraecum nasutum Schltr. in Repert. Sp. Nov. Regni Veg. Beih. 33: 315 (1925), **syn. nov.** TYPE: Madagascar, Mt. Tsaratanana, *H.Perrier* 15307 (holo. B†; iso. P).

DISTRIBUTION: Madagascar: Antsiranana. Mascarenes: Réunion, Mauritius.

HABITAT: lichen-rich and mossy forest on moss- and lichen-covered trees.

ALTITUDE: 2000 – 2200 m.

FLOWERING TIME: January.

LIFE FORM: epiphyte.

PHYTOGEOGRAPHICAL REGION: III, IV.

DESCRIPTION: {896}.

NOTES: tall, erect plant; leaves narrowly ligulate, thick; flowers fleshy; lip concave, folding around the column and broadly oval at the base, then beaked-acuminate; spur slender, cylindrical, thickened and almost club-shaped at the apex, 15 mm long. There is little doubt that *A. nasutum* is the same as *A. pingue* from the Mascarenes; the plant and flower are very similar, the shape of the spur and the lip wich is folded backward at the margins are typical.

Angraecum pinifolium *Bosser* in Adansonia, sér. 2, 10 (1): 99 (1970). TYPE: Madagascar, S of Moramanga, *Bosser* 17633 (holo. P).

DISTRIBUTION: Madagascar: Toamasina.

HABITAT: humid, evergreen forest, on plateau in mossy forest on moss- and lichen-covered trees.

ALTITUDE: 900 m.

FLOWERING TIME: January.

LIFE FORM: epiphyte.

PHYTOGEOGRAPHICAL REGION: III.

DESCRIPTION: {115}.

ILLUSTRATION: {115}.

NOTES: leaves narrowly linear, c. 10 cm long, 2 mm wide; lip ovate-acuminate, 4.5 – 5.5 mm long, with a prominent ridge at the base; spur 6 – 6.5 mm long, 2 mm in diameter, gradually inflated at the tip, with a few thin hairs at the spur entrance.

Angraecum platycornu *Hermans, P.J.Cribb & Bosser* in Orchid Review 110, 1242: 22 – 23 (2002). TYPE: Madagascar, NE, Masoala Peninsula, Ambatoavy escarpment, E of Andranobe, west coast of Masoala, *Lance* 136 (holo. K).

DISTRIBUTION: Madagascar: Toamasina.

HABITAT: humid, lowland forest.

FLOWERING TIME: July.

LIFE FORM: epiphyte.

PHYTOGEOGRAPHICAL REGION: I.

DESCRIPTION: {483}.

ILLUSTRATION: {483}.

NOTES: closest to *A. conchoglossum* but the flower is distinguished by the lip which is much broader than long, very obscurely 3-lobed, and with a broad-mouthed, horn-like spur at its base. Its rostellum is also distinctive in having a distinctive short tooth between the large lateral lobes.

Angraecum popowii *Braem* in Schlechteriana 2 (4): 163 (1991). TYPE: Madagascar, cult. (holo. SCHLE).

DISTRIBUTION: Madagascar: Antananarivo, Fianarantsoa.

HABITAT: sclerophyllous woodland; lichen-rich forest; evergreen highland forest.

ALTITUDE: 1200 – 1500 m.

FLOWERING TIME: November – June.

LIFE FORM: epiphyte.

PHYTOGEOGRAPHICAL REGION: III.

DESCRIPTION: {153}.

ILLUSTRATION: {153}, {281}.

NOTES: this may be no more than a form of *A. teretifolium*. Leaves up to 9 cm long, terete and fleshy; flowers medium-sized; lip 3 × 2 cm; spur up to 14 cm long.

Angraecum potamophilum *Schltr.* in Ann. Mus. Col. Marseille, sér. 3, 1: 199 (56), t.23 (1913). TYPE: Madagascar, Morataitra, Betsiboka, *H.Perrier* 834 (holo. B†; iso. P).

Aerangis potamophila (Schltr.) Schltr. in Beih. Bot. Centralbl. 33 (2): 427 (1915) & 36 (2): 122 (1918).

DISTRIBUTION: Madagascar: Mahajanga.

HABITAT: trees by rivers.

FLOWERING TIME: November.

LIFE FORM: epiphyte on trunks of *Eugenia*.

PHYTOGEOGRAPHICAL REGION: V.

DESCRIPTION: {896}.

ILLUSTRATION: {513}, {896}, {1205}.

CULTIVATION: {513}.

NOTES: short-stemmed plant; leaves thick, up to 15 cm long; lip oval-lanceolate, c. 18 × 8 mm, narrowed-acute towards the tip, a little contracted at the wide base; spur filiform-acute, c. 8 cm long.

Angraecum praestans *Schltr.* in Ann. Mus. Col. Marseille, sér. 3, 1: 200, t.21 (1913). TYPE: Madagascar, Lac Kinkony, *H.Perrier* 1450 (holo. B†; iso. P).

DISTRIBUTION: Madagascar: Antsiranana, Mahajanga, Toliara.

HABITAT: dry coastal forest on tamarind trees; on rock formations; at the base and on trunks of large trees in dry forest; on limestone cliffs.

ALTITUDE: 0 – 100 m.

FLOWERING TIME: November – June.

LIFE FORM: epiphyte or lithophyte.

PHYTOGEOGRAPHICAL REGION: II, V, VI.

DESCRIPTION: {896}.

ILLUSTRATION: {281}, {470}, {529}, {622}, {896}, {1205}, {1223}.

CULTIVATION: {529}.

NOTES: a robust plant with loriform-ligulate, leathery leaves; flowers large; lip obovate-rhomboidal, 4 × 2.2 cm; spur funnel-shaped at the base, then attenuate-filiform, curved and 9 cm long.

Angraecum protensum *Schltr.* in Repert. Sp. Nov. Regni Veg. Beih. 33: 358 (1925). TYPE: Madagascar, Mt. Analamamy, *H.Perrier* 12494 (holo. B†; iso. P).

DISTRIBUTION: Madagascar: Fianarantsoa, only found on the Itremo massif in Central Madagascar.

HABITAT: rocky outcrops.

ALTITUDE: 1600 – 2000 m.

FLOWERING TIME: January – March.

LIFE FORM: lithophyte on quartz.

PHYTOGEOGRAPHICAL REGION: III, IV.

DESCRIPTION: {896}.

ILLUSTRATION: {520}, {529}, {896} the illustration is plate LXXXV 7 – 8 not 1 – 6, {1223}.

CULTIVATION: {529}, {533}, {698}.

NOTES: plant erect, up to 35 cm tall; leaves greyish-green; flowers large; lip oval-concave, acute or apiculate, 4 × 2.5 cm; spur pendent, 14 – 15 cm long.

Angraecum pseudodidieri *H.Perrier* in Notul. Syst. (Paris) 7: 113 (1938). TYPE: Madagascar, Mt. Ambre, *H.Perrier* 18864 (holo. P).

DISTRIBUTION: Madagascar: Antsiranana.

HABITAT: mossy forest on moss- and lichen-covered trees.

ALTITUDE: c. 1000 m.

FLOWERING TIME: November.

LIFE FORM: epiphyte.

PHYTOGEOGRAPHICAL REGION: III.

DESCRIPTION: {896}

ILLUSTRATION: {1269}.

NOTES: plant erect, up to 20 cm; flowers large; lip almost flat, broadly obovate-cuspidate, 2.5 ×
1.8 cm; spur 10 – 11 cm long, narrowed from the entrance to the tip.

Angraecum pseudofilicornu *H.Perrier* in Notul. Syst. (Paris) 7: 108 (1938). TYPE: Madagascar, Mt.
Ambre, *H.Perrier* 18863 (holo. P).

DISTRIBUTION: Madagascar: Antsiranana.
HABITAT: mossy forest on moss- and lichen-covered trees.
ALTITUDE: c. 1000 m.
FLOWERING TIME: November.
LIFE FORM: epiphyte.
PHYTOGEOGRAPHICAL REGION: III.
DESCRIPTION: {896}.
ILLUSTRATION: {3}, {470}, {529}.
CULTIVATION: {529}.
NOTES: similar to *A. scottianum*. Plant elongate; leaves cylindrical, thick-textured; flowers with lip
uppermost; lip very concave, wider (2.5 cm) than long (1.8 cm), cuspidate, with a ridge at
the base; spur 15 cm, compressed and narrowed below the 6 mm wide opening, then
filiform.

Angraecum pterophyllum *H.Perrier* in Notul. Syst. (Paris) 7: 106 (1938). TYPE: Madagascar,
Andasibe, Onive, *H.Perrier* 18647 (holo. P).

DISTRIBUTION: Madagascar: Antananarivo, Fianarantsoa.
HABITAT: mossy forest, on moss- and lichen-covered trees.
ALTITUDE: 1100 – 1200m.
FLOWERING TIME: February.
LIFE FORM: epiphyte.
PHYTOGEOGRAPHICAL REGION: III.
DESCRIPTION: {896}.
NOTES: erroneously thought by Garay {314} to be conspecific with *A. hermannii* (Cordem.)
Schltr. which is endemic to Réunion. Small plant; leaves elliptic, 10 × 3 cm; flowers small; lip
oval-acute, 2.5 × 1.5 mm, very concave; spur obtusely sac-like, 2 × 1.2 mm.

Angraecum pumilio *Schltr.* in Beih. Bot. Centralbl. 34 (2): 33 (1916). TYPE: Madagascar,
Tsaratanana, *H.Perrier* 11356 (= *Schlechter* 143) (holo. B†; iso. P).

DISTRIBUTION: Madagascar: Antsiranana, Fianarantsoa.
HABITAT: montane, ericoid scrubland and lichen-rich forest, on moss- and lichen-covered trees.
ALTITUDE: 1000 – 2400m.
FLOWERING TIME: February – April.
LIFE FORM: epiphyte.
PHYTOGEOGRAPHICAL REGION: IV.
DESCRIPTION: {896}.
ILLUSTRATION: {1223}.
NOTES: differs from *Angraecum acutipetalum* by the wider leaves and petals and the longer,
ascending spur. Plant and flowers small; lip thick, fleshy, suborbicular, cochlear-concave; spur
narrowed from the wide entrance to the middle, then inflated and clavate, 6 – 9 mm long.

Angraecum rhizanthium *H.Perrier* in Notul. Syst. (Paris) 14: 163 (1951). TYPE: Madagascar, W of
Marojejy massif, *Humbert* 23123 (holo. P).

DISTRIBUTION: Madagascar: Antsiranana.
HABITAT: mossy, evergreen forest.
ALTITUDE: 1000 – 1500 m.
FLOWERING TIME: January – February.
LIFE FORM: epiphyte.
PHYTOGEOGRAPHICAL REGION: IV.

DESCRIPTION: {901}.

NOTES: known only from the type. Differs from *A. multiflorum* by the very narrowed conch-shaped lip, the cylindrical and straight spur and the reflexed tip of the dorsal sepal. Stem long; leaves ligulate-linear.

Angraecum rhizomaniacum *Schltr.* in Repert. Sp. Nov. Regni Veg. Beih. 33: 315 (1925). TYPE: Madagascar, Mt. Tsaratanana, *H.Perrier* 15315 (holo. P; iso. K).

DISTRIBUTION: Madagascar: Antsiranana.

HABITAT: lichen-rich forest on moss- and lichen-covered trees.

ALTITUDE: c. 2000 m.

FLOWERING TIME: October, January.

LIFE FORM: epiphyte.

PHYTOGEOGRAPHICAL REGION: II, III, IV.

DESCRIPTION: {896}.

NOTES: similar to *A. zaratananae*, but larger in all parts. Plant up to 20 cm long; flowers small, fleshy; lip very widely rhomboid–suborbicular, 4.5 × 5 mm, concave and shortly apiculate-acuminate; spur slender, cylindric, a little thickened at the obtuse tip, 6.5 mm long.

Angraecum rhynchoglossum *Schltr.* in Repert. Sp. Nov. Regni Veg. Beih. 33: 339 (1925). TYPE: Madagascar, Mt. Vohitrilongo, N Imerina, *H.Perrier* 14973 (holo. P).

Mystacidium viride Ridl. in J. Linn. Soc., Bot. 22: 122 (1886). TYPE: Madagascar, Ankeramadinika, *Fox* s.n. (holo. BM).

Epidorchis viridis (Ridl.) Kuntze, Rev. Gen. Pl. 2: 660 (1891).

Angraecum viride (Ridl.) Schltr. in Beih. Bot. Centralbl. 33 (2): 438 (1915).

Angraecum foxii Summerh. in Kew Bull.: 278 (1948), *nom. superfl.*

DISTRIBUTION: Madagascar: Antananarivo.

HABITAT: mossy forest on moss- and lichen-covered trees.

ALTITUDE: 1000 – 1400 m.

FLOWERING TIME: September – April.

LIFE FORM: epiphyte.

PHYTOGEOGRAPHICAL REGION: III.

DESCRIPTION: {896}.

ILLUSTRATION: {3}, {281}, {896}.

NOTES: small stemless plant; flowers with lip uppermost, yellow-green, 6 × 3 mm, expanded and very rounded at the base; spur funnel-shaped and narrowed below the entrance for 6 mm, then a little contracted, finally a little expanded towards the upper third, 25 mm long.

Angraecum rigidifolium *H.Perrier* in Notul. Syst. (Paris) 7: 116 (1938). TYPE: Madagascar, Ankeramadinika, *E.François* in *H.Perrier* 17292 (holo. P).

DISTRIBUTION: Madagascar: Antananarivo.

HABITAT: mossy forest on moss- and lichen-covered trees.

ALTITUDE: c. 1300m.

FLOWERING TIME: February.

LIFE FORM: epiphyte.

PHYTOGEOGRAPHICAL REGION: III.

DESCRIPTION: {896}.

ILLUSTRATION: {896}.

NOTES: known only from the type. Plant medium-sized; lip 15 × 6 mm, caudate towards the base; spur almost filiform, c. 3 cm long.

Angraecum rostratum *Ridl.* in J. Linn. Soc., Bot. 21: 485 (1885). TYPE: Madagascar, Ankafina, *Hildebrandt* 3976 (holo. BM; iso. P).

Angorchis rostrata (Ridl.) Kuntze, Rev. Gen. Pl. 2: 651 (1891).

DISTRIBUTION: Madagascar: Fianarantsoa.

HABITAT: mossy forest on moss- and lichen-covered trees.
ALTITUDE: c. 1500 m.
FLOWERING TIME: March.
LIFE FORM: epiphyte.
PHYTOGEOGRAPHICAL REGION: III.
DESCRIPTION: {896}.
ILLUSTRATION: {246}, {896}.
NOTES: plant with a 30 cm long stem; flowers greenish; lip 12 mm long, tapering to a point; spur 8 mm long, a little expanded into a club and recurved at the end.

Angraecum rubellum *Bosser* in Adansonia 10 (1): 22 (1988). TYPE: Madagascar, Moramanga-Anosibe Road, *Bosser* 17716 (holo. P).

DISTRIBUTION: Madagascar: Toamasina.
HABITAT: humid, evergreen forest; on *Pandanus*.
ALTITUDE: 500 – 1000 m.
FLOWERING TIME: March.
LIFE FORM: epiphyte.
PHYTOGEOGRAPHICAL REGION: III.
DESCRIPTION: {127}.
ILLUSTRATION: {85} {127}, {281}, {484}.
NOTES: small plant with reddish leaves; flowers tiny, fleshy, pink and carrying red hairs on the outer surface; lip, 2.5 × 2.5 mm, acute at the tip, with rounded edges, raised and narrowly surrounding the column; spur 2.5 – 3 mm long, slightly flattened dorso-ventrally, expanded at the tip.

Angraecum rutenbergianum *Kraenzl.* in Abh. Naturwiss. Vereine Bremen 7: 257 (1882). TYPE: Madagascar, Ankaratra, *Rutenberg* s.n. (holo. B†).
Jumellea rutenbergiana (Kraenzl.) Schltr. in Beih. Bot. Centralbl. 33 (2): 430 (1915).
Angraecum catati Baill. in H.Perrier in Humbert ed., Fl. Mad. Orch. 2: 270 (1941) *nom. nud.*

DISTRIBUTION: Madagascar: Antananarivo, Fianarantsoa.
HABITAT: humid, evergreen forest on trunks and branches of *Agauria salicifolia* on plateau and on shaded wet rocks.
ALTITUDE: 1500 – 2600 m.
FLOWERING TIME: November – February.
LIFE FORM: epiphyte or lithophyte.
PHYTOGEOGRAPHICAL REGION: III, IV.
DESCRIPTION: {896}.
ILLUSTRATION: {3}, {470}, {514}, {517}, {529}, {1329}.
CULTIVATION: {517}, {529}.
NOTES: a variable species. Plant medium-sized, almost stemless; roots glabrous; leaves rigid, thick; flowers large; lip rhombic-elliptic, 30 – 35 × 10 – 18 mm; spur filiform, variable in length, up to 14 cm long.

Angraecum sacculatum *Schltr.* in Repert. Sp. Nov. Regni Veg. Beih. 33: 347 (1925). TYPES: Madagascar, Vohitrilonga, N Imerina, *H.Perrier* 14940 (lecto. P, selected here); Andringitra massif, *H.Perrier* s.n. (syn. P).

DISTRIBUTION: Madagascar: Antananarivo, Fianarantsoa.
HABITAT: mossy forest .
ALTITUDE: 1300 – 1800m.
FLOWERING TIME: September – February.
LIFE FORM: epiphyte.
PHYTOGEOGRAPHICAL REGION: III.
DESCRIPTION: {896}, {1221}.
NOTES: plant with a short stem; flowers small; lip oval-acuminate, up to 5 mm long; spur oblong saccate, obtuse, scarcely 4 mm long.

Angraecum sambiranoense *Schltr.* in Repert. Sp. Nov. Regni Veg. Beih. 33: 329 (1925). TYPE: Madagascar, Sambirano Valley, *H.Perrier* 15699 (holo. P).

DISTRIBUTION: Madagascar: Antananarivo, Antsiranana.
HABITAT: humid, evergreen forest.
ALTITUDE: 700 – 800 m.
FLOWERING TIME: February.
LIFE FORM: epiphyte.
PHYTOGEOGRAPHICAL REGION: II.
DESCRIPTION: {896}.
NOTES: the very long stem is typical; flowers are somewhat similar to those of *A. rutenbergianum*, but the spur is thicker and longer, up to 11 cm.

Angraecum scalariforme *H.Perrier* in Humbert, Mém. Inst. Sci. Mad. sér. B, 6: 265 (1955). TYPE: Madagascar, Mt. Beondroka, *Humbert* 23548 (holo. P).

DISTRIBUTION: Madagascar: Antsiranana.
HABITAT: lichen-rich forest.
ALTITUDE: 1000 – 1400 m.
FLOWERING TIME: March.
LIFE FORM: epiphyte or lithophyte.
PHYTOGEOGRAPHICAL REGION: III.
DESCRIPTION: {902}.
ILLUSTRATION: {902}.
NOTES: plant elongate; leaves lanceolate, small; flowers small; lip oblong above the middle, narrowly acuminate at the tip, 9 mm long; spur 4 mm long, narrowed at the base.

Angraecum scottianum *Rchb.f.* in Gard. Chron. n.s. 10: 556 (1878). TYPE: Comoros, *R.Scott* s.n. (holo. W, Reichenbach Herbarium 46263).
Angraecum reichenbachianum Kraenzl., Xenia Orch. 3: 74, t.239 (1890). TYPE: illn. in Xenia Orchid, t.239 (lecto.).
Angorchis scottiana (Rchb.f.) Kuntze, Rev. Gen. Pl. 2: 652 (1891).

DISTRIBUTION: Comoros: Anjouan, Grande Comore. Records from Madagascar refer to *Angraecum pseudofilicornu*.
HABITAT: rainforest, on bare trunks and lowest branches of *Albizzia*.
ALTITUDE: 700 – 1000 m.
FLOWERING TIME: August – September.
LIFE FORM: epiphyte.
DESCRIPTION: {543}, {896}.
ILLUSTRATION: {3}, {226}, {245}, {407}, {446}, {479}, {529}, {543}, {655}, {896}, {944}, {1019}, {1285}, {1298}, {1313}, {1319}, {1344}, {1492}.
HISTORY: {226}
CULTIVATION: {529}, {761}, {1313}.
NOTES: close to *A. pseudofilicornu*, but differs by the more cylindrical leaves, the sheaths, the peduncle and the pedicellate ovary which are longer, the more elongated sepals and petals, the bigger lip and shorter spur.

Angraecum sedifolium *Schltr.* in Repert. Sp. Nov. Regni Veg. Beih. 33: 316 (1925). TYPES: Madagascar, Mt. Tsiafajavona, Ankaratra, *H.Perrier* 14765 (lecto. P, selected here); Mt. Tsiafajavona, *H.Perrier* 13410 (syn. P).

DISTRIBUTION: Madagascar: Antananarivo.
HABITAT: Evergreen and mossy forest on moss- and lichen-covered trees.
ALTITUDE: 1800 – 2000 m.
FLOWERING TIME: December – May.
LIFE FORM: epiphyte.
PHYTOGEOGRAPHICAL REGION: III, IV.
DESCRIPTION: {896}.
ILLUSTRATION: {281}, {470}, {478}, {529}, {896}.

CULTIVATION: {529}.

NOTES: small plant; leaves resemble those of *Sedum*; flowers green.

Angraecum serpens *(H.Perrier) Bosser* in Adansonia, sér. 2, 10 (1): 109 (1970). TYPE: Madagascar, Fanovana, *Decary* 17986 (holo. P).

Jumellea serpens H.Perrier in Notul. Syst. (Paris) 14: 162 (1951).

DISTRIBUTION: Madagascar: Toamasina.

HABITAT: humid, evergreen forest.

ALTITUDE: c. 600 m.

FLOWERING TIME: July.

LIFE FORM: epiphyte.

PHYTOGEOGRAPHICAL REGION: I.

DESCRIPTION: {901}.

NOTES: in shape resembles *A. penzigianum* but differs in its single-flowered inflorescence, the very large floral sheath, which reaches 2.5 cm in length, the flower with a totally different lip, and the spur which is heavier, twisted and shorter.

Angraecum sesquipedale *Thouars,* Hist. Orch.: t.66 – 67 (1822). TYPE: *Thouars* s.n (P98880). (holo. P).

Aeranthes sesquipedalis (Thouars) Lindl. in Bot. Reg. 10: sub t.817 (1824).

Macroplectrum sesquipedale (Thouars) Pfitzer in Engler & Prantl, Nat. Pflanzenfam. 2 (6): 215, t.234 (1889).

Angorchis sesquipedalis (Thouars) Kuntze, Rev. Gen. Pl. 2: 651 (1891).

Mystacidium sesquipedale (Thouars) Rolfe in Orch. Rev. 12: 47 (1904).

DISTRIBUTION: Madagascar: Fianarantsoa, Toamasina, Toliara.

HABITAT: coastal forest; eastern forest; dry, semi-deciduous forest.

ALTITUDE: sea level – 700m.

FLOWERING TIME: May – November.

LIFE FORM: epiphyte, lithophyte or semi- terrestrial.

PHYTOGEOGRAPHICAL REGION: I.

DESCRIPTION: {96}, {564}, {896}.

ILLUSTRATION: {3}, {6}, {17}, {20}, {30}, {32}, {45}, {50}, {57}, {87}, {96}, {97}, {98}, {152}, {171}, {199}, {208}, {222}, {226}, {243}, {244}, {256}, {259}, {263}, {265}, {281}, {287}, {292}, {302}, {304}, {308}, {311}, {312}, {339}, {342}, {348}, {389}, {402}, {424}, {444}, {451}, {464}, {470}, {478}, {480}, {494}, {509}, {510}, {512}, {523}, {529}, {562}, {591}, {603}, {608}, {615}, {621}, {635}, {669}, {706}, {713}, {714}, {729}, {735}, {737}, {748}, {792}, {797}, {832}, {862}, {867}, {850}, {896}, {904}, {912}, {941}, {944}, {952}, {958}, {962}, {1018}, {1025}, {1026}, {1036}, {1051}, {1055}, {1105}, {1178}, {1206}, {1231}, {1255}, {1259}, {1264}, {1276}, {1281}, {1285}, {1298}, {1319}, {1343}, {1392}, {1398}, {1400}, {1420}, {1423}, {1427}, {1438}, {1439}, {1441}, {1442}, {1453}, {1466}, {1467}, {1468}, {1473}, {1477}, {1480}, {1493}, {1507}, {1510}.

HISTORY: {45}, {54}, {57}, {151}, {183}, {210}, {226}, {233}, {241}, {243}, {275}, {287}, {288}, {302}, {325}, {335}, {338}, {429}, {562}, {571}, {734}, {952}, {1042}, {1105}, {1159}, {1286}, {1420}.

CULTIVATION: {38}, {45}, {96}, {97}, {251}, {326}, {392}, {402}, {473}, {512}, {529}, {603}, {669}, {724}, {737}, {748}, {935}, {1036}, {1230}, {1423}, {1477}.

NOTES: a large plant with ligulate, coriaceous leaves; flowers large, showy, star-shaped; lip concave, ovate-subpandurate, obtusely-acuminate at apex, 6.5 – 8 cm long, 3.5 – 4 cm broad; spur pendent, 30 – 35 cm long, gradually narrowing from the base to the apex.

Angraecum sesquipedale var. **angustifolium** *Bosser & Morat* in Adansonia, sér. 2, 12 (1): 77 (1972). TYPE: Madagascar, Baie d'Italy (W of Fort-Dauphin), *P.Morat* 3834 (holo. P).

Angraecum bosseri Senghas in Orchidee 24: 191 (1973).

DISTRIBUTION: Madagascar: Toliara.

HABITAT: seasonally dry, deciduous forest, woodland and scrub, on exposed scree.

FLOWERING TIME: May – June.

LIFE FORM: terrestrial.

PHYTOGEOGRAPHICAL REGION: VI.

DESCRIPTION: {121}, {1327}.
ILLUSTRATION: {121}, {529}, {1271}.
CULTIVATION: {529}.
NOTES: a smaller xerophytic variant of the above.

Angraecum sesquisectangulum *Kraenzl.* in Mitt. Inst. Allg. Bot. Hamb. 5: 238 (1922). TYPE: Madagascar, Ankor (Sahamabendrana), *Majastre* s.n. (holo. not located).

DISTRIBUTION: Madagascar.
NOTES: it is likely that the type of this species was destroyed, Kraenzlin's description is insufficient to determine its identity.

Angraecum setipes *Schltr.* in Repert. Sp. Nov. Regni Veg. Beih. 33: 340 (1925). TYPE: Madagascar, Mt. Tsiafajavona (Ankaratra), *H.Perrier* 13988 (holo. B†; iso. P).

DISTRIBUTION: Madagascar: Antananarivo, Fianarantsoa, Toamasina, Toliara.
HABITAT: mossy forest on moss- and lichen-covered trees.
ALTITUDE: 600 – 2000 m.
FLOWERING TIME: September – February.
LIFE FORM: epiphyte.
PHYTOGEOGRAPHICAL REGION: III, IV.
DESCRIPTION: {896}.
ILLUSTRATION: {1223}.
NOTES: similar to *A. ochraceum* and *A. brachyrhophalon* but differs in the shape of the leaves and spur. Plant slender, stemless; lip suborbicular to broadly rhombic, 7 × 7 mm, thickly apiculate; spur horizontal, subfiliform-cylindrical, very slightly thickened at the tip, up to 22 mm long.

Angraecum sinuatiflorum *H.Perrier* in Notul. Syst. (Paris) 14: 163 (1951). TYPE: Madagascar, Vondrozo, *Decary* 4868 (holo. P).

DISTRIBUTION: Madagascar: Fianarantsoa.
HABITAT: humid, evergreen forest on plateau.
FLOWERING TIME: September.
LIFE FORM: epiphyte.
PHYTOGEOGRAPHICAL REGION: III.
DESCRIPTION: {901}.
NOTES: known only from the type. Medium-sized plant without a stem; inflorescence in a series of 2 to 4, as in *A. multiflorum*; lip slipper-shaped, 4 × 2 mm, obtusely apiculate; spur pendent, 13 – 14 mm long.

Angraecum sororium *Schltr.* in Repert. Sp. Nov. Regni Veg. Beih. 33: 360 (1925). TYPES: Madagascar, Mt. Ibity, *H.Perrier* 8076 (lecto. P); nr. Ambatofanjena, *H.Perrier* 12405 (syn. P); without locality, *H.Perrier* 12979 (syn. P).

DISTRIBUTION: Madagascar: Antananarivo, Fianarantsoa, Toamasina.
HABITAT: on inselbergs, restricted to the mountainous regions of the island.
ALTITUDE: 1600 – 2200m.
FLOWERING TIME: December – March.
LIFE FORM: epiphyte, lithophyte, terrestrial.
PHYTOGEOGRAPHICAL REGION: III, IV.
DESCRIPTION: {306}, {896}.
ILLUSTRATION: {3}, {135}, {281}, {426}, {463}, {470}, {482}, {510}, {513}, {515}, {524}, {529}, {621}, {638}, {727}, {862}, {896}, {906}, {917}, {944}, {1223}, {1270}.
CULTIVATION: {463}, {515}, {527}, {529}, {533}, {698}, {917}.
NOTES: erect plant up to 1 m tall; flowers large and fleshy; lip suborbicular-ovate, 30 – 33 mm wide; spur 25 – 32 cm long, cylindrical.

Angraecum sterrophyllum *Schltr.* in Beih. Bot. Centralbl. 34 (2): 338 (1916). TYPE: Madagascar, *H.Perrier* XIV (holo. B†).

Angraecum filicornu sensu Kraenzl. in Abh. Naturwiss. Ver. Bremen 7: 257 (1882), *non* Thouars.

DISTRIBUTION: Madagascar: Antananarivo.
HABITAT: on *Uapaca bojeri*, in drier places on small trees, in humid evergreen forest on plateau.
ALTITUDE: c. 1400 m.
FLOWERING TIME: October – January.
LIFE FORM: epiphyte.
PHYTOGEOGRAPHICAL REGION: III.
DESCRIPTION: {896}.
ILLUSTRATION: {1223}.
NOTES: a little-known species said to be related to *A. scottianum*. Leaves c. 4 cm long; lip suborbicular, 12 × 15 mm, wider than long, concave and apiculate, with a central ridge at the base; spur arched-pendent expanded at the entrance, then filiform, 7 – 8 cm long.

Angraecum tamarindicolum *Schltr.* in Repert. Sp. Nov. Regni Veg. Beih. 33: 340 (1925). TYPE: Madagascar, Sambirano Valley, *H.Perrier* 15443 (holo. B†; iso. P).

DISTRIBUTION: Madagascar: Antsiranana.
HABITAT: humid, evergreen forest at low elevation, on *Tamarindus indica*.
ALTITUDE: below 100m.
FLOWERING TIME: February.
LIFE FORM: epiphyte.
PHYTOGEOGRAPHICAL REGION: II.
DESCRIPTION: {896}.
NOTES: known only from the type. The plant has typically very thin, papery leaves reminiscent of species in *Aeranthes*. Flowers small; lip broadly rhombic-oval, 3 × 3 mm; spur 7.5 mm long, narrowed in the basal part, then inflated into a somewhat obtuse club.

Angraecum tenellum *(Ridl.) Schltr.* in Beih. Bot. Centralbl. 33 (2): 438 (1915). TYPE: Madagascar, Ankafana, S Betsileo, *Deans Cowan* s.n. (holo. BM).

Mystacidium tenellum Ridl. in J. Linn. Soc., Bot. 21: 489 (1885).
Epidorchis tenella (Ridl.) Kuntze, Rev. Gen. Pl. 2: 660 (1891).
Angraecum waterlotii H.Perrier in Notul. Syst. (Paris) 7: 122 (1938). TYPE: Madagascar, Manjakandriana, *Waterlot* 993 (holo. P).
Saccolabium micromegas Frapp., Orch Réunion, Cat. Espèces Indig.: 14 (1880), *nom. nud.* TYPE see *S. microphyton*.
Saccolabium microphyton Frapp. in Cordem., Fl. Réunion: 195 (1895). Based upon: ?Réunion, Grande & petite plaine des Palmistes; Salazie; Cilaos (Ilet des Etangs), *Frappier*.
Angraecum microphyton (Frapp.) Schltr. in Beih. Bot Centralbl. 33 (2): 435 (1915).

DISTRIBUTION: Madagascar: Antananarivo, Fianarantsoa.
HABITAT: humid, evergreen forest.
ALTITUDE: 1000 – 1500 m.
FLOWERING TIME: October, February – June.
LIFE FORM: epiphyte.
PHYTOGEOGRAPHICAL REGION: III.
DESCRIPTION: {896}.
ILLUSTRATION: {246}, {896}.
NOTES: tiny plant with small flowers, 2 mm; lip oval-obtuse 1.8 mm long; spur thick, cylindric-obtuse and a little shorter than the ovary.

Angraecum tenuipes *Summerh.* in Kew Bull. 6, 3: 473 (1951). Based upon same type as *A. ischnopus* Schltr. (1916).

Angraecum ischnopus Schltr. in Beih. Bot. Centralbl. 34 (2): 336 (1916), *non* Schltr. (1905). TYPE: Madagascar, Mt. Tsaratanana. *H.Perrier* 140 (holo. B†; iso. P).

DISTRIBUTION: Madagascar: Antananarivo, Antsiranana.
HABITAT: lichen-rich forest on moss- and lichen-covered trees.
ALTITUDE: 1400 – 2500 m.

FLOWERING TIME: September – December.

LIFE FORM: epiphyte.

PHYTOGEOGRAPHICAL REGION: IV.

DESCRIPTION: {896}.

ILLUSTRATION: {1223}.

NOTES: related to *A. caulescens*, but differs in its thin inflorescence and pointed leaves. Flowers small; lip oval, concave, shortly acuminate subacute, c. 5 mm long; spur as long as the lip, curved, gradually inflated from the entrance to the apex into a broad purse-shape.

Angraecum tenuispica *Schltr.* in Repert. Sp. Nov. Regni Veg. Beih. 15: 339 (1918). TYPE: Madagascar, *Laggiara* s.n. (holo. B†); neo. the illustration in Repert. Spec. Nov. Reg. Veg., 68 t.94, 373 (1932), selected here.

DISTRIBUTION: Madagascar.

LIFE FORM: epiphyte.

DESCRIPTION: {896}.

ILLUSTRATION: {1223}.

NOTES: type destroyed with only Schlechter's illustration and description remaining. Said to be related to *A. graminifolium* but with wider and more obtuse leaves, more slender towards the tip, also the thickened spur and more rhombic, blunter lip-blade.

Angraecum teretifolium *Ridl.* in J. Linn. Soc., Bot. 21: 484 (1885). TYPE: Madagascar, Ankafana, *Deans Cowan* s.n. (holo. BM).

Angorchis teretifolia (Ridl.) Kuntze, Rev. Gen. Pl. 2: 652 (1891).

Monixus teretifolius (Ridl.) Finet in Bull. Soc. Bot. France 54, Mém. 9: 16, pl.2 (1907).

DISTRIBUTION: Madagascar: Antananarivo, Fianarantsoa, Toamasina.

HABITAT: shady forest.

ALTITUDE: 1200 – 1300 m.

FLOWERING TIME: October.

LIFE FORM: epiphyte.

PHYTOGEOGRAPHICAL REGION: III.

DESCRIPTION: {130}, {896}.

ILLUSTRATION: {3}, {68}, {130}, {292}, {513}, {529}, {724}, all except {130} and {269} refer to *A. linearifolium.*

NOTES: often confused with *A. linearifolium* but the leaves are a little thicker and shorter, the sepals and petals acuminate, the lip very concave and with a central ridge at the base, and the anther truncate and a little emarginate at the front.

Angraecum triangulifolium *Senghas* in Adansonia sér. 2, 4: 310 (1964). TYPE: Madagascar, Perinet, *Rauh* 7137 (holo. HEID).

DISTRIBUTION: Madagascar: Toamasina.

HABITAT: lower-montane and plateau, humid, evergreen forest.

ALTITUDE: c. 1200 m.

FLOWERING TIME: August.

LIFE FORM: epiphyte.

PHYTOGEOGRAPHICAL REGION: III.

DESCRIPTION: {1239}.

ILLUSTRATION: {1239}.

NOTES: stem long; flowers medium-sized, lip obovate, 3 × 1.8 mm, with a narrow callus in the middle extending into the spur; spur c. 2.5 cm long, fairly wide at the opening and narrowing towards the base.

Angraecum trichoplectron *(Rchb.f.) Schltr.* in Beih. Bot. Centralbl. 33 (2): 438 (1915). TYPE: ?Madagascar, H. Low & Co. *Hamelin* s.n. (holo. W).

Aeranthes trichoplectron Rchb.f. in Gard. Chron. 3: 264 (1888).

Mystacidium trichoplectron (Rchb.f.) T.Durand & Schinz, Consp. Fl. Afric. 5: 54 (1892).

DISTRIBUTION: Madagascar.

DESCRIPTION: {1014}.

NOTES: a dubious species known only from the type, based on a cultivated plant.

Angraecum urschianum *Toill.-Gen. & Bosser* in Adansonia, sér. 2, 1: 100 (1961). TYPE: Madagascar, between Moramanga & Lake Alaotra, *Harmelin* s.n. (holo. P).

DISTRIBUTION: Madagascar: Toamasina.

HABITAT: humid, evergreen forest at intermediate elevation, mossy forest.

ALTITUDE: c. 1100 m.

FLOWERING TIME: September.

LIFE FORM: epiphyte.

PHYTOGEOGRAPHICAL REGION: III.

DESCRIPTION: {1407}.

ILLUSTRATION: {1407}.

NOTES: small plant with characteristic darkly pitted leaves; flowers comparatively large; lip broadly oval-triangular, flat, 10 – 13 mm × 7.5 – 9 mm; spur slender, thread-like, 11 – 12 cm long.

Angraecum verecundum *Schltr.* in Repert. Sp. Nov. Regni Veg. Beih. 33: 348 (1925). TYPE: Madagascar, Mt. Tsiafajavona, Ankaratra, *H.Perrier* 13494 (holo. P).

DISTRIBUTION: Madagascar: Antananarivo.

HABITAT: mossy forest on moss- and lichen-covered trees.

ALTITUDE: c. 2000 m.

FLOWERING TIME: February.

LIFE FORM: epiphyte.

PHYTOGEOGRAPHICAL REGION: III, IV.

DESCRIPTION: {896}, {1221}.

ILLUSTRATION: {1223}.

NOTES: known from the type only. Long-stemmed epiphyte; leaves linear-ligulate up to 14 cm long; flowers medium-sized; lip broadly oval-apiculate, 6.5 × 6 mm; spur subcylindric, a little expanded towards the obtuse tip, c. 10 mm long.

Angraecum vesiculatum *Schltr.* in Repert. Sp. Nov. Regni Veg. Beih. 33: 348 (1925). TYPE: Madagascar, Mt. Tsaratanana, *H.Perrier* s.n. (holo. B†).

DISTRIBUTION: Madagascar: Antsiranana.

HABITAT: lower-montane and plateau, humid, evergreen forest.

ALTITUDE: 1500 – 2000 m.

FLOWERING TIME: October.

LIFE FORM: epiphyte.

PHYTOGEOGRAPHICAL REGION: III, IV.

DESCRIPTION: {896}.

NOTES: the type was destroyed. Related to *A. acutipetalum*; flowers small; lip 7 × 7 mm, oval-elliptic and acuminate; spur sac-shaped, almost as long as the sepals, 8 mm.

Angraecum vesiculiferum *Schltr.* in Repert. Sp. Nov. Regni Veg. Beih. 33: 341 (1925).TYPE: Madagascar, Mt. Tsiafajavona, *H.Perrier* 13988a (holo. P).

DISTRIBUTION: Madagascar: Antananarivo.

HABITAT: mossy forest.

ALTITUDE: 1500 – 2000 m.

FLOWERING TIME: September.

LIFE FORM: epiphyte on moss- and lichen-covered trees.

PHYTOGEOGRAPHICAL REGION: IV.

DESCRIPTION: {896}.

NOTES: known from the type only. Differs from *Angraecum setipes* in its thicker and pointed leaves; lip broadly oval, 4.5 × 3 mm, concave, acuminate; spur horizontally recurved, inflated at the tip, obtuse, 2.5 mm long,

Angraecum viguieri *Schltr.* in Repert. Sp. Nov. Regni Veg. Beih. 18: 326 (1922). TYPE: Madagascar, Analamazaotra, *Viguier & Humbert* 1100 (holo. P).

DISTRIBUTION: Madagascar: Antsiranana, Toamasina.
HABITAT: humid, evergreen forest on plateau, mossy ridge-top forest.
ALTITUDE: 900 – 1100 m.
FLOWERING TIME: October – November.
LIFE FORM: epiphyte.
PHYTOGEOGRAPHICAL REGION: III.
DESCRIPTION: {896}.
ILLUSTRATION: {3}, {73}, {79}, {281}, {426}, {463}, {501}, {514}, {516}, {529}, {579}, {632}, {724}, {850}, {869}, {896}, {1035}, {1333}.
CULTIVATION: {516}, {529}, {869}, {1035}.
NOTES: robust plant with thin, ligulate-linear leaves; flowers large; sepals and petals and spur brownish-orange; lip c. 5.5 cm × 3.5 cm; spur strongly funnel-shaped at the base, then filiform, up to 10 cm long.

Angraecum xylopus *Rchb.f.* in Flora 68: 538 (1885). TYPE: Comoros, Grande Comore, Combani *Humblot* 1451 (451) (holo. W, Reichenbach Herbarium 46135 & 5337).
Macroplectrum xylopus (Rchb.f.) Finet in Bull. Soc. Bot. France 54, Mém. 9: 23 (1907).

DISTRIBUTION: Comoros: Grande Comore.
FLOWERING TIME: December – January.
LIFE FORM: epiphyte {896}.
DESCRIPTION: {896}.
ILLUSTRATION: {3}, {292}.
NOTES: flowers similar to *A. calceolus* but sepals and petals broader.

Angraecum zaratananae *Schltr.* in Repert. Sp. Nov. Regni Veg. Beih. 33: 317 (1925). TYPES: Madagascar, Mt. Tsaratanana, *H.Perrier* 15303 (syn. P), *H.Perrier* 15318 (syn. P).

DISTRIBUTION: Madagascar: Antsiranana.
HABITAT: lichen-rich forest on moss- and lichen-rich-covered trees.
ALTITUDE: 1500 – 2200m.
FLOWERING TIME: January – February.
LIFE FORM: epiphyte.
PHYTOGEOGRAPHICAL REGION: III, IV.
DESCRIPTION: {896}.
NOTES: related to *A. rhizomaniacum* which has smaller leaves and much smaller flowers with a shorter spur; lip very broadly rhomboid; spur a little expanded at the opening, then slender and cylindrical and finally recurved-inflated into a club-shape at the apex, 4 cm long. Perrier de la Bâthie changed Schlechter's spelling of this name to *A. tsaratananae*; there is no basis for this and Schlechter's name stands.

Auxopus madagascariensis *Schltr.* in Repert. Sp. Nov. Regni Veg. Beih. 33: 121 (1924). TYPE: Madagascar, nr. Andranomavo, Ambongo, *H.Perrier* 8013 (holo. P).

DISTRIBUTION: Madagascar: Mahajanga.
HABITAT: seasonally dry, deciduous forest and woodland and rocky outcrops on limestone.
FLOWERING TIME: January.
LIFE FORM: terrestrial.
PHYTOGEOGRAPHICAL REGION: V.
DESCRIPTION: {896}.
NOTES: known only from the type. Saprophyte; inflorescence up to 12 cm; perianth 4 – 5 mm, glabrous; lip 3 × 2.5 mm, shortly unguiculate, sagittate-biauriculate at the base, then broadly rhombic, very obtuse and a little thickened at the apex.

Beclardia grandiflora *Bosser* in Adansonia 19 (2): 185 (1997). TYPE: Madagascar, Perinet, *Bosser* 17957 (holo. P).

DISTRIBUTION: Madagascar: Toamasina.
HABITAT: shaded forest.
ALTITUDE: 900 m.
LIFE FORM: epiphyte.
PHYTOGEOGRAPHICAL REGION: III.
DESCRIPTION: {137}.
ILLUSTRATION: {137}.
NOTES: more robust than *B. macrostachya*, with longer and wider leaves and bigger flowers; lip is 4-lobed, with the terminal lobes bigger than the laterals and with the margins just about undulate, not crispate.

Beclardia macrostachya *(Thouars)* A.Rich., Mon. Orch. Iles France Bourbon: 79, t.2 (1828). TYPE: Réunion, *Thouars* s.n. (holo. P).
Epidendrum macrostachyum Thouars, Hist. Orch.: t.83 (1822).
Aerides macrostachyum (Thouars) Spreng., Syst. Veg. 3: 719 (1826).
Oeonia macrostachya (Thouars) Lindl., Gen. Sp. Orch. Pl.: 245 (1833).
Aeranthes macrostachyus (Thouars) Rchb.f. in Walpers Ann. Bot. Syst. 6: 900 (1861), as *Aeranthus*.
Rhaphidorhynchus macrostachyus (Thouars) Finet in Bull. Soc. Bot. France 54, Mém. 9: 43 (1907).
Epidendrum brachystachyum Thouars, Hist. Orch.: t.84 (1822).
Beclardia brachystachya (Thouars) A.Rich., Mon. Orch. Iles France Bourbon: 70, (1828).
Oeonia brachystachya (Thouars) Lindl., Gen. Sp. Orch. Pl.: 245 (1833).
Aeranthes brachystachyus (Thouars) Bojer, Hort. Maur.: 14 (1837), as *Aeranthus*.
Rhaphidorhynchus macrostachyus var. *brachystachyus* (Thouars) Finet in Bull. Soc. Bot. France 54, Mém. 9: 43 (1907).
Beclardia erostris Frapp., Cat. Orch. Réunion: 12 (1880), *nom. nud.*
Oeonia erostris Cordem., Fl. Réunion: 217 (1895). TYPE: Réunion, Côteau Maigre de la Rivière des Marsouins, *Cordemoy* s.n. (lecto. MARS).
Oeonia erostris var. *egena* Cordem., Fl. Réunion: 219 (1895). TYPE: Réunion *Cordemoy* s.n., not found at MARS.
Oeonia erostris var. *robusta* Cordem., Fl. Réunion: 219 (1895). TYPE: Réunion *Cordemoy* s.n., not found at MARS.

DISTRIBUTION: Madagascar: Antananarivo, Antsiranana, Fianarantsoa, Toamasina, Toliara. Mascarenes: Réunion, Mauritius.
HABITAT: coastal and humid, evergreen forest; seasonally dry, deciduous forest and woodland; lichen-rich forest.
ALTITUDE: 400 – 2000 m.
FLOWERING TIME: December – June.
LIFE FORM: epiphyte.
PHYTOGEOGRAPHICAL REGION: I, III, V.
DESCRIPTION: {896}, {1044}, {1259}.
ILLUSTRATION: {3}, {149b}, {166}, {167}, {246}, {281}, {691}, {896}, {906}, {1028}, {1044}, {1259}, {1289}, {1398}.
CULTIVATION: {1044}.
NOTES: medium-sized plants with flabellate leaves; flowers white, often with a green-tipped spur and a hairy yellow disk on the lip; lip 1.2 – 2 × 1 – 1.8 cm; lateral lobes rounded; terminal lobe sub-entire, bilobed at the tip, the edges undulate-crispate; spur 7 – 12 mm long, the base funnel-shaped, tip suddenly contracted in a cylinder.

Benthamia bathieana *Schltr.* in Repert. Sp. Nov. Regni Veg. Beih. 33: 25 (1924). TYPE: Madagascar, Mt. Ibity, *H.Perrier* 13582 & *loc. cit.* 14 (syn. P).
Benthamia latifolia Schltr. in Beih. Bot. Centralbl. 34 (2): 303 (1916), *non* A.Rich.

DISTRIBUTION: Madagascar: Antananarivo. Mascarenes: Réunion, Mauritius.
HABITAT: near shaded, wet rocks.
ALTITUDE: c. 2000 m.
FLOWERING TIME: February – April.

LIFE FORM: terrestrial.

PHYTOGEOGRAPHICAL REGION: III.

DESCRIPTION: {896}.

ILLUSTRATION: {149b}, {444}, {896}, {1223}.

NOTES: related to *B. rostrata* but differs by the leaves which are closer together and wider; flowers characterised by shape of petals and sepals, the side-lobes of the lip and the shorter spur.

Benthamia calceolata *H.Perrier* in Bull. Soc. Bot. France 81: 32 (1934). TYPE: Madagascar, Andratamarinia, Mananara, *Decary* 24 (holo. P).

DISTRIBUTION: Madagascar: Toamasina, Toliara.

HABITAT: lowland marshland.

FLOWERING TIME: June – August.

LIFE FORM: terrestrial.

PHYTOGEOGRAPHICAL REGION: I.

DESCRIPTION: {896}.

ILLUSTRATION: {896}.

NOTES: the lateral inflorescence and deep calceolate lip are unusual for *Benthamia*; lip entire, very shallowly slipper-shaped and very short; spur scrotiform, 1 mm long, contracted at the base.

Benthamia catatiana *H.Perrier* in Bull. Soc. Bot. France 81: 29 (1934). TYPE: Madagascar, Ambolo Valley (Fort Dauphin), *Catat* 4338 (holo. P).

DISTRIBUTION: Madagascar: Toliara.

HABITAT: lowland marshland.

LIFE FORM: epiphyte.

PHYTOGEOGRAPHICAL REGION: VI.

DESCRIPTION: {896}

NOTES: known only from the type. Related to *B. nigrescens*, but differs in its lip which is slipper-shaped and narrowed into an obtuse point; spur obtuse; ovary remarkably short, 3 mm long.

Benthamia cinnabarina *(Rolfe) H.Perrier* in Bull. Soc. Bot. France 81: 38 (1934). TYPE: Madagascar, Trabonjy, *Hildebrandt* 3450 (holo. K).

Habenaria cinnabarina Rolfe in Kew Bull. 1893: 173 (1893).

Benthamia flavida Schltr. in Beih. Bot. Centralbl. 34 (2): 302 (1916). TYPE: Madagascar, Famaizankova, N of Antsirabe, *H.Perrier* 113 (holo. B†).

Benthamia perrieri Schltr. in Beih. Bot. Centralbl. 34 (2): 302 (1916). TYPE: Madagascar, without locality, *H.Perrier* XV (holo. B†).

DISTRIBUTION: Madagascar: Antananarivo.

HABITAT: marshes, peat-bog and wet rocks.

ALTITUDE: 1500 – 2200m.

FLOWERING TIME: February – May.

LIFE FORM: terrestrial.

PHYTOGEOGRAPHICAL REGION: III, IV.

DESCRIPTION: {896}.

ILLUSTRATION: {1223}.

NOTES: plant terrestrial; flowers relatively large, cinnabar-orange; lip oblong, 3-lobed, the lateral lobes very obtuse; spur scrotiform and a little flattened, 0.8 mm long.

Benthamia cuspidata *H.Perrier* in Bull. Soc. Bot. France 81: 29 (1934). TYPE: Madagascar, Mananara, *Decary* 79 (holo. P).

DISTRIBUTION: Madagascar: Toamasina.

HABITAT: lowland marshes.

FLOWERING TIME: September.

LIFE FORM: epiphyte or terrestrial.

PHYTOGEOGRAPHICAL REGION: I.

DESCRIPTION: {896}.

NOTES: known only from the type. Leaves widely obovate and cuspidate-obtuse; raceme long, cylindrical and densely flowered; and staminodes very long.

Benthamia dauphinensis *(Rolfe) Schltr.* in Repert. Sp. Nov. Regni Veg. Beih. 33: 25 (1924). TYPE: Madagascar, Fort Dauphin, *Scott Elliot* 2867 (holo. K).
Habenaria dauphinensis Rolfe in J. Linn. Soc., Bot. 29: 56 (1891).

DISTRIBUTION: Madagascar: Toliara.
HABITAT: lowland marshes.
FLOWERING TIME: June.
LIFE FORM: terrestrial.
PHYTOGEOGRAPHICAL REGION: I.
DESCRIPTION: {896}.
NOTES: known only from the type. Leaves narrow, almost 10 times longer than wide; spur flattened and almost bidentate at the tip.

Benthamia elata *Schltr.* in Repert. Sp. Nov. Regni Veg. Beih. 15: 324 (1918). TYPE: Madagascar, *Laggiara* s.n. (holo. B†).

DISTRIBUTION: Madagascar: Fianarantsoa.
HABITAT: wet embankments.
ALTITUDE: 1000 – 1200 m.
FLOWERING TIME: June – August.
LIFE FORM: terrestrial.
PHYTOGEOGRAPHICAL REGION: III.
DESCRIPTION: {896}.
ILLUSTRATION: {1223}.
NOTES: the tallest species of the genus; lip oblong, 3-lobed in the upper part, the lateral lobes narrowly lanceolate-ligulate, the middle one ligulate-obtuse, almost twice as big as the laterals; spur semi-oblong, saccate.

Benthamia exilis *Schltr.* in Repert. Sp. Nov. Regni Veg. Beih. 33: 26 (1924). TYPES: Madagascar, Andringitra massif, *H.Perrier* 13652 (lecto. P, selected here); *H.Perrier* 14390 (syn. P).

DISTRIBUTION: Madagascar: Fianarantsoa.
HABITAT: montane, ericaceous scrub.
ALTITUDE: 2000 – 2400 m.
FLOWERING TIME: January – April.
LIFE FORM: terrestrial.
PHYTOGEOGRAPHICAL REGION: IV.
DESCRIPTION: {896}.
ILLUSTRATION: {896}, {1223}.
NOTES: leaves not developed when flowering; plant slender; spur bifid at the apex.

Benthamia exilis var. **tenuissima** *Schltr.* in Repert. Sp. Nov. Regni Veg. Beih. 33: 26 (1924). TYPES: Madagascar, Mt. Tsiafajavona, *H.Perrier* 13522 (lecto. P, selected here); Andringitra massif, *H.Perrier* 14579 (syn. P).

DISTRIBUTION: Madagascar: Antananarivo, Fianarantsoa.
HABITAT: montane, ericaceous scrub.
ALTITUDE: 2000 – 2500 m.
FLOWERING TIME: February – March.
LIFE FORM: terrestrial.
PHYTOGEOGRAPHICAL REGION: III, IV.
DESCRIPTION: {896}, {1221}.
NOTES: differs from the typical form by its more slender, elongated, 60 cm long stem and yellow flowers with brown sepals.

Benthamia glaberrima *(Ridl.)* *H.Perrier* in Bull. Soc. Bot. France 22: 28 (1934). TYPES: Madagascar, Imerina, *Baron* 849 (lecto. BM, selected here), *Fox* 123 (syn. BM), Mahobo, *Scott Elliot* 1944 (syn. BM).
Holothrix glaberrima Ridl. in J. Linn. Soc., Bot. 22: 125 (1886).
Peristylus glaberrima (Ridl.) Rolfe in Orch. Rev. 7 (73): 4 (1899).
Platanthera glaberrima (Ridl.) Kraenzl., Orch. Gen. Sp. 1: 609 (1899).
Habenaria glaberrima (Ridl.) Schltr. in Beih. Bot. Centralbl. 33 (2): 405 (1915).
Rolfeella glaberrima (Ridl.) Schltr. in Repert. Sp. Nov. Regni Veg. Beih. 33: 19 (1924).

DISTRIBUTION: Madagascar: Antananarivo, Fianarantsoa; fairly common in the whole central highlands.
HABITAT: grassland, rocky outcrops and dry scrub.
ALTITUDE: 1501 – 2200 m.
FLOWERING TIME: February – May.
LIFE FORM: terrestrial.
PHYTOGEOGRAPHICAL REGION: III, IV.
DESCRIPTION: {896}.
ILLUSTRATION: {246}, {896}, {1223}.
NOTES: close to *Benthamia madagascariensis*, but is distinguished by its sulphur-yellow flowers and its cylindrical, elongated spur.

Benthamia herminioides *Schltr.* in Repert. Sp. Nov. Regni Veg. Beih. 33: 27 (1924). TYPE: Madagascar, Mt. Ibity, S Antsirabe, *H.Perrier* 8103 (holo. P).
Benthamia herminioides subsp. *typica* H.Perrier in Bull. Soc. Bot. France 81: 36 (1934), *nom. inval.*
Benthamia herminioides forma *sulfurea* H.Perrier in Mem. Inst. Sc. Mad. 6: 253 (1955), *nom. nud.*

DISTRIBUTION: Madagascar: Antananarivo, Antsiranana.
HABITAT: mainly found at high elevation amongst rocky outcrops, in *Philippia* scrub and in lichen-rich forest.
ALTITUDE: 1000 – 2000 m.
FLOWERING TIME: February – March.
LIFE FORM: terrestrial.
PHYTOGEOGRAPHICAL REGION: III, IV.
DESCRIPTION: {896}.
ILLUSTRATION: {896}.
NOTES: one of the smallest species of the genus. Spike laxly-flowered; flowers white, small; lip rhombic, 2.75 × 2.5 mm, 3-lobed in the upper third; spur very short, subglobular.

Benthamia herminioides subsp. **angustifolia** *H.Perrier* in Bull. Soc. Bot. France 81: 37 (1934). TYPE: Madagascar, Mt. Tsaratanana, *H.Perrier* 16513 (holo. P).

DISTRIBUTION: Madagascar: Antsiranana.
HABITAT: shaded, montane, ericaceous scrub.
ALTITUDE: c. 2600 m.
FLOWERING TIME: April.
LIFE FORM: terrestrial.
PHYTOGEOGRAPHICAL REGION: IV.
DESCRIPTION: {896}.
NOTES: differs from the other subspecies by the more slender shape and the shorter, linear-acute leaves, c. 5 × 5 mm, and the distinct stigma.

Benthamia herminioides subsp. **arcuata** *H.Perrier* in Bull. Soc. Bot. France 81: 36 (1934). TYPE: Madagascar, Mt. Tsaratanana, *H.Perrier* 16477 (holo. P).

DISTRIBUTION: Madagascar: Antsiranana.
HABITAT: in recently burnt, montane, *Philippia* scrub.
ALTITUDE: 2600 – 2800m.
FLOWERING TIME: April.
LIFE FORM: terrestrial.
PHYTOGEOGRAPHICAL REGION: IV.
DESCRIPTION: {896}.

NOTES: floral bracts at most equalling the ovary in length; lateral lobes of the lip narrow and acute; anther with the apicule higher, with central lobe of the rostellum rounded.

Benthamia herminioides subsp. **intermedia** *H.Perrier* in Bull. Soc. Bot. France 81: 37 (1934). TYPE: Madagascar, Mt. Tsaratanana, *H.Perrier* 16479 (holo. P).

DISTRIBUTION: Madagascar: Antsiranana.
HABITAT: lichen-rich forest.
ALTITUDE: c. 2000 m.
FLOWERING TIME: April.
LIFE FORM: terrestrial.
PHYTOGEOGRAPHICAL REGION: IV.
DESCRIPTION: {896}.
NOTES: flowers smaller than the other subspecies; lateral lobes of the lip obtuse and shorter, equalling half the size of the midlobe; rostellum with the central tooth narrower and obtuse; ovary longer, 4 mm long.

Benthamia humbertii *H.Perrier* in Bull. Soc. Bot. France 81: 35 (1934). TYPE: Madagascar, Ivohibe Peak (Bara), *Humbert* 3287 (holo. B; iso. P, K, TAN).

DISTRIBUTION: Madagascar: Fianarantsoa, Toliara.
HABITAT: humid, highland forest.
ALTITUDE: 1100 – 2000m.
FLOWERING TIME: November.
LIFE FORM: terrestrial, amongst rock.
PHYTOGEOGRAPHICAL REGION: III.
DESCRIPTION: {896}.
ILLUSTRATION: {896}.
NOTES: close to *B. latifolia* from the Mascarenes. Tall plant with a densely-flowered spike; lip broadly obovate, 5 × 4.5 mm, thick, 2-lobed in the upper third, the lobes short, obtuse, almost equal, the middle one however a little wider; spur compressed, 1 × 2 mm.

Benthamia longecalceata *H.Perrier* in Not. Syst. (Paris) 14: 139 (1951). TYPES: Madagascar, summit E of Marojejy (Manantenina), *Humbert & Cours* 23868 (lecto. P); Mt. Maimborondro (Lokoho basin), *Humbert* 23388 & *Humbert* 23602 (syn. P).

DISTRIBUTION: Madagascar.
HABITAT: peaty marsh in peaty and marshy soil; rocky outcrops; evergreen forest.
ALTITUDE: 500 – 2000 m.
FLOWERING TIME: March – April.
LIFE FORM: terrestrial.
PHYTOGEOGRAPHICAL REGION: III, IV.
DESCRIPTION: {901}.
NOTES: plant slender; flowers small; lip elongate, 2.5 mm long, slipper-shaped; spur pendent, cylindric, the same length as the lip.

Benthamia macra *Schltr.* in Repert. Sp. Nov. Regni Veg. Beih. 33: 28 (1924). TYPE: Madagascar, Andringitra massif, *H.Perrier* 14577 (holo. P).

DISTRIBUTION: Madagascar: Fianarantsoa.
HABITAT: montane, ericaceous scrub.
ALTITUDE: 1100 – 2500 m.
FLOWERING TIME: February.
LIFE FORM: terrestrial.
PHYTOGEOGRAPHICAL REGION: IV.
DESCRIPTION: {896}.
NOTES: the flower is close to those of *B. melanopoda* but the lip does not have the raised callus at the base of the mid-lobe. Plant sparsely leaved, slender; lip oblong, 4 × 2 mm, a little contracted in the middle, 3-lobed in the upper third; spur very short, oblong and obtusely saccate.

Benthamia madagascariensis *(Rolfe) Schltr.* in Repert. Sp. Nov. Regni Veg. Beih. 33: 24 (1924).
TYPES: Madagascar, Fort Dauphin, *Scott Elliot* 2643 (lecto. K, selected here), Vangaindrano, *Scott Elliot* 2257 (syn. K).
Holothrix madagascariensis Rolfe in J. Linn. Soc., Bot. 29: 55, t.12 (1891).
Habenaria madagascariensis (Rolfe) Schltr. in Oesterr. Bot. Z., 49 (1): 23 (1899).
Platanthera madagascariensis (Rolfe) Kraenzl., Orch. Gen. Sp. 1: 609 (1899).
Peristylus madagascariensis (Rolfe) Rolfe in Orch. Rev. 7 (73): 4 (1899).
Peristylus macropetalus Finet in Notul. Syst. (Paris) 2: 24 (1911). TYPE: Madagascar, without locality, *Geay* 1905 (holo. P).

DISTRIBUTION: Madagascar: Toamasina, Toliara.
HABITAT: marshes.
ALTITUDE: sea level – 500 m.
FLOWERING TIME: December – July.
LIFE FORM: terrestrial.
PHYTOGEOGRAPHICAL REGION: I, III.
DESCRIPTION: {896}.
ILLUSTRATION: {896}, {1062} but see errata.
VERNACULAR NAME: Soazombitra.
NOTES: the species has the habit of *B. glaberima*, but is readily distinguished by its laxer raceme of larger flowers, the much larger side-lobes of the lip, and the very short, saccate spur.

Benthamia majoriflora *H.Perrier* in Notul. Syst. (Paris) 14: 140 (1951). TYPE: Madagascar, W of Marojejy (Lokoho Basin), *Humbert & Capuron* 22224 (holo. P).

DISTRIBUTION: Madagascar: Antsiranana.
HABITAT: mossy forest.
ALTITUDE: c. 1500 m.
FLOWERING TIME: December.
LIFE FORM: terrestrial.
PHYTOGEOGRAPHICAL REGION: IV.
DESCRIPTION: {901}.
NOTES: known from the type only. Related to *B. verecunda* and to *B. humberti* but the lip is widely ventricose, concave, the apex 3-lobed, the lobes narrow, acute, and the spur purse-shaped, 1 × 1 mm.

Benthamia melanopoda *Schltr.* in Repert. Sp. Nov. Regni Veg. Beih. 33: 29 (1924). TYPE: Madagascar, Mt. Tsiafajavona, *H.Perrier* 13543 (holo. P).

DISTRIBUTION: Madagascar: Antananarivo, Fianarantoa.
HABITAT: grassland on peaty soil.
ALTITUDE: 1500 – 2000 m.
FLOWERING TIME: March – April.
LIFE FORM: terrestrial.
PHYTOGEOGRAPHICAL REGION: III, IV.
DESCRIPTION: {896}.
ILLUSTRATION: {877}, {896}.
NOTES: flower similar to *B. spiralis* but the lip has a large raised callus at the base of the mid-lobe, the 2 lateral lobes are also larger.

Benthamia misera *(Ridl.) Schltr.* in Repert. Sp. Nov. Regni Veg. Beih. 33: 24 (1924). TYPE: Madagascar, Imerina, *Deans Cowan* (holo. BM).
Habenaria misera Ridl. in J. Linn. Soc., Bot. 21: 503 (1885).

DISTRIBUTION: Madagascar: Antananarivo, Antsiranana, Fianarantsoa.
HABITAT: in grassland; amongst mosses; in marshy areas.
ALTITUDE: 900 – 1200 m.
FLOWERING TIME: September, March – April.
LIFE FORM: terrestrial.
PHYTOGEOGRAPHICAL REGION: III.

DESCRIPTION: {896}.

NOTES: close to *B. spiralis* but distinguished by its oval, cuspidate leaves, shorter ovary and entire lip.

Benthamia monophylla *Schltr.* in Repert. Sp. Nov. Regni Veg. Beih. 33: 30 (1924). TYPE: Madagascar, Andringitra massif, *H.Perrier* 14387 (holo. P).

DISTRIBUTION: Madagascar: Fianarantsoa, Toliara.
HABITAT: montane *Philippia* scrub; high altitude moss; lichen-rich forest, and rock crevices.
ALTITUDE: 1800 – 2600 m.
FLOWERING TIME: March.
LIFE FORM: terrestrial.
PHYTOGEOGRAPHICAL REGION: III, IV.
DESCRIPTION: {896}.
ILLUSTRATION: {896}, {1221}.
NOTES: medium-sized terrestrial with a single leaf; spike subdensely flowered; flowers small; lip entire.

Benthamia nigrescens *Schltr.* in Beih. Bot. Centralbl. 34 (2): 300 (1916). TYPE: Madagascar, Ankaratra, *H.Perrier* IX (holo. B†; iso. P).
Benthamia nigrescens subsp. *typica* H.Perrier in Bull. Soc. Bot. France 81: 30 (1934).

DISTRIBUTION: Madagascar: Antananarivo, Toamasina.
HABITAT: shaded and wet rocks; grassland and evergreen forest.
ALTITUDE: 2000 – 2400 m.
FLOWERING TIME: October – April.
LIFE FORM: lithophyte, terrestrial, epiphyte on moss and lichen-covered trees.
PHYTOGEOGRAPHICAL REGION: III, IV.
DESCRIPTION: {896}.
ILLUSTRATION: {167}, {896}, {1223}, {1289}.
NOTES: spike very densely flowered, obscurely unilateral; lip ligulate, obtuse, c.5 × 1.5 mm, thickened and contracted at the tip; spur scrotiform, a little wider than long.

Benthamia nigrescens subsp. **decaryana** *H.Perrier* in Bull. Soc. Bot. France 81: 31 (1934). TYPE: Madagascar, Ankaizina, *Decary* 1982 (holo. P).

DISTRIBUTION: Madagascar: Mahajanga.
HABITAT: humid, highland forest.
ALTITUDE: c. 1700 m.
FLOWERING TIME: April.
PHYTOGEOGRAPHICAL REGION: III {896}.
DESCRIPTION: {896}.
NOTES: differs by the more unilateral spike, the narrower lip, and the size of the anther.

Benthamia nigrescens subsp. **humblotiana** *H.Perrier* in Bull. Soc. Bot. France 81: 31 (1934). TYPE: Madagascar, Mt. Tsaratanana, *H.Perrier* 16110 (holo. P).

DISTRIBUTION: Madagascar: Antsiranana.
HABITAT: mossy forest.
ALTITUDE: c. 2000 m.
FLOWERING TIME: April.
PHYTOGEOGRAPHICAL REGION: III.
DESCRIPTION: {896}.
ILLUSTRATION: {896}.
NOTES: differs by the obtuse leaves, the denser spike and the flowers facing all directions; lip is thickened from the base; spur smaller.

Benthamia nigro-vaginata *H.Perrier* in Bull. Soc. Bot. France 81: 32 (1934). TYPE: Madagascar, Ikongo massif, *Decary* 5789 (holo. P).

DISTRIBUTION: Madagascar: Fianarantsoa.
HABITAT: river margins and riverine forest.
FLOWERING TIME: October.
LIFE FORM: terrestrial.
PHYTOGEOGRAPHICAL REGION: III.
DESCRIPTION: {896}.
NOTES: known from the type only. Related to *B. spiralis*, but distinguished by its taller size, the big leaves and long black sheaths at the base of the stem.

Benthamia nivea *Schltr.* in Repert. Sp. Nov. Regni Veg. Beih. 33: 31 (1924). TYPE: Madagascar, Mt. Tsiafajavona, *H.Perrier* 13507 (holo. P).

DISTRIBUTION: Madagascar: Antananarivo.
HABITAT: mossy, montane forest on moss- and lichen-covered trees.
ALTITUDE: 2000 – 2200 m.
FLOWERING TIME: February – March, September.
LIFE FORM: epiphyte.
PHYTOGEOGRAPHICAL REGION: IV.
DESCRIPTION: {896}.
ILLUSTRATION: {896}, {1223}, {475}.
NOTES: epiphyte; flowers white; lip narrowly oblong, 8 – 9 × 3.5 – 4 mm, the base a little concave, 3-lobed in their upper third; spur subglobose, saccate, very short.

Benthamia nivea subsp. **parviflora** *H.Perrier* in Bull. Soc. Bot. France 81: 34 (1934). TYPE: Madagascar, Mt. Tsaratanana, *H.Perrier* 16478 (holo. P).

DISTRIBUTION: Madagascar: Antsiranana.
HABITAT: mossy, montane forest.
ALTITUDE: c. 2000 m.
FLOWERING TIME: April.
PHYTOGEOGRAPHICAL REGION: IV.
DESCRIPTION: {896}.
ILLUSTRATION: {896}.
NOTES: flowers smaller than typical variety; lip structure is also different.

Benthamia perfecunda *H.Perrier* in Notul. Syst. (Paris) 14: 140 (1951). TYPES: Madagascar, E summit of Marojejy, *Humbert & Cours* 23754 (lecto. P, selected here); *Humbert* 23755 (syn. P).

DISTRIBUTION: Madagascar: Antsiranana.
HABITAT: mossy, montane forest.
ALTITUDE: 1800 – 2000 m.
FLOWERING TIME: April.
LIFE FORM: terrestrial in peaty soil.
PHYTOGEOGRAPHICAL REGION: IV.
DESCRIPTION: {901}.
NOTES: close to *B. macra* and to *B. verecunda*, but differs from both by its slipper-shaped lip and the grass-like leaves.

Benthamia perularioides *Schltr.* in Repert. Sp. Nov. Regni Veg. Beih. 33: 32 (1924). TYPE: Madagascar, Mt. Tsiafajavona, *H.Perrier* 13523 (holo. P).

DISTRIBUTION: Madagascar: Antananarivo.
HABITAT: grassland.
ALTITUDE: 1200 – 2500m.
FLOWERING TIME: March – April, November.
LIFE FORM: terrestrial.

PHYTOGEOGRAPHICAL REGION: III, IV.

DESCRIPTION: {896}.

ILLUSTRATION: {1223}.

NOTES: related to *B. rostrata* but with smaller, closer together flowers and a narrower lip with a smaller, c.3 mm long spur.

Benthamia praecox *Schltr.* in Beih. Bot. Centralbl. 34 (2): 303 (1916). TYPE: Madagascar, W of Andringitra massif, *H.Perrier* 161 (11481) (holo. B†; iso. P).

DISTRIBUTION: Madagascar: Fianarantsoa.

HABITAT: rocky outcrops of granite and gneiss; high -elevation moss- and lichen-rich forest and ericaceous bush, with *Philippia* dominant.

ALTITUDE: 1000 – 1950m.

FLOWERING TIME: September – January.

LIFE FORM: terrestrial or lithophyte.

PHYTOGEOGRAPHICAL REGION: III.

DESCRIPTION: {896}.

ILLUSTRATION: {281}, {896}, {945}, {1223}.

NOTES: slender plant characterised by its long, cylindrical spur and the short staminodes.

Benthamia procera *Schltr.* in Beih. Bot. Centralbl. 34 (2): 301 (1916). TYPE: Madagascar, Mangoro Basin, *H.Perrier* 119 (holo. P).

DISTRIBUTION: Madagascar: Toamasina.

HABITAT: shaded rock; cool, shady spots in humid evergreen forest at intermediate elevation; river margins.

ALTITUDE: 400 – 800m.

FLOWERING TIME: November – December.

LIFE FORM: terrestrial.

PHYTOGEOGRAPHICAL REGION: I.

DESCRIPTION: {896}.

NOTES: leaves narrow; flowers small; lip oblong-ligulate, 3 × 1.8 mm, 3-lobed in the upper part, the middle lobe thickened, longer than the laterals which are obscure and obtuse; spur wider at the apex than long, 0.8 × 1.2 mm, purse-shaped.

Benthamia rostrata *Schltr.* in Repert. Sp. Nov. Regni Veg. Beih. 33: 33 (1924). TYPE: Madagascar, Mt. Ibity, *H.Perrier* 8101 (holo. B†; iso. P).

Platanthera madagascariensis sensu Schltr. in Beih. Bot. Centralbl. 34 (2): 299 (1916), *non* Schumann (1899) {1227} *nec* Kraenzl. (1901) {657}.

DISTRIBUTION: Madagascar: Antananarivo, Fianarantsoa.

HABITAT: shaded rocky outcrops; grassland.

ALTITUDE: 1500 – 2000 m.

FLOWERING TIME: February – March.

PHYTOGEOGRAPHICAL REGION: III, IV.

DESCRIPTION: {896}.

ILLUSTRATION: {896}.

NOTES: similar in shape to *B. perularioides*, but differs by the 2 radical leaves, the spike which slightly unilateral, the 4 mm long perianth segments which are longer, the lozenge-shaped lip which is expanded near the middle and 3-lobed, and the longer, filiform spur, 5 mm long.

Benthamia spiralis *(Thouars) A.Rich.,* Monogr. Orch. Iles France Bourbon: 44 (1828). TYPE: Réunion, *Thouars* s.n. (holo. P).

Satyrium spirale Thouars, Hist. Orchid.: t.9 (1822).

Spiranthes africana Lindl. in Bot. Reg.: t.823 (1824).

Habenaria spiralis (Thouars) A.Rich., Monogr. Orch. Iles France Bourbon: 22 (1828).

Herminium spirale (Thouars) Rchb.f. in Bonplandia 3: 213 (1855).

Peristylus spiralis (Thouars) S.Moore in Baker, Fl. Mauritius: 335 (1877).

Habenaria minutiflora Ridl. in J. Linn. Soc., Bot. 21: 503 (1885). TYPE: Madagascar, Andrangoloaka, E Imerina, *Hildebrandt* 3729 (holo. BM; iso. HBG).

Habenaria dissimulata Schltr. in Beih. Bot. Centralbl. 33 (2): 404 (1915), *nom. nov. pro B. spiralis* (Thouars) A.Rich.

Benthamia minutiflora (Ridl.) Schltr. in Repert. Sp. Nov. Regni Veg. Beih. 33: 29 (1924).

Benthamia spiralis var. *dissimulata* (Schltr.) H.Perrier in Bull. Soc. Bot. France 81: 34 (1934).

DISTRIBUTION: Madagascar: Antananarivo, Fianarantsoa, Toamasina, Toliara. Mascarenes: Réunion, Mauritius.

HABITAT: marshland; forest; woodland margins; shaded wet rocks.

ALTITUDE: 100 – 1500 m.

FLOWERING TIME: January – December.

LIFE FORM: terrestrial.

PHYTOGEOGRAPHICAL REGION: I, III, IV.

DESCRIPTION: {896}.

ILLUSTRATION: {246}, {1398}.

NOTES: a common and variable species. Plant up to 60 cm tall; flowers small; lip 2 – 3 × 1 – 1.8 mm, 3-lobed at the apex, the middle lobe thicker and at least twice the length of the laterals, which are obscure and sometimes even missing; spur subrectangular, purse-shaped, entire or a little emarginate at the tip, c. 1 mm long. This may be the same as *B. africana* (Lindl.) Garay.

Benthamia verecunda *Schltr.* in Repert. Sp. Nov. Regni Veg. Beih. 33: 34 (1924). TYPE: Madagascar, Manongarivo massif, *H.Perrier* 1944 (19) (holo. P).

Habenaria chlorantha sensu Schltr. in Ann. Mus. Col. Mars., ser.3, I: 14 (1913), *non* Spreng.

DISTRIBUTION: Madagascar: Antsiranana.

HABITAT: montane, humid, evergreen forest; forested mountain-tops, amongst mosses.

ALTITUDE: c. 1200 m.

FLOWERING TIME: April.

LIFE FORM: terrestrial.

PHYTOGEOGRAPHICAL REGION: III.

DESCRIPTION: {896}.

ILLUSTRATION: {1223}.

NOTES: known only from the type. Inflorescence elongate, up to 30 cm long, laxly many-flowered; flowers virtually secund; lip oblong, 3.5 × 2 mm, a little concave at the base, 3-lobed in the upper third, the middle lobe oval, subobtuse; spur almost globular, 1 mm in diameter.

Brachycorythis disoides *(Ridl.) Kraenzl.*, Orch. Gen. Sp.: 1 (9): 543 (1898). TYPE: Madagascar, Ankafana, *Deans Cowan* s.n. (holo. BM).

Habenaria disoides Ridl. in J. Linn. Soc., Bot. 21: 511 (1885).

Diplacorchis disoides (Ridl.) Schltr. in Beih. Bot. Centralbl. 38 (2): 131 (1921).

DISTRIBUTION: Madagascar: Antananarivo, Fianarantsoa.

HABITAT: grassland; marsh.

ALTITUDE: 1500 – 2000 m.

FLOWERING TIME: December – July.

LIFE FORM: terrestrial.

PHYTOGEOGRAPHICAL REGION: III, IV.

DESCRIPTION: {896}, {1370}.

ILLUSTRATION: {896}.

NOTES: plant stout; flowers white marked with crimson, pink or pinkish white, the lip darker; lip oblong, obtuse, obscurely 3-lobed, crenulate, with a ridge in the middle; spur short, 5 mm long and, 2 mm wide, flattened.

Brachycorythis pleistophylla *Rchb.f.*, Otia Bot. Hamburg.: 104 (1881). TYPE: Mozambique, *Meller* s.n. (holo. W).

Platanthera pleistophylla (Rchb.f.) Schltr., Westafr. Kautschuk-Exped.: 274 (1900).

Brachycorythis briartiana Kraenzl. (apud De Wild. & Duv.) in Bull. Soc. Bot. Belg. 38: 219 (1900). TYPE from Zaire.

Brachycorythis pulchra Schltr. in Bot. Jahrb. Syst. 53: 485 (1915), *pro parte.* TYPE from Tanzania.

Brachycorythis perrieri Schltr. in Beih. Bot. Centralbl. 34 (2): 296 (1916). TYPE: Madagascar, Tsaratanana, *H.Perrier* 138 (holo. B).

Brachycorythis macclouniei Braid in Kew Bull. 1925: 357 (1925). TYPE from Malawi.

DISTRIBUTION: Madagascar: Antsiranana, Mahajanga, Toamasina. Also in East & West Africa.

HABITAT: grassland; dry woods; and woodland margins.

ALTITUDE: 1400 – 1500 m.

FLOWERING TIME: October – January.

LIFE FORM: terrestrial.

PHYTOGEOGRAPHICAL REGION: II, III, IV.

DESCRIPTION: {896}, {1370}.

ILLUSTRATION: {281}, {1210}, {1223}, {1345}, {1472}.

NOTES: tall terrestrial; flowers mauve-purple to purple or crimson; lip 11 – 15 mm long; hypochile navicular.

Brownleea coerulea *Harv. ex Lindl.* in Hook. J. Bot. 1: 16 (1842); TYPE: South Africa, Caffraria, Eastern Cape Province, *Brownlee* s.n. (holo. K).

Disa coerulea (Harv. ex Lindl.) Rchb.f., Otia Bot. Hamb. 2: 119 (1881).

Brownleea madagascarica Ridl. in J. Linn. Soc., Bot. 22: 126 (1885). TYPE: Madagascar, Imerina, *Fox* s.n. (holo. BM).

Brownleea nelsonii Rolfe in Fl. Cap. 5, 3: 262 (1913). TYPE from Transvaal.

Brownleea woodii Rolfe in Fl. Cap. 5, 3: 262 (1913). TYPE from Natal.

DISTRIBUTION: Madagascar: Antananarivo, Fianarantsoa. Also in Southern Africa.

HABITAT: forest and woodland remnants; humid, evergreen forest on plateau, mossy forest; and rocky outcrops.

ALTITUDE: 1500 – 2500 m.

FLOWERING TIME: February – June.

LIFE FORM: epiphyte, lithophyte or terrestrial.

PHYTOGEOGRAPHICAL REGION: III, IV.

DESCRIPTION: {549}, {764}, {930}, {768}, {930}.

ILLUSTRATION: {208}, {281}, {549}, {768}, {896}, {906}, {930}, {1298}, {1329} {1505}.

NOTES: inflorescence subimbricate; flowers c. 20 mm, mauve; dorsal sepal galeate; lip very small, linear, erect in front of the stigma, c. 1 mm long; spur slender, cylindrical, 13 – 26 mm long.

Brownleea parviflora *Harv. ex Lindl.* in Hook. J. Bot. 1: 16 (1842); TYPE: South Africa, Caffraria, Eastern Cape Province, *Brownlee* s.n. (holo. K).

Disa parviflora (Harv. ex Lindl.) Rchb.f., Otia Bot. Hamburg. 2: 119 (1881).

Disa alpina Hook. f. in J. Linn. Soc. 7: 22 (1864). TYPE from Cameroon (holo. K).

Brownleea alpina (Hook.f.) N.E. Br. in Fl. Trop. Afr. 7: 287 (1898).

Disa preussii Kraenzl. in Bot. Jahrb. Syst. 17: 64 (1863). TYPE from Cameroon (holo. B).

Disa apetala Kraenzl. in Bot. Jahrb. Syst. 22: 21 (1896). TYPE from Tanzania (holo. B).

Brownleea apetala (Kraenzl.) N.E. Br. in Fl. Trop. Afr. 7: 287 (1898).

Brownleea gracilis Schltr. in Bot. Jahrb. Syst. 53: 545 (1915). TYPE from Tanzania (syn. B).

Brownleea perrieri Schltr. in Repert. Sp. Nov. Regni Veg. Beih. 33: 102 (1924). TYPES: Madagascar, Manondona, *H.Perrier* 11881 (lecto. P, selected here); Andringitra massif, *H.Perrier* 13752 (syn. P).

Brownleea transvaalensis Schltr. in Ann. Transv. Mus. 10: 250 (1924) TYPE from Transvaal (syn. B).

DISTRIBUTION: Madagascar: Antananarivo, Fianarantsoa. Also in East & West Africa.

HABITAT: grassland; marshes; rocky outcrops.

ALTITUDE: 1000 – 2500 m.

FLOWERING TIME: February – April.

LIFE FORM: terrestrial in peaty and marshy soil.

PHYTOGEOGRAPHICAL REGION: III, IV.

DESCRIPTION: {764}.

ILLUSTRATION: {682}, {945}, {1223}.

NOTES: plant tall; sheath hispid at base; inflorescence, densely 20 – 60-flowered; flowers small, white with a slight green or brown tone; lip minute, 1 mm long; spur 3 – 5 mm long.

Bulbophyllum acutispicatum *H.Perrier* in Notul. Syst. (Paris) 14: 152 (1951). TYPE: Madagascar, E of Marojejy massif, affl. of Lokoho, *Humbert* 22531 (holo. P).

DISTRIBUTION: Madagascar: Antsiranana.
HABITAT: humid, evergreen forest on plateau.
ALTITUDE: 1500 – 1700m.
FLOWERING TIME: December.
LIFE FORM: epiphyte.
PHYTOGEOGRAPHICAL REGION: IV.
DESCRIPTION: {901}.
NOTES: known only from the type. Plant 2-leaved, characterised by its approximate flowers, set shallowly within the rachis, covered by the distichous. acute floral bracts, distant on the raceme.

Bulbophyllum afzelii *Schltr.* in Repert. Sp. Nov. Regni Veg. Beih. 15: 328 (1918). TYPE: Madagascar, Andakambararata, *Afzelius* s.n. (holo. B†; iso. P).

DISTRIBUTION: Madagascar: Antsiranana.
HABITAT: humid, evergreen forest at intermediate elevation.
FLOWERING TIME: September.
LIFE FORM: epiphyte {874}.
PHYTOGEOGRAPHICAL REGION: I.
DESCRIPTION: {896}.
ILLUSTRATION: {896}, {1223}.
NOTES: plant and flowers small, 2-leaved; lip fleshy, curved, elliptic, with 2 ridges on top, the stelids triangular, acute, short and erect.

Bulbophyllum afzelii var. **microdoron** *(Schltr.) Bosser* in Adansonia, ser.2, 5: 379 (1965). TYPE: Madagascar, Zaratanana massif, *H.Perrier* 15730 (holo. P).
Bulbophyllum microdoron Schltr. in Repert. Sp. Nov. Regni Veg. Beih. 33: 195 (1924).
Bulbophyllum lichen-richophyllax var. *microdoron* (Schltr.) H.Perrier in Humbert ed., Fl. Mad. Orch. 1: 322 (1939).

DISTRIBUTION: Madagascar: Antananarivo, Antsiranana, Toamasina.
ALTITUDE: c. 1000 m.
HABITAT: humid, evergreen forest on plateau.
FLOWERING TIME: July – October.
LIFE FORM: epiphyte, lithophyte on wet rock.
PHYTOGEOGRAPHICAL REGION: III.
DESCRIPTION: {108}.
NOTES: leaves shorter than typical variety, lanceolate, acute, subacuminate, up to 5mm long, contracted at the base into a very short petiole; petals obovate, rounded at the tip and shortly mucronate.

Bulbophyllum aggregatum *Bosser* in Adansonia, ser.2, 5: 401 (1965). TYPE: Madagascar, Lake Mantasoa, Imerina, *J.-P.Peyrot* 4 (holo. P).

DISTRIBUTION: Madagascar: Antananarivo, Fianarantsoa.
ALTITUDE: 1300 – 1700m.
HABITAT: humid, evergreen forest on plateau, mossy forest.
FLOWERING TIME: April – August.
LIFE FORM: epiphyte.
PHYTOGEOGRAPHICAL REGION: III.
DESCRIPTION: {108}.
ILLUSTRATION: {108}.
NOTES: two-leaved plant close to *B. quadrialatum* which is a more robust species with a laxer spike and different flowers.

Bulbophyllum alexandrae *Schltr.* in Repert. Sp. Nov. Regni Veg. Beih. 33: 249 (1925). TYPE: Madagascar, Manjakandriana, E Imerina, *H.Perrier* 12976 (lecto. P, selected here).

DISTRIBUTION: Madagascar: Antananarivo, Toamasina, Toliara.
HABITAT: seasonally dry, deciduous forest and woodland; humid, evergreen forest on plateau; on granite.
ALTITUDE: 700 – 1500m.
FLOWERING TIME: December – February.
LIFE FORM: epiphyte or lithophyte.
PHYTOGEOGRAPHICAL REGION: III, V.
DESCRIPTION: {896}.
ILLUSTRATION: {896}, {1223}, {281}, {139}, {484}.
NOTES: large, bifoliate plant characterised by its large, pendant flowers with a bright yellow lip nestled between the green sepals.

Bulbophyllum alleizettei *Schltr.* in Repert. Sp. Nov. Regni Veg. Beih. 18: 322 (1922). TYPE: Madagascar, Mandraka forests, *Alleizette* 506 (holo. P).

DISTRIBUTION: Madagascar: Antananarivo, Toamasina.
HABITAT: humid, evergreen forest on plateau.
ALTITUDE: 1000 – 1500m.
FLOWERING TIME: February, October.
LIFE FORM: epiphyte.
PHYTOGEOGRAPHICAL REGION: III.
DESCRIPTION: {896}.
NOTES: pseudobulbs yellow, single-leaved, the sheaths woolly; inflorescence shorter than the plant; flowers small, c. 3 mm long, yellow.

Bulbophyllum ambatoavense *Bosser* in Adansonia sér. 3, 26 (1): 54 (2004). TYPE: Ambatoava, Cap Masoala, 700 m, *Haevermans* 232 (holo. P; iso. P).

DISTRIBUTION: Madagascar: Toamasina.
HABITAT: humid, lowland forest.
ALTITUDE: 700 m.
LIFE FORM: epiphyte.
PHYTOGEOGRAPHICAL REGION: I.
DESCRIPTION: {145}.
ILLUSTRATION: {145}.
NOTES: plant single-leafed; similar to *Bulbophyllum subsessile* but pseudobulbs reddish, spherical and flattened; leaves small and narrow; lip fleshy with 2 wings at the base.

Bulbophyllum ambrense *H.Perrier* in Notul. Syst. (Paris) 6 (2): 58 (1937). TYPE: Madagascar, Mt. d'Ambre, *H.Perrier* 17718 (holo. P).

DISTRIBUTION: Madagascar: Antsiranana.
HABITAT: mossy forest on moss- and lichen-covered trees.
ALTITUDE: c. 1200 m.
FLOWERING TIME: September.
LIFE FORM: epiphyte.
PHYTOGEOGRAPHICAL REGION: III.
DESCRIPTION: {896}.
NOTES: known from the type only. Pseudobulbs single-leafed; sheaths woolly; inflorescence short, laxly-flowered; lip small, 2.6 × 2 mm, a little emarginate at the tip, with a thick rounded heel, the lobes suborbicular and the margins denticulate.

Bulbophyllum amoenum *Bosser* in Adansonia, ser.2, 5: 398 (1965). TYPE: Madagascar, Lake Mantasoa, Imerina, *Bosser* 16210 (holo. P).

DISTRIBUTION: Madagascar: Antananarivo, Toamasina.
HABITAT: humid, evergreen forest on plateau; mossy forest.

ALTITUDE: 1200 – 1400m.

FLOWERING TIME: September – December.

LIFE FORM: epiphyte.

PHYTOGEOGRAPHICAL REGION: III.

DESCRIPTION: {108}.

ILLUSTRATION: {108}.

NOTES: two-leaved plant related to *B. microglossum* but the sepals longer, the petals broader, the lip pale yellow, glabrous and a different structure, and the stelids pyramidical.

Bulbophyllum amphorimorphum *H.Perrier* in Notul. Syst. (Paris) 14 (2): 151 (1951); Bosser in Adansonia, sér. 2, 5: 386 (1965). TYPE: Madagascar, E of Marojejy massif, *Humbert* 22216 (holo. P).

DISTRIBUTION: Madagascar: Antsiranana.

HABITAT: humid, highland forest; lichen-rich forest.

ALTITUDE: c. 1400 m.

FLOWERING TIME: December.

LIFE FORM: epiphyte.

PHYTOGEOGRAPHICAL REGION: IV.

DESCRIPTION: {108}.

NOTES: known only from the type. Plant usually single-leafed; lip sigmoid-curved, 4 mm long, apex widened; stelids long.

Bulbophyllum analamazoatrae *Schltr.* in Repert. Sp. Nov. Regni Veg. Beih. 33: 209 (1924). TYPE: Madagascar, Analamazaotra, *H.Perrier* 11855 (holo. B†; iso. P).

DISTRIBUTION: Madagascar: Antsiranana, Toamasina.

HABITAT: humid, hilltop, evergreen forest; mossy forest.

ALTITUDE: 400 – 1300 m.

FLOWERING TIME: September – January.

LIFE FORM: epiphyte.

PHYTOGEOGRAPHICAL REGION: I.

DESCRIPTION: {896}.

ILLUSTRATION: {505}, {896}, {1223}.

NOTES: plant single-leaved; inflorescence very thin, few-flowered; flowers yellow-orange; lip subspherical; stelids subulate, 1 mm long.

Bulbophyllum andohahelense *H.Perrier* in Notul. Syst. (Paris) 8: 37 (1939). TYPE: Madagascar, Andohahela massif, *Humbert* 13600 (holo. P).

DISTRIBUTION: Madagascar: Fianarantsoa, Toliara.

HABITAT: montane, ericaceous scrub.

ALTITUDE: 1800 – 2000 m.

FLOWERING TIME: November, January.

LIFE FORM: epiphyte.

PHYTOGEOGRAPHICAL REGION: III.

DESCRIPTION: {896}.

NOTES: plant bifoliate, c. 6 cm, single-leafed; inflorescence longer than the leaf; flowers dark violet; petals longly ciliate; lip 2.5 mm long, margins ciliate.

Bulbophyllum anjozorobeense *Bosser* in Adansonia 22 (2): 170 (2000): TYPE: Madagascar, Anjozorobe, *Bosser* 17256 (holo. P).

DISTRIBUTION: Madagascar: Antananarivo, Toamasina.

HABITAT: humid forest remnants at intermediate elevation; evergreen forest.

ALTITUDE: 1000 – 1200 m.

FLOWERING TIME: October.

LIFE FORM: epiphyte.

PHYTOGEOGRAPHICAL REGION: III.

DESCRIPTION: {139}.

ILLUSTRATION: {139}.

NOTES: close to *B. imerinense*, but a more robust bifoliate plant which is also distinct by the presence of caduceus, black glands on the ovary and the back of the floral bracts and sepals.

Bulbophyllum ankaizinense *(Jum. & H.Perrier) Schltr.* in Repert. Sp. Nov. Regni Veg. Beih. 33: 228 (1924). TYPES: Madagascar, Ankaizine, *H.Perrier* 1910 (lecto. P, selected here), Mt. Tsaratanana, *H.Perrier* 15731 (syn. P).

Bulbophyllum ophiuchus var. *ankaizinensis* Jum. & H.Perrier in Ann. Fac. Sci. Marseille 21 (2): 208 (1912).

DISTRIBUTION: Madagascar: Antsiranana, Fianarantsoa, Mahajanga, Toamasina.
HABITAT: mossy forest on moss- and lichen-covered trees.
ALTITUDE: 600 – 2000.
FLOWERING TIME: October – January.
LIFE FORM: epiphyte.
PHYTOGEOGRAPHICAL REGION: II, III, IV.
DESCRIPTION: {896}.
ILLUSTRATION: {896}.
NOTES: pseudobulbs two-leaved; inflorescence long; raceme red, densely-flowered; flowers c. 6 mm long; petals villous; lip suborbicular, 2 × 1.7 mm with a central depression. A number of local forms have been found {888}.

Bulbophyllum ankaratranum *Schltr.* in Repert. Sp. Nov. Regni Veg. Beih. 33: 229 (1924). TYPES: Madagascar, Mt. Triafajavona, *H.Perrier* 13923 (lecto. P; isolecto. K, both selected here); Ankaratrana, *H.Perrier* 14803 (syn. P).

DISTRIBUTION: Madagascar: Antananarivo.
HABITAT: mossy forest on moss- and lichen-covered trees.
ALTITUDE: c. 2000 m.
FLOWERING TIME: February – June.
LIFE FORM: epiphyte.
PHYTOGEOGRAPHICAL REGION: III, IV.
DESCRIPTION: {896}.
ILLUSTRATION: {896}, {1223}.
NOTES: plant 13 cm tall, 2-leaved; flowers greenish-yellow, with the extremities of the lip and the petals blackish-purple, red spots mainly on the sepals, 9 mm long; lip very curved at the base, 3 mm long.

Bulbophyllum antongilense *Schltr.* in Repert. Sp. Nov. Regni Veg. Beih. 33: 230 (1924). TYPE: Madagascar, around Antongil Bay, *H.Perrier* 11346 (holo. P).

DISTRIBUTION: Madagascar: Toamasina.
HABITAT: humid, lowland, evergreen forest.
ALTITUDE: c. 500 m.
FLOWERING TIME: August.
LIFE FORM: epiphyte.
PHYTOGEOGRAPHICAL REGION: I.
DESCRIPTION: {896}.
NOTES: known only from the type. Bifoliate plant close to *B. ankaratranum* but stronger in growth and with larger, broader pseudobulbs, longer leaves, differently coloured flowers and larger floral bracts.

Bulbophyllum approximatum *Ridl.* in J. Linn. Soc., Bot. 22: 117 (1886). TYPE: Madagascar, without locality, *Baron* 4128 (holo. BM; iso. K).

DISTRIBUTION: Madagascar: Antsiranana, Fianarantsoa.
HABITAT: humid, highland forest.
FLOWERING TIME: September.
LIFE FORM: epiphyte.

PHYTOGEOGRAPHICAL REGION: III.

DESCRIPTION: {896}.

NOTES: single-leaved plant allied to *B. baronii*, but a larger plant with larger and fewer flowers; lip thin, with the edges curled inward, and a bilobed apex.

Bulbophyllum aubrevillei *Bosser* in Adansonia, ser.2, 5: 386 (1965). TYPE: Madagascar, S of Moramanga, *Bosser* 16458 (holo. P; iso. BR, TAN).

DISTRIBUTION: Madagascar: Toamasina.

HABITAT: humid, evergreen forest; mossy forest.

ALTITUDE: 500 – 1100 m.

FLOWERING TIME: November – December.

LIFE FORM: epiphyte.

DESCRIPTION: {108}.

ILLUSTRATION: {108}.

NOTES: plant bifoliate, flowers small; lip red-purple, recurved, fleshy, ovate, 4 mm long, 2 mm wide, apex almost concave, base almost bilobed, 'V'-shaped; sepals spotted with purple. Close to *B. amphorimorphum* which has a similar flower but has unifoliate bulbs.

Bulbophyllum auriflorum *H.Perrier* in Notul. Syst. (Paris) 6 (2): 109 (1937). TYPE: Madagascar, Mongoro basin, *H.Perrier* 18281 (holo. P).

DISTRIBUTION: Madagascar: Toamasina.

HABITAT: humid, lowland, evergreen forest.

ALTITUDE: 400 – 800m.

FLOWERING TIME: September – October.

LIFE FORM: epiphyte.

PHYTOGEOGRAPHICAL REGION: I, III.

DESCRIPTION: {896}.

ILLUSTRATION: {896}.

NOTES: plant up to 15 cm tall, 2-leaved; flowers bright orange-yellow.

Bulbophyllum baronii *Ridl.* in J. Linn. Soc., Bot. 21: 463 (1885). TYPES: Madagascar, E Imerina, Andrangaloaka, *Hildebrandt* 3728 (lecto. BM; isolecto. P, K, all selected here), *Baron* 714 (syn. K); *Baron* 1176 (syn. BM); Ankafana, *Deans Cowan* s.n. (syn. BM 25077).
Phyllorchis baronii (Ridl.) Kuntze in Rev. Gen. Pl. 2: 677 (1891).

DISTRIBUTION: Madagascar: Antananarivo, Antsiranana, Fianarantsoa, Toamasina.

HABITAT: humid, evergreen forest at intermediate elevation and on plateau; lichen-rich forest; on rocks.

ALTITUDE: 800 – 2200 m.

FLOWERING TIME: November – June.

LIFE FORM: epiphyte or lithophyte.

PHYTOGEOGRAPHICAL REGION: II, III, IV.

DESCRIPTION: {896}.

ILLUSTRATION: {281}, {479}.

NOTES: a widespread and variable species allied to *B. nutans* but distinguished by its elongate pseudobulbs. Pseudobulbs yellow or sometimes red, single-leafed; flowers pale yellow, small; inflorescence somewhat densely flowered.

Bulbophyllum bathieanum *Schltr.* in Beih. Bot. Centralbl. 34 (2): 330 (1916). TYPE: Madagascar, Masoala, *H.Perrier* 11354 (145) (holo. P).

DISTRIBUTION: Madagascar: Antsiranana, Toamasina.

HABITAT: humid, evergreen forest at intermediate elevation.

ALTITUDE: 500 – 1400 m.

FLOWERING TIME: November – December.

LIFE FORM: epiphyte.

PHYTOGEOGRAPHICAL REGION: I, IV.

DESCRIPTION: {896}.

ILLUSTRATION: {896}, {1223}.

NOTES: plant bifoliate, flower pale; lip dark red, c. 12 mm, very curved-folded, the margins of the upper $^2/_3$ roundly recurved, folded into a furrow in the middle, flanked by 2 prominent lines of papillae.

Bulbophyllum bicoloratum *Schltr.* in Repert. Sp. Nov. Regni Veg. Beih. 33: 218 (1924). Based on same type as *B. bicolor* Jum. & H.Perrier (1912).

Bulbophyllum bicolor auct. non Lindl. (1930), *nec.* Hook. f. (1890): Jum. & H.Perrier in Ann. Fac. Sci. Marseille 21 (2): 202 (1912). Based on: Madagascar, Manongariva massif, *H.Perrier* s.n. (P).

Bulbophyllum theiochlamys Schltr. in Repert. Spec. Nov. Regn. Veg., Beih. 33: 223 (1925). TYPE: Madagascar, Analamahitso forest, Bemarivo, *H.Perrier* 8022 (holo. P).

Bulbophyllum coeruleum H.Perrier in Notul. Syst. (Paris) 14 (2): 154 (1951). TYPE: Madagascar, Mt. S of Tanandava, Mandrere, *Humbert* 20480 (holo. P).

DISTRIBUTION: Madagascar: Antsiranana, Fianarantsoa, Toliara.

HABITAT: humid, lowland, evergreen forest.

ALTITUDE: 300 – 1000m.

FLOWERING TIME: February – May.

LIFE FORM: epiphyte.

PHYTOGEOGRAPHICAL REGION: II.

DESCRIPTION: {896}.

ILLUSTRATION: {896}.

NOTES: plant two-leaved; raceme short; flowers small, hidden underneath the floral bracts; lip thickened, 2.5 mm long, ciliate with black hairs

Bulbophyllum boiteaui *H.Perrier* in Notul. Syst. (Paris) 8: 38 (1939) TYPE: Madagascar, probably Mandraka basin, *Boiteau* 2352 (holo. P).

DISTRIBUTION: Madagascar: Antananarivo, Fianarantsoa.

HABITAT: mossy forest.

ALTITUDE: 1200 – 2000 m.

FLOWERING TIME: January – February.

LIFE FORM: epiphyte.

PHYTOGEOGRAPHICAL REGION: III.

DESCRIPTION: {896}.

NOTES: single-leafed plant, close to *B. nigriflorum*, but differs mainly by the flowers which are twice as large and have a differently shaped lip.

Bulbophyllum brachyphyton *Schltr.* in Repert. Sp. Nov. Regni Veg. Beih. 15: 329 (1918). TYPE: Madagascar, without locality, *Laggiara* s.n. (holo. B†); neo.: the illustration in Repert. Spec. Nov. Reg. Veg., 68: t.64, 254 (1932), selected here.

DISTRIBUTION: Madagascar: Toamasina.

LIFE FORM: epiphyte.

PHYTOGEOGRAPHICAL REGION: I.

DESCRIPTION: {896}.

ILLUSTRATION: {1223}.

NOTES: the only specimen was destroyed and only the description and an illustration by Schlechter remain. Plant bifoliate, very small, similar to *B. sarcorhachis* with the lip curved, very short, 1.25 mm long, subrectangular above a contracted base.

Bulbophyllum brachystachyum *Schltr.* in Repert. Sp. Nov. Regni Veg. Beih. 33: 200 (1924). TYPE: Madagascar, Tsaratanana massif, *H.Perrier* 15459 (holo. P).

Bulbophyllum pseudonutans H.Perrier in Notul. Syst. (Paris) 6 (2): 71 (1937). TYPE: Madagascar, Manongarivo massif, *H.Perrier* 1912 (holo. P).

DISTRIBUTION: Madagascar: Antsiranana.

HABITAT: lichen-rich forest; humid, evergreen forest on plateau on moss- and lichen-covered trees.

ALTITUDE: 1200m – 1600m.

FLOWERING TIME: January, April.

LIFE FORM: epiphyte.

PHYTOGEOGRAPHICAL REGION: II, IV.

DESCRIPTION: {896}.

ILLUSTRATION: {896}.

NOTES: plant bifoliate, small, 5 – 6 cm high, similar in shape to *B. nutans*, with 6 – 10, 7 mm long, yellowish flowers; lip flat, spathulate at the base, with a rather protruding triangular callus.

Bulbophyllum brevipetalum *H.Perrier* in Notul. Syst. (Paris) 6 (2): 88 (1937). TYPE: Madagascar, Ivohibe, *Decary* 5597 (holo. P).

Bulbophyllum brevipetalum subsp. *majus* H.Perrier in Notul. Syst. (Paris) 6 (2): 88 (1937). TYPE: Madagascar, Ankeramadininika forest, *H.Perrier* 19011.

Bulbophyllum brevipetalum subsp. *speculiferum* H.Perrier in Humbert ed., Fl. Mad. Orch. 1: 386 (1939), *nom. nud.* Based on: same specimen as *B. brevipetalum* subsp. *majus*.

DISTRIBUTION: Madagascar: Antananarivo, Fianarantsoa.

HABITAT: humid, evergreen forest on plateau.

ALTITUDE: c. 1500 m.

FLOWERING TIME: February, September.

LIFE FORM: epiphyte.

PHYTOGEOGRAPHICAL REGION: III.

DESCRIPTION: {896}.

NOTES: plant large, 2-leaved; inflorescence up to 45 cm long; flowers yellow, striped with brownish-red, 15 – 18 mm long; lip oblong, 10 × 4 mm, ciliate in the lower part only, the margins folded back and the wings erect and ciliate.

Bulbophyllum bryophilum *Hermans* **nom. nov.**

Bulbophyllum muscicola Schltr. in Ann. Mus. Col. Marseille, sér. 3, 1: 179, t.15 (1913), *non* Rchb.f. in Flora 55, 18 p. 275 (1872), *nec* Schltr. in Repert. Spec. Nov. Regni Veg. Beih. 1: 870 (1913). TYPE: Madagascar: Manongarivo Massif, Mt. Bekolosy, *H.Perrier* 11 (holo. B†; iso. P).

DISTRIBUTION: Madagascar: Antsiranana.

HABITAT: seasonally dry, deciduous woodland.

FLOWERING TIME: March.

LIFE FORM: epiphyte.

PHYTOGEOGRAPHICAL REGION: II.

DESCRIPTION: {896}.

ILLUSTRATION: {1205}.

NOTES: plant small, single-leafed, carpeting its substrate, single-flowered on a 12 mm peduncle; lip 2 × 1 mm, fleshy, oblong-ligulate, a little narrowed in the lower half, obtuse, blackish-violet. Owing to the earlier *B. muscicolum* Rchb.f., a Himalayan species now considered to be conspecific with *B. retusiusculum* Rchb.f., this name cannot be used for the Madagascan plant described by Schlechter in 1913. In the same year that he described the Madagascan species, Schlechter also used the epithet for a plant from New Guinea. I therefore propose the epithet name *bryophilum*, referring to its moss-dwelling habitat, for the Madagascan species.

Bulbophyllum callosum *Bosser* in Adansonia, ser.2, 5: 400 (1965). TYPE: Madagascar, Lake Mantasoa, Imerina, *Bosser* 16400 (holo. P).

DISTRIBUTION: Madagascar: Antananarivo.

HABITAT: humid, mossy, evergreen forest on plateau.

ALTITUDE: 1300 – 1400m.

FLOWERING TIME: September – October.

LIFE FORM: epiphyte.

PHYTOGEOGRAPHICAL REGION: III.

DESCRIPTION: {108}.

ILLUSTRATION: {108}.

NOTES: the lip is very characteristic as is the exterior surface of the dorsal sepal and the lateral sepals, which are downy-papillose. It is close to *B. hirsutiusculum*, but differs by the lip morphology.

Bulbophyllum calyptropus *Schltr.* in Repert. Sp. Nov. Regni Veg. Beih. 33: 178 (1924). TYPE: Madagascar, Mt. Tsaratanana, *H.Perrier* 15729 (holo. P).

DISTRIBUTION: Madagascar: Antsiranana.
HABITAT: lichen-rich-rich forest on moss- and lichen-covered trees.
ALTITUDE: c. 2000m.
FLOWERING TIME: January.
LIFE FORM: epiphyte.
PHYTOGEOGRAPHICAL REGION: III, IV.
DESCRIPTION: {896}.
ILLUSTRATION: {896}.
NOTES: plant 2-leaved, flower glabrous, almost transparent, finely streaked with red and with a purple lip. *Bulbophyllum bryophilum* is a close relative, but larger in all its parts, the shape of the lip is also different.

Bulbophyllum capuronii *Bosser* in Adansonia, sér. 2, 11 (2): 333 (1971). TYPE: Madagascar, S of Farafangana, *Bosser* 20396 (holo. P).

DISTRIBUTION: Madagascar: Fianarantsoa.
HABITAT: coastal forest.
ALTITUDE: sea level – 100 m.
LIFE FORM: epiphyte.
PHYTOGEOGRAPHICAL REGION: I.
DESCRIPTION: {119}.
NOTES: plant single-leafed, several characteristics distinguish this species from *B. implexum*, especially the short, few-flowered inflorescence and the very different shape of the lip.

Bulbophyllum cardiobulbum *Bosser* in Adansonia, sér. 2, 5: 396 (1965). TYPE: Madagascar, Ankeramadinika, *J.-P.Peyrot* 5 (holo. P).

DISTRIBUTION: Madagascar: Fianarantsoa.
HABITAT: humid, mossy, evergreen forest on plateau.
ALTITUDE: 1300 – 1400m.
FLOWERING TIME: November.
LIFE FORM: epiphyte.
PHYTOGEOGRAPHICAL REGION: III.
DESCRIPTION: {108}.
ILLUSTRATION: {108}.
NOTES: characterised by its cordate, flattened, yellow-brown, 2-leaved pseudobulbs; lip yellow-white, spotted with red, 12 mm × 3.5 mm, the base recurved, with two ridges, the basal lobe erect, papillate-ciliolate, the apex lanceolate.

Bulbophyllum cataractarum *Schltr.* in Repert. Sp. Nov. Regni Veg. Beih. 33: 194 (1924). TYPE: Madagascar, Manongarivo, *H.Perrier* 1849 (holo. P).
Bulbophyllum forsythianum sensu Schltr. in Ann. Mus. Col. Mar. 3, I: 35 (1913), *non* Kraenzlin.

DISTRIBUTION: Madagascar: Antsiranana.
HABITAT: on very wet rocks in rivers.
ALTITUDE: c. 1200m.
FLOWERING TIME: April.
LIFE FORM: lithophyte.
PHYTOGEOGRAPHICAL REGION: III.
DESCRIPTION: {896}.
ILLUSTRATION: {1223}.
NOTES: known only from the type. Plant bifoliate, less than 3 cm high, single-flowered; peduncle thread-like, 2 – 2.5 cm long; lip 2.5 × 1.5 mm, curved, oval, cordate at the base, carrying a large, rounded, verrucose callus.

Bulbophyllum ceriodorum *Boiteau* in Bull. Acad. Malgache n.s. 24: 89 (1942). TYPE: Madagascar, S Betsileo, Mt. Tsitonoroina, *Boiteau* in Jard. Bot. Tananarive 4850 (holo. P).

DISTRIBUTION: Madagascar: Fianarantsoa.
HABITAT: mossy, montane forest.
ALTITUDE: c. 1700m.
LIFE FORM: epiphyte.
PHYTOGEOGRAPHICAL REGION: III.
DESCRIPTION: {100}.
NOTES: known only from the type. Plant 2-leaved; flowers c. 8 mm long; lip strongly S-shaped in outline, slightly biauriculate at the base, with 2 parallel ridges in the upper part.

Bulbophyllum ciliatilabrum *H.Perrier* in Notul. Syst. (Paris) 6 (2): 79 (1937). TYPE: Madagascar, Analamazaotra forest, *H.Perrier* 11850 (holo. P).

DISTRIBUTION: Madagascar: Toamasina.
HABITAT: humid, evergreen forest at intermediate altitude and on plateau.
ALTITUDE: 800 – 900m.
FLOWERING TIME: January.
LIFE FORM: epiphyte.
PHYTOGEOGRAPHICAL REGION: III.
DESCRIPTION: {896}.
NOTES: plant single-leafed, 10 cm high; flowers deep red, small; lip oblanceolate, 1.2 mm long, with an obtuse protruding callus on top, the upper third a little enlarged-spathulate, ciliate with long white hairs at the margins.

Bulbophyllum cirrhoglossum *H.Perrier* in Notul. Syst. (Paris) 14: 154 (1951). TYPE: Madagascar, Moramanga, Lake Alaotra, cult. Jard. Bot. Tan. 453 (holo. P).

DISTRIBUTION: Madagascar: Toamasina.
HABITAT: humid, evergreen forest on plateau.
LIFE FORM: epiphyte.
PHYTOGEOGRAPHICAL REGION: I.
DESCRIPTION: {901}.
NOTES: plant bifoliate, inflorescence c. 20 cm long; lip saddle-shaped, 3 mm long, dark red, with long violet hairs in the middle.

Bulbophyllum coccinatum *H.Perrier* in Notul. Syst. (Paris) 7: 141 (1938). TYPE: Madagascar, Lac Nossy-ve, *Humblot* 387 (holo. P; iso. K, W).

DISTRIBUTION: Madagascar: Toamasina.
HABITAT: humid, lowland forest.
LIFE FORM: epiphyte or terrestrial.
PHYTOGEOGRAPHICAL REGION: I.
DESCRIPTION: {896}.
NOTES: plant 2-leaved; inflorescence c. 8 cm long, as long as the leaves; raceme red; lip oval, 2.2 × 1.2 mm, a little narrowed at the tip, biauriculate at the base with 2 swellings.

Bulbophyllum comorianum *H.Perrier* in Notul. Syst. (Paris) 7: 140 (1938). TYPE: Grande Comore, *Pobéguin* 14.6.1809 (holo. P).

DISTRIBUTION: Comoros: Grande Comore.
FLOWERING TIME: February, June.
LIFE FORM: epiphyte.
DESCRIPTION: {896}.
NOTES: plant bifoliate, c. 20 cm high; flowers small, c. 4.5 mm; lip broadly elliptic-suborbicular with a rounded speculum in the middle, the stelids are divided in 2 big obtuse teeth, the front one thin, the back one very thick.

Bulbophyllum complanatum *H.Perrier* in Notul. Syst. (Paris) 6 (2): 47 (1937). TYPE: Madagascar, around Ambanja, *H.Perrier* 16502 (holo. P).

Bulbophyllum sigilliforme H.Perrier in Notul. Syst. (Paris) 6 (2): 66 (1937). TYPE: Madagascar, Maromandia (Sambirano), *Decary* 921 (holo. P).

DISTRIBUTION: Madagascar: Antsiranana.
HABITAT: seasonally dry, deciduous forest and woodland.
ALTITUDE: sea level – 100 m.
FLOWERING TIME: February – April, August – September.
LIFE FORM: epiphyte.
PHYTOGEOGRAPHICAL REGION: II.
DESCRIPTION: {896}.
ILLUSTRATION: {896}.
NOTES: close to *B. humblotii* which is single-leaved; lip 3 × 1.5 mm, curved, narrowly oblong and obtuse.

Bulbophyllum conchidioides *Ridl.* in J. Linn. Soc., Bot. 22: 117 (1886). TYPE: Madagascar, without locality, *Baron* 4471 (holo. BM; iso. K).

Bulbophyllum pleurothalloides Schltr. in Repert. Sp. Nov. Regni Veg. Beih. 33: 185 (1924). TYPE: Madagascar, Analamazaotra, *H.Perrier* 11856 (holo. P).

DISTRIBUTION: Madagascar: Antananarivo, Toamasina.
HABITAT: humid evergreen forest on plateau, on the tops of hills in forest on moss- and lichen-covered trees or rocks.
ALTITUDE: 900 – 1400m.
FLOWERING TIME: October – February.
LIFE FORM: epiphyte or lithophyte.
PHYTOGEOGRAPHICAL REGION: III.
DESCRIPTION: {896}.
ILLUSTRATION: {896}.
NOTES: a somewhat variable species in the size of the leaves and flowers. Plant tiny, single-leafed; flower pale, small on a setiform peduncle; lip 3 mm long, pinkish-red, ovate.

Bulbophyllum coriophorum *Ridl.* in J. Linn. Soc., Bot. 22: 119 (1886). TYPE: Comoros, without locality, *Humblot* 337 (holo. BM; iso. P, W).

Bulbophyllum compactum Kraenzl. in Bot. Jahrb. Syst. 17: 48 (1893).

Bulbophyllum crenulatum Rolfe in Curtis's Bot. Mag. 131: t.8000 (1905). TYPE: Madagascar, *Braun* s.n. (holo. K).

Bulbophyllum robustum Rolfe in Curtis's Bot. Mag. 131: sub t.8000 (1905) & in Curtis' Bot. Mag. 145: t.8792 (1919). TYPES: Madagascar, without locality, *Baron* 2324 & *Baron* 2723 (both syn. K).

Bulbophyllum mandrakanum Schltr. in Repert. Sp. Nov. Regni Veg. Beih. 33: 235 (1924). TYPE: Madagascar, Mandraka, *H.Perrier* 14880 (holo. P).

DISTRIBUTION: Madagascar: Antananarivo, Antsiranana, Fianarantsoa, Toamasina. Comoros: Grande Comore.
HABITAT: humid, evergreen forest; dry forest.
ALTITUDE: 95 – 1600m.
FLOWERING TIME: October – June.
LIFE FORM: epiphyte or terrestrial.
PHYTOGEOGRAPHICAL REGION: I, III.
DESCRIPTION: {142}, {896}, {1129}.
ILLUSTRATION: {105}, {142}, {281}, {672}, {896}, {1129}, {1298}.
NOTES: robust bifoliate plant up to 50 cm high; rachis thickened and spongy; flowers dark red, very numerous, inserted in cavities within; lip orbicular, 3 mm long, thick especially at the base, dotted with warts. Siegerist {1288} erroneously put *Bulbophyllum cyclanthum* Schltr. (1916) as a synonym.

Bulbophyllum crassipetalum *H.Perrier* in Notul. Syst. (Paris) 6 (2): 114 (1937). TYPE: Madagascar, Sambirano valley, *H.Perrier* 15198 (holo. P).

DISTRIBUTION: Madagascar: Antsiranana, Fianarantosa.
HABITAT: humid, evergreen forest.
ALTITUDE: 400 – 1750 m.
FLOWERING TIME: November – December.
LIFE FORM: epiphyte.
PHYTOGEOGRAPHICAL REGION: II.
DESCRIPTION: {896}.
NOTES: bifoliate plant up to 24 cm tall; raceme dark red, many-flowered; flowers small, 5 mm long; petals papillose; lip curved, fleshy, suborbicular, 2 × 1.5 mm, papillose and with a wide speculum.

Bulbophyllum cryptostachyum *Schltr.* in Repert. Sp. Nov. Regni Veg. Beih. 33: 219 (1924). TYPES: Madagascar, Sambirano Mts, *H.Perrier* 15195 (lecto. P, selected here); Ananalamahitso, *H.Perrier* 11884 (syn. P).

DISTRIBUTION: Madagascar: Antsiranana, Mahajanga.
HABITAT: humid, evergreen forest.
ALTITUDE: c. 1000 m.
FLOWERING TIME: August – October.
LIFE FORM: epiphyte.
PHYTOGEOGRAPHICAL REGION: II, III.
DESCRIPTION: {896}.
ILLUSTRATION: {896}.
NOTES: two-leaved plant, related to *B. occultum*; flowers small, 6 mm long, a little scaly on the exterior; lip curved, oblong, very obtuse, c. 2 mm long, base with a small swelling.

Bulbophyllum curvifolium *Schltr.* in Beih. Bot. Centralbl. 34 (2): 329 (1917). TYPE: Madagascar, Masoala, *H.Perrier* 151 (11402) (holo. B†; iso. P).

DISTRIBUTION: Madagascar: Antsiranana.
HABITAT: humid, evergreen forest.
ALTITUDE: c. 500 m.
FLOWERING TIME: October.
LIFE FORM: epiphyte.
PHYTOGEOGRAPHICAL REGION: I.
DESCRIPTION: {896}.
ILLUSTRATION: {1223}.
NOTES: single-leafed plant related to *B. subsecundum*, but with shorter inflorescence and differently shaped flowers; lip 2.5 mm long, very obtuse, rectangular-suborbicular, the base narrower.

Bulbophyllum cyclanthum *Schltr.* in Beih. Bot. Centralbl. 34 (2): 327 (1917). TYPE: Madagascar, around Tsaratanana, *H.Perrier* 11362. (holo. B†; iso. P).

DISTRIBUTION: Madagascar: Antsiranana.
HABITAT: humid, evergreen forest on plateau.
ALTITUDE: c. 1300 m.
FLOWERING TIME: December.
LIFE FORM: epiphyte.
PHYTOGEOGRAPHICAL REGION: III.
DESCRIPTION: {896}.
ILLUSTRATION: {281}, {1223}.
NOTES: bifoliate plant up to 15 cm, 2-leaved; floral bracts rigid, almost as long as the flower; flowers small, finely papillose on the exterior; lip orbicular, curved, very obtuse.

Bulbophyllum cylindrocarpum var. **andringitrense** *Bosser* in Adansonia 22 (2): 169 (2000): TYPE: Madagascar, Andringitra massif, *Bosser* 19510 (holo. P).

Bulbophyllum cylindrocarpum auct. non Frapp. ex Cordem. referring to the Madagascan material.

DISTRIBUTION: Madagascar: Fianarantsoa.
HABITAT: higher elevation forest remnants.
ALTITUDE: 1500 – 1800 m.
FLOWERING TIME: April.
LIFE FORM: epiphyte.
PHYTOGEOGRAPHICAL REGION: IV.
DESCRIPTION:{139}.
NOTES: known only from the type. Bifoliate plant close to *Bulbophyllum cylindrocarpum* Frapp. ex Cordem. which is only found on Réunion, distinct by the shape of the pseudobulbs and by the larger flower with more elongate petals.

Bulbophyllum debile *Bosser* in Adansonia 11 (4): 378 (1989). TYPE: Madagascar, Crete Forest, Lakato Road, *Bosser* 22583 (holo. P).

DISTRIBUTION: Madagascar: Toamasina.
HABITAT: mossy and lichen-rich forest on moss- and lichen-covered trees.
ALTITUDE: 900 – 1000m.
FLOWERING TIME: February.
LIFE FORM: epiphyte.
DESCRIPTION: {130}.
ILLUSTRATION: {130}, {672}.
NOTES: known only from the type. Two-leaved plant, close to *B. percorniculatum* but differs by the non-acuminate sepals and the shape of the lip.

Bulbophyllum decaryanum *H.Perrier* in Notul. Syst. (Paris) 6 (2): 61 (1937). TYPE: Madagascar, around Moramanga, *Decary* 7257 (holo. P).

DISTRIBUTION: Madagascar: Toamasina.
HABITAT: humid, evergreen forest.
ALTITUDE: 900 – 1000m.
LIFE FORM: epiphyte.
PHYTOGEOGRAPHICAL REGION: III.
DESCRIPTION: {896}.
NOTES: known only from the type. Plant 12 cm long; inflorescence about the same length of the single leaf; flowers small, c. 5 mm long.

Bulbophyllum discilabium *H.Perrier* in Notul. Syst. (Paris) 14: 150 (1951). TYPES: Madagascar, Mt. Marojejy, *Humbert* 22573 (lecto. P, selected here); Beandroka, Lokono Valley, *Humbert* 23562 (syn. P).

DISTRIBUTION: Madagascar: Antsiranana.
HABITAT: humid, evergreen forest on plateau.
ALTITUDE: 1500 – 2000 m.
FLOWERING TIME: December – March.
LIFE FORM: epiphyte.
PHYTOGEOGRAPHICAL REGION: III.
DESCRIPTION: {901}.
NOTES: small single-leafed plant; inflorescence 25 mm, 2- to 3-flowered; base of the lip thickened then clavate, 4 mm long, apex longly ciliate.

Bulbophyllum divaricatum *H.Perrier* in Notul. Syst. (Paris) 6 (2): 105 (1937). TYPE: Madagascar, Mt. Tsaratanana, *H.Perrier* 16514 (holo. P).

DISTRIBUTION: Madagascar: Antsiranana.
HABITAT: lichen-rich forest on moss- and lichen-covered trees.

ALTITUDE: 1500 – 2000 m.

FLOWERING TIME: April, October.

LIFE FORM: epiphyte.

PHYTOGEOGRAPHICAL REGION: II, IV.

DESCRIPTION: {896}.

ILLUSTRATION: {896}.

NOTES: plant two-leaved, up to 30 cm high; inflorescence shorter than the leaves; flowers papillose on the exterior, c. 7 mm long; lip glabrous, thick, elliptical, tongue-shaped, 2 × 1.5 mm, with three furrows in the lower half.

Bulbophyllum edentatum *H.Perrier* in Notul. Syst. (Paris) 6 (2): 92 (1937). TYPE: Madagascar, Mt. Papanga, nr. Befotaka, Itomampy basin, *Humbert* 6929bis (holo. P).

DISTRIBUTION: Madagascar: Fianarantsoa.

HABITAT: humid, evergreen forest on plateau.

ALTITUDE: 1200 – 1800m.

FLOWERING TIME: November – December.

LIFE FORM: epiphyte.

PHYTOGEOGRAPHICAL REGION: III.

DESCRIPTION: {139}, {896}.

ILLUSTRATION: {896}.

NOTES: plant bifoliate, characterised by the narrow sepals, the linear petals as long as the column, and the slender column with subulate stelids.

Bulbophyllum elliotii *Rolfe* in J. Linn. Soc., Bot. 29: 51 (1891). TYPE: Madagascar, nr. Fort Dauphin, *Scott Elliot* s.n. (holo. K; iso. P).

Bulbophyllum malawiense B.Morris in Proc. Linn. Soc. London 179: 63 (1968). TYPE: Malawi, *Morris* (holo. K).

DISTRIBUTION: Madagascar: Toamasina, Toliara. Also in East and South-central Africa.

HABITAT: humid, evergreen forest.

ALTITUDE: sea level – 1200 m.

FLOWERING TIME: April – June.

LIFE FORM: epiphyte.

PHYTOGEOGRAPHICAL REGION: I, III.

DESCRIPTION: {896}.

ILLUSTRATION: {384}, {676}, {677}, {1431}.

NOTES: habit of *B. pervillei*, bifoliate, the pseudobulbs more distant and the leaves larger, whereas the inflorescence resembles that of *B. conitum* with dull maroon flowers but with a strongly recurved, very fimbriate lip.

Bulbophyllum erectum *Thouars*, Hist. Orch.: t.96 (1822). TYPE: *Thouars* s.n. (holo. P).

Bulbophyllum lobulatum Schltr. in Repert. Sp. Nov. Regni Veg. Beih. 33: 220 (1924). TYPE: Madagascar, Sambirano, *H.Perrier* 15442 (holo. P).

Bulbophyllum calamarioides Schltr. in Repert. Sp. Nov. Regni Veg. Beih. 33: 218 (1924). TYPE: Madagascar, nr. Maivoto, *H.Perrier* 11333 (holo. P).

Phyllorchis erecta (Thouars) Kuntze, Rev. Gen. Pl. 2: 675 (1891).

DISTRIBUTION: Madagascar: Antsiranana, Toamasina. Mascarenes: Mauritius.

HABITAT: coastal dry forests, not far from the sea; evergreen forest on *Sarcolaenaceae.*

ALTITUDE: sea level – 100 m.

FLOWERING TIME: February – April, August, October.

LIFE FORM: epiphyte.

PHYTOGEOGRAPHICAL REGION: I.

DESCRIPTION: {896}.

ILLUSTRATION: {246}, {1223}, {1398}.

NOTES: plant robust, 1-leafed; inflorescence up to 25 cm long, laxly-flowered; flowers greenish-brown; lip reddish, distinctly ciliate and a little narrowed towards the obtuse apex, stelids long.

Bulbophyllum erythroglossum *Bosser* in Adansonia 22, 2: 179 (2000) TYPE: Madagascar, Moramanga – Anosibe road, *Bosser* 18379 (holo. P).

> DISTRIBUTION: Madagascar: Toamasina.
> HABITAT: humid forest.
> ALTITUDE: 900 – 1000 m.
> FLOWERING TIME: May.
> LIFE FORM: epiphyte.
> PHYTOGEOGRAPHICAL REGION: III.
> DESCRIPTION: {139}.
> ILLUSTRATION: {139}.
> NOTES: known only from the type, plant 2-leaved, distinct by the reflexed flowers and the stelids which are broadened and denticulate at the apex.

Bulbophyllum erythrostachyum *Rolfe* in Orch. Rev. 11: 200 (!903). TYPE: Madagascar, without locality, cult. Glasnevin BG (holo. K).

> DISTRIBUTION: Madagascar.
> DESCRIPTION: {1118}.
> NOTES: known only from a cultivated plant, single-leafed, said to be allied to *B. clavatum* from the Mascarenes.

Bulbophyllum ferkoanum *Schltr.* in Repert. Sp. Nov. Regni Veg. Beih. 15: 329 (1918). TYPE: Madagascar, *Laggiara* s.n. (holo. B†); neo.: the illustration in Repert. Spec. Nov. Reg. Veg., 68 t.65, 260 (1932), selected here.

> DISTRIBUTION: Madagascar: Toamasina.
> LIFE FORM: epiphyte.
> PHYTOGEOGRAPHICAL REGION: I.
> DESCRIPTION: {896}.
> ILLUSTRATION: {1223}.
> NOTES: known only from the description and Schlechter's drawing. Two-leaved plant related to *B. sarcorhachis* but differs in its longer, rectangular pseudobulbs, longer spike, and somewhat bigger flowers, with wider petals and a distinct lip and column.

Bulbophyllum florulentum *Schltr.* in Repert. Sp. Nov. Regni Veg. Beih. 33: 189 (1924). TYPE: Madagascar, Mt. Tsaratanana, *H.Perrier* 155333 (holo. P).

> DISTRIBUTION: Madagascar: Antsiranana.
> HABITAT: lichen-rich forest on moss- and lichen-covered trees.
> ALTITUDE: 1500 – 2000m.
> FLOWERING TIME: January, October.
> LIFE FORM: epiphyte.
> PHYTOGEOGRAPHICAL REGION: III, IV.
> DESCRIPTION: {896}.
> ILLUSTRATION: {896}.
> NOTES: plant single-leafed, up to 20 cm high; inflorescence similar in length, densely-flowered; flowers small, 5 mm long, yellowish-white; lip very curved, suborbicular, with the lower half wide and with the edges folded back.

Bulbophyllum forsythianum *Kraenzl.* in Bot. Jahrb. Syst. 28: 163 (1900). TYPE: Madagascar, Ambohimitombo, *Forsyth-Major* 480 (holo. BM; iso. G).

> DISTRIBUTION: Madagascar: Fianarantsoa, Toamasina, Toliara.
> HABITAT: shaded and humid rocks in river-bed.
> ALTITUDE: 300 – 1400m.
> FLOWERING TIME: October – February.
> LIFE FORM: lithophyte.
> PHYTOGEOGRAPHICAL REGION: I, III.
> DESCRIPTION: {896}.

ILLUSTRATION: {896}.

NOTES: plant bifoliate, small, carpeting rocks; inflorescence filiform, single-flowered; lip 3 × 0.8 mm, very curved, shortly angular at the apex, with 2 callosities on top joining together to form a central ridge.

Bulbophyllum françoisii *H.Perrier* in Notul. Syst. (Paris) 6 (2): 76 (1937). TYPES: Madagascar, Ambilo, *E.François* 7 (lecto. P, selected here); Analamazaotra, *H.Perrier* 8133 (syn. P).

DISTRIBUTION: Madagascar: Toamasina.

HABITAT: coastal and humid, evergreen forest.

ALTITUDE: sea level – 800 m.

FLOWERING TIME: September – November, February.

LIFE FORM: epiphyte.

PHYTOGEOGRAPHICAL REGION: I.

DESCRIPTION: {896}.

ILLUSTRATION: {505}, {896}.

NOTES: two-leaved plant, characterised by its c. 5 mm long, oval-oblong, oscillating lip, set in the middle of the foot, the back part a little widened and ciliate and the front half a little narrowed and glabrous.

Bulbophyllum françoisii var. **andrangense** *(H.Perrier) Bosser* in Adansonia, sér. 2, 5: 384 (1965). TYPE: Madagascar, Andrangovalo massif, *Humbert* 17760 (holo. P).
Bulbophyllum andrangense H.Perrier in Notul. Syst. (Paris) 8: 39 (1939).

DISTRIBUTION: Madagascar: Fianarantsoa.

HABITAT: in very shady, humid, evergreen forest on plateau.

ALTITUDE: 1200 – 1400m.

FLOWERING TIME: October.

LIFE FORM: epiphyte.

PHYTOGEOGRAPHICAL REGION: III.

DESCRIPTION: {108}, {896}.

NOTES: two-leaved plant, the inflorescence has a very short peduncle, much shorter than the leaves, the petals are oblong, less elongate, and the anther is also distinct.

Bulbophyllum hamelinii *W.Watson* in Garden & Forest 6 (285): 336 (1893). TYPE: Madagascar, cult. Sander & Co. (holo. K).

DISTRIBUTION: Madagascar: Toamasina.

HABITAT: evergreen, humid forest.

ALTITUDE: 800 – 1000m.

FLOWERING TIME: February.

LIFE FORM: epiphyte.

PHYTOGEOGRAPHICAL REGION: I, III.

DESCRIPTION: {365}, {1113}, {1171}, {1449}.

ILLUSTRATION: {365}, {1171}, {1298}.

CULTIVATION: {365}, {1171}.

NOTES: the largest *Bulbophyllum* in Madagascar; pseudobulbs flattened, discoid, up to 11 cm in diameter, 2-leaved; flowers in a 10 – 15 cm tight raceme, pinkish-purple and malodorous.

Bulbophyllum hapalanthos *Garay* in Harv. Papers Bot. 4, 1: 302 (1999). TYPE: Madagascar, without locality, hort. University of Vienna Botanical Garden, *A.Sieder* 097016 – 2 (holo. AMES).

DISTRIBUTION: Madagascar.

LIFE FORM: epiphyte.

DESCRIPTION: {320}.

ILLUSTRATION: {320}.

NOTES: only known from the type, plant single-leafed, similar in habit and column structure to *B. lakatoese* but differs in the completely glabrous flowers and in the shape of the petals and lip.

Bulbophyllum henrici *Schltr.* in Repert. Sp. Nov. Regni Veg. Beih. 33: 232 (1924). TYPE: Madagascar, nr. Befaona, *H.Perrier* 14297 (holo. P).

DISTRIBUTION: Madagascar: Toamasina.
HABITAT: humid, evergreen forest.
ALTITUDE: c. 700m.
FLOWERING TIME: September.
LIFE FORM: epiphyte.
PHYTOGEOGRAPHICAL REGION: I.
DESCRIPTION: {896}.
ILLUSTRATION: {896}, {1223}.
NOTES: bifoliate plant, related to *B. ophiuchus* & *B. moramangensis*, but distinguished by the broad leaves, thick spike and rugose flowers with long sepals.

Bulbophyllum henrici var. **rectangulare** *H.Perrier ex Hermans* **var. nov.** TYPE: Madagascar, Farafangana, Ikongo massif, *Decary* 5785 (holo.P). See Appendix .
Bulbophyllum henrici var. *rectangulare* H.Perrier in Notul. Syst. (Paris) 6 (2): 113 (1937), *nom. nud.*

DISTRIBUTION: Madagascar: Fianarantsoa.
FLOWERING TIME: October.
LIFE FORM: epiphyte.
PHYTOGEOGRAPHICAL REGION: III.
DESCRIPTION: {896}.
NOTES: differs from the typical variety by the synsepal that is bi-apiculate at the apex, the smaller subrectangular lip, lacking a circular depression, and the papillae which are less pronounced.

Bulbophyllum hildebrandtii *Rchb.f.*, Otia Bot. Hamburg.: 74 (1881). TYPE: Madagascar, Beravi Mt, *Hildebrandt* 2988a (holo. W).
Phyllorchis hildebrandtii (Rchb.f.) Kuntze, Rev. Gen. Pl. 2: 677 (1891).
Bulbophyllum maculatum Jum. & H.Perrier in Ann. Fac. Sci. Marseille 21 (2): 215 (1912), *non* Boxall. TYPE: Madagascar, Berizoka, nr. Mevatanana, *Jumelle & H.Perrier* s.n. (holo. P).
Bulbophyllum johannum H.Perrier in sched. Based on: Rés. Nat. 1400, Andringitra, *Jean de Dieu Rakoto* s.n. (P).
Bulbophyllum madagascariense Schltr. in Beih. Bot. Centralbl. 33 (2): 418 (1915). Based on same type as *B. maculatum* Jum. & H.Perrier.
Bulbophyllum melanopogon Schltr. in Repert. Sp. Nov. Regni Veg. Beih. 15: 330 (1918). TYPE: Madagascar, St Marie de Marrovay, *Afzelius* s.n. (holo. B†).

DISTRIBUTION: Madagascar: Antsiranana, Fianarantsoa, Mahajanga, Toamasina.
HABITAT: on *Eugenia* and other large trees by rivers, abundant; seasonally dry, deciduous forest and woodland; river margins.
ALTITUDE: sea-level – 600 m.
FLOWERING TIME: July – December.
LIFE FORM: epiphyte.
PHYTOGEOGRAPHICAL REGION: I, II, V.
DESCRIPTION: {896}.
ILLUSTRATION: {896}, {1223}.
NOTES: plant up to 40 cm, normally 2-leaved; inflorescence c. 17 cm long; flowers 7 – 10 mm, greenish-yellow with a red lip, ciliate in the upper part.

Bulbophyllum hirsutiusculum *H.Perrier* in Notul. Syst. (Paris) 6 (2): 114 (1937). TYPE: Madagascar, Mt. Papanga, Itomampy basin, *Humbert* 6902 (holo. P).

DISTRIBUTION: Madagascar: Fianarantsoa.
HABITAT: humid, evergreen forest on plateau.
ALTITUDE: 1300 – 1700m.
FLOWERING TIME: December.
LIFE FORM: epiphyte.
PHYTOGEOGRAPHICAL REGION: III.
DESCRIPTION: {896}.

NOTES: known only from the type. Plant bifoliate, substantial, up to 60 cm tall, recognised by the short stiff hairs which cover the outside of the flower.

Bulbophyllum horizontale *Bosser* in Adansonia, ser.2, 5: 394 (1965). TYPE: Madagascar, S of Moramanga, Anosibe road, *Bosser* 17170 (holo. P).

DISTRIBUTION: Madagascar: Antananarivo, Toamasina.
HABITAT: humid, evergreen and mossy forest.
ALTITUDE: c. 900m.
FLOWERING TIME: March – July.
LIFE FORM: epiphyte.
DESCRIPTION: {108}.
ILLUSTRATION: {108}, {906}.
NOTES: plant has typical shiny red-brown tetragonal pseudobulbs and 2 leathery leaves; lip dark red, very fleshy, with a central rounded ridge on top, the edges are raised into two wings at the base, extended by two papillose cushions on either side, lateral margins hairy, the tip has longer ligulate hairs.

Bulbophyllum hovarum *Schltr.* in Repert. Sp. Nov. Regni Veg. Beih. 33: 233 (1924). TYPE: Madagascar, Manongarivo massif, *H.Perrier* 1906 (holo. P).

DISTRIBUTION: Madagascar: Antsiranana.
HABITAT: lichen-rich forest on moss- and lichen-covered trees.
ALTITUDE: 1000 – 2000m.
FLOWERING TIME: May.
LIFE FORM: epiphyte.
PHYTOGEOGRAPHICAL REGION: III.
DESCRIPTION: {896}.
ILLUSTRATION: {1223}.
NOTES: plant 2-leaved, c. 15 cm tall; inflorescence shorter, densely-flowered; flowers small, blackish-violet, finely verrucose on the exterior.

Bulbophyllum humbertii *Schltr.* in Repert. Sp. Nov. Regni Veg. Beih. 18: 322 (1922). TYPE: Madagascar, Andramasina, Antananarivo, *Viguier & Humbert* 1925 (holo. P).

DISTRIBUTION: Madagascar: Antananarivo.
HABITAT: humid, evergreen forest on plateau.
ALTITUDE: c. 1700 m.
FLOWERING TIME: December.
LIFE FORM: epiphyte.
PHYTOGEOGRAPHICAL REGION: III.
DESCRIPTION: {896}.
ILLUSTRATION: {1223}.
NOTES: two-leaved plant, related to *B. baronii*, characterised by the long and narrow floral bracts in the lower half of the raceme.

Bulbophyllum humblotii *Rolfe* in J. Linn. Soc., Bot. 29: 50 (1891). TYPES: Madagascar, *Humblot* 378 (lecto. K; isolecto. P, both selected here); nr. Fort Dauphin, *Scott Elliot* 2771; & Hirondio, *Meller* 1862; (both syn. K).
Bulbophyllum album Jum. & H.Perrier in Ann. Fac. Sci. Marseille 21 (2): 213 (1912). TYPE: Madagascar, Manongarivo, *H.Perrier* 1908 (holo. P).
Bulbophyllum laggiarae Schltr. in Repert. Sp. Nov. Regni Veg. Beih. 15: 330 (1918). TYPE: Madagascar, *Laggiara* s.n. (holo. B†).
Bulbophyllum luteolabium H.Perrier in Notul. Syst. (Paris) 14: 155 (1951). TYPES: Madagascar, Foulpointe, N of Toamasina, *Decary* 16996 & 17013 (both iso. P).
Bulbophyllum linguiforme P.J.Cribb in Kew Bull. 32: 157 (1977). TYPE from Zimbabwe.

DISTRIBUTION: Madagascar: Antsiranana, Toamasina, Toliara; Seychelles. Also in East Africa.

HABITAT: seasonally dry, deciduous forest and woodland; humid, lowland, evergreen forest.
ALTITUDE: sea-level – 1300 m.
FLOWERING TIME: March – October.
LIFE FORM: epiphyte.
PHYTOGEOGRAPHICAL REGION: I, II.
DESCRIPTION: {896}, {901}.
ILLUSTRATION: {34}, {676}, {896}, {1431}.
NOTES: plant small, single-leafed; flowers creamy-white with a yellow, fleshy, 2.5 × 1 – 1.5 mm, tongue-like lip, geniculate at base, the tip rounded, the margins entire; stelids well-developed.

Bulbophyllum hyalinum *Schltr.* in Repert. Sp. Nov. Regni Veg. Beih. 33: 184 (1924). TYPES: Madagascar, Sambirano valley, *H.Perrier* 15733 (lecto. P, selected here); Manongarivo massif, *H.Perrier* 8036 & (syn. P).

DISTRIBUTION: Madagascar: Antsiranana. Comoros: Grande Comore.
HABITAT: humid, evergreen forest from lowlands to plateau.
ALTITUDE: 400 – 1200m.
FLOWERING TIME: January – May.
LIFE FORM: epiphyte.
PHYTOGEOGRAPHICAL REGION: II, III.
DESCRIPTION: {896}.
ILLUSTRATION: {896}.
NOTES: plant small, single-leafed; inflorescence slender, with 8 – 15 white flowers; lip curved, oblong, ciliate at the margins.

Bulbophyllum ikongoense *H.Perrier* in Notul. Syst. (Paris) 6 (2): 70 (1937). TYPE: Madagascar, Ikongo massif, *Decary* 5820 (holo. P).

DISTRIBUTION: Madagascar: Fianarantsoa.
HABITAT: humid, lowland forest.
FLOWERING TIME: October – December.
LIFE FORM: epiphyte.
PHYTOGEOGRAPHICAL REGION: III.
DESCRIPTION: {896}.
NOTES: plant bifoliate, up to 20 cm tall; raceme strongly reclining, 4.5 – 6 cm long, not very densely many-flowered; lip very small, 1 × 1 mm, with 5 obtuse lobules, and a narrow transverse callus.

Bulbophyllum imerinense *Schltr.* in Repert. Sp. Nov. Regni Veg. Beih. 33: 251 (1925). TYPE: Madagascar, Mt. Vohitralongo, N Imerina, *H.Perrier* 14938 (holo. P).

DISTRIBUTION: Madagascar: Antananarivo, Fianarantsoa, Toamasina.
HABITAT: humid, evergreen forest on plateau.
ALTITUDE: 1000 – 1300 m.
FLOWERING TIME: September.
LIFE FORM: epiphyte.
PHYTOGEOGRAPHICAL REGION: III.
DESCRIPTION: {896}.
NOTES: related to *B. viguieri*, but with much longer, 2-leaved pseudobulbs, a longer inflorescence, narrower sepals, and a 3.5 – 4 mm, thick, ligulate lip with 2 short erect ridges at the base, and 2 lateral rows of white hairs.

Bulbophyllum insolitum *Bosser* in Adansonia, ser.2, 11 (2): 325 (1971). TYPE: Madagascar, Sambava region, cult. Jard. Bot. Tananarive 1405 (holo. P).

DISTRIBUTION: Madagascar: Antsiranana.
HABITAT: humid, lowland forest.
LIFE FORM: epiphyte.
PHYTOGEOGRAPHICAL REGION: I.

DESCRIPTION: {119}.

ILLUSTRATION: {119}.

NOTES: known only from the type, distinguished by its flattened, 2-leaved pseudobulbs and short single-flowered inflorescence.

Bulbophyllum johannis *H.Wendl. & Kraenzl.* in Gard. Chron. sér. 3, 16: 592 (1894) (as *"Bolbophyllum"*). TYPE: Madagascar, *Braun* s.n. (holo. not located).

DISTRIBUTION: Madagascar {647}.

NOTES: very little is known about this species; the plant is single-leafed, the flowers are small and white.

Bulbophyllum jumelleanum *Schltr.* in Ann. Mus. Col. Marseille, ser.3, 1: 178, t.15 (1913). TYPE: Madagascar, Manongarivo, *H.Perrier* 20 (1945) (holo. B†; iso. P).

DISTRIBUTION: Madagascar: Antsiranana.

HABITAT: humid, evergreen forest on plateau, on tree trunks.

ALTITUDE: c. 1200m.

FLOWERING TIME: October.

LIFE FORM: epiphyte.

PHYTOGEOGRAPHICAL REGION: III, IV.

DESCRIPTION: {896}.

ILLUSTRATION: {896}, {1205}.

NOTES: related to *B. pervillei.* Plant small, 2-leaved; flower small, white, the lip greenish or yellow; lip, 2 mm, pandurate, furrowed, obtuse and shortly indented at the apex.

Bulbophyllum kainochiloides *H.Perrier* in Notul. Syst. (Paris) 6 (2): 60 (1937). TYPE: Madagascar, along Mandraka River, *E.François* in *H.Perrier* 17587 (holo. P).

DISTRIBUTION: Madagascar: Antananarivo.

HABITAT: humid, evergreen forest on plateau.

ALTITUDE: 1200 – 1400 m.

FLOWERING TIME: October, January.

LIFE FORM: epiphyte.

PHYTOGEOGRAPHICAL REGION: III.

DESCRIPTION: {896}.

NOTES: plant 20 – 30 cm tall; pseudobulbs conical, single-leafed; peduncle c. 12 cm long; lip curved, the margins truncate and a little expanded in the upper third and with a thick heel; stelids narrow, triangular, acute.

Bulbophyllum kieneri *Bosser* in Adansonia, ser.2, 11 (2): 328 (1971). TYPE: Madagascar, Toamasina region, *Kiener*, cult. Jard. Bot. Tan. 526 (holo. P).

DISTRIBUTION: Madagascar: Toamasina.

HABITAT: humid, lowland forest.

LIFE FORM: epiphyte.

PHYTOGEOGRAPHICAL REGION: V.

DESCRIPTION: {119}.

NOTES: known only from the type. Plant 2-leaved, c. 12 cm tall; pseudobulbs cylindrical, canaliculated; inflorescence short, slender; flower fleshy, whitish spotted with red; lip fleshy, curved, whitish, 2.5 × 2 mm wide.

Bulbophyllum labatii *Bosser* in Adansonia sér. 3, 26 (1): 53 (2004). TYPE: Madagascar, Toamasina, Maroantsetra, Masoala peninsula, Ambodiforaha, 200 m, *Labat, Andrianjafy, Breteler, Rabevohitra & Randrianorivo* 3367 (holo. P; iso. K, MO, TAN, TEF).

DISTRIBUTION: Madagascar: Tamatave.

HABITAT: humid, lowland forest.

ALTITUDE: 200 m.

FLOWERING TIME: October.

LIFE FORM: epiphyte.

PHYTOGEOGRAPHICAL REGION: I.

DESCRIPTION: {145}.

ILLUSTRATION: {145}.

NOTES: small bifoliate plant; lateral sepals fused together; distinct by the long and slender inflorescence, the small yellow flowers and the morphology of the lip.

Bulbophyllum lakatoense *Bosser* in Adansonia, ser.2, 9 (1): 135 (1969). TYPE: Madagascar, Lakato road, E of Moramanga, *Bosser* 19298 (holo. P).

DISTRIBUTION: Madagascar: Toamasina.

HABITAT: lichen-rich forest on moss- and lichen-covered trees.

ALTITUDE: c. 1000 m.

FLOWERING TIME: June.

LIFE FORM: epiphyte.

PHYTOGEOGRAPHICAL REGION: III.

DESCRIPTION: {110}.

ILLUSTRATION: {110}.

NOTES: two-leaved plant, similar to *B. lichenophylax* but distinguished by the lip which is ciliate at the edges.

Bulbophyllum lancisepalum *H.Perrier* in Notul. Syst. (Paris) 7: 142 (1938). TYPE: Madagascar, Belavenoka, nr. Fort Dauphin, *Decary* 10585bis (holo. P).

DISTRIBUTION: Madagascar: Toliara.

HABITAT: humid, lowland forest.

LIFE FORM: epiphyte.

PHYTOGEOGRAPHICAL REGION: I.

DESCRIPTION: {896}.

NOTES: known only from the type. Differs from *B. minax* by the distinctly angular, bifoliate pseudobulbs, the yellow flowers, the anterior piece of the lateral sepals with winged ridges, the petals velvety at the top, and the flat lip.

Bulbophyllum latipetalum *H.Perrier* in Notul. Syst. (Paris) 14: 150 (1951). TYPE: Madagascar, E of Marojejy massif, *Humbert* 22656 (holo. P).

DISTRIBUTION: Madagascar: Antsiranana.

HABITAT: humid, evergreen forest on plateau; lichen-rich forest.

ALTITUDE: c. 1500 m.

FLOWERING TIME: December.

LIFE FORM: epiphyte.

PHYTOGEOGRAPHICAL REGION: IV.

DESCRIPTION: {901}.

NOTES: known only from the type. Plant single-leafed, small; inflorescence short, single-flowered; lip incurved, 3 × 2 mm, red, with short hairs at the apex.

Bulbophyllum leandrianum *H.Perrier* in Notul. Syst. (Paris) 6 (2): 54 (1937). TYPE: Madagascar, Analamazaotra forest, *Leandri* 727 (holo. P; iso. K, TAN).

DISTRIBUTION: Madagascar: Antananarivo, Toamasina.

HABITAT: humid, evergreen forest.

ALTITUDE: c. 900 m.

FLOWERING TIME: December – January.

LIFE FORM: epiphyte.

PHYTOGEOGRAPHICAL REGION: III.

DESCRIPTION: {896}.

ILLUSTRATION: {281}, {478}, {896}.

NOTES: pseudobulbs long; single leaf 22 cm long; flowers greenish yellow; lip curved wider towards the base, the edges very finely denticulate.

Bulbophyllum lecouflei *Bosser* in Adansonia 11 (1): 30 (1989). TYPE: Madagascar, Ambodirafia (Ambohitralanana), *M.Lecoufle* 15762 (holo. P).

DISTRIBUTION: Madagascar: Antsiranana.
HABITAT: coastal and lowland forest.
ALTITUDE: sea level – 100 m.
FLOWERING TIME: November (in cultivation).
LIFE FORM: epiphyte.
PHYTOGEOGRAPHICAL REGION: I.
DESCRIPTION: {128}.
ILLUSTRATION: {128}, {501}, {567}, {722}, {724}, {726}.
NOTES: bifoliate plant, close to *B. cirrhoglossum*, but differs by its linear, narrow, glabrous petals, the lip with a long, narrow mid-lobe, glabrous at the margins, and the longer stelids which curl at the tip.

Bulbophyllum lemuraeoides *H.Perrier* in Notul. Syst. (Paris) 6 (2): 106 (1937). TYPE: Madagascar, Mandraka valley, *E.François* in *H.Perrier* 17583 (holo. P).

DISTRIBUTION: Madagascar: Antananarivo, Fianarantsoa.
HABITAT: humid, evergreen forest on plateau.
ALTITUDE: 800 – 2000 m.
FLOWERING TIME: November – February.
LIFE FORM: epiphyte.
PHYTOGEOGRAPHICAL REGION: III.
DESCRIPTION: {896}.
ILLUSTRATION: {896}.
NOTES: plant small, 2-leaved; inflorescence slender, elongate; lateral sepals fused, broad, bicarinate; lip 2.3 × 1.2 – 2 mm, thick, curved and heart-shaped at the base.

Bulbophyllum lemurense *Bosser & P.J.Cribb* in DuPuy *et al.*, Orch. Mad (1999).
Bulbophyllum clavigerum H.Perrier in Notul. Syst. (Paris) 14: 153 (1951), *nom. illeg. non* F. Muell. (1875). TYPE: Madagascar, Mt. Beandroka (N of Maroambihy), *Humbert* 23600 (holo. P).

DISTRIBUTION: Madagascar: Antsiranana.
HABITAT: on rocks and in humid, medium-elevation forest.
ALTITUDE: c. 1400 m.
FLOWERING TIME: March.
LIFE FORM: epiphyte.
PHYTOGEOGRAPHICAL REGION: II.
DESCRIPTION: {901}.
NOTES: related to *B. edentatum* which has 2-leaved pseudobulbs and a different lip.

Bulbophyllum leoni *Kraenzl.* in Bot. Jahrb. Syst. 28: 164 (1900), as "*Bolbophyllum*". TYPE: Comoros, Grande Comore, *Humblot* 1530 (holo. BM; iso. P).
Bulbophyllum humblotianum Kraenzl. in Bot. Jahrb. Syst. 33: 71 (1904), *non* Rolfe (1891).

DISTRIBUTION: Comoros: Grande Comore.
LIFE FORM: terrestrial {896}.
FLOWERING TIME: May.
DESCRIPTION: {896}.
ILLUSTRATION: {896}.
NOTES: plant up to 18 cm high; pseudobulbs angular, 2-leaved; flowers distant, yellow, 9 – 11 mm long; lip curved, 3 – 4 mm long, the edges folded back.

Bulbophyllum leptochlamys *Schltr.* in Repert. Sp. Nov. Regni Veg. Beih. 33: 234 (1924). TYPE: Madagascar, Mt. Tsaratanana, *H.Perrier* 15739 (holo. P).

DISTRIBUTION: Madagascar: Antsiranana.
HABITAT: humid, evergreen forest on plateau; lichen-rich forest.

ALTITUDE: c. 1500 m.

FLOWERING TIME: January.

LIFE FORM: epiphyte.

PHYTOGEOGRAPHICAL REGION: III.

DESCRIPTION: {896}.

ILLUSTRATION: {896}.

NOTES: known only from the type. Two-leaved plant, close to *B. cyclanthum* and *B. paleiformis* but characterised by its oval flowers, the sickle-shaped lip and curved stelids.

Bulbophyllum leptostachyum *Schltr.* in Repert. Sp. Nov. Regni Veg. Beih. 18: 323 (1922). TYPE: Madagascar, Analamazaotra forest, *Alleizette* 713 (holo. B†); Central, Mandraka valley, *H.Perrier* 16718. (neo. P), selected here.

DISTRIBUTION: Madagascar: Antananarivo, Fianarantsoa, Toamasina.

HABITAT: humid, evergreen and mossy forest.

ALTITUDE: 400 – 1600 m.

FLOWERING TIME: October – March.

LIFE FORM: epiphyte.

PHYTOGEOGRAPHICAL REGION: I, III.

DESCRIPTION: {896}.

NOTES: Schlechter's type was destroyed and *H.Perrier* 16718 is selected here as the neotype. Related to *B. baronii*: plant single-leafed, c. 10 cm tall; flowers 5 – 7 mm long, yellow; lip recurved, wide at the base and the front oblong-ligulate; stelids short, triangular, acute.

Bulbophyllum lichenophylax *Schltr.* in Repert. Sp. Nov. Regni Veg. Beih. 33: 194 (1924); Bosser in Adansonia, sér. 2, 5: 381 (1965). TYPES: Madagascar, Mt. Tsaratanana, *H.Perrier* 15305 (lecto. P, selected here) & *H.Perrier* 15328 (syn. P).

Bulbophyllum quinquecornutum H.Perrier in Notul. Syst. (Paris) 6 (2): 48 (1937). TYPE: Madagascar, Ankeramadinika forest, H.Perrier 18297bis (holo. P).

DISTRIBUTION: Madagascar: Antananarivo, Antsiranana, Toamasina.

HABITAT: mossy and lichen-rich forest.

ALTITUDE: 800 – 2000 m.

FLOWERING TIME: January – March, August.

LIFE FORM: epiphyte on moss- and lichen-covered trees.

PHYTOGEOGRAPHICAL REGION: III, IV.

DESCRIPTION: {108}, {896}.

NOTES: two-leaved plant, small with bristle-like, few-flowered inflorescences; lip without a callus; stelids narrow and acute.

Bulbophyllum lineariligulatum *Schltr.* in Repert. Sp. Nov. Regni Veg. Beih. 33: 190 (1924). TYPES: Madagascar, Mt. Tsaratanana, *H.Perrier* 11334 (lecto. P, selected here); *H.Perrier* 15723; *H.Perrier* 15725; (both syn. P).

DISTRIBUTION: Madagascar: Antsiranana, Fianarantsoa, Toamasina.

HABITAT: on moss- and lichen-covered trees in lichen-rich forest and on ericaceous shrubs.

ALTITUDE: 1000 – 2000m.

FLOWERING TIME: September – April.

LIFE FORM: epiphyte.

PHYTOGEOGRAPHICAL REGION: III, IV.

DESCRIPTION: {896}.

ILLUSTRATION: {896}.

NOTES: closely related to *B. curvifolium*, but differs by its smaller pseudobulbs, shorter leaf; the blunter and differently shaped petals and the lip which is the same width at apex and base but narrowed in the middle.

Bulbophyllum liparidioides *Schltr.* in Repert. Sp. Nov. Regni Veg. Beih. 33: 201 (1924). TYPE: Madagascar, Analamazaotra, *H.Perrier* 8132 (holo. P).

DISTRIBUTION: Madagascar: Toamasina.
HABITAT: humid, evergreen forest at intermediate elevation and on plateau.
ALTITUDE: 800 – 1200m.
FLOWERING TIME: February.
LIFE FORM: epiphyte.
PHYTOGEOGRAPHICAL REGION: III.
DESCRIPTION: {896}.
NOTES: resembles some species of *Liparis* in habit. Two-leaved plant, small, few-flowered; inflorescence 4 – 7 cm long; flowers 5 – 6 mm long; lip recurved, glabrous and obtuse.

Bulbophyllum longiflorum *Thouars*, Hist. Orchid.: t.98 (1822). TYPE: *Thouars* s.n. (holo. P).
Epidendrum umbellatum Forst., Fl. Insul. Aust. Prodr.: 60 (1786), *non* Sw. (1788).
Zygoglossum umbellatum (Forst.) Reinw. ex Blume, Cat. Gew. Buitenz.: 100 (1823).
Cirrhopetalum thouarsii Lindl. in Bot. Reg.: sub t.832 (1824), *nom. illeg.*
Cymbidium umbellatum (Forst.) Spreng., Syst. Veg. 3: 723 (1826).
Cirrhopetalum thomasii Otto & A.Dietr. in Allg. Gart. Zeit. 6 (22): 174 (1838), sphalm. for "*thouarsii*".
Cirrhopetalum clavigerum Fitzg. in J. Bot. 21: 204 (1883). TYPE: not traced.
Phyllorkis longiflora (Thouars) Kuntze, Rev. Gen. Pl. 2: 675 (1891).
Cirrhopetalum umbellatum (Forst.) Frapp. ex Cordem. in Fl. Réunion: 177 (1895).
Cirrhopetalum longiflorum (Thouars) Schltr. in Beih. Bot. Centralbl. 33 (2): 420 (1915).
Cirrohopetalum africanum Schltr. in Bot. Jahrb. Syst. 53: 573 (1915). TYPE from Tanzania.
Bulbophyllum clavigerum (Fitzg.) Dockrill in Orchadian 1: 106 (1964), *nom. illeg. non B. clavigerum* F. Muell (1875).
For further synonymy see Seidenfaden in Dansk Bot. Ark. 29: 126 (1973).

DISTRIBUTION: Madagascar: Antananarivo, Antsiranana, Fianarantsoa, Mahajanga, Toamasina, Toliara. Comoros: Grande Comore, Mayotte, Moheli. Mascarenes: Réunion, Mauritius. Seychelles. Also in East Africa, SE Asia and SW Pacific Islands.
HABITAT: mossy forest; coastal forest; humid, evergreen forest.
ALTITUDE: sea level – 1200 m.
FLOWERING TIME: October – April.
LIFE FORM: epiphyte.
PHYTOGEOGRAPHICAL REGION: I, II, III.
DESCRIPTION: {582}, {896}, {768}.
ILLUSTRATION: {34}, {50}, {166}, {219}, {319}, {384}, {504}, {547}, {582}, {672}, {676}, {677}, {682}, {724}, {768}, {896}, {906}, {922}, {943}, {1398}, {1328}, {1500}.
CULTIVATION: {582}, {782}.
VERNACULAR NAME: Solofa.
NOTES: widespread and easily recognisable species; pseudobulbs one-leafed; inflorescence subumbellate; flowers purple; lateral sepals c. 27 × 6 mm.

Bulbophyllum longivaginans *H.Perrier* in Notul. Syst. (Paris) 6 (2): 75 (1937). TYPES: Madagascar, Andasibe forest, *H.Perrier* 17122 (lecto. P, selected here); *H.Perrier* 17120 (syn. P).

DISTRIBUTION: Madagascar: Antananarivo, Toamasina.
HABITAT: humid, evergreen forest on plateau.
ALTITUDE: c. 1000 m.
FLOWERING TIME: January – February.
LIFE FORM: epiphyte.
PHYTOGEOGRAPHICAL REGION: III.
DESCRIPTION: {896}.
ILLUSTRATION: {896}.
NOTES: two-leaved plant, up to 25 cm high; inflorescence shorter than the leaves, few-flowered; flowers red, glabrous, c. 7 mm long; lip recurved, obtuse at the apex and auriculate at the base.

Bulbophyllum lucidum *Schltr.* in Repert. Sp. Nov. Regni Veg. Beih. 33: 234 (1924). TYPE: Madagascar, Sambirano Mts, *H.Perrier* 15199 (holo. P).

Bulbophyllum rictorium Schltr. in Repert. Sp. Nov. Regni Veg. Beih. 33: 241 (1925). TYPE: Madagascar, Mt. Tsaratanana, *H.Perrier* 11344 (holo. P).

DISTRIBUTION: Madagascar: Antsiranana.
HABITAT: coastal forest; mossy forest on moss- and lichen-covered trees.
ALTITUDE: sea level – 1500 m.
FLOWERING TIME: October.
LIFE FORM: epiphyte.
PHYTOGEOGRAPHICAL REGION: II, III.
DESCRIPTION: {896}.
NOTES: two-leaved plant, known only from the type. The wide leathery leaves are characteristic, and the greenish flowers are similar to those of *B. ankaratranum*.

Bulbophyllum luteobracteatum *Jum. & H.Perrier* in Ann. Fac. Sci. Marseille 21 (2): 204 (1912). TYPE: Madagascar, Manongariva, *H.Perrier* 1905 (holo. P).

DISTRIBUTION: Madagascar: Antsiranana.
HABITAT: lichen-rich forest.
ALTITUDE: c. 1500 m.
FLOWERING TIME: May.
LIFE FORM: epiphyte on moss- and lichen-covered trees.
PHYTOGEOGRAPHICAL REGION: III.
DESCRIPTION: {896}.
ILLUSTRATION: {896}.
NOTES: known only from the type. Plant small, 2-leaved; inflorescence c. 30 cm tall; flowers c. 16 mm long, yellowish and spotted by dull brown, the lip red, curved and truncate at the base, then extended into a thick velvety, ciliate lump.

Bulbophyllum lyperocephalum *Schltr.* in Repert. Sp. Nov. Regni Veg. Beih. 33: 182 (1924). TYPE: Madagascar, Analamazaotra, *H.Perrier* 8039 (holo. P).

DISTRIBUTION: Madagascar: Antsiranana, Toamasina.
HABITAT: humid, evergreen forest.
ALTITUDE: 800 – 900m.
FLOWERING TIME: February.
LIFE FORM: epiphyte.
PHYTOGEOGRAPHICAL REGION: III.
DESCRIPTION: {896}.
ILLUSTRATION: {281}, {896}.
NOTES: small single-leafed plant, on a creeping rhizome; flowers greenish, 5 mm long, glabrous; lip 2 × 1.25 mm, folded-curved, with a thick keel, the lower half sunk and with 2 thick, pronounced horns, the upper part rounded.

Bulbophyllum lyperostachyum *Schltr.* in Repert. Sp. Nov. Regni Veg. Beih. 33: 183 (1924). TYPE: Madagascar, Mt. Tsaratanana, *H.Perrier* 15317 (holo. P).

DISTRIBUTION: Madagascar: Antsiranana.
HABITAT: lichen-rich forest.
ALTITUDE: c. 2000 m.
FLOWERING TIME: January.
LIFE FORM: epiphyte on moss- and lichen-covered trees.
PHYTOGEOGRAPHICAL REGION: III, IV.
DESCRIPTION: {896}.
ILLUSTRATION: {896}.
NOTES: known only from the type. Plant single-leafed, compact, up to 17 cm tall; inflorescence shorter than the leaf, laxly-flowered; flower yellow with darker orange-brown lines, 6.5 – 7 mm; lip 3.5 × 2 mm, curved, obtuse, glabrous, a small obsolete swelling in the middle; stelids well-developed, bidentate at the apex.

Bulbophyllum maleolens *Kraenzlin* in Repert. Sp. Nov. Regni. Veg. 25, 4 – 6: 55 (1928). TYPE: Madagascar, without locality, cult. Heidelberg Botanic Garden (holo. HBG).

DISTRIBUTION: Madagascar.
DESCRIPTION: {666}.
NOTES: known only from the type. Two-leaved plant, small; inflorescence 10 cm long; flowers unpleasantly scented; dorsal sepal yellow, with 3 purple lines; lateral sepals red, extensively spotted; lip yellow, suborbicular, base indented, with 2 ridges at the base of the disc, papillose 2.3 mm long and wide.

Bulbophyllum mananjarense *Poiss.*, Recherch. Fl. Merid. Mad.: 173 (1912). TYPE: Madagascar, Mananjary, *Geay* 8744 (holo. P, not located).

DISTRIBUTION: Madagascar.
HABITAT: humid, lowland forest.
LIFE FORM: epiphyte.
DESCRIPTION: {928}.
NOTES: according to Perrier de la Bâthie {888} the type has been lost.

Bulbophyllum mangenotii *Bosser* in Adansonia, sér. 2, 5: 378 (1965). TYPE: Madagascar, Ankeramadinika (Ambatoloana), *Bosser* 16224 (holo. P).

DISTRIBUTION: Madagascar: Fianarantsoa.
HABITAT: humid, evergreen forest on plateau; mossy forest.
ALTITUDE: 1200 – 1300m.
FLOWERING TIME: August.
LIFE FORM: epiphyte.
PHYTOGEOGRAPHICAL REGION: III.
DESCRIPTION: {108}.
ILLUSTRATION: {108}.
NOTES: small bifoliate plant, flower tiny, lip recurved, dark red, with a central depression on the upper surface that branches into two lateral horns, two winged crests edge this depression.

Bulbophyllum marojejiense *H.Perrier* in Notul. Syst. (Paris) 14: 156 (1951). TYPES: Madagascar, E of Marojejy massif, *Humbert* 22612 (lecto. P, selected here); W of Marojejy massif, *Humbert* 22223 (syn. P).

DISTRIBUTION: Madagascar: Antsiranana.
HABITAT: humid evergreen forest on plateau; lichen-rich forest.
ALTITUDE: 1000 – 1600m.
FLOWERING TIME: November – December.
LIFE FORM: epiphyte.
PHYTOGEOGRAPHICAL REGION: III, IV.
DESCRIPTION: {901}.
NOTES: plant small; pseudobulbs with 1 leaf; inflorescence 6- to 10-flowered; flowers small, 4 – 5 mm; lip curved, 4 × 2 mm.

Bulbophyllum marovoense *H.Perrier* in Notul. Syst. (Paris) 14: 155 (1951). TYPE: Madagascar, nr. Marovoay, N of Moramanga, *François* 2 (holo. P).

DISTRIBUTION: Madagascar: Toamasina.
HABITAT: humid, evergreen forest on plateau.
LIFE FORM: epiphyte.
PHYTOGEOGRAPHICAL REGION: III {901}.
DESCRIPTION: {901}.
NOTES: single-leafed plant known only from the type. Distinct by the lanceolate-linear sepals and the very short semi-rounded petals; stelids triangular, acute.

Bulbophyllum masoalanum *Schltr.* in Beih. Bot. Centralbl. 34 (2): 328 (1916). TYPE: Madagascar, Masoala, *H.Perrier* 148 (11399) (holo. B†; iso. P).

DISTRIBUTION: Madagascar: Antsiranana, Toamasina.
HABITAT: humid, lowland, evergreen forest.
ALTITUDE: c. 300 m.
FLOWERING TIME: October – November.
LIFE FORM: epiphyte.
PHYTOGEOGRAPHICAL REGION: I.
DESCRIPTION: {896}.
NOTES: close to *B. henrici* but differs by the less robust plant, closer together, shorter and relatively wider pseudobulbs, smaller, shorter leaves (2), a thinner raceme, smaller and differently coloured, red-veined flowers, and a column toothed at the edge of the not acute but obtuse.

Bulbophyllum maudeae *A.D.Hawkes* in Phytologia 13: 309 (1966). Based on same type as *B. nigrilabium* H.Perrier.
Bulbophyllum nigrilabium H.Perrier in Notul. Syst. (Paris) 14: 150 (1951), *non* Schltr. (1913). TYPE: Madagascar, E of Marojejy massif, *Humbert* 22532 (holo. P).

DISTRIBUTION: Madagascar: Antsiranana.
HABITAT: lichen-rich forest.
ALTITUDE: c. 1700 m.
FLOWERING TIME: December.
LIFE FORM: epiphyte.
PHYTOGEOGRAPHICAL REGION: IV.
DESCRIPTION: {901}.
NOTES: single-leafed plant known only from the type. Inflorescence 8 – 10 cm long, three times the length of the plant, 5- to 8-flowered; lip dark, narrow, 3 mm long, base much thickened, from there extended and spathulate, apex with black hairs.

Bulbophyllum megalonyx *Rchb.f.*, Otia Bot. Hamburg. 2: 74 (1881) & in Bot. Zeit. 39: 449 (1881). TYPE: Comoros, *Hildebrandt* 1707 (holo. W).

DISTRIBUTION: Comoros: Anjouan.
HABITAT: forest.
ALTITUDE: 1000 – 1500 m.
FLOWERING TIME: June – August.
LIFE FORM: epiphyte on tree trunks.
DESCRIPTION: {896}.
NOTES: a poorly known species; the type has an inflorescence only, said to be close to *B. cupreum* from Burma.

Bulbophyllum melleum *H.Perrier* in Notul. Syst. (Paris) 6 (2): 54 (1937).TYPES: Madagascar, Rienana valley, *Humbert* 3500 (lecto. P; isolecto. K, both selected here); Ikongo massif, *Decary* 5739 (syn. P).

DISTRIBUTION: Madagascar: Fianarantsoa, Toamasina.
HABITAT: humid, evergreen forest on plateau.
ALTITUDE: 1000 – 1400 m.
FLOWERING TIME: October – November.
LIFE FORM: epiphyte.
PHYTOGEOGRAPHICAL REGION: III.
DESCRIPTION: {896}.
NOTES: pseudobulbs small with 1 leaf; inflorescence slender, laxly flowered; flowers yellow, 10 – 12 mm long, honey-scented.

Bulbophyllum metonymon *Summerh.* in Kew Bull: 470 (1951). TYPE: Madagascar, Mt. Tsaratanana, *H.Perrier* 15732 (holo. P).
Bulbophyllum zaratananae Schltr. in Repert. Sp. Nov. Regni Veg. Beih. 33: 246 (1925), *non* Schltr. (1924).

Bulbophyllum schlechteri Kraenzl. in Repert. Spec. Nov. Regni Veg. 25: 56 (1928), *non* De Wildem. (1921).

DISTRIBUTION: Madagascar: Antsiranana.
HABITAT: lichen-rich forest on moss- and lichen-covered trees.
ALTITUDE: c. 1200 m.
FLOWERING TIME: January.
LIFE FORM: epiphyte.
PHYTOGEOGRAPHICAL REGION: III.
DESCRIPTION: {896}.
NOTES: bifoliate plant known from the type only which is fragmented. Similar to *B. clavatum* and *B. rubrilabium* but with larger completely red flowers and a thickened pale green rachis, the shape of the lip is also different.

Bulbophyllum minax *Schltr.* in Repert. Sp. Nov. Regni Veg. Beih. 33: 202 (1924). TYPE: Madagascar, Mt. Isomaino, nr. Fianarantsoa, *H.Perrier* 12580 (holo. P).

DISTRIBUTION: Madagascar: Fianarantsoa.
HABITAT: mossy, montane forest on moss- and lichen-covered trees.
ALTITUDE: c. 1700 m.
FLOWERING TIME: May.
LIFE FORM: epiphyte.
PHYTOGEOGRAPHICAL REGION: III.
DESCRIPTION: {896}.
NOTES: known only from the type. Two-leaved plant, similar to *B. brachystachyum*; inflorescence slender, longer than the leaves; peduncle c. 15 cm, laxly-flowered; flowers yellow, c. 8 mm long; lip small, 2.3 mm long, curved, obtuse, emarginate in the lower half, with 2 pronounced ridges on top.

Bulbophyllum minutilabrum *H.Perrier* in Notul. Syst. (Paris) 6 (2): 58 (1937). TYPE: Madagascar, Mt. Tsaratanana, *H.Perrier* 16475 (holo. P).

DISTRIBUTION: Madagascar: Antsiranana.
HABITAT: lichen-rich forest on moss- and lichen-covered trees.
ALTITUDE: c. 2400 m.
FLOWERING TIME: April.
LIFE FORM: epiphyte.
PHYTOGEOGRAPHICAL REGION: IV {896}.
DESCRIPTION: {896}.
NOTES: known from the type only. Large plant with two wide leaves; lip expanded at the base; anther small with an obtuse lip at the front.

Bulbophyllum minutum *Thouars,* Hist. Orch.: t.110 (1822). TYPE: Madagascar, *Thouars* s.n. (holo. P).
Phyllorchis minuta (Thouars) Kuntze, Rev. Gen. Pl. 2: 677 (1891).
Bulbophyllum implexum Jum. & H.Perrier in Ann. Fac. Sci. Marseille 21 (2): 212 (1912), TYPE: Madagascar, Sambirano, *H.Perrier* 1914 (holo. P).

DISTRIBUTION: Madagascar: Antananarivo, Antsiranana.
HABITAT: semi-deciduous forest, on *Chlaenaceae*, in seasonally dry deciduous woodland.
ALTITUDE: sea level – 800 m.
FLOWERING TIME: September – April.
LIFE FORM: epiphyte.
PHYTOGEOGRAPHICAL REGION: I, II, III.
DESCRIPTION: {896}.
ILLUSTRATION: {505}, {896}, {1398}.
NOTES: two-leaved plant, very small; rachis zig-zag, with 14 – 15 small dark red flowers; stelids subulate.

Bulbophyllum mirificum *Schltr.* in Repert. Sp. Nov. Regni Veg. Beih. 15: 331 (1918). TYPE: Madagascar, Ile St Marie, *Laggiara* s.n. (holo. B†).

DISTRIBUTION: Madagascar: Toamasina.
HABITAT: eastern forest.
FLOWERING TIME: February.
LIFE FORM: epiphyte.
PHYTOGEOGRAPHICAL REGION: I.
DESCRIPTION: {139}, {896}.
ILLUSTRATION: {139}
NOTES: small bifoliate plant with a long inflorescence; flowers 15 – 20, not reflexed, laxly grouped; lip 4 – 5 mm, spathulate, broadened and rounded at the apex; base narrow, with 2 short, rounded raised keels, in-between them a central keel extending to the apex, carrying a small area of branched hairs and below the apex a transversal row of numerous ligulate, mobile, caducous appendages.

Bulbophyllum moldekeanum *A.D.Hawkes* in Phytologia 13 (5): 309 (1966). TYPE: Madagascar, E of Marojejy massif, *Humbert* 22657bis (holo. P.).
Bulbophyllum microglossum H.Perrier in Notul. Syst. (Paris) 14: 151 (1951), *non* Ridl. (1908).

DISTRIBUTION: Madagascar: Antsiranana.
HABITAT: lichen-rich forest on moss- and lichen-covered trees.
ALTITUDE: 1500 – 1700 m.
FLOWERING TIME: November – December.
LIFE FORM: epiphyte.
PHYTOGEOGRAPHICAL REGION: III, IV.
DESCRIPTION: {445}.
NOTES: small, single-leafed plant, 5- to 7-flowered; pseudobulbs sub-confluent, flattened; lip very small, ovate-lanceolate, ciliate fimbriate on the lower side, 4 – 6 × 1.5 mm.

Bulbophyllum molossus *Rchb.f.* in Flora 71: 155 (1888). TYPE: Central Madagascar, *Baron* s.n. (holo. W).

DISTRIBUTION: Madagascar: Antananarivo, Toamasina.
HABITAT: humid, evergreen forest on plateau.
ALTITUDE: 1200 – 1500 m.
FLOWERING TIME: November – February.
LIFE FORM: epiphyte.
PHYTOGEOGRAPHICAL REGION: III.
DESCRIPTION: {896}.
ILLUSTRATION: {896}.
NOTES: plant with two leaves, up to 7 cm long, papery sheaths at the base of the pseudobulbs; inflorescence about as long as the leaves; peduncle 6 – 12 cm, few-flowered; flowers red-pink, 12 – 13 mm long.

Bulbophyllum moramanganum *Schltr.* in Repert. Sp. Nov. Regni Veg. Beih. 18: 324 (1922) (as "*maromanganum*"). TYPE: Madagascar, Analamazaotra forest, Moramanga, *Viguier & Humbert* 1137 (holo. P).

DISTRIBUTION: Madagascar: Toamasina.
HABITAT: humid, evergreen forest on plateau.
ALTITUDE: 900 m.
FLOWERING TIME: October – November.
LIFE FORM: epiphyte.
PHYTOGEOGRAPHICAL REGION: III.
DESCRIPTION: {896}.
NOTES: related to *B. humbertii* but differs in the plant and in the sepals which are verrucose on the exterior.

Bulbophyllum moratii *Bosser* in Adansonia 11 (1): 29 (1989). TYPE: Madagascar, nr. Iaraka, Masoala peninsula, *Morat* 4947 (holo. P).

DISTRIBUTION: Madagascar: Antsiranana.
HABITAT: humid, evergreen forest at intermediate elevation and on plateau.
ALTITUDE: c. 1000 m.
FLOWERING TIME: May.
LIFE FORM: epiphyte.
PHYTOGEOGRAPHICAL REGION: III.
DESCRIPTION: {128}.
ILLUSTRATION: {128}, {672}.
NOTES: bifoliate plant, known only from the type. One of the biggest *Bulbophyllum* species of the island, up to 29 cm tall; raceme thick, cylindrical, with numerous densely grouped flowers; flowers largely covered by the bracts; lip 5 × 3 mm, yellow, ligulate with 2 pronounced narrow crenulate crests.

Bulbophyllum multiflorum *Ridl.* in J. Linn. Soc., Bot. 21: 463 (1885), *non* (Breda) Kraenzl. TYPE: Madagascar, Imerina, Ankafana, *Deans Cowan* s.n. (holo. BM).
Phyllorchis multiflora (Ridl.) Kuntze, Rev. Gen. Pl. 2: 677 (1891).
Bulbophyllum ridleyi Kraenzl. in Gard. Chron. sér. 3, 19: 294 (1896), *nom. superfl.*

DISTRIBUTION: Madagascar: Antananarivo, Fianarantsoa.
HABITAT: humid, evergreen forest on plateau, covering tree trunks.
ALTITUDE: 900 – 1500 m.
FLOWERING TIME: March, October.
LIFE FORM: epiphyte.
PHYTOGEOGRAPHICAL REGION: III.
ILLUSTRATION: {246}.
DESCRIPTION: {896}.
NOTES: pseudobulbs single-leafed, c. 10 cm high; inflorescence slender, c. 25 cm long, many-flowered; flowers white, c. 5 cm long; lip very curved, hardly 2 mm long; stelids acute and long.

Bulbophyllum multiligulatum *H.Perrier* in Notul. Syst. (Paris) 6 (2): 92 (1937). TYPES: Madagascar, Mt. Tsaratanana, *H.Perrier* 16490 (lecto. P, selected here); Ankaizina, c. 1200 m, *R.Decary* 2071 (syn. P).

DISTRIBUTION: Madagascar: Antsiranana, Mahajanga.
HABITAT: humid evergreen forest on plateau and at medium-elevation.
ALTITUDE: 1000 – 1500m.
FLOWERING TIME: April.
LIFE FORM: epiphyte.
PHYTOGEOGRAPHICAL REGION: III.
DESCRIPTION: {139}, {896}.
ILLUSTRATION: {896}.
NOTES: two-leaved plant characterised by its laxly several-flowered raceme, longer than the peduncle and by the lip which is fleshy, clavate, 6 – 7 mm long, with a lateral row of branched hairs and below the apex a transversal row of ligulate appendices, mobile, caducous.

Bulbophyllum multivaginatum *Jum. & H.Perrier* in Ann. Fac. Sci. Marseille 21, 2: 211 (1912). TYPE: Madagascar, Antsatotra, Analalava, *H.Perrier* 1913 (holo. P).

DISTRIBUTION: Madagascar: Antsiranana.
HABITAT: humid, lowland, evergreen forest.
ALTITUDE: 1100 – 1600m.
FLOWERING TIME: May, October.
LIFE FORM: epiphyte.
PHYTOGEOGRAPHICAL REGION: II.
DESCRIPTION: {896}.
NOTES: two-leaved plant up to 7 cm tall; inflorescence 18 – 23 cm tall; flowers thick, c. 8 mm long; sepals reddish brown; lateral sepals greenish on the outside, red-brown on the inside; disc of the lip purple-black; lip 2.5 mm long, curved, obtuse, velvety; stelids broadly triangular.

Bulbophyllum myrmecochilum *Schltr.* in Repert. Sp. Nov. Regni Veg. Beih. 33: 237 (1924). TYPE: Madagascar, Mt. Tsaratanana, *H.Perrier* 15308 (holo. P).

DISTRIBUTION: Madagascar: Antsiranana.
HABITAT: lichen-rich forest on moss- and lichen-covered trees.
ALTITUDE: c. 2000 m.
FLOWERING TIME: January.
LIFE FORM: epiphyte.
PHYTOGEOGRAPHICAL REGION: III, IV.
DESCRIPTION: {896}.
ILLUSTRATION: {896}.
NOTES: two-leaved plant known from the type only. Plant and inflorescence slender, up to 20 cm tall; flowers 8 – 9 mm long, rugose on the exterior; lip green.

Bulbophyllum namoronae *Bosser* in Adansonia, sér. 2, 11 (2): 327 (1971). TYPE: Madagascar, along the Namorona, Ifanadiana, *Bosser* 17637 (holo. P).

DISTRIBUTION: Madagascar: Fianarantsoa.
HABITAT: humid, evergreen forest.
FLOWERING TIME: January.
LIFE FORM: epiphyte.
PHYTOGEOGRAPHICAL REGION: III.
DESCRIPTION: {119}.
ILLUSTRATION: {119}.
NOTES: typical by the cylindrical pseudobulbs with two leaves; flowers fleshy; lip 2 × 1.8 mm long, violet, curved, concave at the base and the margins upright, with in the lower part, on both sides a central depression; stelids bidentate.

Bulbophyllum neglectum *Bosser* in Adansonia, sér. 2, 5: 379 (1965). TYPE: Madagascar, S of Moramanga, *Bosser* 15585 (holo. P).

DISTRIBUTION: Madagascar: Toamasina.
HABITAT: humid, evergreen forest, plant partly immersed in moss and liverworts.
ALTITUDE: 800 – 1000 m.
LIFE FORM: epiphyte.
PHYTOGEOGRAPHICAL REGION: III.
DESCRIPTION: {108}.
ILLUSTRATION: {108}.
NOTES: two-leaved plant tiny; peduncle thin, up to 1.7 mm long, single-flowered; lip glabrous, violet-red, the base with two keels, pale yellow; stelids bilobed.

Bulbophyllum nigriflorum *H.Perrier* in Notul. Syst. (Paris) 6 (2): 80 (1937). TYPE: Madagascar, Andasibe forest, Onive, *H.Perrier* 17126 (holo. P).

DISTRIBUTION: Madagascar: Antananarivo, Toamasina.
HABITAT: humid, evergreen forest on plateau.
ALTITUDE: c. 1000 m.
FLOWERING TIME: February.
LIFE FORM: epiphyte.
PHYTOGEOGRAPHICAL REGION: III.
DESCRIPTION: {896}.
ILLUSTRATION: {896}.
NOTES: plant small, single-leafed; inflorescence filiform, up to 10 cm long; petals ciliate; lip oblanceolate, 2 × 0.8 mm, very obtuse, ciliate with long bristles.

Bulbophyllum nitens *Jum.* & *H.Perrier* in Ann. Fac. Sci. Marseille 21 (2): 209 (1912). Type: Madagascar, Manongarivo massif, *H.Perrier* 1869 (holo. P).
Bulbophyllum nitens var. *typica* H.Perrier in Notul. Syst. (Paris) 6 (2): 111 (1937), *nom. inval.*
Bulbophyllum nitens var. *minus* H.Perrier in Notul. Syst. (Paris) 6 (2): 112 (1937), *nom. nud.* Based upon: Madagascar, Ankeramdinika forest, March 1928, *François* in *H.Perrier* 18521 (P).

DISTRIBUTION: Madagascar: Antananarivo, Antsiranana.

HABITAT: lichen-rich forest; humid evergreen forest on plateau, on moss- and lichen-covered trees.

ALTITUDE: 1000 – 2000 m.

FLOWERING TIME: March, August – June.

LIFE FORM: epiphyte.

PHYTOGEOGRAPHICAL REGION: III.

DESCRIPTION: {896}.

ILLUSTRATION: {896}.

VERNACULAR NAME: Tsiakondroakandro.

NOTES: this is a widespread and variable species from Northern, Central and East-Central Madagascar. Plant 2-leaved, rarely single leafed, up to 8 cm tall; peduncle slender, up to 30 cm long, densely-many flowered; flowers c. 3 mm long, yellowish green; lip with 2 slightly oblique raised ridges. The variety *minus* is smaller than the typical variety. Perrier de la Bâthie (1937) invalidly published several other varieties, based on flower size, minor variations in shape and the presence of papillae on floral parts, viz. var. *pulverulentum*, *intermedium* and *majus*, plus an unpublished variety *ioloides* indicated on a herbarium sheet *Humbert* 18602 – P97443 (P). These may not belong to this species but require further examination.

Bulbophyllum nitens var. **intermedium** *H.Perrier* in Notul. Syst. (Paris) 6 (2): 112 (1937), *nom. nud.* Based upon: Madagascar, Mt. Tsaratanana, *H.Perrier* 16510 (P, K).

DISTRIBUTION: Madagascar: Antsiranana.

HABITAT: mossy forest.

ALTITUDE: 1501 – 2000 m.

FLOWERING TIME: April.

PHYTOGEOGRAPHICAL REGION: III.

DESCRIPTION: {896}.

NOTES: this is likely to be a distinct species.

Bulbophyllum nitens var. **majus** *H.Perrier* in Notul. Syst. (Paris) 6 (2): 112 (1937), *nom. nud.* Based upon: Madagascar, Mandraka river, *H.Perrier* 18007 (P).

DISTRIBUTION: Madagascar: Antananarivo.

HABITAT: mossy forest.

FLOWERING TIME: September.

PHYTOGEOGRAPHICAL REGION: III.

DESCRIPTION: {896}.

NOTES: this is likely to be a distinct species.

Bulbophyllum nitens var. **pulverulentum** *H.Perrier* in Notul. Syst. (Paris) 6 (2): 113 (1937), *nom. nud.* Based upon: Madagascar, Analabe (N of Antananarivo), *H.Perrier* 18462 (P).

DISTRIBUTION: Madagascar: Antananarivo.

HABITAT: evergreen forest.

FLOWERING TIME: February.

LIFE FORM: epiphyte.

PHYTOGEOGRAPHICAL REGION: III.

DESCRIPTION: {896}.

NOTES: this is likely to be a distinct species.

Bulbophyllum nutans *(Thouars) Thouars*, Hist. Orchid.: t.107 (1822). TYPE: *Thouars* s.n. (holo. P).
Phyllorkis nutans Thouars, Cahier de six planches: t.4 (1819).
Phyllorkis nuphyllis Thouars, Cahier de six planches: t.4 (1819), *nom. superfl.*
Bulbophyllum nutans nanum Cordem., Fl. Réunion 171 (1895).
Bulbophyllum nutans genuinum Cordem., Fl. Réunion 171 (1895).
Bulbophyllum nutans flavum Cordem., Fl. Réunion 171 (1895).
Bulbophyllum nutans rubellum Cordem., Fl. Réunion 171 (1895).

Bulbophyllum nutans pictum Cordem., Fl. Réunion 171 (1895).

Bulbophyllum andringitranum Schltr. in Repert. Sp. Nov. Regni Veg. Beih. 33: 199 (1924). TYPE: Madagascar, Andringitra massif, *H.Perrier* 14389 (holo. P).

Bulbophyllum tsinjoarivense H.Perrier in Notul. Syst. (Paris) 6 (2): 72 (1937). TYPE: Madagascar, Tsinjoarivo, *H.Perrier* 16954 (holo. P).

Bulbophyllum chrysobulbum H.Perrier in Notul. Syst. (Paris) 14 (2): 158 (1951). TYPE: Herbier de l'Exposition Coloniale (holo. P).

DISTRIBUTION: Madagascar: Antananarivo, Fianarantsoa. Comoros: Grande Comore. Mascarenes: Réunion, Mauritius.
HABITAT: shaded rocks; humid, medium-elevation forest; highland forest.
ALTITUDE: 800 – 2200m.
FLOWERING TIME: December – June.
LIFE FORM: epiphyte or lithophyte.
PHYTOGEOGRAPHICAL REGION: III, IV.
DESCRIPTION: {108}, {896}.
ILLUSTRATION: {167}, {896}, {1223}, {1289}, {1398}.
NOTES: a widespread and variable species; plant 2-leaved, small; pseudobulbs ovoid, yellow; raceme reclining, a long as the peduncle, with 9 – 14 distant flowers; flowers c. 4 mm, yellowish; lip very mobile.

Bulbophyllum nutans var. **variifolium** *(Schltr.) Bosser* in Adansonia, sér. 2, 5: 389 (1965). TYPE: Madagascar, Mt. Tsaratanana, *H.Perrier* 15327 (holo. P).

Bulbophyllum variifolium Schltr. in Repert. Sp. Nov. Regni Veg. Beih. 33: 206 (1924).

Bulbophyllum ambohitrense H.Perrier in Notul. Syst. (Paris) 6 (2): 72 (1937). TYPE: Madagascar, Mt. Ambre, *H.Perrier* 17716 (holo. P).

DISTRIBUTION: Madagascar: Antsiranana.
HABITAT: lichen-rich forest on moss- and lichen-rich-covered trees.
ALTITUDE: 1500 – 2000 m.
FLOWERING TIME: January.
LIFE FORM: epiphyte.
PHYTOGEOGRAPHICAL REGION: III, IV.
DESCRIPTION: {108}, {896}.
NOTES: distinct from the typical variety by the more slender inflorescence, longer peduncle, leaves which are contracted at the base into a longer petiole; the flowers are similar but a little larger, whereas the shape of the stelids is slightly different.

Bulbophyllum obscuriflorum *H.Perrier* in Notul. Syst. (Paris) 6 (2): 78 (1937). TYPE: Madagascar, Sambirano valley, *H.Perrier* 15733 (holo. P).

DISTRIBUTION: Madagascar: Antsiranana.
HABITAT: seasonally dry, deciduous forest.
ALTITUDE: c. 400 m.
FLOWERING TIME: January.
LIFE FORM: epiphyte.
PHYTOGEOGRAPHICAL REGION: II.
DESCRIPTION: {896}.
NOTES: known only from the type. Plant single-leafed, small; inflorescence 4 – 5 cm long, laxly-flowered; flowers dark red, 5 mm long.

Bulbophyllum obtusatum *(Jum. & H.Perrier) Schltr.* in Repert. Sp. Nov. Regni Veg. Beih. 33: 222 (1924), *non* Lindl. TYPE: Madagascar, Manongarivo massif, *H.Perrier* s.n. (holo. P).

Bulbophyllum obtusum Jum. & H.Perrier in Ann. Fac. Sci. Marseille 21 (2): 203 (1912). TYPE: Madagascar, Manongarivo, *H.Perrier* 1907 (holo. P).

DISTRIBUTION: Madagascar: Antsiranana, Mahajanga.
HABITAT: humid, evergreen forest on plateau; dry forests.
ALTITUDE: sea level – 1400 m.
FLOWERING TIME: April – May.

LIFE FORM: epiphyte.

PHYTOGEOGRAPHICAL REGION: III.

DESCRIPTION: {896}.

NOTES: related to *B. occultum*, bifoliate; lip narrow, very thick, the edges recurved, ciliate with many violet-red hairs in the lower half.

Bulbophyllum obtusilabium *W. Kittr.* in Bull. Mus. Leafl. Harvard. Univ. 30 (2): 100 (1984). TYPE: Madagascar, Mt. Tsaratanana, *H.Perrier* 15735 (holo. P).

Bulbophyllum rhizomatosum Schltr. in Repert. Sp. Nov. Regni Veg. Beih. 33: 240 (1924), *non B. rhizomatosum* Ames & C. Schweinf. (1920).

DISTRIBUTION: Madagascar: Toamasina.

HABITAT: humid, mossy, evergreen forest on plateau.

ALTITUDE: 800 – 1600m.

FLOWERING TIME: January.

LIFE FORM: epiphyte.

PHYTOGEOGRAPHICAL REGION: III.

DESCRIPTION: {896}.

ILLUSTRATION: {896}.

NOTES: known only from the type. Bifoliate plant, related to *B. ankaratranum*, but more compact, the spike somewhat longer, and the lip longer with short parallel lamellae.

Bulbophyllum occlusum *Ridl.* in J. Linn. Soc., Bot. 21: 464 (1885). TYPE: Madagascar, Ankafana, *Hildebrandt* 3985 (holo. BM; iso. K).

Phyllorchis occlusa (Ridl.) Kuntze, Rev. Gen. Pl. 2: 677 (1891).

DISTRIBUTION: Madagascar: Antananarivo, Antsiranana, Fianarantsoa, Toamasina. Mascarenes: Réunion.

HABITAT: humid, mossy, evergreen forest.

ALTITUDE: 600 – 1600m.

FLOWERING TIME: August – March.

LIFE FORM: epiphyte.

PHYTOGEOGRAPHICAL REGION: I, II, III, IV.

DESCRIPTION: {896}.

ILLUSTRATION: {167}, {281}, {896}, {906}.

NOTES: plant bifoliate, large; peduncle c. 20 cm long, almost completely covered by papery sheaths, covering the almost purple flowers; lip broad; column-teeth long.

Bulbophyllum occultum *Thouars,* Hist. Orchid.: t.93 (1822). TYPE: *Thouars* s.n. (holo. P).

Dendrochilum occultum (Thouars) Lindl., Gen. Sp. Orch. Pl.: 34 (1830).

Diphyes occulta (Thouars) Kuntze, Rev. Gen. Pl. 2: 675 (1891).

DISTRIBUTION: Madagascar: Antananarivo, Antsiranana, Fianarantsoa, Mahajanga, Toliara. Mascarenes: Réunion, Mauritius, Rodriguez.

HABITAT: coastal forest; humid, evergreen forest.

ALTITUDE: sea level – 1500 m.

FLOWERING TIME: September – May.

LIFE FORM: epiphyte.

PHYTOGEOGRAPHICAL REGION: I, II, III, IV.

DESCRIPTION: {896}.

ILLUSTRATION: {165}, {166}, {167}, {168}, {312}, {457}, {896}, {906}, {1398}.

NOTES: plant two-leaved, up to 15 cm high; peduncle 8 – 14 cm long, about as long as the raceme, with some lax brownish-yellow sheaths, 1.5 – 1.8 cm apart; flowers hidden, small, reddish, 4 mm long.

Bulbophyllum onivense *H.Perrier* in Notul. Syst. (Paris) 6 (2): 81 (1937). TYPE: Madagascar, Tsinjoarivo, *H.Perrier* 16923 (holo. P).

DISTRIBUTION: Madagascar: Antananarivo, Toamasina.

HABITAT: humid, evergreen forest on plateau.

ALTITUDE: 1300 – 1500 m.

FLOWERING TIME: February.

LIFE FORM: epiphyte.

PHYTOGEOGRAPHICAL REGION: III.

DESCRIPTION: {896}.

ILLUSTRATION: {896}.

NOTES: plant single-leafed, similar to *B. pleurothallopsis* but has more acute floral bracts, shorter and more acute petals, and a pale lip.

Bulbophyllum ophiuchus *Ridl.* in J. Linn. Soc., Bot. 22: 118 (1886). TYPE: Madagascar, Ankeramadinika, *Fox* s.n. (holo. BM; iso. K).

Bulbophyllum mangoroanum Schltr. in Beih. Bot. Centralbl. 34 (2): 328 (1917). TYPE: Madagascar, Mangoro, *H.Perrier* 118 (11381) (holo. B).

DISTRIBUTION: Madagascar: Toamasina.

HABITAT: humid, evergreen forest on plateau.

ALTITUDE: 900 – 1400 m.

FLOWERING TIME: November – February.

LIFE FORM: epiphyte.

PHYTOGEOGRAPHICAL REGION: III.

DESCRIPTION: {896}.

NOTES: plant bifoliate, related to *B. masoalanum*; inflorescence c. 50 cm long; flowers green, turning red on fading; lip curved-folded, subrectangular, 2 × 1 mm, with a broad central speculum surrounded by a wide, finely verrucose margin.

Bulbophyllum ophiuchus var. **baronianum** *H.Perrier ex Hermans* **var. nov.** TYPE: Madagascar, without locality, *Baron* 4468 (holo. K; iso. BM, P). See Appendix .

Bulbophyllum ophiuchus var. *baronianum* H.Perrier in Humbert ed., Fl. Mad. Orch. 1: 374 (1939), *nom. nud.*

DISTRIBUTION: Madagascar.

PHYTOGEOGRAPHICAL REGION: III.

DESCRIPTION: {896}.

NOTES: differs from the typical variety by its smaller flowers, its dorsal sepal which is obtuse and as wide as the anterior part of the almost orbicular synsepal, its suborbicular lip and the foot of the column which is not dilated in front of the stigmatic surface.

Bulbophyllum oreodorum *Schltr.* in Repert. Sp. Nov. Regni Veg. Beih. 33: 238 (1924). TYPE: Madagascar, Andringitra massif, *H.Perrier* 13702 (holo. P).

DISTRIBUTION: Madagascar: Fianarantsoa.

HABITAT: rocky outcrops amongst mosses.

ALTITUDE: 2000 – 2400m.

FLOWERING TIME: April.

LIFE FORM: lithophyte.

PHYTOGEOGRAPHICAL REGION: IV.

DESCRIPTION: {896}.

ILLUSTRATION: {896}.

NOTES: plant bifoliate, compact, up to 13 cm long; peduncle up to 9 cm long, very densely many-flowered; flowers fleshy, red and finely papillose on the outside; stelids short, obtusely sub-bilobed.

Bulbophyllum ormerodianum *Hermans* **nom. nov.**

Bulbophyllum abbreviatum Schltr. in Repert. Sp. Nov. Regni Veg. Beih. 33: 198 (1924), *non* Rchb.f. in Gard. Chron. ns. 16: 70 (1881). TYPE: Madagascar, Analamazaotra, *H.Perrier* 8054 (holo. P!).

DISTRIBUTION: Madagascar: Toamasina.

HABITAT: humid, evergreen forest at mid-elevation and on plateau.
ALTITUDE: 700 – 1000m.
FLOWERING TIME: February.
LIFE FORM: epiphyte.
PHYTOGEOGRAPHICAL REGION: I, III.
DESCRIPTION: {896}.
NOTES: plant, 2-leaved, small, similar to *B. nutans*; inflorescence as *B. pseudonutans*; flowers c. 5 mm long, yellowish; lip 3-lobed, curved, the margins folded downwards. The earlier *B. abbreviatum* Rchb.f. (1881), precludes the use of that name for the Madagascan plant described by Schlechter in 1924. I therefore propose the epithet *ormerodianum*, in honour of Paul Ormerod who has critically commented on many of the name changes proposed here.

Bulbophyllum oxycalyx *Schltr.* in Repert. Sp. Nov. Regni Veg. Beih. 33: 212 (1924). TYPE: Madagascar, Mt. Tsaratanana, *H.Perrier* 15228 (holo. P).
Bulbophyllum rubescens var. *meizobulbon* Schltr. in Repert. Sp. Nov. Regni Veg. Beih. 33: 213 (1924). TYPE: Madagascar, Mt. Tsaratanana, *H.Perrier* 15229 (holo. P).

DISTRIBUTION: Madagascar: Antananarivo, Antsiranana.
HABITAT: lichen-rich forest on moss- and lichen-covered trees.
ALTITUDE: c. 1500 m.
FLOWERING TIME: March – February, October.
LIFE FORM: epiphyte.
PHYTOGEOGRAPHICAL REGION: III, IV.
DESCRIPTION: {896}.
NOTES: plant slender, 2-leaved; inflorescence longer than the leaves; peduncle very slender, filiform, up to 11 cm long; flowers 7 – 7.5 mm long, red, striped or spotted; lip dark greenish-red, recurved, narrowly oblong-linguiform, with a small obsolete callus, and obscurely 2-ridged in the lower half.

Bulbophyllum oxycalyx var. **rubescens** *(Schltr.) Bosser* in Adansonia, sér. 2, 5: 383 (1965). TYPE: Madagascar, Mt. Tsaratanana, *H.Perrier* 11337 (holo. P).
Bulbophyllum rubescens Schltr. in Repert. Sp. Nov. Regni Veg. Beih. 33: 213 (1924).
Bulbophyllum caeruleolineatum H.Perrier in Notul. Syst. (Paris) 14 (2): 158 (1951). TYPE: Marojejy massif, *Humbert* 22572 (holo. P).
Bulbophyllum loxodiphyllum H.Perrier in Notul. Syst. (Paris) 14 (2): 159 (1951). TYPE: Madagascar, Marojejy massif, *Humbert* 22613 (holo. P).
Bulbophyllum rostriferum H.Perrier in Notul. Syst. (Paris) 14 (2): 157 (1951). TYPE: Madagascar, cult. Jardin Bot. Antananarivo (holo. P).

DISTRIBUTION: Madagascar: Antsiranana.
HABITAT: lichen-rich forest.
ALTITUDE: 2000 – 2500 m.
FLOWERING TIME: November – December.
LIFE FORM: epiphyte.
PHYTOGEOGRAPHICAL AREA: III, IV.
DESCRIPTION: {108}.
NOTES: petals are larger and more elongate than in the typical variety and do not have a red violet-red spot near the tip.

Bulbophyllum pachypus *Schltr.* in Repert. Sp. Nov. Regni Veg. Beih. 33: 208 (1924). TYPES: Madagascar, Analamazaotra forest, *H.Perrier* 11874 (lecto. P, selected here); *H.Perrier* 11854 (syn. P).

DISTRIBUTION: Madagascar: Antananarivo, Fianarantsoa, Toamasina.
HABITAT: humid, evergreen forest.
ALTITUDE: 800 – 1400 m.
FLOWERING TIME: January – April.
LIFE FORM: epiphyte.
PHYTOGEOGRAPHICAL REGION: III.
DESCRIPTION: {896}.

ILLUSTRATION: {281} (= *B. molossus*), {505}, {896}, {906}.

NOTES: plant bifoliate, up to 20 cm tall; peduncle 10 – 12 cm long, slender and with a few distant sheaths, laxly 4 – 8 flowered; flowers 11 – 13 mm long, tinted violet with a few darker stripes; lip red and yellow.

Bulbophyllum paleiferum *Schltr.* in Repert. Sp. Nov. Regni Veg. Beih. 33: 239 (1924). TYPE: Madagascar, Analamazaotra, *H.Perrier* 11858 (holo. P).

DISTRIBUTION: Madagascar: Toamasina.
HABITAT: humid, evergreen forest.
ALTITUDE: 500 – 900 m.
FLOWERING TIME: January – March.
LIFE FORM: epiphyte.
PHYTOGEOGRAPHICAL REGION: I, III.
DESCRIPTION: {896}.
ILLUSTRATION: {896}.
NOTES: two-leaved plant, related to *B. cyclanthum* and from the same area but differs in its more cylindrical, longer pseudobulbs and the larger bracts that cover the flowers completely.

Bulbophyllum pallens *(Jum. & H.Perrier) Schltr.* in Repert. Sp. Nov. Regni Veg. Beih. 33: 240 (1924). TYPE: Madagascar, Ankaizina, *H.Perrier* 1911 (holo. P).
Bulbophyllum ophiuchus var. *pallens* Jum. & H.Perrier in Ann. Fac. Sci. Marseille 21 (2): 209 (1912). TYPE: Madagascar, Ankaizina, *Jumelle & H.Perrier* s.n. (holo. P).

DISTRIBUTION: Madagascar: Mahajanga.
HABITAT: humid, evergreen forest on plateau; on on moss- and lichen-rich-covered trees by streams.
ALTITUDE: c. 1000 m.
FLOWERING TIME: August.
LIFE FORM: epiphyte.
PHYTOGEOGRAPHICAL REGION: III.
DESCRIPTION: {896}.
NOTES: known only from the type. Plant bifoliate, related to *B. ankaizinense* and *B. ophiuchus* but they differ by the hairless petals and lip.

Bulbophyllum pandurella *Schltr.* in Repert. Sp. Nov. Regni Veg. Beih. 33: 196 (1924); TYPE: Madagascar, Mt. Tsaratanana, *H.Perrier* 15727 (holo. P).

DISTRIBUTION: Madagascar: Antsiranana.
HABITAT: lichen-rich forest on moss- and lichen-covered trees.
ALTITUDE: c. 2000 m.
FLOWERING TIME: January.
LIFE FORM: epiphyte.
PHYTOGEOGRAPHICAL REGION: III, IV.
DESCRIPTION: {896}.
NOTES: plant 2-leaved, inflorescence single-flowered; lip is similar to *B. oxycalyx*, the column has long aciculate stelids.

Bulbophyllum pantoblepharon *Schltr.* in Repert. Sp. Nov. Regni Veg. Beih. 33: 180 (1924). TYPE: Madagascar, Mt. Tsaratanana, *H.Perrier* 15738 (holo. P).

DISTRIBUTION: Madagascar: Antananarivo, Antsiranana.
HABITAT: lichen-rich forest on moss- and lichen-covered trees.
ALTITUDE: c. 1600 m.
FLOWERING TIME: January.
LIFE FORM: epiphyte.
PHYTOGEOGRAPHICAL REGION: III, IV.
DESCRIPTION: {896}.
ILLUSTRATION: {281}, {459}, {896}.

NOTES: single-leafed plant, related to *B. pleurothallopsis* but with smaller pseudobulbs and leaves, a fewer flowered raceme and sepals of the same length; flowers purplish-red; lip ciliate at the margins.

Bulbophyllum pantoblepharon var. **vestitum** *H.Perrier ex Hermans* **var. nov.** TYPE: Madagascar, Ankeramadinika, *E.François* 4 in *H.Perrier* 18646 (holo. P). See Appendix .

Bulbophyllum pantoblepharon var. *vestitum* H.Perrier in Humbert, Fl. Mad. Orch. 1: 324 (1939), *nom. nud.*

DISTRIBUTION: Madagascar: Antananarivo, Fianarantsoa.
HABITAT: humid, evergreen forest on plateau.
PHYTOGEOGRAPHICAL REGION: III.
DESCRIPTION: {896}.
NOTES: differs from the typical variety by its peduncle that is completely covered by 5 – 6 thick, lax sheaths, that open at the apex and are keeled at the back, its elliptical sepals, smaller, 0.6 mm long, completely orbicular petals, the 2 small, triangular, thick bracteoles on the outside of the perianth, the acute stelids that are the same height as the erect part of the column, and the anther that lacks an apicule.

Bulbophyllum papangense *H.Perrier* in Notul. Syst. (Paris) 6 (2): 59 (1937). TYPE: Madagascar, Mt. Papanga, Beampingaratra massif, *Humbert* 6348 (holo. P).

DISTRIBUTION: Madagascar: Fianarantsoa, Toliara.
HABITAT: humid, evergreen forest on plateau; ericaceous scrub.
ALTITUDE: c. 1500 m.
FLOWERING TIME: November.
LIFE FORM: epiphyte.
PHYTOGEOGRAPHICAL REGION: IV.
DESCRIPTION: {896}.
NOTES: known only from the type. Plant up to 20 cm, single-leafed; flowers yellowish-white c. 6 mm long; lip very recurved, 2 mm long, a little indented at the apex and the margins reflexed.

Bulbophyllum pentastichum *(Pfitzer ex Kraenzl.) Schltr.* in Beih. Bot. Centralbl. 33 (2): 419 (1915). TYPE: Madagascar, without locality, cult. Hamburg Botanic Garden (holo. HBG).

Bulbophyllaria pentasticha Pfitzer ex Kraenzl. in Orchis 2: 135 (1908).

Bulbophyllum pentastichum Rolfe in Orch. Rev. 23: 181 (1915). Type: Madagascar, hort. Sir Jeremiah Colman, May 1915 (holo. Icon. Matilda Smith del. K169/346).

Bulbophyllum matitanense H.Perrier in Notul. Syst. (Paris) 6 (2): 85 (1937), **syn. nov.** Type: Madagascar, at the mouth of the Matitana (= Matitanana ?), *H.Perrier* 8004 (holo. P).

DISTRIBUTION: Madagascar: Fianarantsoa, Toliara.
HABITAT: coastal forest.
ALTITUDE: sea level – 100 m.
FLOWERING TIME: October.
PHYTOGEOGRAPHICAL REGION: I.
LIFE FORM: epiphyte.
DESCRIPTION: {896}.
ILLUSTRATION: {896}.
NOTES: Pseudobulbs 2-leaved, rachis red, thick, up to18 cm long, floral bracts in 3 rows, flowers shorter than the bracts c. 4 × 3 mm, set within the rachis, lip almost semi-circular, the margins fimbriate.

    Bulbophyllum pentastichum is a characteristic species from coastal south-eastern Madagascar. Its nomenclature is somewhat confused. The genus *Bulbophyllaria* was established by H.G. Reichenbach in 1852 (Bot. Zeit. 10: 934). Based upon Pfitzer's unpublished material, Kraenzlin (1908) clearly placed this species in Reichenbach's genus. Schlechter (1915) transferred it to the genus *Bulbophyllum*. Rolfe (1915) made the same transfer but somewhat later in the same year. A herbarium sheet annotated by Rolfe and a watercolour by Mathilda Smith showing the spike from a plant in the collection of Sir Jeremiah Colman both at K, correspond well with the Kraenzlin description. It is possible that this cultivated plant and the type grown at the Hamburg Botanic Garden originated from the same source. Perrier de la Bâthie (1937) described *B. matitanense* which is, without doubt, conspecific with *B. pentastichum.*

Bulbophyllum pentastichum *(Pfitzer ex Kraenzl.) Schltr.* subsp. **rostratum** *H.Perrier ex Hermans* **subsp. nov.** TYPE: Madagascar, Fort Dauphin district, Vinanibe, *Decary* 10865 (holo. P). See Appendix. *Bulbophyllum matitanense* subsp. *rostratum* H.Perrier in Humbert, Fl. Mad. Orch. 1: 418 (1939), *nom. nud.*

DISTRIBUTION: Madagascar: Toliara.
HABITAT: humid, lowland forest.
FLOWERING TIME: October.
LIFE FORM: epiphyte.
PHYTOGEOGRAPHICAL REGION: I.
DESCRIPTION: {896}.
NOTES: this subspecies differs from the typical one by its more markedly curved, often longer raceme, its flowers that have reddish warts scattered on the outside, the 3-veined sepals, the narrower dorsal sepal and especially the thicker anther, with a verrucose, conical rostrum at the front.

Bulbophyllum percorniculatum *H.Perrier* in Notul. Syst. (Paris) 14: 149 (1951). TYPE: Madagascar, Mt. Vohimavo, Manampanihy Valley, *Humbert* 20688 (holo. P).

DISTRIBUTION: Madagascar: Toliara.
HABITAT: humid, evergreen forest.
ALTITUDE: c. 600 m.
FLOWERING TIME: March.
LIFE FORM: epiphyte.
PHYTOGEOGRAPHICAL REGION: I.
DESCRIPTION: {901}.
NOTES: known only from the type. Small bifoliate plant; pseudobulbs disc-like; floral bracts extended at the tip; lip ligulate, obtuse, porrect, somewhat concave, oblong, 3.5 × 1.5 mm.

Bulbophyllum perpusillum *H.Wendl. & Kraenzl.* in Gard. Chron. sér. 3, 16: 592 (1894), (as "*Bolbophyllum*"). TYPE: Madagascar, without locality, cult. Herrenhausen B. G., *Braun* s.n. (holo. HBG).

DISTRIBUTION: Madagascar.
FLOWERING TIME: June (in cultivation).
PHYTOGEOGRAPHICAL REGION: I.
DESCRIPTION: {896}.
NOTES: known only from cultivated material. Plant very small, 2-leaved; flowers small, white-pink; lip with a velvety callus.

Bulbophyllum perreflexum *Bosser & P.J.Cribb* in Adansonia sér. 3, 23 (1): 129 (2001). TYPE: Madagascar, Antsiranana, Marojejy reserve, N of Mandena, *Miller & Randrianasolo* 4652 (holo. P; iso. MO, TAN).

DISTRIBUTION: Madagascar: Antsiranana.
HABITAT: pre-montane forest.
ALTITUDE: 900 – 1300 m.
FLOWERING TIME: December.
LIFE FORM: epiphyte.
PHYTOGEOGRAPHICAL REGION: IV.
DESCRIPTION: {142}.
ILLUSTRATION: {142}.
NOTES: bifoliate plant, known only from the type. Flowers brown, strongly reflexed against the axis, close to *B. coriophorum* and *B. divaricatum,* but the flowers of these species are very different, particularly in the shape of the lip.

Bulbophyllum perrieri *Schltr.* in Ann. Mus. Col. Marseille, ser.3, 1: 180, t.15 (1913). TYPE: Madagascar, Manongarivo, *H.Perrier* 131850 (holo. B†; iso. P).

DISTRIBUTION: Madagascar: Antsiranana.
HABITAT: humid, evergreen forest on plateau; moss- and lichen-rich forest.
ALTITUDE: 1500 – 1600 m.
FLOWERING TIME: January.
LIFE FORM: epiphyte.
PHYTOGEOGRAPHICAL REGION: II, III, IV.
DESCRIPTION: {896}.
ILLUSTRATION: {1205}.
NOTES: two-leaved plant, related to *B. variegatum*; inflorescence c. 20 cm long, somewhat longer than the leaves; mentum very pronounced, protruding by 4 mm.

Bulbophyllum perseverans *Hermans* **nom. nov.**
Bulbophyllum graciliscapum H.Perrier in Notul. Syst. (Paris) 6 (2): 107 (1937), *non* Schltr. in K.Schum. & Lauterb., Nachtr. Fl. Deutsch. Sudsee 203, (1905), *nec* Ames & Rolfe, Orchidaceae 5: 175 (1915), *nec B. graciliscapum* Summerh. in Kew Bull., 579 (1954). TYPE: Madagascar, Mt. Tsaratanana, *H.Perrier* 16513 (holo. P).

DISTRIBUTION: Madagascar: Antsiranana, Toamasina.
HABITAT: lichen-rich forest on moss- and lichen-covered trees.
ALTITUDE: c. 2000 m.
FLOWERING TIME: February.
LIFE FORM: epiphyte.
PHYTOGEOGRAPHICAL REGION: III, IV.
DESCRIPTION: {896}.
ILLUSTRATION: {896}.
NOTES: plant 2-leaved, 20 cm high; peduncle up to 18 cm, many-flowered; flowers 9 mm long; lip flat, very obtuse, oval, 3 × 2 mm, the base almost auriculate, with two distinct spots below the middle. Owing to the earlier *B. graciliscapum* Schltr., a species from New Guinea, the name cannot be used for the Madagascan plant described by H.Perrier. The same epithet was also used by Ames and Rolfe (1915) for a plant from the Philippines and later still by Summerhayes (1954) for a plant from Gabon. Summerhayes changed the epithet for the last species to *flexiliscapum* (in Kew Bull. 11, 2: 232, 1956), when he noted that Perrier's 1937 name had precedence but without realising that Schlechter had used the same name many years previously. I therefore propose the epithet *perseverans*, referring to the way this epithet *graciliscapum* recurs in the literature, for the Madagascan species.

Bulbophyllum pervillei *Rolfe* in J. Linn. Soc., Bot. 29: 51 (1891). TYPES: Madagascar, nr. Fort Dauphin, *Scott Elliot* s.n. (lecto. K; isolecto. P, both selected here); Nossi Be, *Perville* 136 (syn. K, L) & *Humblot* 380 (syn. K, P).

DISTRIBUTION: Madagascar: Antsiranana, Toamasina, Toliara.
HABITAT: coastal forest.
ALTITUDE: sea level – 1000 m.
FLOWERING TIME: October – July.
LIFE FORM: epiphyte.
PHYTOGEOGRAPHICAL REGION: I, II.
DESCRIPTION: {896}.
ILLUSTRATION: {246}, {505}.
NOTES: pseudobulbs flattened, suborbicular, 2-leaved; peduncle subfiliform, 10 – 15 cm long; flowers small; lip long and narrow, 4 × 0.6 mm, papillose and with 2 ridges on the upper surface.

Bulbophyllum peyrotii *Bosser* in Adansonia, sér. 2, 5: 405 (1965) & in Adansonia 11 (1): 32 (1989). TYPE: Madagascar, nr. Marovoay, Moramanga, *E.François* 403 (holo. P).
Bulbophyllum fimbriatum H.Perrier in Notul. Syst. (Paris) 9 (4): 145 (1941), *non* (Lindl.) Rchb.f. (1861)

Bulbophyllum flickingerianum Hawkes in Phytologia 13 (5): 309 (1966), *nom. superfl.*

Bulbophyllum mayae Hawkes in Phytologia 13 (5): 309 (1966), *nom. superfl.*

DISTRIBUTION: Madagascar: Antananarivo, Toamasina.

HABITAT: humid, evergreen mossy forest, on tree trunks.

ALTITUDE: 900 – 1000 m.

FLOWERING TIME: January – August.

LIFE FORM: epiphyte or lithophyte.

PHYTOGEOGRAPHICAL AREA: III.

DESCRIPTION: {108}.

ILLUSTRATION: {108}.

NOTES: distinguished by its small flowers, which are greenish-yellow scattered with warts and dark red dots, and the deeply fimbriate-dentate margins of the big blade formed by the joined lateral sepals.

Bulbophyllum platypodum *H.Perrier* in Notul. Syst. (Paris) 6 (2): 103 (1937). TYPE: Madagascar, Mt. Analampanga, Mangoro, *H.Perrier* 18296 (holo. P).

DISTRIBUTION: Madagascar: Antananarivo, Fianarantsoa, Toamasina.

HABITAT: humid, evergreen forest.

ALTITUDE: 400 – 1200 m.

FLOWERING TIME: October.

LIFE FORM: epiphyte.

PHYTOGEOGRAPHICAL REGION: I, III.

DESCRIPTION: {896}.

ILLUSTRATION: {105}, {896}.

NOTES: plant robust, up to 32 cm high; pseudobulbs 2-leaved; peduncle flattened, laxly many-flowered; lip curved-geniculate, oblong, obtuse, bicarinate above the base and with a slightly concave speculum.

Bulbophyllum pleiopterum *Schltr.* in Orchis: 114, t.25 (1912). TYPE: Madagascar, without locality, cult. Herrenhausen Botanic Garden, *Braun* s.n. (holo. B†); East, Masoala, east coast, *H.Perrier* 11353 (146) (neo. P), selected here.

DISTRIBUTION: Madagascar: Antsiranana, Toamasina.

HABITAT: humid, lowland, evergreen forest.

ALTITUDE: c. 300 m.

FLOWERING TIME: October.

LIFE FORM: epiphyte.

PHYTOGEOGRAPHICAL REGION: I.

DESCRIPTION: {896}.

ILLUSTRATION: {896}, {1204}.

NOTES: two-leaved plant, rachis with expanded 3 mm broad wings below each flower. The type of this species was destroyed in Berlin, the neotype chosen is the one illustrated by Perrier de la Bâthie in {896}.

Bulbophyllum pleurothallopsis *Schltr.* in Repert. Sp. Nov. Regni Veg. Beih. 33: 180 (1924). TYPE: Madagascar, Mt. Tsaratanana, *H.Perrier* 15728 (lecto. P selected here).

DISTRIBUTION: Madagascar: Antsiranana.

HABITAT: lichen-rich forest.

ALTITUDE: c. 1400 m.

FLOWERING TIME: January – February.

LIFE FORM: epiphyte on moss- and lichen-covered trees.

PHYTOGEOGRAPHICAL REGION: III.

DESCRIPTION: {896}.

NOTES: plant single-leafed, small; flowers 1 to 3, small, greenish or reddish; petals and lip ciliate.

Bulbophyllum protectum *H.Perrier* in Notul. Syst. (Paris) 6 (2): 105 (1937). TYPE: Madagascar, around Onive & Mangoro, *H.Perrier* 17139 (holo. P).

DISTRIBUTION: Madagascar: Antananarivo, Toamasina.
HABITAT: humid, lowland, evergreen forest.
ALTITUDE: c. 700 m.
FLOWERING TIME: February, October.
LIFE FORM: epiphyte.
PHYTOGEOGRAPHICAL REGION: I, III.
DESCRIPTION: {896}.
ILLUSTRATION: {896}.
NOTES: plant medium-sized, 2-leaved; flowers c. 1 cm long; raceme thickened, white or pink; petals extended by a filiform acumen; lip short, truncate at the tip, triangular.

Bulbophyllum ptiloglossum *H.Wendl. & Kraenzl.* in Gard. Chron. ser. 3, 21: 330 (1897) (as "*Bolbophyllum*"). TYPE: Madagascar, without locality, cult. Herrenhausen Botanic Garden, *Braun* s.n. (type not located in HBG, B?).

DISTRIBUTION: Madagascar.
LIFE FORM: terrestrial.
DESCRIPTION: {896}.
NOTES: only the description of the cultivated plant has been found. Said to be a small single-leafed plant with a slender inflorescence; flowers green, spotted with red; lip 3-lobed, fimbriate and mobile.

Bulbophyllum quadrialatum *H.Perrier* in Notul. Syst. (Paris) 8: 39 (1939). TYPE: Madagascar, Tsaratanana massif, *Humbert* 18165 (holo. P).

DISTRIBUTION: Madagascar: Antsiranana.
HABITAT: mossy, montane forest on moss- and lichen-covered trees.
ALTITUDE: c. 1800 m.
FLOWERING TIME: November.
LIFE FORM: epiphyte.
PHYTOGEOGRAPHICAL REGION: II.
DESCRIPTION: {896}.
NOTES: known only from the type. Plant 2-leaved, related to *B. paleiferum*, *B. leptochlamys* and *B. cyclanthum*, but distinguished by its yellow, 4-angular, winged pseudobulbs, the less dense raceme, and the flowers not completely covering the rachis.

Bulbophyllum quadrifarium *Rolfe* in Orch. Rev. 11 (126): 190 (1903). TYPE: Madagascar, without locality, cult. Glasnevin Botanic Garden, *Peeters* s.n. (holo. K).

DISTRIBUTION: Madagascar.
DESCRIPTION: {1118}.
NOTES: a poorly known species described from a cultivated plant. This may be an older name for *Bulbophyllum pentastichum*.

Bulbophyllum ranomafanae *Bosser & P.J.Cribb* in Adansonia sér. 3, 23 (1): 130 (2001). TYPE: Madagascar, Fianarantsoa, Ranomafana reserve, *Turk, Randriamanantenina & Solo* 353 (holo. P; iso. MO, TAN).

DISTRIBUTION: Madagascar: Fianarantsoa.
HABITAT: moist, montane forest.
ALTITUDE: 950 – 1150 m.
FLOWERING TIME: March.
LIFE FORM: epiphyte.
PHYTOGEOGRAPHICAL REGION: III.
DESCRIPTION: {142}.
ILLUSTRATION: {142}.

NOTES: close to *B. acutispicatum*, but the pseudobulb smaller and conical, the leaves (2) shorter and wider and especially the flower with a fleshy lip, 3 × 1.3 mm, its margins finely denticulate, upper surface with 2 depressions at the base of the basal lobes, and 2 short fine central crests.

Bulbophyllum rauhii *Toill.-Gen. & Bosser* in Nat. Malg. 12: 18 (1960). TYPE: Madagascar, Mandraka Forest, *Rauh* s.n. (holo. P).

DISTRIBUTION: Madagascar: Antananarivo.
HABITAT: humid, evergreen forest on plateau.
ALTITUDE: 1200 – 1300m.
FLOWERING TIME: February.
LIFE FORM: epiphyte.
PHYTOGEOGRAPHICAL REGION: III.
DESCRIPTION: {1406}.
ILLUSTRATION: {1406}.
NOTES: known only from the type. Bifoliate plant, close to *B. oxycalyx* but differs by the more slender, bare, rhizome, the linear narrow leaves, the smaller flowers and especially the size and shape of the lip.

Bulbophyllum rauhii var. **andranobeense** *Bosser* in Adansonia, sér. 2, 11 (2): 334 (1971). TYPE: Madagascar, Andranobe forest, SW of Lake Alaotra, *Bosser* 19848 (holo. P).

DISTRIBUTION: Madagascar: Toamasina.
HABITAT: humid evergreen forest.
FLOWERING TIME: March.
LIFE FORM: epiphyte.
PHYTOGEOGRAPHICAL REGION: III.
DESCRIPTION: {119}.
ILLUSTRATION: {119}.
NOTES: leaves (2) shorter and wider than in the typical variety; pseudobulbs smaller; flowers smaller; the dorsal sepal has 5 veins instead of 3.

Bulbophyllum reflexiflorum *H.Perrier* in Notul. Syst. (Paris) 6 (2): 93 (1937). TYPE: Madagascar, Mt. Tsaratanana, *H.Perrier* 16480 (holo. P).
Bulbophyllum inauditum Schltr. in Repert. Sp. Nov. Regni Veg. Beih. 33: 250 (1925), *nom. illeg. non B. inauditum* Schltr. in Repert. Sp. Nov. Regni Veg. Beih. 1: 815 (1913).
Bulbophyllum bosseri K.Lemcke in Die Orchidee 50, 6: 663 (2000), *nom. superfl.* TYPE: Madagascar, Mt. Tsiafajavona, Ankaratra, *H.Perrier* 13992 (lecto. P).

DISTRIBUTION: Madagascar: Antsiranana, Antananarivo.
HABITAT: montane, ericaceous scrub; lichen-rich forest on moss- and lichen-covered trees.
ALTITUDE: 2000 – 2400 m.
FLOWERING TIME: March – April, September.
LIFE FORM: epiphyte.
PHYTOGEOGRAPHICAL REGION: III, IV.
DESCRIPTION: {896}.
ILLUSTRATION: {139}, {896}.
NOTES: plant two-leaved, up to 20 cm high; raceme 10 – 16 cm long, with 15 – 20 flowers; flowers violet-red; lip fleshy, obconical, 6.5 – 7 × 2.5 mm, with rows of hairs and ligulate appendages, the upper surface has 2 short upright rounded keels at the back, with a slightly raised central keel, lateral sides rounded, papillose.

Bulbophyllum reflexiflorum subsp. **pogonochilum** *Summerh.* in Adansonia 22 (2): 174 (2000). TYPE: Madagascar, Mt. Papanga (Itomampy basin), *Humbert* 6929 (holo. P).
Bulbophyllum pogonochilum Summerh. in Kew Bull. 6: 471 (1951).
Bulbophyllum comosum H.Perrier in Notul. Syst. (Paris) 6 (2): 94 (1937), *non* Collett & Hemsley (1890).

DISTRIBUTION: Madagascar: Fianarantsoa.

HABITAT: humid evergreen forest on plateau, humid forest.

ALTITUDE: 1200 – 1700 m.

FLOWERING TIME: November – December.

LIFE FORM: epiphyte.

PHYTOGEOGRAPHICAL REGION: III.

DESCRIPTION: {896}.

NOTES: similar in general shape to the typical subspecies but lip more elongate, and the petals longer and of a different shape. Plant up to 20 cm tall; raceme laxly 5 – 8-flowered, rachis a little verrucose; flowers 22 mm long, dark red, verrucose on the exterior; lip carrying long, mobile, linear appendages.

Bulbophyllum rhodostachys *Schltr.* in Beih. Bot. Centralbl. 34 (2): 326 (1917). TYPE: Madagascar, Masoala, *H.Perrier* 147 (11352) (holo. P).

DISTRIBUTION: Madagascar: Antsiranana, Toamasina.

HABITAT: humid, lowland, evergreen forest, along the edge of streams.

ALTITUDE: c. 300 m.

FLOWERING TIME: November.

LIFE FORM: epiphyte.

PHYTOGEOGRAPHICAL REGION: I.

DESCRIPTION: {896}.

NOTES: bifoliate plant, close to *B. rubrilabium*, but differs in the thickening of the rachis and the larger flowers.

Bulbophyllum rienanense *H.Perrier* in Notul. Syst. (Paris) 6 (2): 56 (1937). TYPE: Madagascar, Rienana Valley, *Humbert* 3500 (holo. P).

DISTRIBUTION: Madagascar: Fianarantsoa.

HABITAT: humid, evergreen forest on plateau.

ALTITUDE: c. 1200 m.

FLOWERING TIME: November.

LIFE FORM: epiphyte.

PHYTOGEOGRAPHICAL REGION: III.

DESCRIPTION: {896}.

NOTES: known only from the type. Pseudobulbs very close together; leaf (1) up to 22 mm long; inflorescence slender, 7 – 10 cm long; flowers 13 mm long; stelids very short and subacute.

Bulbophyllum rubrigemmum *Hermans* **nom. nov.**

Bulbophyllum simulacrum Schltr. in Repert. Sp. Nov. Regni Veg. Beih. 33: 243 (1925), *non* Ames in Orchid. 5: 189 (1915). TYPE: Madagascar, *H.Perrier* 14297*bis* (holo. B†; iso. P).

DISTRIBUTION: Madagascar: Toamasina.

HABITAT: humid, evergreen forest.

ALTITUDE: c. 700 m.

FLOWERING TIME: January.

LIFE FORM: epiphyte.

PHYTOGEOGRAPHICAL REGION: I.

DESCRIPTION: {896}.

NOTES: known only from the type. Similar to *B. lucidum* but with smaller and narrower pseudobulbs, with two much smaller leaves, smaller flowers, more pointed petals, a different lip structure and stelids shorter. Owing to the earlier *B. simulacrum* Ames, based on a Philippine collection, this name cannot be used for the Madagascan plant described by Schlechter in 1925. I therefore propose for this species the epithet *rubrigemmum*, referring to the red raceme.

Bulbophyllum rubiginosum *Schltr.* in Repert. Sp. Nov. Regni Veg. Beih. 33: 242 (1925). TYPE: Madagascar, Analamazaotra, *H.Perrier* 11853 (holo. P).

DISTRIBUTION: Madagascar: Toamasina.
HABITAT: humid, evergreen forest.
ALTITUDE: 800 – 900 m.
FLOWERING TIME: January.
LIFE FORM: epiphyte.
PHYTOGEOGRAPHICAL REGION: I, III.
DESCRIPTION: {896}.
ILLUSTRATION: {896}.
NOTES: related to *B. brachyphyton* but with a less leathery leaf, narrower petals and a differently structured lip which is verrucose.

Bulbophyllum rubrolabium *Schltr.* in Beih. Bot. Centralbl. 34 (2): 326 (1917). TYPE: Madagascar, Tsarakanana, *H.Perrier* 135 (11364) (holo. P).

DISTRIBUTION: Madagascar: Antsiranana, Toliara.
HABITAT: humid, evergreen forest.
ALTITUDE: 1100 – 1500 m.
FLOWERING TIME: November – December.
LIFE FORM: epiphyte.
PHYTOGEOGRAPHICAL REGION: III.
DESCRIPTION: {896}.
NOTES: plant bifoliate, up to 15 cm high; inflorescence short; rachis thickened; flowers c. 7 mm; lip red, oval, verrucose.

Bulbophyllum rubrum *Jum. & H.Perrier* in Ann. Fac. Sci. Marseille 21 (2): 215 (1912). TYPE: Madagascar, Manongarivo, Ambongo, *H.Perrier* 956 (holo. P).
Bulbophyllum ambongense Schltr. in Repert. Sp. Nov. Regni Veg. Beih. 33: 217 (1924). TYPES: Madagascar, Mahavovy, Ambongo, *H.Perrier* 13866 & Mananjeba, *H.Perrier* 11329 (both syn. P).

DISTRIBUTION: Madagascar: Antsiranana, Mahajanga.
HABITAT: rain forest; coastal forest; dry forest.
ALTITUDE: sea level – 300 m.
FLOWERING TIME: July – November.
LIFE FORM: epiphyte.
PHYTOGEOGRAPHICAL REGION: V.
DESCRIPTION: {108}, {896}.
ILLUSTRATION: {896}, {1223}.
NOTES: plant bifoliate, up to 20 cm tall; inflorescence c. 27 cm long, with 15 – 30 flowers; flowers yellowish with a dark red lip carrying 2 rows of hairs.

Bulbophyllum ruginosum *H.Perrier* in Notul. Syst. (Paris) 6 (2): 65 (1937). TYPE: Madagascar, Mt. Ambre, *H.Perrier* 18832 (holo. P).

DISTRIBUTION: Madagascar: Antsiranana.
HABITAT: humid, evergreen forest on plateau.
ALTITUDE: c. 1000 m.
FLOWERING TIME: November.
LIFE FORM: epiphyte.
PHYTOGEOGRAPHICAL REGION: III.
DESCRIPTION: {896}.
ILLUSTRATION: {896}.
NOTES: pseudobulbs confluent and flattened, 2-leaved; inflorescence long, with 15 – 30 flowers; flowers 8 – 9 mm, warty – scaly on the exterior.

Bulbophyllum rutenbergianum *Schltr.* in Repert. Sp. Nov. Regni Veg. Beih. 33: 204 (1924). TYPES: Madagascar, Mt. Tsaratanana, *H.Perrier* 8026 (lecto. P, selected here); Ankaratra, *Rutenberg* s.n. (syn. P).

Bulbophyllum peniculus Schltr. in Repert. Sp. Nov. Regni Veg. Beih. 33: 204 (1924). TYPE: Madagascar, Mt. Tsaratanana, *H.Perrier* 15726 (holo. P).

Bulbophyllum spathulifolium H.Perrier in Notul. Syst. (Paris) 14 (2): 158 (1951). TYPE: Madagascar, Marojejy massif, *Humbert* 22641 (holo. P).

Bulbophyllum coursianum H.Perrier in Mém. Inst. Sci. Mad. sér. B, 6: 264 (1955), *nom. nud.*

DISTRIBUTION: Madagascar: Antananarivo, Antsiranana, Fianarantsoa.

HABITAT: lichen-rich forest on moss- and lichen-covered trees.

ALTITUDE: 1400 – 2500m.

FLOWERING TIME: October – February.

LIFE FORM: epiphyte.

PHYTOGEOGRAPHICAL REGION: III, IV.

DESCRIPTION: {108}, {888}, {896}.

ILLUSTRATION: {896}.

NOTES: a variable species close to *B. nutans*. Plant bifoliate, flowers numerous, small, 4 – 6 mm long, yellowish white; lip curved, small, c. 2 mm long, emarginate at the apex.

Bulbophyllum sambiranense *Jum. & H.Perrier* in Ann. Fac. Sci. Marseille 21 (2): 214 (1912). TYPE: Madagascar, Manongarivo massif, *H.Perrier* 1916 (holo. P).

Bulbophyllum sambiranense var. *typicum* H.Perrier in Notul. Syst. (Paris) 6 (2): 86 (1937), *nom. inval.*

Bulbophyllum sambiranense var. *ankeranense* H.Perrier in Notul. Syst. (Paris) 6 (2): 86 (1937), *nom. nud.*

DISTRIBUTION: Madagascar: Antananarivo, Antsiranana, Mahajanga.

HABITAT: seasonally dry, deciduous forest and woodland; humid, evergreen forest on plateau.

ALTITUDE: 100 – 1900 m.

FLOWERING TIME: September – April.

LIFE FORM: epiphyte.

PHYTOGEOGRAPHICAL REGION: II, III, V.

DESCRIPTION: {896}.

ILLUSTRATION: {1223}.

NOTES: plant scrambling, up to 15 cm tall; inflorescence 15 – 20 cm long; floral bracts and flowers rugose, red; flowers small, 7.5 – 8 mm long; lip curved, ligulate, obtuse, 2 mm long, ciliate with thick hairs. The specimens labelled by Perrier de la Bâthie as var. *ankeranense* differ from the typical variety only in having pseudobulbs that are broader than long, its smaller and broader leaves and its shorter sepals. *Bulbophyllum sambiranense* is a widespread species in north and north-western Madagascar. All specimens examined show some variation from the type.

Bulbophyllum sambiranense var. **latibracteatum** *H.Perrier ex Hermans* **var. nov.** TYPE: Madagascar, Centre, Tampoketsa between the Ikopa and the Betsiboka, forest of the eastern slopes, *H.Perrier* 16724 (holo. P!). See Appendix .

Bulbophyllum sambiranense var. *latibracteatum* H.Perrier in Notul. Syst. (Paris) 6 (2): 86 (1937), *nom. nud.*

DISTRIBUTION: Madagascar: Antananarivo, Antsiranana.

HABITAT: humid, highland forest.

ALTITUDE: 600 – 1500 m.

FLOWERING TIME: April, August.

LIFE FORM: epiphyte.

PHYTOGEOGRAPHICAL REGION: II, III.

DESCRIPTION: {896}.

NOTES: distinguished by its orbicular, approximate pseudobulbs, smaller, wider leaves, broadly ovate floral bracts, and flowers that are completely hidden by the bracts. It is represented by two H.Perrier collections that show consistent characteristics.

Bulbophyllum sandrangatense *Bosser* in Adansonia, sér. 2, 5: 394 (1965). TYPE: Madagascar, S of Moramanga, *Bosser* 16769 (holo. P).

DISTRIBUTION: Madagascar: Toamasina.

HABITAT: humid, mossy, evergreen forest.

ALTITUDE: 900 – 1100 m.

FLOWERING TIME: February.
LIFE FORM: epiphyte.
PHYTOGEOGRAPHICAL REGION: III.
DESCRIPTION: {108}.
ILLUSTRATION: {108}, {505}.
NOTES: plant 2-leaved, distinguished by the fibres enveloping the pseudobulbs, the triangular, subcordate lip with a papillose-ciliate surface, and the column with very short stelids reflexed at the back.

Bulbophyllum sanguineum *H.Perrier* in Notul. Syst. (Paris) 6 (2): 110 (1937). TYPE: Madagascar, Andasibe forest, Onive, *H.Perrier* 17105 (holo. P).

DISTRIBUTION: Madagascar: Toamasina.
HABITAT: humid, evergreen forest on plateau.
ALTITUDE: c. 1000 m.
FLOWERING TIME: February.
LIFE FORM: epiphyte.
PHYTOGEOGRAPHICAL REGION: III.
DESCRIPTION: {896}.
NOTES: known only from the type. Close to *B. rhodostachys*, but differs by the inflorescence which is much longer than the leaves (2), the thinner rachis, less than 2 times the thickness of the apex of the peduncle, the smaller flowers, acute dorsal sepal, and entire stelids.

Bulbophyllum sarcorhachis *Schltr.* in Repert. Sp. Nov. Regni Veg. Beih. 15: 331 (1918). TYPE: Madagascar, without locality, *Laggiara* s.n. (holo. B†).
Bulbophyllum sarcorhachis var. *typicum* H.Perrier in Notul. Syst. (Paris) 6 (2): 109 (1937), *nom. inval.*

DISTRIBUTION: Madagascar: Fianarantsoa, Toamasina.
HABITAT: humid, lowland, evergreen forest.
FLOWERING TIME: September – December.
LIFE FORM: epiphyte.
PHYTOGEOGRAPHICAL REGION: I.
DESCRIPTION: {896}.
NOTES: the type was destroyed. Two leaved plant, said to be close to *B. rubrilabium* and *B. rhodostachyum*; further research is needed.

Bulbophyllum sarcorhachis var. **befaonense** *(Schltr.)* *H.Perrier* in Notul. Syst. (Paris) 6 (2): 110 (1937). TYPE: Madagascar, near Befaona (=Beforona), *H.Perrier* 14272 (holo. P).
Bulbophyllum befaonense Schltr. in Repert. Sp. Nov. Regni Veg. Beih. 33: 230 (1924).

DISTRIBUTION: Madagascar: Toamasina.
HABITAT: humid, lowland, evergreen forest.
ALTITUDE: c. 700 m.
FLOWERING TIME: September.
LIFE FORM: epiphyte.
PHYTOGEOGRAPHICAL REGION: I.
DESCRIPTION: {896}.
ILLUSTRATION: {1223}.
NOTES: differs from the typical form by the narrower and longer pseudobulb, longer leaves two, 9 – 14 cm × 8 – 12 mm long, the inflorescence almost twice the length, 16 – 19 cm, raceme greenish yellow, the sepals a little wider and the petals more acute.

Bulbophyllum sarcorhachis var. **flavomarginatum** *H.Perrier ex Hermans* **var. nov.** TYPE: Madagascar, Mt. Tsaratanana, *H.Perrier* 16511 (P). See Appendix .
Bulbophyllum sarcorhachis var. *flavomarginatum* H.Perrier in Notul. Syst. (Paris) 6 (2): 110 (1937), *nom. nud.*

DISTRIBUTION: Madagascar: Antsiranana.
HABITAT: lichen-rich forest on moss- and lichen-covered trees.

ALTITUDE: 1100 – 2600 m.

FLOWERING TIME: April – July.

LIFE FORM: epiphyte.

PHYTOGEOGRAPHICAL REGION: II, IV.

DESCRIPTION: {896}.

ILLUSTRATION: {1223}.

NOTES: the habit and inflorescence of this variety are similar to those of var. *befoana* (Schltr.) H.Perrier, but its pseudobulbs are wider and oval-oblong and its leaves are wider; raceme greyish-green; dorsal sepal slightly obovate, narrowed below the middle; lateral sepals narrower, connate, with light yellow margins, contrasting with the brown colouration of its sepals.

Bulbophyllum sciaphile *Bosser* in Adansonia, sér. 2, 5: 391 (1965). TYPE: Madagascar, S of Lake Mantasoa, *Bosser* 16166 (holo. P).

DISTRIBUTION: Madagascar: Antananarivo.

HABITAT: humid, mossy, evergreen forest on plateau.

ALTITUDE: 1300 – 1400 m.

FLOWERING TIME: August.

LIFE FORM: epiphyte.

PHYTOGEOGRAPHICAL REGION: III.

DESCRIPTION: {108}.

ILLUSTRATION: {108}.

NOTES: two-leaved plant, related to *B. jumelleanum* but distinguished by the few-flowered inflorescence and the floral morphology, especially the lip and the column.

Bulbophyllum septatum *Schltr.* in Repert. Sp. Nov. Regni Veg. Beih. 33: 205 (1924). TYPE: Madagascar, Mt. Tsaratanana, *H.Perrier* 15200 (holo. P).

Bulbophyllum serratum H.Perrier in Notul. Syst. (Paris) 6 (2): 104 (1937), TYPE: Madagascar, Mt. Tsaratanana, *H.Perrier* 15308 (holo. P).

Bulbophyllum ambreae H.Perrier in Notul. Syst. (Paris) 7: 142 (1938), TYPE: Madagascar, Mt. Ambre, *H.Perrier* 18873 (holo. P).

DISTRIBUTION: Madagascar: Antsiranana.

HABITAT: lichen-rich and mossy forests on moss- and lichen-covered trees.

ALTITUDE: 1500 – 2000 m.

FLOWERING TIME: November January.

LIFE FORM: epiphyte.

PHYTOGEOGRAPHICAL REGION: II, III, IV.

DESCRIPTION: {896}.

ILLUSTRATION: {896}.

NOTES: leaves two, oblong-ligulate, up to 7 cm, shortly contracted at the base; flowers yellowish, c. 1 cm long, typically with pointed sepals, and a tongue-shaped lip with a distinct mentum.

Bulbophyllum sphaerobulbum *H.Perrier* in Notul. Syst. (Paris) 6 (2): 57 (1937). TYPE: Madagascar, Mt. Analampanga, Mangoro, *H.Perrier* 18295 (holo. P).

DISTRIBUTION: Madagascar: Toamasina.

HABITAT: humid, lowland, evergreen forest.

ALTITUDE: c. 400 m.

FLOWERING TIME: October.

LIFE FORM: epiphyte.

PHYTOGEOGRAPHICAL REGION: I.

DESCRIPTION: {896}.

ILLUSTRATION: {896}.

NOTES: known only from the type. Pseudobulbs globose; single leaf up to 5 cm long; flowers 6 – 8 mm long; lip very small, 1 mm long, with the edges a little expanded in the lower half.

Bulbophyllum subapproximatum *H.Perrier* in Notul. Syst. (Paris) 6 (2): 60 (1937). TYPE: Madagascar, Manongarivo massif, *H.Perrier* 1915 (holo. P).

DISTRIBUTION: Madagascar: Antsiranana.
HABITAT: humid, evergreen forest on plateau.
ALTITUDE: c. 1000 m.
FLOWERING TIME: May.
LIFE FORM: epiphyte.
PHYTOGEOGRAPHICAL REGION: II.
DESCRIPTION: {896}.
NOTES: known only from the type. Plant small, with a single leaf; inflorescence 12 – 15 cm long; flowers 6 – 8 mm long; lip subrectangular; stelids short and acute.

Bulbophyllum subclavatum *Schltr.* in Repert. Sp. Nov. Regni Veg. Beih. 33: 244 (1925). TYPE: Madagascar, Manankazo, NE of Ankazobe, *H.Perrier* 11316 (holo. P).

DISTRIBUTION: Madagascar: Antananarivo, Fianarantsoa.
HABITAT: humid, evergreen forest on plateau.
ALTITUDE: 1200 – 1700 m.
FLOWERING TIME: September – December.
LIFE FORM: epiphyte.
PHYTOGEOGRAPHICAL REGION: III {896}.
DESCRIPTION: {896}.
NOTES: leaves two, up to 11 cm long, leathery; inflorescence much longer than the leaves, densely many-flowered; flowers c. 7 mm long; lip curved at the base, suborbicular, 2 × 2.3 mm and subapiculate.

Bulbophyllum subcrenulatum *Schltr.* in Repert. Sp. Nov. Regni Veg. Beih. 33: 245 (1925). TYPE: Madagascar, right bank of the Mangoro river, *H.Perrier* 11382 (holo. P).

DISTRIBUTION: Madagascar: Antsiranana, Toamasina.
HABITAT: humid, evergreen forest.
ALTITUDE: 800 – 900 m.
FLOWERING TIME: August.
LIFE FORM: epiphyte.
PHYTOGEOGRAPHICAL REGION: III.
DESCRIPTION: {896}.
NOTES: leaves two, ligulate, 13 – 15 × 1.2 – 1.5 cm; flowers greenish, becoming red with age, c. 8 mm long; petals papillose; lip with the margins expanded, broadly oval or slightly obovate, 2.5 × 1.8 mm.

Bulbophyllum subsecundum *Schltr.* in Beih. Bot. Centralbl. 34 (2): 329 (1917). TYPE: Madagascar, source of the Androranga, *H.Perrier* 134 (11365) (holo. B†; iso. P).

DISTRIBUTION: Madagascar: Antsiranana, Toamasina.
HABITAT: humid, evergreen forest.
ALTITUDE: 800 – 1400 m.
FLOWERING TIME: November – January.
LIFE FORM: epiphyte.
PHYTOGEOGRAPHICAL REGION: III.
DESCRIPTION: {896}.
NOTES: single leaf oblong-ligulate, 4 – 8 × 0.8 – 1.5 cm, longly narrowed towards the base; flowers yellowish-white, 4 – 5 mm long; lip obtuse; stelids narrow and acute.

Bulbophyllum subsessile *Schltr.* in Repert. Sp. Nov. Regni Veg. Beih. 33: 186 (1924). TYPE: Madagascar, eastern forests, *H.Perrier* 14046 (holo. P).

DISTRIBUTION: Madagascar: Toamasina.
HABITAT: humid, lowland, evergreen forest.

ALTITUDE: 500 – 1000 m.

FLOWERING TIME: September.

LIFE FORM: epiphyte.

PHYTOGEOGRAPHICAL REGION: I.

DESCRIPTION: {896}.

NOTES: known only from the type. Plant with a single leaf, c. 10 cm tall; leaf oblong-ligulate; flowers small, 5 mm long; lip wide and rounded.

Bulbophyllum sulfureum *Schltr.* in Repert. Sp. Nov. Regni Veg. Beih. 33: 225 (1924). TYPE: Madagascar, Analamazaotra, *H.Perrier* 8053 (holo. P).

Bulbophyllum brevipetalum subsp. *majus* H.Perrier in Notul. Syst. (Paris) 6 (2): 89 (1937), *nom. nud.* Based upon: Madagascar, Ankeramadanika, *H.Perrier* 19011 (P).

DISTRIBUTION: Madagascar: Antananarivo, Toamasina.

HABITAT: humid, evergreen forest.

ALTITUDE: 800 – 1700 m.

FLOWERING TIME: December – February.

LIFE FORM: epiphyte.

PHYTOGEOGRAPHICAL REGION: III.

DESCRIPTION: {896}.

ILLUSTRATION: {896}.

NOTES: related to *B. occlusum*; pseudobulbs ovoid carrying 2 leaves; bracts yellow, c. 3 cm long; flowers small; petals and the whole surface of the lip hairy; stelids large and blunt.

Bulbophyllum tampoketsense *H.Perrier* in Notul. Syst. (Paris) 6 (2): 107 (1937). TYPE: Madagascar, Manerinerina, Tampoketsa d'Ankazobe, *H.Perrier* 16747 (holo. P).

DISTRIBUTION: Madagascar: Antananarivo.

HABITAT: humid, evergreen forest on plateau.

ALTITUDE: c. 1600 m.

FLOWERING TIME: December.

LIFE FORM: epiphyte.

PHYTOGEOGRAPHICAL REGION: III.

DESCRIPTION: {896}.

NOTES: known only from the type. Plant small, 2-leaved; flowers red, 5 mm long, papillose on the exterior; lip oval-tongue-shaped, 3 × 2 mm, with 2 coloured spots below the middle.

Bulbophyllum teretibulbum *H.Perrier* in Notul. Syst. (Paris) 6 (2): 108 (1937). TYPE: Madagascar, Analamazaotra forest, *Viguier & Humbert* 920 (holo. P).

DISTRIBUTION: Madagascar: Toamasina.

HABITAT: humid, evergreen forest on plateau.

ALTITUDE: 900 – 1080 m.

FLOWERING TIME: October.

LIFE FORM: epiphyte.

PHYTOGEOGRAPHICAL REGION: III.

DESCRIPTION: {896}.

NOTES: plant small, two-leaved; peduncle 8 – 9 cm, laxly-flowered; flowers glabrous, yellowish, c. 9 mm long; lip flattened, obovate, 4 × 2.5 mm, very obtuse.

Bulbophyllum therezienii *Bosser* in Adansonia, ser.2, 11 (2): 330 (1971). TYPE: Madagascar, Ifanadiana, *Therezien* s.n. (holo. P).

DISTRIBUTION: Madagascar: Fianarantsoa.

HABITAT: mossy forest {119}.

FLOWERING TIME: February.

LIFE FORM: epiphyte {119}.

PHYTOGEOGRAPHICAL REGION: III.

DESCRIPTION: {119}.

NOTES: known only from the type. Plant smallish, 2-leaved; inflorescence 10 – 12 cm long, laxly 8- to 12-flowered; flowers reddish, slightly fleshy; stelids erect, triangular, acute, 0.2 – 0.3 mm long.

Bulbophyllum thompsonii *Ridl.* in J. Linn. Soc., Bot. 21: 464 (1885). TYPE: Madagascar, *Thompson* s.n. (holo. BM).
Phyllorchis thompsonii (Ridl.) Kuntze in Rev. Gen. Pl. 2: 678 (1891).

DISTRIBUTION: Madagascar.
LIFE FORM: epiphyte.
DESCRIPTION: {896}.
NOTES: known only from the type. Two leaved; an ambiguous species very close to *Bulbophyllum incurvum* Thouars from the Mascarenes.

Bulbophyllum toilliezae *Bosser* in Adansonia, sér. 2, 5: 403 (1965). TYPE: Madagascar, nr. Perinet, *Bosser* 16802 (holo. P).

DISTRIBUTION: Madagascar: Toamasina.
HABITAT: humid, evergreen forest.
ALTITUDE: c. 900 m.
FLOWERING TIME: March.
LIFE FORM: epiphyte.
PHYTOGEOGRAPHICAL REGION: III.
DESCRIPTION: {108}.
ILLUSTRATION: {108}.
NOTES: known only from the type. Plant up to 25 cm tall, 2-leaved; rachis 18- to 22-flowered, thickened, 13 – 14 cm long; flowers not opening much, yellow-green; lip fleshy, much recurved, two oval, deep depressions near the base, ovate c. 4 mm × 2 mm, upper surface very papillose.

Bulbophyllum trichochlamys *H.Perrier* in Notul. Syst. (Paris) 6 (2): 58 (1937). TYPE: Madagascar, Mandraka forest, *E.François* in *H.Perrier* 17584 (holo. P).

DISTRIBUTION: Madagascar: Antananarivo.
HABITAT: humid; evergreen forest on plateau.
ALTITUDE: c. 1200 m.
FLOWERING TIME: February.
LIFE FORM: epiphyte.
PHYTOGEOGRAPHICAL REGION: III.
DESCRIPTION: {896}.
NOTES: known only from the type. Pseudobulbs oval-conical, c. 12 mm long, single-leafed; leaf oval; flowers 8 mm long; lip relatively large, 3 mm long.

Bulbophyllum trifarium *Rolfe* in Kew Bull. 1910: 280 (1910). TYPE: Madagascar, cult. T. Lawrence (holo. K).

DISTRIBUTION: Madagascar.
DESCRIPTION: {1146}, {1150}.
NOTES: known only from the cultivated type, a poorly known species.

Bulbophyllum trilineatum *H.Perrier* in Notul. Syst. (Paris) 6 (2): 48 (1937). TYPE: Madagascar, Beampingaratra massif, Bekoha summit, *Humbert* 6456 (holo. P; iso. BR, K, TAN).

DISTRIBUTION: Madagascar: Toliara.
HABITAT: humid, evergreen forest on plateau.
ALTITUDE: c. 1500 m.
FLOWERING TIME: November.
LIFE FORM: epiphyte.

PHYTOGEOGRAPHICAL REGION: III.

DESCRIPTION: {108}, {896}.

NOTES: plant very small, c. 5 cm tall, 2-leaved: inflorescence 2 – 4 cm long, filiform, 1- to 3-flowered; flowers pinkish-white; stelids very small and obtuse.

Bulbophyllum turkii *Bosser & P.J.Cribb* in Adansonia sér. 3, 23 (1): 132 (2001). TYPE: Madagascar, Fianarantsoa, Ranomafana reserve, *Turk & Solo* 263 (holo. P; iso. MO, TAN).

DISTRIBUTION: Madagascar: Fianaranatsoa.

HABITAT: moist, montane forest.

ALTITUDE: 950 – 1150 m.

FLOWERING TIME: March.

LIFE FORM: epiphyte.

PHYTOGEOGRAPHICAL REGION: III.

DESCRIPTION: {142}.

ILLUSTRATION: {142}.

NOTES: two-leaved plant, close to *B. perreflexum* by its shape and size of the flower and the strong reflection of the ovary, but distinct by the shape of the lip and petals.

Bulbophyllum vakonae *Hermans* **nom. nov.**

Bulbophyllum ochrochlamys Schltr. in Repert. Sp. Nov. Regni Veg. Beih. 33: 190 (1924), *non* Schltr. in Repert. Sp. Nov. Regni Veg. Beih. 1: 856 (1913). Type: Madagascar, Analamazaotra forests, *H.Perrier* 11852 (holo. P; iso. K).

DISTRIBUTION: Madagascar: Antananarivo, Toamasina.

HABITAT: humid, evergreen forest.

ALTITUDE: 800 – 900m.

FLOWERING TIME: January.

LIFE FORM: epiphyte.

PHYTOGEOGRAPHICAL REGION: III.

DESCRIPTION: {896}.

ILLUSTRATION: {896}.

NOTES: two-leaved plant, related to *B. florulentum*, but with a thick rhizome, shorter ovoid pseudobulbs, wider leaves and somewhat wider sepals, the shape of the petals and lip are also different. Schlechter (1913) had already used the epithet *ochrochlamys* for a New Guinea species of *Bulbophyllum* before he used it for a Madagascan species. I therefore propose the epithet name *vakonae,* referring to a lodge in the Analamazaotra forest where this species was first found.

Bulbophyllum variegatum *Thouars*, Hist. Orch. t.105 & t.106 (1822). TYPE: lecto. Thouars, Hist. Orch., t.105 & also t.106 (1822).

Phyllorchis variegata (Thouars) Kuntze, Rev. Gen. Pl.: 675 (1891).

DISTRIBUTION: Madagascar: Toamasina. Mascarenes: Mauritius, Réunion.

HABITAT: humid, forest at low-elevation.

PHYTOGEOGRAPHICAL REGION: I.

DESCRIPTION: {103}.

ILLUSTRATION: {166}, {167}, {169}.

NOTES: sporadic and undefined records for this species exist for Madagascar.

Bulbophyllum ventriosum *H.Perrier* in Notul. Syst. (Paris) 6 (2): 55 (1937). TYPE: Madagascar, Tsaratanana massif, *H.Perrier* 16476 (holo. P).

DISTRIBUTION: Madagascar: Antsiranana.

HABITAT: lichen-rich and evergreen forests, on moss- and lichen-covered trees.

ALTITUDE: c. 2000 m.

FLOWERING TIME: April, August – September.

LIFE FORM: epiphyte.

PHYTOGEOGRAPHICAL REGION: III, IV.

DESCRIPTION: {896}.

ILLUSTRATION: {896}.

NOTES: single-leafed plant, distinct by the wide, rounded mentum which is formed by the fusion of the two lateral sepals and by its lip with spreading margins.

Bulbophyllum verruculiferum *H.Perrier* in Notul. Syst. (Paris) 14: 152 (1951). TYPE: Madagascar, E of Marojejy massif, *Humbert* 22642 (holo. P).

DISTRIBUTION: Madagascar: Antsiranana, Fianarantsoa.
HABITAT: lichen-rich forest on moss- and lichen-covered trees.
ALTITUDE: 1400 – 1700 m.
FLOWERING TIME: November – December.
LIFE FORM: epiphyte.
PHYTOGEOGRAPHICAL REGION: IV.
DESCRIPTION: {901}.
NOTES: plant bifoliate; distinct by the large dorsal sepal and the tiny warts which cover the exterior and also the crests and margins of the joined lateral sepals; stelids shortly denticulate.

Bulbophyllum vestitum *Bosser* in Adansonia, sér. 2, 11 (2): 331 (1971). TYPE: Madagascar, road to Lakato, *Bosser* 19277 (holo. P).

DISTRIBUTION: Madagascar: Antsiranana, Toamasina.
HABITAT: humid, mossy, evergreen forest.
ALTITUDE: 500 – 1100 m.
FLOWERING TIME: June – September.
LIFE FORM: epiphyte.
PHYTOGEOGRAPHICAL REGION: I, III.
DESCRIPTION: {119}.
ILLUSTRATION: {119}, {906}.
NOTES: plant 2-leaved, close to *B. sandrangatense* with similar fibres at the base of the bulb, but different in floral morphology, especially in the lip; also similar to *B. longivaginans* but differs in the plant and shape of the petals and lip.

Bulbophyllum vestitum var. **meridionale** *Bosser* in Adansonia, sér. 2, 11 (2): 331 (1971). TYPE: Madagascar, Andohahela massif, cult. Jard. Bot. Tan. 457 (holo. P).

DISTRIBUTION: Madagascar: Toliara.
HABITAT: evergreen forest.
LIFE FORM: epiphyte.
PHYTOGEOGRAPHICAL REGION: III.
DESCRIPTION: {119}.
NOTES: petals ovoid, acutely lanceolate at tip and are also much wider and larger, c.3 × 2 mm; inflorescence laxly few-flowered.

Bulbophyllum viguieri *Schltr.* in Repert. Sp. Nov. Regni Veg. Beih. 18: 324 (1922). TYPE: Madagascar, 3 km E of Ambatolaona, *Viguier & Humbert* 1217 (holo. B†).

DISTRIBUTION: Madagascar: Antananarivo.
HABITAT: humid, evergreen forest on plateau.
ALTITUDE: c. 1500 m.
FLOWERING TIME: November.
LIFE FORM: epiphyte.
PHYTOGEOGRAPHICAL REGION: III.
DESCRIPTION: {896}.
NOTES: no herbarium material of this species has been found and the description lacks precision, no specimens can be recognised with any certainty as this species.

Bulbophyllum vulcanorum *H.Perrier* in Notul. Syst. (Paris) 7: 140 (1938). TYPE: Madagascar, Mt. d'Ambre, *H.Perrier* 18877 (holo. P).

DISTRIBUTION: Madagascar: Antsiranana.
HABITAT: humid evergreen forest on plateau.
ALTITUDE: c. 1000 m.
FLOWERING TIME: November.
LIFE FORM: epiphyte.
PHYTOGEOGRAPHICAL REGION: III.
DESCRIPTION: {896}.
NOTES: known only from the type. Plant bifoliate, up to 40 cm tall; leaves elliptical, c. 9 cm long; flowers 7.5 – 8 mm long; lip short and wide, 3 × 1.6 mm, very rounded at the tip.

Bulbophyllum xanthobulbum *Schltr.* in Repert. Sp. Nov. Regni Veg. Beih. 15: 332 (1918). TYPE: Madagascar, *Laggiara* s.n. (holo. B†), East-Central, Analamazaotra area, *H.Perrier* 11849 (neo. P, selected here).

DISTRIBUTION: Madagascar: Toamasina.
HABITAT: coastal forest; humid evergreen forest.
ALTITUDE: sea level – 800m.
FLOWERING TIME: January.
LIFE FORM: epiphyte.
PHYTOGEOGRAPHICAL REGION: I.
DESCRIPTION: {896}.
ILLUSTRATION: {896}.
NOTES: the type was destroyed. Plant small, single-leafed; flowers c. 4.5 mm long; lip very recurved, with a wide heal, the margins of thepart upright and those of the upper part reflexed.

Bulbophyllum zaratananae *Schltr.* in Repert. Sp. Nov. Regni Veg. Beih. 33: 192 (1924) *non* Schltr. (1925). TYPE: Madagascar, Sambirano, *H.Perrier* 15201 (holo. P).

DISTRIBUTION: Madagascar: Antsiranana.
HABITAT: humid, evergreen forest; lichen-rich-rich forest; ericaceous scrub.
ALTITUDE: 400 – 2000m.
FLOWERING TIME: November – January.
LIFE FORM: epiphyte.
PHYTOGEOGRAPHICAL REGION: II, III.
DESCRIPTION: {896}.
NOTES: single-leafed plant, related to *B. baronii* but with a thicker rhizome and longer leaves, also a distinct lip.

Bulbophyllum zaratananae subsp. **disjunctum** *H.Perrier* ex *Hermans* **subsp. nov.** TYPE: Madagascar, Farafangana, Ikongo massif, *Decary* 5611 (holo. P; iso. L). See Appendix .
Bulbophyllum tsaratananae subsp. *disjunctum H.Perrier* in Notul. Syst. (Paris) 6 (2): 59 (1937), *nom. nud.*

DISTRIBUTION: Madagascar: Fianarantsoa.
HABITAT: humid, mid-elevation forest.
FLOWERING TIME: October.
LIFE FORM: epiphyte.
PHYTOGEOGRAPHICAL REGION: III.
DESCRIPTION: {896}.
NOTES: the variety has slightly larger flowers up to 9 mm long, with the margins of the column straight below the rostellum, a column-foot that is flat at the front and acute stelids.

Calanthe humbertii *H.Perrier* in Humbert, Mém. Inst. Sci. Mad. sér. B, 6: 258 (1955). TYPE: Madagascar, W of Marojejy, *Humbert* 23693 (holo. P).

DISTRIBUTION: Madagascar: Antsiranana.
HABITAT: lichen-rich forest on granite, gneiss and limestone.

ALTITUDE: 1700 – 2000 m.

FLOWERING TIME: March.

LIFE FORM: epiphyte.

PHYTOGEOGRAPHICAL REGION: IV.

DESCRIPTION: {902}.

ILLUSTRATION: {902}.

NOTES: flowers violet; lip entire, 12 × 10 mm, the disc with a tridentate callus; spur cylindrical up to 3 cm long; ovary more or less pilose-glandulose.

Calanthe madagascariensis *Watson ex Rolfe* in Kew Bull.: 84 (1906); Watson in Gard. Chron. ser. 3, 28: 335 (1900), *nom. nud.* TYPES: Madagascar, E Betsileo, *Warpur* s.n.; Fort Dauphin, *Scott Elliot* 2357; without locality, *Baron* 254; *Baron* 328; *Baron* 2697; *Baron* 6555 (all syn. K).

Calanthe sylvatica sensu Rolfe in J. Linn. Soc. 29: 52 (1891), *non* Lindl.

Calanthe warpuri Watson ex Rolfe in Kew Bull.: 85 (1906). TYPE: Madagascar, *Warpur* s.n. (holo. K).

DISTRIBUTION: Madagascar: Antananarivo, Antsiranana, Fianarantsoa, Toliara.

HABITAT: in shady places, in moist clay in humid, evergreen forest on plateau; river margins; on *Asplenium nidus*; on rocks in humus.

ALTITUDE: 1001 – 1500 m.

FLOWERING TIME: February – September.

LIFE FORM: terrestrial, lithophyte, or rarely epiphyte.

PHYTOGEOGRAPHICAL REGION: III.

DESCRIPTION: {216}, {554}, {896}.

ILLUSTRATION: {216}, {554}, {896}, {906}, {1298}.

CULTIVATION: {216}, {413}, {1452}.

NOTES: plant small; leaves undulate; flowers small;lobes of the lip narrow, 5- to 6-times longer than wide.

Calanthe millotae *Ursch & Genoud ex Bosser* in Adansonia, sér. 2, 6 (3): 399 (1966); Ursch & Genoud in Nat. Malg. 3 (2): 104 (1951), *nom. nud.* TYPE: Madagascar, Marojejy massif, *Humbert* 944 (holo. P).

DISTRIBUTION: Madagascar: Antsiranana.

HABITAT: humid, eastern forests.

FLOWERING TIME: March – April.

LIFE FORM: terrestrial or lithophyte.

PHYTOGEOGRAPHICAL REGION: III, IV.

DESCRIPTION: {109}, {1414}.

ILLUSTRATION: {109}, {1414}.

NOTES: inflorescence short; spur shorter than the ovary; flower pure white, including the lip; lip with a red callus. Similar to the widespread Asiatic species *C. triplicata* (Willem.) Ames.

Calanthe repens *Schltr.* in Repert. Spec. Nov. Regni Veg. Beih. 33: 164 (1924). TYPE: Madagascar, Mt. Tsaratanana, *H.Perrier* 15278 (holo. P).

Calanthe repens var. *pauliani* Ursch & Genoud in Nat. Malg. 3 (2): 102 (1951), *nom. nud.* Based upon: Madagascar, Andringitra, *Millot* 938a (P).

DISTRIBUTION: Madagascar: Antsiranana.

HABITAT: mossy, evergreen forest.

ALTITUDE: 1200 – 1800 m.

FLOWERING TIME: February – March, October.

PHYTOGEOGRAPHICAL REGION: II, III, IV.

DESCRIPTION: {896}, {1414}.

ILLUSTRATION: {281}, {896}, {906}, {1414}.

NOTES: lip not 3-lobed, just expanded-rounded on each side towards the base, contracted in the middle, then a little widened at the apex.

Calanthe sylvatica *(Thouars) Lindl.,* Gen. Sp. Orch. Pl.: 250 (1833). TYPE: Mauritius, *Thouars* s.n. (holo. P).

Centrosis sylvatica Thouars, Hist. Orch.: t.35 & 36 (1822).

Bletia silvatica (Thouars) Spreng., Syst. Veg. 3: 743 (1826).

Centrosia auberti A.Rich., Monogr. Orch. Iles Fr. Bourbon: 45, t.7 (1828), *nom. illeg.* Based upon same type as *Calanthe sylvatica.*

Alismorkis centrosis Steud., Nomencl. Bot., ed. 2, 1: 49 (1840).

Calanthe sylvatica var. *alba* Cordem., Fl. Réunion: 225 (1895).

Calanthe sylvatica var. *purpurea* Cordem., Fl. Réunion: 225 (1895).

Calanthe sylvatica var. *lilacina* Cordem., Fl. Réunion: 225 (1895).

Calanthe sylvatica var. *iodes* Cordem., Fl. Réunion: 225 (1895), *nom. nud.*

Calanthe violacea Rolfe in Kew Bull. 1913: 29 (1913). TYPE: Madagascar, cult. England (holo. K, not found).

Calanthe sylvatica var. *pallidipetala* Schltr. in Repert. Spec. Nov. Regni Veg. Beih. 33: 166 (1924). TYPE: Madagascar, NE of Inanatonana, Andrantsay bassin, *H.Perrier* 8104 (holo. P).

Calanthe durani Ursch & Genoud in Nat. Malg. 3 (2): 104 (1951), *nom. nud.* Based upon: Madagascar, Betampona, *Duran* 809 (P).

Calanthe perrieri Ursch & Genoud in Nat. Malg. 3 (2): 102 (1951), *nom. nud.* Based upon: Madagascar, North Tamatave, *Duran* 811 (P).

Calanthe natalensis Rchb.f. (Rchb.f.) in Bonplandia 4: 322 (1856). TYPE from South Africa (Natal).

Calanthe corymbosa Lindl. in J. Linn. Soc., Bot. 6: 129 (1862). TYPE from Bioko (Fernando Po).

Calanthe sanderiana Rolfe in Gard. Chron., Ser. 3: 396 (1892). TYPE not located.

Calanthe delphinioides Kraenzl. in Bot Jahrb. Syst. 17: 55 (1893). TYPE from Cameroon.

Calanthe volkensii Rolfe in F.T.A. 7: 46 (1897). TYPE from Tanzania.

Calanthe neglecta Schltr. in Bot. Jahrb. Syst. 53: 570 (1915). TYPE from Tanzania.

Calanthe stolzii Schltr. in Bot. Jahrb. Syst. 53: 569 (1915). TYPE from Tanzania.

Calanthe schliebenii Mansfeld in Notizbl. Bot. Gart. Berlin 11: 808 (1933). TYPE from Tanzania.

Calanthe sylvatica forma *imerina* Ursch & Genoud in Nat. Malg. 3 (2): 108 (1951), *nom. nud.* TYPE: Madagascar, Mandraka, Ambohitantely, *Ursch* 24 (holo. TAN).

Calanthe sylvatica forma *humberti* Ursch & Genoud in Nat. Malg. 3 (2): 108 (1951), *nom. nud.* TYPE: Madagascar, cult. Bot. Garden Antananarivo 808, *Humbert* (holo. TAN).

DISTRIBUTION: Madagascar, Antananarivo, Antsiranana, Fianarantsoa, Toamasina, Toliara. Comoros: Anjouan, Grande Comore, Mayotte, Moheli. Mascarenes: Réunion, Mauritius. Seychelles. Also in East and South Africa.

HABITAT: humid, evergreen forest from sea level to plateau.

ALTITUDE: 300 – 2000m.

FLOWERING TIME: December – July.

LIFE FORM: terrestrial.

PHYTOGEOGRAPHICAL REGION: I, II, III, IV.

DESCRIPTION: {896}, {768}.

HISTORY: {571}, {790}, {990}.

ILLUSTRATION: {50}, {135}, {149b}, {166}, {167}, {168}, {219}, {246}, {308}, {417}, {677}, {682}, {691}, {768}, {862}, {906}, {957}, {1028}, {1264}, {1289}, {1298}, {1329}, {1345}, {1398}, {1505}.

NOTES: a widespread and variable species. Plant large, 40 – 80 cm tall; leaves generally wider and longer than the other species; flowers large; lip 3-lobed with the lobes 2 times longer than wide, bearing a basal callus of 3 verrucose ridges.

Chamaeangis gracilis *(Thouars) Schltr.,* in Beih. Bot. Centrbl., 33, 2: 426 (1915). TYPE Madagascar?, Thouars s.n. (holo. P).

Angraecum gracile Thouars, Orch. Iles Afr. t.77 (1822).

Oeceoclades gracilis (Thouars) Lindl., Gen. Sp. Orch. Pl. 237: 237 (1833).

Aerobion gracile (Thouars) Spreng., Syst. 3: 716 (1826).

Angorchis gracilis (Thouars) Kuntze, Rev. Gen. Pl.: 651 (1891).

Mystacidium gracile Finet, in Bull. Soc. France 54, Mém. 9: 63 (index) (1907).

DISTRIBUTION: Madagascar, Mascarenes, Mauritius.

DESCRIPTION: {896}.

ILLUSTRATION: {1398}.

Cheirostylis nuda *(Thouars) Ormerod* in Lindleyana 17(4): 193 (2002). TYPE: Mauritius *Thouars* s.n. (syn. P).

Goodyera nuda Thouars, Orch. Iles Austral. Afr.: t.29 & 30B (1822).

Cheirostylis gymnochiloides (Ridl.) Rchb.f. in Flora 68: 537 (1885). TYPE: Central Madagascar, *Baron* 2842 (holo. BM).

Cheirostylis humblotii Rchb.f. in Linnaea 41: 60 (1876). TYPE: Comoros, *Humblot* s.n. (holo. W).

Cheirostylis sarcopus Schltr. in Bot. Jahrb. Syst. 53: 558 (1915). TYPE from Tanzania.

Cheirostylis micrantha Schltr. in Repert. Spec. Nov. Regni Veg. Beih. 33: 127 (1924). TYPE: Madagascar, Mt. Vtovavy, *H.Perrier* 1930 (4) (holo. B†; iso. P).

Erporkis gymnerpis Thouars, Orch. Iles Austral. Afr.: t.29 & 30B (1822), *nom. alt. Gymnochilus nudum* (Thouars) Blume, Orch. Arch. Ind.: 108 (1858).

Gymnochilus recurvum Blume, Coll. Orch. Arch. Ind. 1: 109, t.32, fig. 2 (1858). TYPE: Madagascar, without locality "*Ophrys spica recurva radice decurrente*" in herb Bruguieres s.n. (holo. L).

Monochilus boryi Rchb.f., Linnaea 41: 60 (1877). TYPE: Réunion, *Bory du Saint Vincent* s.n. (holo. W).

Monochilus gymnochiloides Ridl. in J. Linn. Soc., Bot. 21: 499 (1885). Types: Madagascar, Imerina, *Hildebrandt* 3800 (syn. BM; iso. K); central part, *Baron* 2842 (syn. BM).

Orchiodes nudum (Thouars) Kuntze, Rev. Gen. Pl. 2: 671 (1891).

Zeuxine boryi (Rchb.f.) Schltr. in Beih. Cot. Centralbl. 33: 2: 410 (1915). TYPE: Bourbon, *Bory* s.n. (holo. P?).

Zeuxine gymnochiloides Schltr. in Beih. Bot. Centralbl. 34 (2): 320 (1916). TYPE: Madagascar, Simiana Basin, 100 m, *H.Perrier* 11936 (102) (holo. B†; iso. P).

Zeuxine sambiranoensis Schltr. in Repert. Sp. Nov. Regni Veg. Beih. 33: 129 (1924). TYPE: Madagascar, Ramena Valley, *H.Perrier* 11559 (holo. P).

DISTRIBUTION: Madagascar: Antananarivo, Antsiranana, Fianarantsoa, Toliara. Mascarenes: Réunion, Mauritius. Comoros.

HABITAT: humid, medium-elevation forest; seasonally dry, deciduous forest and woodland; humid evergreen forest, in humus and detritus.

ALTITUDE: sea level – 1700 m.

FLOWERING TIME: August – March.

LIFE FORM: terrestrial.

PHYTOGEOGRAPHICAL REGION: I, III, V.

DESCRIPTION: {896}, {768}.

ILLUSTRATION: {94}, {219}, {672}, {682}, {768}, {1028}, {1205}, {1223}, {1398}.

NOTES: nomenclature is discussed by Ormerod {853}. Plant small with erect leafy stem arising from fleshy rhizomes; flowers small, mostly whitish; sepals joined for about half their length; petals adnate to the dorsal sepal; lip saccate.

Corymborkis corymbis *Thouars* in Nouv. Bull. Sci. Soc. Philom. Paris 1: 318 (1809); P.J.Cribb, Fl. Trop. E. Africa Orch. 2: 243 (1984). TYPE: Réunion, *Thouars* s.n. (holo. P).

Corymbis thouarsii Rchb.f. in Bot. Zeit. 7: 868 (1849), *nom. illeg.*

Corymborkis disticha Lindl., Fol. Orch. 1: 14 (1854), *nom. illeg.*

Corymborkis thouarsii (Rchb.f.) Blume, Coll. Orch. Arch. Ind. 1: 126 (1857).

Corymbis welwitschii Rchb.f. in Flora 48 (12): 183 (1865). TYPE: Angola, Cazengo, *Welwitsch* 668 (holo. W).

Corymbis corymbosa Ridl. in J. Linn. Soc., Bot. 21: 498 (1885).

Corymborkis corymbosa (Ridl.) Kuntze in Rev. Gen. Pl. 2: 658 (1891).

Corymborkis welwitschii (Rchb.f.) Kuntze in Rev. Gen. Pl. 2: 658 (1891).

Corymbis leptantha Kraenzl. ex Engl. in Abh. Preuss. Akad. Wiss.: 45 & 53 (1894), *nom. nud.*

Corymbis bolusiana Bolus ex Schltr. in Repert. Spec. Nov. Regni Veg. Beih. 33: 131 (1924), *nom. nud.*

DISTRIBUTION: Madagascar: Antsiranana, Toamasina, Toliara. Mascarenes: Réunion, Mauritius. Also in East & southern Africa.

HABITAT: humid; evergreen forest and lowland forest, often at base of trees in densely shady undergrowth, often forming large colonies.

ALTITUDE: 100 – 1300 m.

FLOWERING TIME: January – April.

LIFE FORM: terrestrial.

PHYTOGEOGRAPHICAL REGION: I, II.

DESCRIPTION: {219}, {312}, {768}, {896}, {954}.

ILLUSTRATION: {94}, {168}, {219}, {256b}, {312}, {384}, {448}, {677}, {682}, {768}.

NOTES: plant rhizomatous with tall, erect leafy stems; flowers white to greenish; lip 60 – 80 mm long, linear; column erect, terete, clavate towards the apex.

Cryptopus brachiatus *H.Perrier* in Notul. Syst. (Paris) 7: 137 (1938). TYPE: Madagascar, confl. Mangoro & Onive, *H.Perrier* 17005 (holo. P).

DISTRIBUTION: Madagascar: Fianarantsoa, Toamasina.
HABITAT: humid, evergreen forest.
ALTITUDE: 600 – 1200 m.
FLOWERING TIME: February – April.
LIFE FORM: epiphyte.
PHYTOGEOGRAPHICAL REGION: I, III.
DESCRIPTION: {896}.
ILLUSTRATION: {896}.
NOTES: basal lobes of the lip oblong, rounded at the tip and contracted at the base into a short tooth; midlobe T-shaped, truncate at the tip extended laterally into 2 elongated filiform appendices; petals T-shaped, similar to the midlobe of the lip.

Cryptopus dissectus *(Bosser) Bosser* in Adansonia, sér. 2, 20 (3): 261 (1980). TYPE: Madagascar, Ifanadiana Road, Fort-Carnot, *Peyrot* 31 (holo. P).
Cryptopus elatus subsp. *dissectus* Bosser in Adansonia, ser.2, 5: 407, (1965) & in Adansonia, sér. 2, 20 (3): 261 (1980).

DISTRIBUTION: Madagascar: Fianarantsoa.
HABITAT: humid, mossy, evergreen forest.
ALTITUDE: 500 – 1000 m.
FLOWERING TIME: December.
PHYTOGEOGRAPHICAL REGION: III.
DESCRIPTION: {108}, {1259}.
ILLUSTRATION: {108}.
NOTES: plant 30 – 40 cm tall; flowers yellow-green; sepals and petals 13 – 15 mm long; terminal lobes of the lip and the petals deeply divided into narrow lobules.

Cryptopus paniculatus *H.Perrier* in Notul. Syst. (Paris) 7: 136 (1938). TYPE: Madagascar, Moramanga, *Decary* 7171 (holo. P).

DISTRIBUTION: Madagascar: Toamasina.
HABITAT: humid, medium altitude forest, on rocks, littoral forest.
ALTITUDE: sea level – 1000 m.
FLOWERING TIME: February – March.
LIFE FORM: epiphyte.
PHYTOGEOGRAPHICAL REGION: III.
DESCRIPTION: {896}.
ILLUSTRATION: {114}, {185}, {710}, {896}, {906}, {1440}.
NOTES: lip 5-lobed, basal lobes very small and rounded, petals expanded at the tip into 2 curved lobes, spur 2 – 3 mm.

Cymbidiella falcigera *(Rchb.f.) Garay* in Orch. Digest 40 (5): 192 (1976). TYPE: Madagascar, without locality, *Humblot* 491 (holo. W).
Grammangis falcigera Rchb.f. in Flora 68: 541 (1885).
Cymbidium humblotii Rolfe in L'Orchidophile 12 (132): 161 (1892). TYPE: Madagascar, *Humblot* s.n. & cult. Ingram (syn. K).
Cymbidium loise-chauvierii Hort. in Gard. Chron. sér. 3, 12: 8 (1892).
Caloglossum humblotii (Rolfe) Schltr. in Repert. Spec. Nov. Regni Veg. Beih. 15: 213 (1918).
Caloglossum magnificum Schltr. in Repert. Spec. Nov. Regni Veg. Beih. 15: 214 (1918). TYPE: Madagascar, without locality, *Chr. Bang* s.n. (holo. B).
Cymbidiella humblotii (Rolfe) Rolfe in Orch. Rev. 26: 58 (1918).

DISTRIBUTION: Madagascar: Antsiranana, Fianarantsoa, Toamasina. Comoros?
HABITAT: river margins, littoral forest, marshland.

ALTITUDE: sea level – 400 m.

FLOWERING TIME: December – March.

LIFE FORM: epiphyte on *Raphia, Afzelia* sp. and *Vonitra thouarsiana.*

PHYTOGEOGRAPHICAL REGION: I, III.

DESCRIPTION: {50}, {896}, {1326}, {1356}.

ILLUSTRATION: {50}, {192}, {277}, {281}, {317}, {452}, {482}, {499}, {612}, {620}, {720}, {731}, {818}, {896}, {904}, {906}, {1169}, {1266}, {1298}, {1326}, {1344}, {1356}, {1476}, {1484}.

HISTORY: {24}, {1140}, {1152}, {1154}, {1161}, {1162}, {1169}, {1356}, {1476}, {1484}.

CULTIVATION: {452}, {470}, {620}, {724}, {731}, {818}, {1152}, {1326}.

NOTES: plant very large with a stout creeping rhizome; inflorescence branched; flowers apple-green to yellow-green with dark blackish-maroon spots on lip; lateral sepals dorsally keeled; lip recurved, 3-lobed, the lateral lobes bigger than the middle one.

Cymbidiella flabellata *(Thouars) Lindl.,* Gen. Sp. Orch. Pl.: 167 (1833). TYPE: Madagascar, *Thouars* s.n. (holo. P).

Limodorum flabellatum Thouars, Hist. Orch.: t.39 – 40 (1822).

Cymbidium flabellatum (Thouars) Spreng., Syst. Veg. 3: 724 (1826).

Caloglossum flabellatum (Thouars) Schltr. in Repert. Spec. Nov. Regni Veg. Beih. 15: 213 (1918).

Cymbidiella perrieri Schltr. in Repert. Spec. Nov. Regni Veg. Beih. 33: 272 (1925). TYPE: Madagascar, source of Androranga, *H.Perrier* 11336 (holo. B†; iso. P).

DISTRIBUTION: Madagascar: Antsiranana, Fianarantsoa, Mahajanga, Toamasina, Toliara.

HABITAT: coastal forest; shaded and humid quartz rocks; marshland on peat; in *Philippia* scrub; in peaty sphagnum moss and sand saturated with water; coarse acid quartzite sand, constantly humid and with black humus, often in the shade of ericaceous scrub on laterite.

ALTITUDE: sea level – 1500 m.

FLOWERING TIME: September – February.

LIFE FORM: terrestrial.

PHYTOGEOGRAPHICAL REGION: I, III.

DESCRIPTION: {131}, {896}.

ILLUSTRATION: {3}, {131}, {281}, {317}, {452}, {499}, {510}, {612}, {724}, {726}, {733}, {906}, {1248}, {1398}.

HISTORY: {571}, {1064}.

CULTIVATION: {131}.

NOTES: a slender terrestrial, 1 to 1.5 m tall, on a long thin rhizome; flowers somewhat similar in appearance to those of *C. pardalina* but more slender and much smaller; sepals and petals yellow-green; lip red with black spots.

Cymbidiella pardalina *(Rchb.f.) Garay* in Orch. Digest 40 (5): 192 (1976). TYPE: Madagascar, without locality, *Humblot* s.n. (holo. W).

Grammangis pardalina Rchb.f. in Flora 68: 541 (1885).

Cymbidium rhodochilum Rolfe in Orch. Rev. 10: 184 (1902). TYPE: Madagascar, Tananbe, cult. Kew, *Warpur* s.n. (holo. K).

Caloglossum rhodochilum (Rolfe) Schltr. in Repert. Spec. Nov. Regni Veg. Beih. 15: 213 (1918).

Cymbidiella rhodochila (Rolfe) Rolfe in Orch. Rev. 26: 58 (1918).

DISTRIBUTION: Madagascar: Fianarantsoa, Toamasina.

HABITAT: humid, evergreen forest, growing in association with *Platycerium madagascariensis*, on *Albizzia fastigiata.*

ALTITUDE: 500 – 2000 m.

FLOWERING TIME: October – December.

LIFE FORM: epiphyte.

PHYTOGEOGRAPHICAL REGION: I.

DESCRIPTION: {50}, {132}, {447}, {703}, {896}.

ILLUSTRATION: {50}, {132}, {135}, {222}, {267}, {317}, {332}, {412}, {416}, {425}, {447}, {452}, {470}, {499}, {509}, {529}, {566}, {587}, {612}, {638}, {700}, {703}, {724}, {727}, {850}, {818}, {862}, {905}, {906}, {944}, {958}, {1166}, {1176}, {1209}, {1235}, {1264}, {1280}, {1281}, {1285}, {1298}.

HISTORY: {332}, {425}, {571}, {606}, {1116}, {1132}.

CULTIVATION: {132}, {238}, {452}, {703}, {818}, {846}, {1132}.

NOTES: plant large on a short stout rhizome; inflorescence not branched; flowers large; tepals yellowish-green, carrying dark spots; lip green with black spots, the lobes red.

Cynorkis alborubra *Schltr.* in Repert. Spec. Nov. Regni Veg. Beih. 33: 39 (1924). TYPE: Madagascar, Andringitra massif, *H.Perrier* 14356 (holo. P).

DISTRIBUTION: Madagascar: Fianarantsoa.
HABITAT: montane, ericaceous scrub.
ALTITUDE: c. 2400 m.
FLOWERING TIME: January.
LIFE FORM: terrestrial.
PHYTOGEOGRAPHICAL REGION: IV.
DESCRIPTION: {896}.
ILLUSTRATION: {896}, {1223}.
NOTES: plant slender with 2 basal spreading leaves; lip 3-lobed, 6.5 mm long, 4.75 mm wide, the lateral lobes lanceolate, subacute, the middle one oblong-ligulate and obtuse, almost 3 times longer than the laterals; spur 2.5 – 3 mm, conical and acute.

Cynorkis ambondrombensis *Boiteau* in Bull. Acad. Malgache n. s. 24: 90 (1942). TYPE: Madagascar, S Betsileo, Ambondrombe, *Boiteau* in Jard. Bot. Tananarive 4920 (holo. P).

DISTRIBUTION: Madagascar: Fianarantsoa.
HABITAT: lichen-rich forest on wet rock.
ALTITUDE: 1800 m.
LIFE FORM: terrestrial.
PHYTOGEOGRAPHICAL REGION: III.
DESCRIPTION: {100}.
NOTES: known only from the type. Plant slender with a small leaf; lip lozenge-shaped, the front margin finely fimbrillate, 18 × 15 mm; spur cylindrical, 3.5 cm long.

Cynorkis ampullacea *H.Perrier ex Hermans* **sp. nov.** TYPE: Madagascar, Lake Andraikibo, nr. Antsirabe, *H.Perrier* 17895 (holo. P). See Appendix .
Cynorkis ampullacea (H.Perrier) H.Perrier in Humbert ed., Fl. Mad. Orch. 1: 112 (1939), *comb. inval.*
Cynosorchis cuneilabia subsp. *ampullacea* H.Perrier in Arch. Bot. Bull. Mens. 5: 51 (1931), *nom. inval.*

DISTRIBUTION: Madagascar: Antananarivo, Fianarantsoa.
HABITAT: rocky outcrops.
ALTITUDE: c. 1400 m.
FLOWERING TIME: January – March.
LIFE FORM: terrestrial.
PHYTOGEOGRAPHICAL REGION: III.
DESCRIPTION: {896}.
ILLUSTRATION: {1433}.
NOTES: plant with a single (rarely 2), radical, linear-lanceolate leaf; raceme glandular-hairy, laxly 4- to 12-flowered; flowers violet; sepals obtuse, the dorsal hood-shaped, ovate; the laterals oblong; petals a little asymmetric, slightly narrowed from the middle towards the base; lip obtriangular-flabellate, the base thickened and hollowed into a gutter, gradually expanded above; spur 4 mm, swollen into a very obtuse flask-shape; anther spherical; staminodes small and spathulate; rostellum shortly tridentate at the front.

Cynorkis ampullifera *H.Perrier* in Notul. Syst. (Paris) 14: 145 (1951). TYPES: Madagascar; Lokoho Valley, Mt. Beondroka, *Humbert* 23570 (lecto. P, selected here); Mt. Maimborondro, *Humbert* 23380; *Humbert* 23387 (both syn. P).

DISTRIBUTION: Madagascar: Antsiranana, Toamasina.
HABITAT: humid evergreen forest; lichen-rich forest; boggy ground.
ALTITUDE: 200 – 1450 m.
FLOWERING TIME: March, September.
LIFE FORM: terrestrial.
PHYTOGEOGRAPHICAL REGION: I.
DESCRIPTION: {901}.
NOTES: close to *C. cardiophylla* but the leaves shortly petiolate and the lip 3-lobed.

Cynorkis andohahelensis *H.Perrier* in Notul. Syst. (Paris) 8 (1): 34 (1939). TYPE: Madagascar, Andohahela massif, *Humbert* 13615 (holo. B; iso. K, P, TAN).

DISTRIBUTION: Madagascar: Toliara.
HABITAT: montane, ericaceous scrub in humus and in peat.
ALTITUDE: 1800 – 2000 m.
FLOWERING TIME: January.
LIFE FORM: terrestrial.
PHYTOGEOGRAPHICAL REGION: III.
DESCRIPTION: {896}.
NOTES: similar to *C. filiformis* but the raceme is shorter, denser; flowers white, edged with yellow; lip yellow, lip 3-lobed.

Cynorkis andringitrana *Schltr.* in Repert. Spec. Nov. Regni Veg. Beih. 33: 40 (1924). TYPE: Madagascar, Andringitra massif, *H.Perrier* 13755 (syn. P) & *H.Perrier* 14570 (syn. P).

DISTRIBUTION: Madagascar: Antananarivo, Fianarantsoa.
HABITAT: mossy forest; wet areas.
ALTITUDE: 1600 – 2200 m.
FLOWERING TIME: February – April.
LIFE FORM: terrestrial.
PHYTOGEOGRAPHICAL REGION: III, IV.
DESCRIPTION: {896}.
ILLUSTRATION: {896}, {1223}.
NOTES: the lip shape is similar to that of *C. tryphioides*, but the plant is taller, has only one basal leaf and much larger flowers.

Cynorkis angustipetala *Ridl.* in J. Linn. Soc., Bot. 21: 514 (1885). TYPE: Madagascar, without locality, *Hilsenberg & Bojer* s.n. (holo. BM).
Cynosorchis angustipetala (Ridley) T.Durand & Schinz, Consp. Fl. Afr. 5: 90 (1892).
Cynorkis angustipetala var. *typica* H.Perrier in Arch. Bot. Bull. Mens. 5: 64 (1931), *nom. inval.*

DISTRIBUTION: Madagascar: Antananarivo, Fianarantsoa, Toamasina.
HABITAT: high-elevation grassland; forest and woodland margins; marshland in peaty soil; rocky outcrops.
ALTITUDE: 900 – 2000 m.
FLOWERING TIME: November – June.
LIFE FORM: terrestrial.
PHYTOGEOGRAPHICAL REGION: III, IV.
DESCRIPTION: {896}.
ILLUSTRATION: {3}, {165}, {281}, {896}, {906}, {1274}.
NOTES: flowers large, 5 – 7 cm long; lip 2 – 3 cm long, with 4 almost equal lobes, the front edge crenulate; spur 2.2 – 3 cm long, narrowed towards the tip.

Cynorkis angustipetala var. **amabilis** *(Schltr.)* *H.Perrier* in Humbert ed., Fl. Mad. Orch. 1: 145 (1939). TYPE: Madagascar, Ambatofinandrahana, W Betsileo, *H.Perrier* 12437 (holo. P).
Cynosorchis amabilis Schltr. in Repert. Spec. Nov. Regni Veg. Beih. 33: 40 (1924).

DISTRIBUTION: Madagascar: Fianarantsoa.
HABITAT: marshland; grassland; rocky outcrops.
ALTITUDE: 1200 – 1800 m.
FLOWERING TIME: January – February.
LIFE FORM: terrestrial.
PHYTOGEOGRAPHICAL REGION: III.
DESCRIPTION: {896}.
ILLUSTRATION: {1223}.
NOTES: flowers with a light violet-red lip, the disc darker; spur reddish, up to 4 cm long.

Cynorkis angustipetala var. **bella** *(Schltr.)* *H.Perrier* in Humbert ed., Fl. Mad. Orch. 1: 145 (1939). TYPE: Madagascar, Andringitra massif, *H.Perrier* 14586 (holo. P).
Cynosorchis bella Schltr. in Repert. Spec. Nov. Regni Veg. Beih. 33: 42 (1924).

> DISTRIBUTION: Madagascar: Fianarantsoa.
> HABITAT: near forest and on rocky outcrops.
> ALTITUDE: 1000 – 2000 m.
> FLOWERING TIME: February – March.
> LIFE FORM: terrestrial.
> PHYTOGEOGRAPHICAL REGION: III.
> DESCRIPTION: {896}.
> ILLUSTRATION: {1223}.
> NOTES: flowers larger; lip 3 – 3.5 cm long.

Cynorkis angustipetala var. **moramangensis** *H.Perrier ex Hermans* **var. nov.** TYPE: Madagascar, Moramanga, *H.Perrier* 16890 (holo. P). See Appendix .
Cynosorchis angustipetala var. *moramangensis* H.Perrier in Humbert ed., Fl. Mad. Orch. 1: 144 (1939), *nom. nud.*

> DISTRIBUTION: Madagascar: Antananarivo.
> HABITAT: grassland and rocky outcrops.
> FLOWERING TIME: January – February, July.
> LIFE FORM: terrestrial.
> PHYTOGEOGRAPHICAL REGION: III.
> DESCRIPTION: {896}.
> NOTES: differs from the typical variety in having the midlobe of the rostellum longer than the arms and not apiculate. It has many of the characteristics of the *C. angustipetala* var. *oxypetala*, but the anther is not apiculate and the flowers are shortly glandular on their exterior surface.

Cynorkis angustipetala var. **oligadenia** *(Schltr.)* *H.Perrier* in Humbert ed., Fl. Mad. Orch. 1: 145 (1939). TYPE: Madagascar, Antananarivo, *H.Perrier* 15876 (holo. P).
Cynosorchis oligadenia Schltr. in Repert. Spec. Nov. Regni Veg. Beih. 33: 62 (1924).

> DISTRIBUTION: Madagascar: Antananarivo.
> HABITAT: grassland and rocky outcrops.
> FLOWERING TIME: February.
> LIFE FORM: terrestrial.
> PHYTOGEOGRAPHICAL REGION: III.
> DESCRIPTION: {896}.
> NOTES: plant with two radical leaves; anther without apicule; middle lobe of the rostellum with the front margin entire.

Cynorkis angustipetala var. **oxypetala** *(Schltr.)* *H.Perrier* in Humbert ed., Fl. Mad. Orch. 1: 144 (1939). TYPE: Madagascar, Antsirabe, *H.Perrier* 8112 (holo. B†; iso. P).
Cynosorchis oxypetala Schltr. in Repert. Spec. Nov. Regni Veg. Beih. 33: 63 (1924).

> DISTRIBUTION: Madagascar: Antananarivo.
> HABITAT: grassland and rocky outcrops.
> ALTITUDE: 1500 – 1900 m.
> FLOWERING TIME: February, June.
> LIFE FORM: terrestrial.
> PHYTOGEOGRAPHICAL REGION: III.
> DESCRIPTION: {896}.
> ILLUSTRATION: {1223}.
> NOTES: spur 20 mm long; anther apiculate, the apicule furrowed; middle lobe of the rostellum a little longer than the arms.

Cynorkis angustipetala var. **speciosa** *(Ridl.) H.Perrier* in Humbert ed., Fl. Mad. Orch. 1: 144 (1939). TYPE: Madagascar, Ambatovory, Imerina, *Fox* 18 (holo. BM).
Cynosorchis speciosa Ridl. in J. Linn. Soc., Bot. 22: 122 (1886).

DISTRIBUTION: Madagascar: Antananarivo.
HABITAT: grassland and rocky outcrops.
LIFE FORM: terrestrial.
PHYTOGEOGRAPHICAL REGION: III.
DESCRIPTION: {896}.
NOTES: leaves two; spur shorter than the ovary; anther apiculate; rostellum with the middle lobe rectangular, shorter than the arms.

Cynorkis angustipetala var. **tananarivensis** *H.Perrier* ex *Hermans* **var. nov.** TYPE: Madagascar, Antananarivo area, *H.Perrier* 16896 (holo. P). See Appendix .
Cynorchis angustipetala var. *tananarivensis* H.Perrier in Humbert, Fl. Mad. Orch. 1: 143 (1939), *nom. nud.*

DISTRIBUTION: Madagascar: Antananarivo.
HABITAT: grassland and rocky outcrops.
FLOWERING TIME: February – April.
LIFE FORM: terrestrial.
PHYTOGEOGRAPHICAL REGION: III.
DESCRIPTION: {896}.
NOTES: this variety differs from the typical one in having flowers with a few small glands and an anther which has an obtuse apicule. It is similar to *C. angustipetala* var. *moramangensis* but the flowers have a few glands, the anther has an obtuse not furrowed apicule and the midlobe of the rostellum is bi-sinuate at the front.

Cynorkis aphylla *Schltr.* in Ann. Mus. Col. Marseille, ser.3, 1: 150 (1913). TYPE: Madagascar, Mt. Kalabenono, *H.Perrier* 76 (1867) (holo. P).

DISTRIBUTION: Madagascar: Antsiranana, Mahajanga.
HABITAT: rocky outcrops; on sandstone and sand.
ALTITUDE: sea level – 200 m.
FLOWERING TIME: September – November, January.
LIFE FORM: terrestrial or lithophyte.
PHYTOGEOGRAPHICAL REGION: II.
DESCRIPTION: {896}.
ILLUSTRATION: {877}, {896}, {1205}, {1223}.
NOTES: leaves not developed at flowering time; flower large, dark pink; lip 4-lobed, wider than long, 3 × 2.4 cm; spur filiform, 4 – 5 cm long.

Cynorkis aurantiaca *Ridl.* in J. Linn. Soc., Bot. 22: 123 (1886). TYPE: Madagascar, Ankeramadinika, *Fox* 12A. (holo. BM; iso. K).

DISTRIBUTION: Madagascar: Antananarivo.
HABITAT: humid, evergreen forest on plateau, epiphyte in moss; marshes on plateau.
ALTITUDE: 1200 – 1800 m.
FLOWERING TIME: January – May.
LIFE FORM: terrestrial.
PHYTOGEOGRAPHICAL REGION: III.
DESCRIPTION: {896}.
ILLUSTRATION: {896}.
NOTES: plant slender, single-leafed; flowers yellowish-white with the lip yellow-orange; lip 6 mm long, 3-lobed, the mid lobe finely papillose; spur c. 4 mm, a little narrowed-obtuse.

Cynorkis bardotiana *Bosser* in Adansonia 20 (2): 281 (1998). TYPE: Madagascar, Diego-Suarez prov, Ankarana massif, *Bardot-Vaucoulon* 716 (holo. P).

DISTRIBUTION: Madagascar: Antsiranana.

HABITAT: in clay in deciduous forest on calcareous rock.
FLOWERING TIME: July.
LIFE FORM: terrestrial.
PHYTOGEOGRAPHICAL REGION: V.
DESCRIPTION: {138}.
ILLUSTRATION: {138}.
NOTES: known only from the type. Lip 3-lobed, 12 × c. 20 mm; spur a little curved, 2 – 2.5 cm long, filiform, inflated at the tip; rostellum with an entire horizontal blade.

Cynorkis baronii *Rolfe* in J. Linn. Soc., Bot. 29: 58 (1891). TYPES: Madagascar, Andringitra Mts, *Scott Elliot* 1853 (lecto. K), Central, *Baron* 725 (syn. K), NW, *Baron* 5246 (syn. K, P).
Cynosorchis pauciflora Rolfe in J. Linn. Soc., Bot. 29: 58 (1891) {896}. TYPE: Madagascar, Ankaratra Mts, *Scott Elliot* 1983 (holo. K).
Cynosorchis baronii (Rolfe) T.Durand & Schinz, Consp. Fl. Afr. 5: 90 (1892).
Cynosorchis nigrescens Schltr. in Beih. Bot. Centralbl. 34 (2): 311 (1916). TYPE: Madagascar, Mt. Ibity, *H.Perrier* XVII (8019) (holo. B†; iso. P).
Cynosorchis nigrescens var. *jumelleana* Schltr. in Beih. Bot. Centralbl. 34 (2): 311 (1916). TYPE: Madagascar, Ankaratra, *H.Perrier* VII (8086) (holo. B†; iso. P).

DISTRIBUTION: Madagascar: Antananarivo, Fianarantsoa.
HABITAT: montane, ericaceous scrub; rocky outcrops in debris; high-elevation grassland.
ALTITUDE: 1500 – 2500 m.
FLOWERING TIME: January – April.
LIFE FORM: terrestrial or lithophyte.
PHYTOGEOGRAPHICAL REGION: III, IV.
DESCRIPTION: {896}.
ILLUSTRATION: {504}, {896}, {906}, {1205}, {1223}.
NOTES: a relatively common species in the highlands; leaves spreading on the soil; flowers small, 12 – 15 mm long, white, pink or pink spotted with red, hirsute-glandular on the exterior; lip obscurely expanded-sublobulate above the narrow base, then 3-lobed, 4.5 – 6 mm long; spur cylindrical, up to 4 mm long.

Cynorkis bathiei *Schltr.* in Repert. Spec. Nov. Regni Veg. Beih. 33: 41 (1924). TYPE: Madagascar, Andringitra massif, *H.Perrier* 14569 (holo. P).
Cynosorchis inversa Schltr. in Repert. Spec. Nov. Regni Veg. Beih. 33: 55 (1924). TYPES: Madagascar, Andringitra massif, *H.Perrier* 14352 & *H.Perrier* 13754 (both syn. P).

DISTRIBUTION: Madagascar: Fianarantsoa.
HABITAT: montane, ericaceous scrub; rocky outcrops.
ALTITUDE: c. 2400 m.
FLOWERING TIME: February – April.
LIFE FORM: terrestrial.
PHYTOGEOGRAPHICAL REGION: IV.
DESCRIPTION: {896}.
ILLUSTRATION: {896}, {1223}.
NOTES: related to *Cynorkis lilacina* and *C. sororia*, but with slightly smaller flowers, also distinguished by the shape of the sepals and sepals and the long side lobes of the lip and the short spur, the waviness at the margin of the lip is very characteristic.

Cynorkis betsileensis *Kraenzl.*, Orch. Gen. Sp. 1: 492 (1898). TYPE: Madagascar, Betsileo, *Baron* 285 (holo. HBG).
DISTRIBUTION: Madagascar: Fianarantsoa.
PHYTOGEOGRAPHICAL REGION: III.
DESCRIPTION: {896}.
NOTES: known from the type only. Lip linear, c.10 × 2 mm, a little expanded towards the apex; spur slender, filiform, 2 times longer than the ovary.

Cynorkis bimaculata *(Ridl.)* *H.Perrier* in Humbert ed., Fl. Mad. Orch. 1: 114 (1939). TYPE: Madagascar, Inkagasoa, Imerina, *Deans Cowan* s.n. (holo. BM).
Habenaria bimaculata Ridl. in J. Linn. Soc., Bot. 21: 506 (1885).

DISTRIBUTION: Madagascar: Antananarivo.
HABITAT: humid, highland forest.
FLOWERING TIME: March.
PHYTOGEOGRAPHICAL REGION: III, IV.
DESCRIPTION: {896}.
ILLUSTRATION: {246}.
NOTES: flowers lilac, 15 – 17 mm long; lip broadly obovate, with two oblong dark purple spots in the centre,

Cynorkis boinana *Schltr.* in Ann. Mus. Col. Marseille, ser.3, 1: 150 (1913). TYPE: Madagascar, Boina, *H.Perrier* 72 (1875) (holo. B†; iso. P).

DISTRIBUTION: Madagascar: Antsiranana, Mahajanga.
HABITAT: seasonally dry, deciduous forest or woodland; rocky limestone outcrops; grassland.
ALTITUDE: sea level – 500 m.
FLOWERING TIME: September – February.
LIFE FORM: terrestrial or lithophyte.
PHYTOGEOGRAPHICAL REGION: V.
DESCRIPTION: {896}.
ILLUSTRATION: {1163}, {1223}.
NOTES: flowers mauve or pinkish-violet, c. 2.5 cm long; lip 3-lobed above; spur short, 5 – 6 mm long, expanded into a wide oval sphere at the tip.

Cynorkis brachycentra *(Frapp. ex A.Rich.)* *Kraenzl.*, Orch. Gen. Sp. 1, 8: 484 (1898). TYPE: Réunion, Saint-Denis, *J.M.C.Rich* (holo.?) or Comoros, Grand Benard, *Boivin* 1072 (cfr. Kraenzl. 1898).
Hemiperis brachycentra Frapp. ex. A.Rich. in Cordemoy, Fl. Réunion: 250 (1895).
DISTRIBUTION: Mascarenes: Réunion. Comoros.
NOTES: a poorly known species dubiously reported from the Comoros.

Cynorkis brachyceras *Schltr.* in Repert. Spec. Nov. Regni Veg. Beih. 33: 43 (1924). TYPE: Madagascar, Manankazo (NE Ankazobe), *H.Perrier* 11327 (holo. P).

DISTRIBUTION: Madagascar: Antananarivo.
HABITAT: marshland.
ALTITUDE: c. 1500 m.
FLOWERING TIME: November.
LIFE FORM: terrestrial in peaty and marshy soil.
PHYTOGEOGRAPHICAL REGION: III.
DESCRIPTION: {896}.
ILLUSTRATION: {896}, {1223}.
NOTES: flowers small, c. 12 mm long, white flushed with pink, the ovary and exterior of the dorsal sepal reddish, spur very short, globular, 2 × 2 mm.

Cynorkis brachystachya *Bosser* in Adansonia, sér. 2, 20 (3): 260 (1980). TYPE: Madagascar, Marivorahona massif, *Humbert & Capuron* 25772 (holo. P).

DISTRIBUTION: Madagascar: Antsiranana.
HABITAT: rocky outcrops.
ALTITUDE: 1500 – 2000 m.
FLOWERING TIME: March.
LIFE FORM: terrestrial.
PHYTOGEOGRAPHICAL REGION: IV.
DESCRIPTION: {124}.
ILLUSTRATION: {124}.
NOTES: known only from the type. In many aspects similar to plants of the genus *Physoceras*, but the column is typical for *Cynorkis*.

Cynorkis brauniana *Kraenzl.* in Bot. Jahrb. Syst. 17: 62 (1893). TYPE: Madagascar, without locality, *Joh. Braun,* not designated by Kraenzlin (holo. B?).

DISTRIBUTION: Madagascar.
DESCRIPTION: {1208}.
NOTES: type not located. Kraenzlin's description is insufficient to identify this species.

Cynorkis brevicornu *Ridl.* in J. Linn. Soc., Bot. 21: 516 (1885). TYPE: Madagascar, Ankafana, *Deans Cowan* (holo. BM).
Cynosorchis brevicornu (Ridley) T.Durand & Schinz, Consp. F. Afr. 5, 90 (1892).

DISTRIBUTION: Madagascar: Fianarantsoa.
HABITAT: damp places amongst rocks.
PHYTOGEOGRAPHICAL REGION: III.
DESCRIPTION: {896}.
ILLUSTRATION: {246}.
NOTES: flowers small, lip linear, narrow, spur c. 2 mm long, short and cylindrical.

Cynorkis cardiophylla *Schltr.* in Beih. Bot. Centralbl. 34 (2): 307 (1916). TYPE: Madagascar, Mt. Ibity, S of Antsirabe, *H.Perrier* 8089 (VI). (holo. B†; iso. K, P).

DISTRIBUTION: Madagascar: Antananarivo, Fianarantsoa.
HABITAT: montane, ericaceous scrub; rocky outcrops, in scree.
ALTITUDE: 1500 – 2400 m.
FLOWERING TIME: January – April.
LIFE FORM: terrestrial.
PHYTOGEOGRAPHICAL REGION: III, IV.
DESCRIPTION: {896}.
ILLUSTRATION: {896}, {1223}.
NOTES: plant with a single radical leaf spreading on the ground, suborbicular or broadly oval, c.4 × 3 cm; spur slender, 8 – 10 mm long.

Cynorkis catatii *Bosser* in Adansonia, sér. 2, 9 (3): 352 (1969). TYPE: Madagascar, road from Mandritsara, *Catat* 3207 B (holo. P).

DISTRIBUTION: Madagascar: Toamasina.
HABITAT: humid, evergreen forest.
FLOWERING TIME: September.
LIFE FORM: terrestrial in humus.
PHYTOGEOGRAPHICAL REGION: I.
DESCRIPTION: {113}.
ILLUSTRATION: {113}.
NOTES: known only from the type. Lip 6 – 8 mm long, 5-lobed, the 2 basal lobes minutely triangular; spur filiform, 13 – 14 mm long; ovary long and slender.

Cynorkis coccinelloides *(Frapp.* ex *Cordem.) Schltr.* in Beih. Bot. Centralbl. 33 (2): 399 (1915). TYPE: Réunion, cliffs of the Verdures river, *Richard* (lecto. MARS).
Camilleugenia coccinelloides Frapp., Orch. Réunion, Cat.: 10 (1880), *nom. nud.*; Frapp. ex Cordem., Fl. Réunion, 234 (1895).

DISTRIBUTION: Madagascar: Antsiranana. Mascarenes: Réunion.
HABITAT: montane, ericaceous scrub amongst sphagnum and peat moss; amongst rocks.
ALTITUDE: 1600 – 2400 m.
FLOWERING TIME: March – April.
LIFE FORM: terrestrial.
PHYTOGEOGRAPHICAL REGION: III, IV.
DESCRIPTION: {896}.
ILLUSTRATION: {896}.
NOTES: the plants from Madagascar differ from those from Réunion by their white flowers.

Cynorkis comorensis *Bosser* in Adansonia sér. 3, 24 (1): 21 (2002). TYPE: Comoros: Grande Comore, Lake Hantsongoma, *Floret* 1283 (holo. P).

DISTRIBUTION: Comoros: Grande Comore.
HABITAT: evergreen forest remnants on large canopy trees.
FLOWERING TIME: March.
LIFE FORM: epiphyte.
DESCRIPTION: {143}.
ILLUSTRATION: {143}.
NOTES: known only from the type. Flowers white, not opening completely; lip flat, c. 3 mm long, 5-lobed; spur c. 5 mm long.

Cynorkis confusa *H.Perrier* in Notul. Syst. (Paris) 14: 147 (1951). TYPE: Madagascar, E summit of Mt. Marojejy, *Humbert* 23756 (lecto. P; isolecto. K, both selected here); Mt. Beondroka (Lokoho Valley), *Humbert* 26601 (syn. P).

DISTRIBUTION: Madagascar: Antsiranana.
HABITAT: mossy forest; shaded and humid rocks; marshland.
ALTITUDE: 1400 – 2000 m.
FLOWERING TIME: March – April.
LIFE FORM: terrestrial in peaty and marshy soil.
PHYTOGEOGRAPHICAL REGION: III.
DESCRIPTION: {901}.
NOTES: close to *C. petiolata* but differs by the shape of the leaf, the petals and the lip in the shape of a halberd.

Cynorkis cuneilabia *Schltr.* in Repert. Spec. Nov. Regni Veg. Beih. 33: 45 (1924). TYPE: Madagascar, Mt. Tsiafajavona, *H.Perrier* 13503 (holo. P).

DISTRIBUTION: Madagascar: Antananarivo.
HABITAT: in the lighter parts of mossy forest.
ALTITUDE: c. 2000 m.
FLOWERING TIME: March.
LIFE FORM: terrestrial.
PHYTOGEOGRAPHICAL REGION: IV.
DESCRIPTION: {896}.
ILLUSTRATION: {896}, {1223}.
NOTES: known only from the type. Flowers pale pink; lip 8 × 5.5 mm, obovate, very shortly 3-lobulate at the tip, the middle lobule tri-crenulate; spur cylindrical, 3 mm long.

Cynorkis decaryana *H.Perrier* ex *Hermans* **sp. nov.** TYPE: Madagascar, Ranolania (nr. Mananara), *Decary* 74 (holo. P). See Appendix .
Cynosorkis decaryana H.Perrier in Humbert ed., Fl. Mad. Orch. 1: 105, fig. VIII, 26 – 31 (1939), *nom. nud.*

DISTRIBUTION: Madagascar: Toamasina.
ALTITUDE: sea level – 500 m.
FLOWERING TIME: September.
PHYTOGEOGRAPHICAL REGION: I.
DESCRIPTION: {896}.
ILLUSTRATION: {896}.
NOTES: known only from the type. A terrestrial herb with 5 radical ovate-lanceolate leaves; raceme, elongate, laxly 15-flowered; flowers small, lilac dorsal sepal obtusely hood-shaped; lateral sepals semi-obovate, the outer margin expanded-rounded below the middle, the opposite margin straight and shortly angular towards the tip; petals obovate, the internal edge expanded-rounded, then angular-obtuse; lip base narrow and concave, 3-lobed above the middle, the lateral lobes narrowly triangular, acute, a little shorter than the mid-lobe, which is very broadly transverse; spur recurved, as long as the sepals, very inflated towards the tip; anther retuse; viscidium as long as the caudicles.

Cynorkis disperidoides *Bosser* in Adansonia, ser.2, 9 (3): 345 (1969). TYPE: Madagascar, Ambondrombe, cult. Jard. Bot. Tananarive 4627 (holo. P).

DISTRIBUTION: Madagascar: Fianarantsoa.
HABITAT: moaay forest; on wet rocks; marshland.
ALTITUDE: 1500 – 2000 m.
FLOWERING TIME: April.
LIFE FORM: terrestrial or lithophyte.
PHYTOGEOGRAPHICAL REGION: III.
ILLUSTRATION: {113}.
NOTES: similar in appearance to *C. papilio* but differs by floral characteristics, in general appearance it reminds one of certain *Disperis*.

Cynorkis elata *Rolfe* in J. Linn. Soc., Bot. 29: 58 (1891). TYPE: Madagascar, Fort Dauphin, *Scott Elliot* 2477 (holo. K; iso. P, BM).

DISTRIBUTION: Madagascar: Fianarantsoa, Toamasina, Toliara.
HABITAT: coastal forest.
ALTITUDE: sea level – 50 m.
FLOWERING TIME: April – May.
LIFE FORM: terrestrial in sand and in humus.
PHYTOGEOGRAPHICAL REGION: I.
DESCRIPTION: {896}.
ILLUSTRATION: {896}, {906}.
NOTES: plant small with 2 marbled leaves spreading on the ground; flowers c. 2 cm; lip 8 mm long, 5-lobed; spur 13 mm long, a little contracted in the middle and very slightly expanded toward the tip.

Cynorkis elegans *Rchb.f.* in Flora: 150 (1888) & in Gard. Chron. ser. 3, 3: 424 (1888). TYPE: Madagascar, cult. R.B.G. Kew, *Low* s.n. (holo. W).
Gymnadenia muricata Brogn. ex Kraenzl., Orch. Gen. Sp.: 482 (1898), in sched.

DISTRIBUTION: Madagascar.
DESCRIPTION: {896}.
NOTES: known only from some detailed sketches of the type by Reichenbach, likely to be conspecific with another plant.

Cynorkis ericophila *H.Perrier* ex *Hermans* **sp. nov.** TYPE: Madagascar, Mt. Tsaratanana, *H.Perrier* 16517 (holo. P). See Appendix .
Cynosorchis ericophila H.Perrier in Fl. Mad. Orch. 1: 157 (1939), *nom. nud.*

DISTRIBUTION: Madagascar: Antsiranana.
HABITAT: montane, ericaceous scrub.
ALTITUDE: c. 2400 m.
FLOWERING TIME: April.
PHYTOGEOGRAPHICAL REGION: IV.
DESCRIPTION: {896}.
NOTES: known only from the type. Plant slender with a lanceolate, acute leaf narrowing below into a long petiole; flowers pink, papillose; dorsal sepal hooded; the laterals larger, very asymmetric, broadly ovate, very obtuse; lip oblong, with an obtuse tooth on each side in the third; spur 6 mm long, a little expanded towards the tip; anther obtuse, the caudicles short, broadened below the pollinia; rostellum with thick arms.

Cynorkis fastigiata *Thouars*, Hist. Orch.: t.13 (1822), as *Cynorchis fastigiata.* TYPE: *Thouars* s.n. (holo. P).
Orchis obcordata Willem. in Usteri, Ann. Bot. 18: 52 (1796), *non* Buch.-Ham. ex D. Don, Prodr. Fl. Nepal. 23 (1825).
Cynorkis obcordata (Willem.) Schltr. in Beih. Bot. Centralbl. 33, 2: 401 (1915). Type: Mauritius, *Willemet* s.n. (holo. not located).

Cynorchis isocynis Thouars *loc. cit., nom. alt.*

Orchis fastigiata (Thouars) Spreng., Syst. Veg. 3: 687 (1826) .

Orchis mauritiana Sieber ex Lindl., Gen. Sp. Orch. Pl.: 332 (1835). Type: Mauritius, without locality. *Sieber* 169 (holo. K; iso. P).

Cynosorchis hygrophila Schltr. in Beih. Bot. Centralbl. 34 (2): 309 (1916). Type: Madagascar, along the river Fandrarazana (NE), wet areas, *H.Perrier* 11398 (*Schlechter* 100) (holo. B†; iso. P).

Cynosorchis diplorhyncha Schltr. in Repert. Spec. Nov. Regni Veg. Beih. 15: 325 (1918). TYPE: Madagascar, St Marie, *Laggiara* s.n.(holo. B†).

Cynosorchis laggiarae Schltr. in Repert. Spec. Nov. Regni Veg. Beih. 15: 326 (1918). Type: Madagascar, *Laggiara* s.n. (holo. B†).

Cynosorchis laggiarae var. *ecalcarata* Schltr. in Repert. Spec. Nov. Regni Veg. Beih. 15: 326 (1918). Type: Madagascar, *Laggiara* s.n. (holo. B†).

Cynosorchis decolorata Schltr. in Repert. Spec. Nov. Regni Veg. Beih. 33: 46 (1924). TYPE: Madagascar, Sambirano mountains, *H.Perrier* 15713 (holo. P).

Gymnadenia fastigiata (Thouars) A.Rich., Monogr. Orch. Iles France Bourbon: 25 (1928).

Habenaria cynosorchidacea C.Schweinf. in Bull. Bishop. Mus. Honolulu 141; 18 (1936).

Cynorkis fastigiata var. *typica* H.Perrier in Humbert, Fl. Mad. Orch. 1: 141 (1939), *nom. inval.*

Cynorchis fastigiata var. *decolorata* (Schltr.) H.Perrier in Humbert, Fl. Mad. Orch. 1: 141 (1939), *nom. inval.*

Cynorchis fastigiata var. *diplorhyncha* (Schltr.) H.Perrier, in Humbert, Fl. Mad. Orch. 1: 141 (1939), *nom. inval.*

Cynorchis fastigiata var. *hygrophila* (Schltr.) H.Perrier in Humbert, Fl. Mad. Orch. 1: 140 (1939), *nom. inval.*

Cynorchis fastigiata var. *laggiarae* (Schltr.) H.Perrier in Humbert, Fl. Mad. Orch. 1: 141 (1939), *nom. inval.*

DISTRIBUTION: Madagascar: Antananarivo, Antsiranana, Fianarantsoa, Toamasina, Toliara. Comoros: Anjouan, Mayotte, Moheli. Mascarenes: Mauritius. Seychelles. Naturalised in several other countries.

HABITAT: on edges of woods; riverine forest; littoral forest; evergreen forest, on base of trees; amongst rocks; in grassland; river margins.

ALTITUDE: sea level – 2000 m.

FLOWERING TIME: November – June.

LIFE FORM: terrestrial, occasionally epiphyte.

PHYTOGEOGRAPHICAL REGION: I, II, III, IV.

DESCRIPTION: {779}, {896}.

HISTORY: {1139}.

ILLUSTRATION: {4}, {253}, {281}, {473}, {504}, {779}, {906}, {944}, {1223}, {1319}, {1398}.

VERNACULAR NAME: Sofinampory.

NOTES: plant with 1 – 2 leaves; flowers c. 3 cm long, creamy white variably toned reddish-purple; lip 10 – 15 mm long, unequally 4-lobed; spur cylindrical-narrow, 18 – 30 mm long. It is likely that the earliest name for the well-known *Cynorkis fastigiata* Thouars was overlooked. Ormerod (1996) concluded that the description of *C. obcordata* in Willemet (1796), based on a plant from Mauritius, matched that of Thouars' plant. There is little doubt that this is the case but without a type specimen there will never be any certainty. The name *C. fastigiata* has been used extensively in the literature and in horticulture. We therefore propose to conserve the later name and a note is being prepared for *Taxon* to keep it as a *nomen conservandum*. *Cynorkis fastigiata* is a variable and widespread species throughout Madagascar, the Comoros, the Mascarenes and the Seychelles, a number of varieties were described by Perrier, most of them fall within a phenotypically plastic species. Two distinct varieties are recognised here.

Cynorkis fastigiata var. **ambatensis** Hermans **var. nov.**: Madagascar, Sambirano nr. Ambato peninsula, *H.Perrier* 1857 (holo. P). See Appendix .

Cynorchis fastigiata var. *ambatensis* H.Perrier, Fl. Mad. Orch. 1: 141 (1939), *nom. nud.*

DISTRIBUTION: Madagascar: Antsiranana.

HABITAT: dry slopes, in the shade of scrub.

FLOWERING TIME: January.

PHYTOGEOGRAPHICAL REGION: II.

NOTES: differs by its small flowers, 4 mm long sepals and petals, 5 mm long lip, 9 mm long spur, glabrous ovary and entire middle lobe of the rostellum.

Cynorkis fastigiata var. **triphylla** *(Thouars) S.Moore* in Baker, Fl. Mauritius: 337 (1877). TYPE: Mauritius, *Thouars* s.n. (holo. P).

Cynorkis triphylla Thouars, Hist. Orch.: t.14 (1822), as *Cynosorchis*. Type: Mauritius, *Thouars* s.n. (holo. P).

Orchis triphylla (Thouars) Spreng., Syst. Veg. 3: 687 (1826).

Gymnadenia triphylla (Thouars) A.Rich., Monogr. Orch. Iles France Bourbon: 26 (1828).

DISTRIBUTION: Comoros: Mayotte. Mascarenes: Mauritius.
DESCRIPTION: {896}.
ILLUSTRATION: {1398}.
NOTES: leaves 3, radical; inflorescence often single-flowered.

Cynorkis filiformis *Schltr.* in Repert. Spec. Nov. Regni Veg. Beih. 33: 48 (1924), *non* (Kraenzl.) H.Perrier. TYPE: Madagascar, Andringitra massif, *H.Perrier* 14354 (holo. P).

Cynorkis hirtula H.Perrier in Arch. Bot. Bull. Mens. 5: 49 (1931), *nom. inval.* TYPE: Madagascar, Andringitra massif, *H.Perrier* 14554 (holo. P).

DISTRIBUTION: Madagascar: Antananarivo, Fianarantsoa.
HABITAT: montane, ericaceous scrub, in sphagnum moss.
ALTITUDE: 1100 – 2400 m.
FLOWERING TIME: January – March.
LIFE FORM: terrestrial.
PHYTOGEOGRAPHICAL REGION: IV.
DESCRIPTION: {896}.
ILLUSTRATION: {281}, {896}, {1223}.
NOTES: plant slender with a single linear-acute leaf; flowers small, c. 14 mm long; lip 4 × 5 mm, the base narrow, the lateral lobes very small and short; spur narrowly cylindrical, 4.5 mm long.

Cynorkis fimbriata *H.Perrier* ex *Hermans* **sp. nov.** TYPE: Madagascar, Mt. Tsaratanana, *H.Perrier* 16492 (holo. P). See Appendix .

Cynosorchis fimbriata H.Perrier in Fl. Mad. Orch. 1: 122, fig. X, 12 – 16 (1939), *nom. nud.*

DISTRIBUTION: Madagascar: Antsiranana.
HABITAT: shaded, amongst wet rocks.
ALTITUDE: 2000 – 2200 m.
FLOWERING TIME: April.
LIFE FORM: terrestrial.
PHYTOGEOGRAPHICAL REGION: III, IV.
DESCRIPTION: {896}.
ILLUSTRATION: {896}.
NOTES: stem and spike enclosed at the base by a long sheath; leaf solitary, well-developed; raceme short, 4- to 8-flowered; floral bracts ample, a third or half the length of the ovary; flowers bright purple; sepals obtuse, the dorsal sepal ovate, concave; the laterals almost semi-orbicular; petals narrow, curved, obtuse; lip 5-lobed with the mid-lobe divided, fimbriate-dentate at the ends of the lobes; spur narrowly cylindrical, then clearly inflated in the apical third; anther retuse; rostellum divided above in 2 parallel projections.

Cynorkis flabellifera *H.Perrier* in Notul. Syst. (Paris) 14: 144 (1951). TYPE: Madagascar, E of Marojejy massif, *Humbert* 22669 (holo. P).

DISTRIBUTION: Madagascar: Antsiranana.
HABITAT: montane, ericaceous scrub.
ALTITUDE: 1500- 2000 m.
FLOWERING TIME: December.
LIFE FORM: terrestrial.
PHYTOGEOGRAPHICAL REGION: IV.
DESCRIPTION: {901}.
NOTES: known from the type only. Similar to *C. henrici* but with smaller leaves and flowers, the spur c. 5 mm long, inflated into a sphere at the tip.

Cynorkis flexuosa *Lindl.,* Gen. Sp. Orch. Pl.: 331 (1835). TYPE: Madagascar, *Lyall* s.n. (holo. K).

Gymnadenia lyallii Steud., Nom. Bot., ed. 2, 1: 712 (1840), *nom. illeg.,* based upon *Cynorkis flexuosa.*

Cynorkis fallax Schltr. in Beih. Bot. Centralbl. 34 (2): 308 (1916). TYPE: Madagascar, nr. Antsirabe, *H.Perrier* 8108 (XLVI) (holo. B†; iso. P).

Cynosorchis flexuosa subsp. *fallax* (Schltr.) H.Perrier in Arch. Bot. Bull. Mens. 5: 64 (1931), *nom. inval.*

DISTRIBUTION: Madagascar: Antananarivo, Antsiranana, Fianarantsoa, Mahajanga, Toamasina. Comoros: Mayotte.

HABITAT: grassland; rocky outcrops.

ALTITUDE: sea level – 1700 m.

FLOWERING TIME: October – August.

LIFE FORM: terrestrial.

PHYTOGEOGRAPHICAL REGION: I, III, V.

ILLUSTRATION: {281}, {906}, {945}, {1282}.

DESCRIPTION: {896}.

NOTES: a common and somewhat variable species; lip yellow, generally with two small reddish spots at base, rest of perianth green. *C. fallax* differs from the typical form of this common and variable species by its smaller size.

Cynorkis flexuosa var. **bifoliata** *Schltr.* in Beih. Bot. Centralbl. 34 (2): 308 (1916). TYPE: Madagascar, nr. Soalala (Ambongo), *H.Perrier* XLII (1682) (holo. B†; iso. P).

Cynosorchis flexuosa var. *ambongensis* H.Perrier in Arch. Bot. Bull. Mens. 5: 64 (1931), *nom. inval.* Based upon: Madagascar, Manongarivo (Ambonga) *H.Perrier* 1682 (holo. P).

DISTRIBUTION: Madagascar: Mahajanga.

HABITAT: marshes in peaty soil.

FLOWERING TIME: January.

LIFE FORM: terrestrial.

PHYTOGEOGRAPHICAL REGION: V.

DESCRIPTION: {896}.

NOTES: differs by the greater number of radical leaves, 2 to 3, and the larger, white or pink lip, with the spot on the disc a rich violet-red; spur 15 mm.

Cynorkis formosa *Bosser* in Adansonia, sér. 2, 9 (3): 356 (1969). TYPE: Madagascar, Marivorahona massif, *Humbert & Capuron* 25770 (holo. P; iso. K, TAN).

DISTRIBUTION: Madagascar: Antsiranana.

HABITAT: montane, ericaceous scrub on granite and gneiss; lichen-rich forest.

ALTITUDE: 2000 – 2500 m.

FLOWERING TIME: March.

LIFE FORM: terrestrial.

PHYTOGEOGRAPHICAL REGION: IV.

DESCRIPTION: {113}.

ILLUSTRATION: {113}.

NOTES: close to *C. bathiei* but differs by its much smaller flowers and sepals, petals and lip of a different shape; flower pinkish white with red spots.

Cynorkis gaesiformis *H.Perrier* in Notul. Syst. (Paris) 14: 146 (1951). TYPE: Madagascar, S of Moramanga, *Decary* 18419 (holo. P).

DISTRIBUTION: Madagascar: Toamasina.

HABITAT: humid, evergreen forest.

ALTITUDE: 600 – 800 m.

FLOWERING TIME: January.

LIFE FORM: terrestrial.

PHYTOGEOGRAPHICAL REGION: I.

DESCRIPTION: {901}.

NOTES: close to *Cynorkis nutans,* also related to *C. spathulata* but differs by the single radical leaf and the lip in the shape of a halberd.

Cynorkis galeata *Rchb.f.* in Flora 68: 536 (1885). TYPE: Comoros, Grande Comore, *Humblot* 1209 (209) (holo. W).

DISTRIBUTION: Madagascar? Comoros: Anjouan, Grande Comore.
HABITAT: rocky outcrops.
FLOWERING TIME: September – May.
LIFE FORM: epiphyte or lithophyte.
DESCRIPTION: {896}.
ILLUSTRATION: {896}.
NOTES: very close to *C. nutans*. Flowers lilac-violet, c. 3 cm; lip 7 – 8 mm long, ligulate.

Cynorkis gibbosa *Ridl.* in J. Linn. Soc., Bot. 20: 331 (1883). TYPE: Madagascar, Imerina, Ankafana, *Deans Cowan* (holo. BM).

DISTRIBUTION: Madagascar: Antananarivo, Fianarantsoa, Toamasina, Toliara.
HABITAT: edge of forest; shaded, wet rocks in open but sheltered situations; generally in peaty or humus-rich soil in areas of seepage; on granite rock outcrop.
ALTITUDE: 600 – 1800 m.
FLOWERING TIME: August – May.
LIFE FORM: lithophyte or terrestrial.
PHYTOGEOGRAPHICAL REGION: I, III, IV.
DESCRIPTION: {278}, {896}.
ILLUSTRATION: {69}, {81}, {278}, {281}, {464},{487}, {504}, {896}, {906}.
CULTIVATION: {69}, {278}, {488}, {700}.
NOTES: plant large, single-leafed with the young growths carrying dark spots; flowers big, orange-red; lip 3-lobed, up to 29 mm long. A yellow-white form is also known.

Cynorkis gigas *Schltr.* in Repert. Spec. Nov. Regni Veg. Beih. 33: 50 (1924). TYPE: Madagascar, Andringitra massif, *H.Perrier* 14330 (holo. P).

DISTRIBUTION: Madagascar: Antananarivo, Fianarantsoa, Toliara.
HABITAT: rocky outcrops.
ALTITUDE: 800 – 2000 m.
FLOWERING TIME: February – April.
LIFE FORM: lithophyte or terrestrial.
PHYTOGEOGRAPHICAL REGION: III, IV.
DESCRIPTION: {896}.
ILLUSTRATION: {896}, {906}.
NOTES: somewhat similar to *C. uniflora* but larger and with a 10 cm long spur.

Cynorkis glandulosa *Ridl.* in J. Linn. Soc., Bot. 22: 123 (1886). TYPE: Madagascar, Ambatovory, Imerina, *Fox* (holo. K).

DISTRIBUTION: Madagascar: Antananarivo.
HABITAT: rocky outcrops.
FLOWERING TIME: March.
PHYTOGEOGRAPHICAL REGION: III.
DESCRIPTION: {896}.
NOTES: known from the type only. Allied to *C. lilacina* but distinguished by its lanceolate leaves, smaller flowers, bifid lip, and short points to the anther.

Cynorkis globifera *H.Perrier* in Notul. Syst. (Paris) 14: 146 (1951). TYPE: Madagascar, Mandrare valley, Manampanihy, *Humbert* 20479 (holo. P; iso. K).

DISTRIBUTION: Madagascar: Toliara.
HABITAT: humid, lowland forest.
FLOWERING TIME: March.
LIFE FORM: terrestrial.
PHYTOGEOGRAPHICAL REGION: I.

DESCRIPTION: {901}.

NOTES: plant with 2 – 3 petiolate leaves; flowers small; spur globular.

Cynorkis globosa *Schltr.* in Bot. Jahrb. Syst. 38: 145 (1906). TYPE: Madagascar, Inala Bay, *Bang* s.n. (holo. B†); neo. illustration in Repert. Spec. Nov. Reg. Veg., 68 t.16, 63 (1932), selected here.

DISTRIBUTION: Madagascar: Toamasina ?
LIFE FORM: terrestrial.
DESCRIPTION: {896}.
ILLUSTRATION: {1223}.
NOTES: known only from Schlechter's short description and illustration.

Cynorkis graminea *(Thouars) Schltr.* in Repert. Spec. Nov. Regni Veg. Beih. 33: 51 (1924). TYPE: Madagascar, *Thouars* s.n. (holo. P).
Satyrium gramineum Thouars, Hist. Orch.: t.6 (1822).
Habenaria graminea (Thouars) Spreng., Syst. Veg. 3: 690 (1826).
Platanthera graminea (Thouars) Lindl., Gen. Sp. Orch. Pl.: 292 (1835).
Bicornella longifolia Lindl., Gen. Sp. Orch. Pl.: 335 (1838). TYPE: Madagascar, in Herb Lehmann s.n. (holo. K).
Peristylus gramineus (Thouars) S.Moore in Baker, Fl. Mauritius: 336 (1877).
Bicornella parviflora Ridl. in J. Linn. Soc., Bot. 21: 500 (1885). TYPES: Madagascar, Imerina, *Deans Cowan* s.n., *Hildebrandt* 3820 & *Lyall* 308 (all syn. BM).
B. similis Schltr. in Beih. Bot. Centralbl. 34 (2): 305 (1916). TYPE: Madagascar, Antsirabe, *H.Perrier* 8125A (XXXIII) (holo. B†).
Benthamia graminea (Thouars) Schltr. in Repert. Spec. Nov. Regni Veg. Beih. 33: 24 (1924).
Cynosorchis longifolia (Lindl.) Schltr. in Repert. Spec. Nov. Regni Veg. Beih. 33: 59 (1924).
Cynosorchis similis (Schltr.) Schltr. in Repert. Spec. Nov. Regni Veg. Beih. 33: 71 (1924).

DISTRIBUTION: Madagascar: Antananarivo, Fianarantsoa, Toamasina, Toliara. Mascarenes: Réunion, Mauritius.
HABITAT: marshland in peaty soil; wet grassland, beside streams.
ALTITUDE: sea level – 2000 m.
FLOWERING TIME: August – March.
LIFE FORM: terrestrial.
PHYTOGEOGRAPHICAL REGION: I, III.
DESCRIPTION: {896}.
ILLUSTRATION: {165}, {1223}, {1398}.
NOTES: a common species in wet areas; roots fleshy; flowers small, pink or purple with darker spots; lip entire, a little expanded in thepart; spur c. 4 mm long, a little contracted in the middle and widened at the tip.

Cynorkis guttata *Hermans & P.J. Cribb*
This new species is in press to be described in The Orchid Review. It was illustrated as *Cynorkis uncinata* by a colour photograph on plate 23 in the first edition of this work (Du Puy *et al.* 1999).

Related to *C. purpurascens* but differs by its larger leaf, the greater number of flowers on the raceme, its distinctly coloured flowers, larger and differently proportioned lip and shorter spur. It is fairly common in the highlands of Madagascar.

Cynorkis gymnochiloides *(Schltr.) H.Perrier* in H.Perrier, Fl. Mad. Orch. 1: 229 (1939) ex. H.Perrier Arch. Bot. Bull. Mens. 5: 66 (1931). TYPE: Madagascar, Mt. Ibity, S Antsirabe, *H.Perrier* 8080 (XVI) (holo. P).
Habenaria gymnochiloides Schltr. in Beih. Bot. Centralbl. 34 (2): 313 (1916).
Lemuranthe gymnochiloides (Schltr.) Schltr. in Repert. Spec. Nov. Regni Veg. Beih. 33: 85 (1924).

DISTRIBUTION: Madagascar: Antananarivo.
HABITAT: rocky outcrops.
ALTITUDE: 1700 – 2200 m.
FLOWERING TIME: February – March.

LIFE FORM: lithophyte or terrestrial.
PHYTOGEOGRAPHICAL REGION: III, IV.
DESCRIPTION: {896}.
ILLUSTRATION: {1223}.
NOTES: raceme short, densely-flowered; flowers glabrous, pink or violet; lip entire, narrowly ligulate and obtuse; spur cylindrical, obtuse, 4 mm long.

Cynorkis henricii *Schltr.* in Repert. Spec. Nov. Regni Veg. Beih. 33: 52 (1924). TYPE: Madagascar, Bay of Antongil, *H.Perrier* 11345 (holo. P).

DISTRIBUTION: Madagascar: Toamasina.
HABITAT: humid evergreen forest.
ALTITUDE: c. 500 m.
FLOWERING TIME: August.
PHYTOGEOGRAPHICAL REGION: I.
DESCRIPTION: {896}.
ILLUSTRATION: {896}.
NOTES: close to *C. ridleyi* but with 3 – 4 leaves; flowers up to 20 mm long; lip 3-lobed; spur hook-shaped.

Cynorkis hispidula *Ridl.* in J. Linn. Soc., Bot. 21: 517 (1885). TYPES: Madagascar, Imerina, *Deans Cowan* s.n. (lecto. BM, selected here), Ankafana, *Hildebrandt* 3981 (syn. BM).
Cynosorchis hispidula (Ridl.) T.Durand & Schinz, Consp. Fl. Afr: 5: 91 (1892).

DISTRIBUTION: Madagascar: Antananarivo, Fianarantsoa.
HABITAT: shaded and humid granite and gneiss rocks.
ALTITUDE: c. 2000 m.
FLOWERING TIME: February – March.
LIFE FORM: lithophyte.
PHYTOGEOGRAPHICAL REGION: III, IV.
DESCRIPTION: {896}.
ILLUSTRATION: {896}.
NOTES: plant small; flowers white; lip spathulate, 4 mm long; spur less than 5 mm long.

Cynorkis hologlossa *Schltr.* in Repert. Spec. Nov. Regni Veg. Beih. 33: 54 (1924). TYPE: Madagascar, Mt. Tsaratanana, *H.Perrier* 15306 (holo. P).

DISTRIBUTION: Madagascar: Antsiranana.
HABITAT: lichen-rich, montane forest.
ALTITUDE: c. 2000 m.
FLOWERING TIME: January.
LIFE FORM: terrestrial.
PHYTOGEOGRAPHICAL REGION: III, IV.
DESCRIPTION: {896}.
NOTES: known only from the type; plant single-leafed; flower pink or lilac, 16 mm long; lip entire; spur narrowly cylindrical, obtuse, 6.5 mm long.

Cynorkis hologlossa var. **angustilabia** *H.Perrier* ex *Hermans* **var. nov.** TYPE: Madagascar, Mt. Tsaratanana, *H.Perrier* 16516 (holo. P). See Appendix.
Cynosorchis hologlossa var. *angustilabia* H.Perrier in Fl. Mad. Orch. 1: 155 (1939), *nom. nud.*

DISTRIBUTION: Madagascar: Antsiranana.
HABITAT: montane, ericaceous scrub.
ALTITUDE: c. 2600 m.
FLOWERING TIME: April.
LIFE FORM: terrestrial.
PHYTOGEOGRAPHICAL REGION: IV.
DESCRIPTION: {896}.

NOTES: differs from the typical form in being taller, and in having shorter floral bracts which are about a third the length of the ovary, a shorter dorsal sepal, a shorter lip that is slightly expanded in thequarter, a shorter spur, and a column with its stigmatic processes free in the upper two-thirds, and a rostellum with markedly shorter arms.

Cynorkis hologlossa var. **gneissicola** *Bosser* in Adansonia, sér. 2, 9 (3): 357 (1969). TYPE: Madagascar, Marivorahona Mts, SW of Manambato (upper Mahavavy), *Humbert & Capuron* 25771 (holo. P; iso. K).

DISTRIBUTION: Madagascar: Antsiranana.
HABITAT: lichen-rich forest; on granite and gneiss.
ALTITUDE: 2000 – 2200 m.
FLOWERING TIME: March.
LIFE FORM: terrestrial or lithophyte.
PHYTOGEOGRAPHICAL REGION: IV.
NOTES: differs from the typical form by the shape of the rostellum.

Cynorkis humbertii *Bosser* in Adansonia, sér. 2, 9 (3): 350 (1969). TYPE: Madagascar, Mts. N of Mangindrano, Upper Maevarano, near the summit of Ambohimirahavavy, *Humbert & Capuron* 24901 (holo. P).

DISTRIBUTION: Madagascar: Mahajanga.
HABITAT: lichen-rich, evergreen forest.
ALTITUDE: 1600 – 2000 m.
LIFE FORM: terrestrial.
PHYTOGEOGRAPHICAL REGION: IV.
DESCRIPTION: {113}.
ILLUSTRATION: {113}.
NOTES: known only from the type. Close to *C. brachyceras* but different by the rostellum and lip which is 3-lobed; spur short, up to 1.5 mm.

Cynorkis humblotiana *Kraenzl.* in Bot. Jahrb. Syst. 28: 176 (1901). TYPE: Comoros, *Humblot* s.n. (holo. not located).

DISTRIBUTION: Comoros.
DESCRIPTION: {896}.
NOTES: a poorly known species, said to be close to *C. fastigiata* and *C. gracilis.*

Cynorkis jumelleana *Schltr.* in Repert. Spec. Nov. Regni Veg. Beih. 33: 56 (1924). TYPE: Madagascar, Andringitra massif, *H.Perrier* 14399 (holo. P).

DISTRIBUTION: Madagascar: Fianarantsoa.
HABITAT: in ericaceous shrub.
ALTITUDE: c. 2500 m.
FLOWERING TIME: February.
LIFE FORM: terrestrial.
PHYTOGEOGRAPHICAL REGION: IV.
DESCRIPTION: {896}.
ILLUSTRATION: {896}, {1223}.
NOTES: related to *C. tenerrima* but the lip is expanded above the narrow base and deeply 3-lobed.

Cynorkis jumelleana var. **gracillima** *Schltr.* in Repert. Spec. Nov. Regni Veg. Beih. 33: 57 (1924). TYPE: Madagascar, Andringitra massif, *H.Perrier* 14353 (holo. P).

DISTRIBUTION: Madagascar: Fianarantsoa.
HABITAT: montane, ericaceous scrub.
ALTITUDE: c. 2400 m.
FLOWERING TIME: January.
LIFE FORM: terrestrial.
PHYTOGEOGRAPHICAL REGION: IV.

DESCRIPTION: {896}.

NOTES: plant more slender, fewer-flowered; flowers smaller; lip wider; the staminodes narrower.

Cynorkis laeta *Schltr.* in Repert. Spec. Nov. Regni Veg. Beih. 33: 57 (1924). TYPES: Madagascar, nr. Itremo, W Betsileo, *H.Perrier* 12464 (lecto. P selected here); Andringitra massif, *H.Perrier* 14452 (syn. P).

DISTRIBUTION: Madagascar: Fianarantsoa.
HABITAT: shaded, humid, granite and gneiss rocks.
ALTITUDE: 1000 – 2000 m.
FLOWERING TIME: February – May.
LIFE FORM: terrestrial or lithophyte .
PHYTOGEOGRAPHICAL REGION: III.
DESCRIPTION: {896}.
ILLUSTRATION: {896}, {1223}.
NOTES: similar to *C. gracilis*; flowers purple; lip up to 10 mm, 3-lobed; spur 9 mm long, cylindrical-obtuse.

Cynorkis laeta var. **angavoensis** *H.Perrier* ex *Hermans* **var. nov.**: TYPE: Madagascar, Angavo massif, *R.Decary* 7344 (lecto. P; isolecto. TAN, both selected here). See Appendix.
Cynorkis laeta var. *angavoensis* H.Perrier in Humbert, Fl. Mad. Orch. 1: 131 (1939), *nom. nud.*

DISTRIBUTION: Madagascar: Antananarivo.
FLOWERING TIME: March – May.
LIFE FORM: terrestrial.
PHYTOGEOGRAPHICAL REGION: III.
DESCRIPTION: {896}
NOTES: differs by its more numerous radical leaves (7 – 8) that are only half the width (4 – 8 mm) of those in the typical form.

Cynorkis lancilabia *Schltr.* in Repert. Spec. Nov. Regni Veg. Beih. 33: 58 (1924). TYPE: Madagascar, Mt. Tsaratanana, *H.Perrier* 15711 (holo. P).

DISTRIBUTION: Madagascar: Antsiranana.
HABITAT: forest.
ALTITUDE: c. 2000 m.
FLOWERING TIME: January.
LIFE FORM: terrestrial.
PHYTOGEOGRAPHICAL REGION: IV.
DESCRIPTION: {896}.
NOTES: known from the type only. Plant single-leafed; flowers violet; lip 7 × 2.3 mm, entire, lanceolate, acute; spur recurved in a complete curl, c. 8 mm long.

Cynorkis latipetala *H.Perrier* in Notul. Syst. (Paris) 14: 147 (1951). TYPES: Madagascar, E summit of Marojejy massif, *Humbert & Cours* 23877 (lecto. P, selected here); *Humbert* 23569 (syn. P).

DISTRIBUTION: Madagascar: Antsiranana.
HABITAT: marshes in peaty soil; mossy and lichen-rich forest; rocky outcrops.
ALTITUDE: 1200 – 2050 m.
LIFE FORM: terrestrial.
PHYTOGEOGRAPHICAL REGION: III, IV.
DESCRIPTION: {901}.
NOTES: flowers green, tinted purple-red; lip 3-lobed, 15 × 15 mm, divided and denticulate in front; spur up to 10 mm long, inflated towards the tip.

Cynorkis lilacina *Ridl.* in J. Linn. Soc., Bot. 21: 515 (1885). TYPES: Madagascar, Ankafana, *Deans Cowan* s.n. (syn. BM); *Baron* s.n (syn. K); & *Lyall* s.n. (syn. BM).
Cynosorchis lilacina (Ridl.) T.Durand & Schinz, Consp. Fl. Afr. 5: 92 (1892).
Cynorkis lilacina var. *typica* H.Perrier in Humbert ed., Fl. Mad. Orch. 1: 90 (1939), *nom. inval.*

DISTRIBUTION: Madagascar: Antananarivo, Fianarantsoa.

HABITAT: grassland; marshes; on inselbergs.

ALTITUDE: 1300 – 1800 m.

FLOWERING TIME: November – April.

LIFE FORM: terrestrial.

PHYTOGEOGRAPHICAL REGION: III.

DESCRIPTION: {896}.

ILLUSTRATION: {246}.

NOTES: a fairly common and variable species; leaf single with a distinct petiole; flowers violet or lilac, small; lip 3-lobed, 1 cm long; spur broad at the base then narrowed and inflated and obtuse-truncate at the tip, 7 mm long.

Cynorkis lilacina var. **boiviniana** *(Kraenzl.) H.Perrier* in Humbert ed., Fl. Mad. Orch. 1: 92 (1939).
TYPE: Comoros, Grande Comore, *Boivin* s.n. (holo. HBG; iso. P).
Cynorkis boiviniana Kraenzl., Orch. Gen. Sp. 1: 483 (1901).

DISTRIBUTION: Comoros: Grande Comore.

FLOWERING TIME: May.

LIFE FORM: terrestrial.

DESCRIPTION: {896}.

NOTES: spur similar to the typical form, but longer; lip narrower.

Cynorkis lilacina var. **comorensis** *H.Perrier ex Hermans* **var. nov.** TYPE: Comoros, Grande Comore, *Boivin* s.n. (holo. P). See Appendix .
Cynorkis lilacina var. *comorensis* H.Perrier in Humbert, Fl. Mad. Orch. 1: 92 (1939), *nom. nud.*

DISTRIBUTION: Comoros: Grande Comore.

FLOWERING TIME: May.

LIFE FORM: terrestrial.

DESCRIPTION: {896}.

NOTES: differs from the typical variety in having a straight spur, gradually narrowing from its mouth to its obtuse apex and more than half the length of the ovary; lip thick with narrow lobes, the lateral ones being one-veined.

Cynorkis lilacina subsp. **curvicalcar** *(H.Perrier) H.Perrier* in Humbert ed., Fl. Mad. Orch. 1: 92 (1939). TYPE: Madagascar, Mt. Tsaratanana, *H.Perrier* 16518 (holo. P).
Cynorkis andringitrana subsp. *curvicalcar* H.Perrier in Arch. Bot. Bull. Mens. 5: 53 (1931).

DISTRIBUTION: Madagascar: Antsiranana.

HABITAT: in 'Savoka' (*Philippia*) scrub.

ALTITUDE: c. 2400 m.

LIFE FORM: terrestrial.

PHYTOGEOGRAPHICAL REGION: IV.

DESCRIPTION: {896}.

NOTES: based on *C. andringitrana* subsp. *curvicalcar* which was never published (Perrier, 1939). Differs by the petals with the outer edges often expanded-angular, the short, 5 mm long spur and the rostellum which is much shorter (1 mm).

Cynorkis lilacina var. **laxiflora** *(Schltr.) H.Perrier* in Humbert ed., Fl. Mad. Orch. 1: 91 (1939).
TYPE: Madagascar, Andringitra massif, *H.Perrier* 14450 (holo. B†; iso. P).
Cynorkis pulchra var. *laxiflora* Schltr. in Repert. Spec. Nov. Regni Veg. Beih. 33: 67 (1924).

DISTRIBUTION: Madagascar: Fianarantsoa.

HABITAT: mossy forest.

ALTITUDE: c. 1600 m.

FLOWERING TIME: February.

LIFE FORM: terrestrial.

PHYTOGEOGRAPHICAL REGION: III.

DESCRIPTION: {896}.

NOTES: flowers much smaller than the typical form; raceme more lax; anther canals short.

Cynorkis lilacina var. **pulchra** *(Schltr.) H.Perrier* in Humbert ed., Fl. Mad. Orch. 1: 91 (1939). TYPE: Madagascar, Mt. Tsaratanana, *H.Perrier* 16520 (lecto. P, selected here); Mt. Ibity, *H.Perrier* 8088 (syn. P); Ambohimanga (nr. Antananarivo), *H.Perrier* 7053 (syn. P); nr. Ambatofinandrahana, *H.Perrier* 12421 (syn. P); Andringitra massif, *H.Perrier* 13619 (syn. P); Mt. Tsiafajavona, *H.Perrier* 13502 (syn. P); Manjakatompo, *H.Perrier* 13501 (syn. P, TAN); Ankaizina, *Decary* 2002 (syn. P); Antananarivo, *Waterlot* s.n. (syn. P); Ilafy (nr. Antananarivo), *Decary* s.n. (syn. P).

Bicornella pulchra Kraenzl. ex Schltr. in Beih. Bot. Centralbl. 33 (2): 398 (1915).

Cynosorchis pulchra Kraenzl. ex Schltr. in Beih. Bot. Centralbl. 34 (2): 311 (1916) & in Repert. Spec. Nov. Regni Veg. Beih. 33: 66 (1924).

Forsythmajoria pulchra Kraenzl., mss. in Herb. Berol.

DISTRIBUTION: Madagascar: Antananarivo, Antsiranana, Fianarantsoa.
HABITAT: in *Philippia* scrub; humid, evergreen forest on plateau.
ALTITUDE: 1600 – 2200 m.
FLOWERING TIME: January – April.
LIFE FORM: terrestrial.
PHYTOGEOGRAPHICAL REGION: III, IV.
DESCRIPTION: {896}.
ILLUSTRATION: {1223}.
NOTES: spur cylindrical then obtuse, 5 – 6.5 mm long; anther erect, small, the anther canals slender; stigmatic processes shorter than the anther canals.

Cynorkis lilacina var. **tereticalcar** *H.Perrier ex Hermans* **var. nov.** TYPE: Madagascar, Andringitra Massif, c. 1800 m, forest and undergrowth, *H.Perrier* 14571 (holo. P). See Appendix .

Cynorkis lilacina var. *tereticalcar* H.Perrier in Humbert, Fl. Mad. Orch. 1: 91 (1939), *nom. nud.*

DISTRIBUTION: Madagascar: Antsiranana, Fianarantsoa.
HABITAT: humid, evergreen forest on plateau; in *Philippia* scrub.
ALTITUDE: 1500 – 2000 m.
FLOWERING TIME: February – March.
LIFE FORM: terrestrial.
PHYTOGEOGRAPHICAL REGION: III.
DESCRIPTION: {896}.
NOTES: differs from the typical variety by the spur which is subcylindrical from the base to the apex, more or less elongate, and at times as long as the pedicellate ovary.

Cynorkis lindleyana *Hermans* **nom. nov.**

Bicornella gracilis Lindl., Gen. Sp. Orch. Pl.: 334 (1835). TYPE: Madagascar, Imerina, *Lyall* 308 (holo. K).

Cynorkis gracilis (Lindl.) Schltr. in Repert. Spec. Nov. Regni Veg. Beih. 33: 51 (1924), *non* (Blume) Kraenzl., Orch. Gen. Sp. 1 (8): 488 (1898).

DISTRIBUTION: Madagascar: Antananarivo, Fianarantsoa, Toamasina.
HABITAT: grassland; marshes in peaty soil; dry river-beds.
ALTITUDE: 1000 – 2500 m.
FLOWERING TIME: September – April.
PHYTOGEOGRAPHICAL REGION: III, IV.
LIFE FORM: terrestrial.
DESCRIPTION: {896}.
ILLUSTRATION: {246}, {1031}.
NOTES: plant slender, with fleshy roots; flowers purple-pink; lip entire, a little pandurate; spur almost cylindrical, wide at the base and a little contracted towards the middle, 7 – 9 mm long. *Cynorkis gracilis* (Blume) Kraenzl. was based on a Korean collection and, thus, the name cannot be used for the Madagascan plant described by Lindley. I therefore propose the epithet *lindleyana* for the Madagascan species.

Cynorkis lowiana *Rchb.f.* in Flora: 150 (1888) & in Gard. Chron. sér. 3, 3: 424 (1888). TYPE: Madagascar, without locality, cult. Low s.n. (holo. W, Reichenbach Herbarium 46503).

Cynosorchis purpurascens Hook.f. in Curtis's Bot. Mag.: t.7551 (1897), *non* Thouars.

DISTRIBUTION: Madagascar: Antsiranana, Fianarantsoa, Toamasina.

HABITAT: humid, evergreen forest in moss; shaded and humid rocks.

ALTITUDE: 200 – 1200

FLOWERING TIME: August – May.

LIFE FORM: epiphyte, terrestrial or lithophyte.

PHYTOGEOGRAPHICAL REGION: I, III, IV.

DESCRIPTION: {551}, {896}.

ILLUSTRATION: {50}, {281}, {366}, {488}, {551}, {556}, {896}, {1298}, {1451}.

CULTIVATION: {366}, {551}.

NOTES: plant single-leafed, few-flowered; flowers large; lip reddish-purple with a darker spot on the disk; lip 3-lobed, 2.2 × 2.7 cm; spur up to 4.5 cm long.

Cynorkis × madagascarica *(Schltr) Hermans* **comb. nov.**

Microtheca madagascarica Schltr. in Repert. Spec. Nov. Regni Veg. Beih. 33: 77 (1924). TYPE: Madagascar, Mt. Tsiafajavona, *H.Perrier* 13504 (holo. P).

Cynosorchis lilacina × ridleyi H.Perrier in Arch. Bot. Bull. Mens. 5: 52 (1931) & in Humbert, Fl. Mad. Orch. 1: 92 (1939).

DISTRIBUTION: Madagascar: Antananarivo, Toamasina.

HABITAT: amongst *Philippia* on wooded slopes.

ALTITUDE: 1400 – 2000 m.

FLOWERING TIME: February – March.

LIFE FORM: terrestrial.

PHYTOGEOGRAPHICAL REGION: III, IV.

DESCRIPTION: {896}.

ILLUSTRATION: {1223}.

NOTES: a very variable natural hybrid, flowers glabrous, violet, lip oval, without a spur. Perrier de la Bâthie's original combination remains unpublished. The hybrid formula *Cynorkis lilacina × ridleyi* was used by him in the Flore de Madagascar (1939).

Cynorkis marojejyensis *Bosser* in Adansonia, sér. 2, 9 (3): 348 (1969). TYPE: Madagascar, Marojejy massif, *Humbert & Cours* 23756 (holo. P).

DISTRIBUTION: Madagascar: Antsiranana.

HABITAT: marshes in peaty soil.

ALTITUDE: 2000 – 2500 m.

FLOWERING TIME: March – April.

LIFE FORM: terrestrial.

PHYTOGEOGRAPHICAL REGION: IV.

DESCRIPTION: {113}.

ILLUSTRATION: {113}.

NOTES: known only from the type. The column and the rostellum with the mid-lobe developed into a flattened blade are characteristic.

Cynorkis melinantha *Schltr.* in Repert. Spec. Nov. Regni Veg. Beih. 33: 60 (1924). TYPE: Madagascar, betw. Mania & Ivato, *H.Perrier* 12383 (holo. P).

DISTRIBUTION: Madagascar: Fianarantsoa.

HABITAT: rocky outcrops.

ALTITUDE: 1500 – 1700 m.

FLOWERING TIME: February.

LIFE FORM: terrestrial on quartz soil.

PHYTOGEOGRAPHICAL REGION: III.

DESCRIPTION: {896}.

ILLUSTRATION: {1223}.

NOTES: related to *C. jumelleana* but differs in shape of petals and lip, the longer spur and anther canal, and the orange-yellow flowers.

Cynorkis mellitula *Toill.-Gen. & Bosser* in Adansonia, sér. 2, 1: 103 (1961). TYPE: Madagascar, Manjakatompo, Ankaratra, *Bosser* 14829 (holo. P).

DISTRIBUTION: Madagascar: Antananarivo, Fianarantsoa.
HABITAT: humid, mossy, evergreen forest.
ALTITUDE: 1900 – 2000 m.
FLOWERING TIME: January.
LIFE FORM: epiphyte or terrestrial.
PHYTOGEOGRAPHICAL REGION: III, IV.
DESCRIPTION: {1407}.
ILLUSTRATION: {1407}.
NOTES: known from the type only. Lip pale yellow, spotted violet; spur green flattened and bilobed at the tip.

Cynorkis minuticalcar *Toill.-Gen. & Bosser* in Nat. Malg. 13: 25 (1962). TYPE: Madagascar, Mt. Tsiafajavona, Ankaratra, *Bosser* 10867 (holo. P; iso. TAN).

DISTRIBUTION: Madagascar: Antananarivo.
HABITAT: grassland; marshes.
ALTITUDE: 2000 – 2500 m.
FLOWERING TIME: February.
LIFE FORM: terrestrial.
PHYTOGEOGRAPHICAL REGION: IV.
DESCRIPTION: {1408}.
ILLUSTRATION: {1408}.
NOTES: close to *C. brachyceras* but the lip shape is different, it has a much shorter spur and column wings.

Cynorkis mirabile *Hermans & P.J. Cribb*
This new species is in press to be described in *The Orchid Review*. It has similarities to *Cynorkis lowiana* Rchb.f and *C. gibbosa* Ridl, but is distinct by the 3-lobed lip and the distinct spotted leaf and bracts. It is probable that this is a natural hybrid between the two species mentioned above.

Cynorkis monadenia *H.Perrier* in Notul. Syst. (Paris) 8: 33 (1939). TYPES: Madagascar, nr. Ambatofinandrahana, *Decary* 13162 (lecto. P selected here); *Decary* 13269 (syn. P).

DISTRIBUTION: Madagascar: Fianarantsoa.
HABITAT: quartz rock, rocky outcrops.
ALTITUDE: 1600 – 1800 m.
FLOWERING TIME: February.
LIFE FORM: terrestrial.
PHYTOGEOGRAPHICAL REGION: III.
DESCRIPTION: {896}.
NOTES: plant with 2 leaves; lip entire, 7 × 2 mm; spur 6 mm long; rostellum entire.

Cynorkis muscicola *Bosser* in Adansonia, sér. 2, 9 (3): 346 (1969). TYPE: Madagascar, Ambondrombe, cult. Jard. Bot. Tananarive 4627 (holo. P).

DISTRIBUTION: Madagascar: Toliara.
HABITAT: mossy forest in humus.
ALTITUDE: 1500 – 2000 m.
LIFE FORM: terrestrial.
PHYTOGEOGRAPHICAL REGION: III.
DESCRIPTION: {113}.
ILLUSTRATION: {113}.
NOTES: known only from the type. Inflorescence long, laxly-flowered; lip ligulate, entire; spur cylindrical and narrow at the base, enlarged and globulose at the tip.

Cynorkis nutans *(Ridl.)* *H.Perrier* in Bull. Soc. Bot. France 83: 582 (1936). TYPES: Madagascar, Central, *Baron* 1703 (lecto. BM; isolecto. K, both selected here); Imerina, *Deans Cowan* s.n.; Andrangoloaka, *Parker* s.n. (all syn. BM).

Habenaria nutans Ridl. in J. Linn. Soc., Bot. 21: 507 (1885).

Cynosorchis nutans var. *campenoni* H.Perrier in Humbert, Fl. Mad. Orch. 1: 156 (1939), *nom. nud.* TYPE: Madagascar, Imerina, *R.P.Campenon* s.n. (holo. P).

> DISTRIBUTION: Madagascar: Antananarivo, Toamasina, Toliara.
> HABITAT: humid, highland, evergreen forest, in moss.
> ALTITUDE: 600 – 1400 m.
> FLOWERING TIME: January – October.
> LIFE FORM: terrestrial or epiphyte.
> PHYTOGEOGRAPHICAL REGION: III.
> DESCRIPTION: {896}.
> ILLUSTRATION: {281}, {896}.
> NOTES: flowers c. 3 cm wide, variable in colour; lateral sepals broadly semi-orbicular, spreading; lip narrow and entire, ligulate, 8 × 1.3 mm; spur c. 12 mm long, a little inflated from the middle. *Cynorkis nutans* var. *campenoni* is just a smaller form of this variable species.

Cynorkis ochroglossa *Schltr.* in Repert. Spec. Nov. Regni Veg. Beih. 33: 61 (1924). TYPE: Madagascar, Andringitra massif, *H.Perrier* 14583 (holo. P).

> DISTRIBUTION: Madagascar: Fianarantsoa.
> HABITAT: rocky outcrops.
> ALTITUDE: c. 2400 m.
> FLOWERING TIME: February.
> LIFE FORM: terrestrial.
> PHYTOGEOGRAPHICAL REGION: IV.
> DESCRIPTION: {896}.
> NOTES: known only from the type. Flowers small, 12 mm long; lip 3-lobed; spur conical and very short, 1 mm long.

Cynorkis orchioides *Schltr.* in Ann. Mus. Col. Marseille, ser.3, 1: 152 (9) (1913). TYPE: Madagascar, nr. Maevatanana, *H.Perrier* 416 (holo. B†; iso. P).

> DISTRIBUTION: Madagascar: Mahajanga.
> HABITAT: seasonally dry, deciduous forest or woodland; rocky forests and rocky outcrops.
> FLOWERING TIME: November – January.
> LIFE FORM: terrestrial or lithophyte.
> PHYTOGEOGRAPHICAL REGION: V.
> DESCRIPTION: {896}.
> ILLUSTRATION: {1205}, {1223}.
> NOTES: close to *C. flexuosa* but differs in its wider leaf and the smaller flowers.

Cynorkis papilio *Bosser* in Adansonia, sér. 2, 9 (3): 343 (1969). TYPE: Madagascar, road to Lakato, *Bosser* 18384 (holo. P; iso. TAN).

> DISTRIBUTION: Madagascar: Fianarantsoa, Toamasina.
> HABITAT: montane *Philippia* scrub on quartz soil.
> ALTITUDE: 1100 – 1200 m.
> FLOWERING TIME: November, March – June.
> LIFE FORM: terrestrial.
> PHYTOGEOGRAPHICAL REGION: III.
> DESCRIPTION: {113}.
> ILLUSTRATION: {113}.
> NOTES: distinguished by its linear lip which is curled up at the tip and rostellum with bidentate arms.

Cynorkis papillosa *(Ridl.) Summerh.* in Kew Bull. 2: 461 (1951). TYPE: Madagascar, Ankaratra, *Hildebrandt* 3860 (holo. BM; iso. K, P, W).

Peristylus filiformis Kraenzl. in Abh. Naturwiss. Ver. Bremen 7: 258 (1882). TYPE: Madagascar, Antananarivo, *Rutenberg* s.n. (holo. HBG).

Habenaria filiformis (Kraenzl.) Ridl. in J. Linn. Soc., Bot. 21: 504 (1885).

Habenaria papillosa Ridl. in J. Linn. Soc., Bot. 21: 504 (1885).

Helorchis filiformis (Kraenzl.) Schltr. in Repert. Spec. Nov. Regni Veg. Beih. 33: 36 (1924).

Cynorkis filiformis (Kraenzl.) H.Perrier in Arch. Bot. Bull. Mens. 5: 48 (1931), *non* Schltr.

DISTRIBUTION: Madagascar: Antananarivo, Fianarantsoa.
HABITAT: ericaceous scrub; marshes in peaty soil.
ALTITUDE: 1500 – 2400 m.
FLOWERING TIME: December – April.
LIFE FORM: terrestrial.
PHYTOGEOGRAPHICAL REGION: III.
DESCRIPTION: {896}.
ILLUSTRATION: {896}.
NOTES: locally common; plant slender; flowers white; lip red; spur conical.

Cynorkis parvula *Schltr.* in Beih. Bot. Centralbl. 33 (2): 401 (1915). TYPES: Comoros, *Schmidt* 157 & 158 (both syn. B†); *Schmidt* 157 (isosyn. K).

Bicornella schmidtii Kraenzl. in Bot. Jahrb. Syst. 17: 67 (1893).

DISTRIBUTION: Comoros.
FLOWERING TIME: June.
DESCRIPTION: {896}.
NOTES: poorly known species; lip 3-lobed; spur obtuse and short.

Cynorkis perrieri *Schltr.* in Beih. Bot. Centralbl. 34 (2): 310 (1916). TYPE: Madagascar, Mt. Ibity, S Antsirabe, *H.Perrier* 8120 (XXXVI) (holo. B†; iso. P).

DISTRIBUTION: Madagascar: Antananarivo.
HABITAT: rocky outcrops, in moss.
ALTITUDE: 1600 – 2000 m.
FLOWERING TIME: February – May.
LIFE FORM: terrestrial or lithophyte.
PHYTOGEOGRAPHICAL REGION: III, IV.
DESCRIPTION: {896}.
ILLUSTRATION: {896}.
NOTES: close to *C. hispidula* but lip shortly unguiculate at the base, 3-lobed in the third, 5.5 – 6 mm long; spur cylindrical, obtuse, 4.5 – 5 mm long.

Cynorkis petiolata *H.Perrier* in Notul. Syst. (Paris) 14: 148 (1951). TYPE: Madagascar, E of Marojejy massif, *Humbert* 22534 (holo. P).

DISTRIBUTION: Madagascar: Antsiranana.
HABITAT: lichen-rich forest.
ALTITUDE: c. 1500 m.
FLOWERING TIME: December.
LIFE FORM: terrestrial.
PHYTOGEOGRAPHICAL REGION: III.
DESCRIPTION: {901}.
NOTES: perianth white, pink spots on the petals; lip sulphur yellow with 3 dark red spots on each side, 3-lobed; spur c. 4 mm long.

Cynorkis peyrotii *Bosser* in Adansonia, sér. 2, 9 (3): 353 (1969). TYPE: Madagascar, S of Lake Mantasoa, *Bosser & Peyrot* 16157 (holo. P).

DISTRIBUTION: Madagascar: Antananarivo, Toamasina.
HABITAT: humid, evergreen forest on moss- and lichen-covered trees.

ALTITUDE: 1000 – 1500 m.

FLOWERING TIME: July – August.

LIFE FORM: epiphyte.

PHYTOGEOGRAPHICAL REGION: III.

DESCRIPTION: {113}.

ILLUSTRATION: {113}, {281}.

NOTES: distinguished by its hairy, non-resupinate flowers and patterned leaves. The typical form has a 3-lobed lip and an dilated, elongated spur, another form has a sub-entire lip and short, 1 – 2.5 mm long, cylindrical spur. The two forms coexist in the same populations.

Cynorkis pinguicularioides *H.Perrier* ex *Hermans* **sp. nov.** TYPE: Madagascar, Pic Saint Louis, nr. Fort Dauphin, *Decary* 10008 (holo. P). See Appendix .

Cynosorchis pinguicularioides H.Perrier in Fl. Mad. Orch. 1: 96, fig. VII, 16 – 18 (1939), *nom. nud.*

DISTRIBUTION: Madagascar: Toliara.

HABITAT: rocky outcrops.

ALTITUDE: 200 – 500 m.

FLOWERING TIME: May – July.

PHYTOGEOGRAPHICAL REGION: III.

DESCRIPTION: {896}.

ILLUSTRATION: {896}.

NOTES: plant small with 3 – 4, radical, lanceolate-acute leaves; raceme rather laxly 4- to 10-flowered; floral bracts acute, shorter than the ovary; flowers small, purplish, papillose on the outside; dorsal sepal concave, obtuse, a little shorter than the laterals which are falcate; petals the same size as the lateral sepals; lip 5-lobed, the lobes obtuse, the three front ones more or less equal, much larger than the others; spur scrotiform; anther retuse; rostellum 0.7 – 1 mm high, subentire at the front.

Cynorkis pseudorolfei *H.Perrier* in Notul. Syst. (Paris) 14: 145, (1951), as *pseudorolfer*. TYPE: Madagascar, Mt. Vohimavo, Manampnihy, *Humbert* 20644 (holo. P; iso. K, TAN).

DISTRIBUTION: Madagascar: Toliara.

HABITAT: coastal forest.

FLOWERING TIME: March.

ALTITUDE: 100 – 150 m.

LIFE FORM: terrestrial.

PHYTOGEOGRAPHICAL REGION: I.

DESCRIPTION: {901}.

NOTES: similar to *C. rolfei*, but the inflorescence is not hairy, the perianth is twice as long and clearly gibbose at the base, the lip is different and the spur is curved and cylindrical.

Cynorkis purpurascens *Thouars*, Hist. Orch.: t.15 (1822). TYPE: Mascarenes, *Thouars* s.n. (holo. P).

Orchis purpurascens (Thouars) Spreng., Syst. Veg. 3: 687 (1826).

Gymnadenia purpurascens (Thouars) A.Rich., Monogr. Orch. Iles France Bourbon: 29, t.6 (1828).

Cynosorchis purpurascens var. *praecox* (Schltr.) Schltr. in Ann. Mus. Col. Marseille, sér. 3, 1: 153 (1913).

Cynosorchis praecox Schltr. in Repert. Spec. Nov. Regni Veg. Beih. 33: 65 (1924). TYPES: Madagascar, Mt. Tsitondroina near Maevatanana *H.Perrier* 425 & Bemarivo, *H.Perrier* 1938 (33 ?) (both syn. P).

DISTRIBUTION: Madagascar: Antananarivo, Antsiranana, Mahajanga. Mascarenes: Réunion, Mauritius. There are also unverified records from the Comoros.

HABITAT: shaded and humid rocks; humid forest on moss- and lichen-rich-covered trees, or *Pandanus* and *Asplenium nidus.*

ALTITUDE: sea level – 1500 m.

FLOWERING TIME: September – May.

LIFE FORM: epiphyte or terrestrial.

PHYTOGEOGRAPHICAL REGION: I, II, III, V.

DESCRIPTION: {896}.

ILLUSTRATION: {166}, {167}, {281}, {556}, {906}, {921}, {922}, {1028}, {1136}, {1148}, {1206}, {1298}, {1398}, {1454}.

HISTORY: {366}, {484}, {555}, {556}, {667}, {921}, {1102}, {1144}, {1454}.

CULTIVATION: {556}, {1102}, {1148}, {1452}.

VERNACULAR NAMES: Katsakandrango, Fitsotsoka, Tsimpelany.

NOTES: plant single-leaved; flowers mauve, large; lip 18 – 25 mm long, the base narrow and hollowed into a gutter, 4-lobed; spur up to 4 cm.

Cynorkis purpurea *(Thouars) Kraenzl.*, Orch. Gen. Sp. 1: 482 (1901). TYPE: Madagascar, *Thouars* s.n. (holo. P).

Habenaria purpurea Thouars, Hist. Orch.: t.17 (1822).

Peristylus purpureus (Thouars) S.Moore in Baker, Fl. Mauritius: 335 (1877).

DISTRIBUTION: Madagascar: Antananarivo, Fianarantsoa. Mauritius {826}.

HABITAT: wet rocks bordering streams; river margins.

ALTITUDE: 800 – 2000 m.

FLOWERING TIME: January – June.

LIFE FORM: terrestrial or lithophyte.

PHYTOGEOGRAPHICAL REGION: III, IV.

DESCRIPTION: {896}.

ILLUSTRATION: {246}, {823}, {1398}.

NOTES: raceme laxly 9- to 13-flowered; lip 6 mm long, angular at the base, 3-lobed in the upper quarter; spur hook-shaped.

Cynorkis quinqueloba *H.Perrier* ex *Hermans* **sp. nov.** TYPE: Madagascar, Mt. Papanga, *Humbert* 6373 (holo. P). See Appendix .

Cynosorchis quinqueloba H.Perrier in Fl. Mad. Orch. 1: 77 (1939), *nom. nud.*

DISTRIBUTION: Madagascar: Fianarantsoa, Toliara.

HABITAT: high–elevation ericaceous scrub; high-elevation moss- and lichen-rich rain forest; ericaceous bush, with *Philippia* dominant.

ALTITUDE: 1400 – 1900 m.

FLOWERING TIME: November, January.

LIFE FORM: terrestrial.

PHYTOGEOGRAPHICAL REGION: III.

DESCRIPTION: {896}.

NOTES: plant slender, terrestrial, with 3 – 4 basal leaves, elliptical or oblong, cuspidate-acute, narrowed towards the base; spike with 5 – 6 small sheaths on the peduncle and hairy towards the tip; raceme short, with 7 – 10 flowers; floral bracts hairy, about a third the length of the flower; flowers very finely papillose on the outside; dorsal sepal broadly ovate, obtuse; lateral sepals larger; petals broadly ovate, very obtuse; lip obtusely 5-lobed, the lateral lobes at least half the size of the middle one; staminodes spathulate and obtuse, half the length of the anther; rostellum tridentate, the middle tooth a little bigger.

Cynorkis quinquepartita *H.Perrier* ex *Hermans* **sp. nov.** TYPE: Madagascar, Mt. Ambre, *H.Perrier* 17738 (holo. P). See Appendix .

Cynosorchis quinquepartita H.Perrier in Fl. Mad. Orch. 1: 78, fig. VI, 8 – 9 (1939), *nom. nud.*

DISTRIBUTION: Madagascar: Antsiranana.

HABITAT: mossy, montane forest.

ALTITUDE: c. 1200 m.

FLOWERING TIME: September.

LIFE FORM: terrestrial.

PHYTOGEOGRAPHICAL REGION: III.

DESCRIPTION: {896}.

ILLUSTRATION: {896}.

NOTES: herb slender, terrestrial, lacking tubers but with bunched, downy, slightly fleshy roots; leaf radical, one (rarely 2), linear-lanceolate, acute, tapering into an elongate petiole; flowers violet, with a few glands on the outside; dorsal sepal concave, orbicular; lateral sepals larger, obovate-oblong; petals ovate, obtuse, adnate to the edge of the lip for a third of their length; lip 5-lobed, the basal lobes bigger than the intermediate ones, but a little smaller than middle one; spur cylindrical, but broad; rostellum tridentate; stigmatic processes free over their upper half, much shorter than the rostellum.

Cynorkis raymondiana *H.Perrier* ex *Hermans* **sp. nov.** TYPE: Madagascar, nr. Fort Dauphin, *Decary* 10142 & 10019 (holo. P).
Cynosorchis raymondiana H.Perrier in Fl. Mad. Orch. 1: 97 (1939), *nom. nud.*

DISTRIBUTION: Madagascar: Toliara.
HABITAT: coastal forest.
ALTITUDE: sea level – 1200 m.
FLOWERING TIME: April, July.
PHYTOGEOGRAPHICAL REGION: I.
LIFE FORM: terrestrial in sand.
DESCRIPTION: {896}.
NOTES: herb glabrous, terrestrial; tubers 2, ovoid; roots villous; leaves radical, 2, rarely 3, oblong-lanceolate, cuspidate-acute; spike slender, with 3 – 4 distant sheaths; raceme rather laxly 6- to 12-flowered; floral bracts acute 3 times shorter than the ovary; flowers very small, pinkish-white; dorsal sepal ovate, obtuse, concave; laterals semi-ovate; petals semi-ovate, obtuse; lip narrowed at the base, very unequally 5-lobed;lobes small, triangular-acute and diverging, the intermediate ones larger and obovate, the middle one as long as the intermediates, but oblong-acute and narrower; spur club-shaped; rostellum tridentate at the front.

Cynorkis ridleyi *T.Durand & Schinz*, Consp. Fl. Afric.: 92 (1895). TYPE: Madagascar, E Imerina, Ankeramadinika, *Hildebrandt* 3730 (holo. BM.; iso. K, P).
Amphorchis lilacina Ridl. in J. Linn. Soc., Bot. 21: 518 (1885). TYPE: based on same type as *C. ridleyi*.
Cynosorchis heterochroma Schltr. in Beih. Bot. Centralbl. 34 (2): 307 (1916). TYPE: Madagascar, Analamagoatra, *H.Perrier* 132 (holo. B†; iso. P).
Cynosorchis moramangana Schltr. in Repert. Spec. Nov. Regni Veg. Beih. 18: 321 (1922). TYPE: Madagascar, Analamazaotra forest, Moramanga, *Viguier & Humbert* 879 (holo. B†; iso. P).
Cynorkis rhomboglossa Schltr. in Repert. Spec. Nov. Regni Veg. Beih. 33: 68 (1924), **syn. nov.** TYPE: Madagascar, Mt. Tsiafajavona, *H.Perrier* 14000 (holo. P).

DISTRIBUTION: Madagascar: Antananarivo, Antsiranana, Fianarantsoa, Toamasina. Comoros: Anjouan, Grande Comore. There are unconfirmed records from Réunion.
HABITAT: humid, mossy, evergreen forest on plateau amongst mosses.
ALTITUDE: 600 – 2000 m.
FLOWERING TIME: September – April.
LIFE FORM: terrestrial or rarely epiphyte.
PHYTOGEOGRAPHICAL REGION: III, IV.
DESCRIPTION: {896}.
ILLUSTRATION: {281}, {1025} misidentified, {1223}, {1433}.
NOTES: leaves often darker coloured underneath; flowers 12- 17 mm long, purple-pink; lip uppermost, spotted with purple, with small lateral lobes, the mid-lobe very variable; spur a little recurved and slightly swollen at the tip. *C. rhomboglossa* has all the characteristics of this somewhat variable species and is here reduced to the synonymy of *C. ridleyi*.

Cynorkis rolfei *Hochr.* in Ann. Conserv. Jard. Bot. Genève 11/12, 2: 54 (1908). TYPE: Madagascar, Vatomandry, *Guillot* 6 (holo. P; iso. G, K?).

DISTRIBUTION: Madagascar: Toamasina.
HABITAT: coastal forest in sandy soil.
ALTITUDE: sea level – 500 m.
FLOWERING TIME: July.
LIFE FORM: terrestrial.
PHYTOGEOGRAPHICAL REGION: I.
DESCRIPTION: {896}.
ILLUSTRATION: {537}.
NOTES: raceme few-flowered; flowers 10 – 12 mm long; lip 5-lobed; spur tubular, c. 2 mm long.

Cynorkis rosellata *(Thouars) Bosser* in Adansonia 19 (2): 188 (1997). TYPE: Thouars, Orch. Isles Aust. Afr.: t.8 (1822) (lecto. P, selected here).

Satyrium rosellatum Thouars, Orch. Isles Aust. Afr.: t.8 (1822).

Habenaria mascarenensis Spreng., Syst. Veg. 3: 690 (1826), *nom. illeg.* based on *Satyrium rosellatum* Thouars.

Gymnadenia rosellata (Thouars) A.Rich. in Mem. Soc. Hist. Nat., Paris 4: 27 (1828).

Habenaria imerinensis Ridl. in J. Linn. Soc., Bot. 21: 505 (1885). TYPE: Madagascar, Andrangaloaka, E Imerina, *Hildebrandt* 3731 (holo. BM; iso. K, P).

Cynorkis imerinensis (Ridl.) Kraenzl., Orch. Gen. Sp. 1: 490 (1901).

Habenaria rosellata (Thouars) Schltr. in Beih. Bot. Centralbl. 33: 406 (1915).

DISTRIBUTION: Madagascar: Antsiranana, Fianarantsoa, Toamasina, Toliara. Mascarenes: Réunion. Dubious records from Mauritius.

HABITAT: mossy forest; river margins; shaded and humid rocks.

ALTITUDE: 500 – 1500 m.

FLOWERING TIME: August – November, March.

LIFE FORM: terrestrial or lithophyte.

PHYTOGEOGRAPHICAL REGION: I, III.

DESCRIPTION: {896}.

ILLUSTRATION: {896}.

NOTES: plant with several leaves, laxly-flowered; flowers small, lilac, spotted with red; lip lanceolate, 4 – 6 × 1.3 mm; spur 6 – 9 mm long, cylindrical.

Cynorkis sacculata *Schltr.* in Beih. Bot. Centralbl. 34 (2): 310 (1916). TYPE: Madagascar, Mt. Ibity, *H.Perrier* 8119 (holo. P).

DISTRIBUTION: Madagascar: Antananarivo.

HABITAT: montane, ericaceous scrub.

ALTITUDE: 1600 – 2300 m.

FLOWERING TIME: February – March.

LIFE FORM: terrestrial.

PHYTOGEOGRAPHICAL REGION: IV.

DESCRIPTION: {896}.

ILLUSTRATION: {896}, {1223}.

NOTES: related to *C. tryphioides* but differs by the 2 – 3 mm long spur which is conical saccate and obtuse, as wide at the base as long.

Cynorkis sagittata *H.Perrier* in Notul. Syst. (Paris) 14: 146 (1951). TYPE: Madagascar, between Vohemar & Ambilobe, *Decary* 14667 (holo. P).

DISTRIBUTION: Madagascar: Antsiranana.

HABITAT: coastal forest.

FLOWERING TIME: July.

LIFE FORM: terrestrial.

PHYTOGEOGRAPHICAL REGION: V.

DESCRIPTION: {901}.

NOTES: known only from the type. Close to *C. andringitrana* but differs by the hairs on the inflorescence, the smaller flowers, and the arrow-shaped lip.

Cynorkis sambiranoensis *Schltr.* in Repert. Spec. Nov. Regni Veg. Beih. 33: 69 (1924). TYPE: Madagascar, Sambirano Valley, *H.Perrier* 15196 (holo. P).

DISTRIBUTION: Madagascar: Antsiranana, Mahajanga.

HABITAT: on rocks; on hills; in wet grassland.

ALTITUDE: 300 – 500 m.

FLOWERING TIME: October – January.

LIFE FORM: terrestrial.

PHYTOGEOGRAPHICAL REGION: II.

DESCRIPTION: {896}.

ILLUSTRATION: {896}.

NOTES: related to *C. orchioides*, and to *C. purpurascens* in habit, but the flowers are smaller and the shape of the spur is different.

Cynorkis saxicola *Schltr.* in Repert. Spec. Nov. Regni Veg. Beih. 33: 70 (1924). TYPE: Madagascar, Andringitra massif, *H.Perrier* 14328 (holo. P).

DISTRIBUTION: Madagascar: Fianarantsoa.
HABITAT: shaded and wet rocks.
ALTITUDE: c. 2000 m.
FLOWERING TIME: January.
LIFE FORM: terrestrial.
PHYTOGEOGRAPHICAL REGION: III, IV.
DESCRIPTION: {896}.
ILLUSTRATION: {896}, {1223}.
NOTES: plant three-leaved; flowers small, 16 mm long, red; lip 5 – 6 × 3 mm, expanded above the base; spur cylindrical, curved at the front and thickened at the tip, 6 – 8 mm long.

Cynorkis schlechteri *H.Perrier* in Arch. Bot. Bull. Mens. 3: 198 (1929). TYPE: Madagascar, Anove basin, *H.Perrier* 11393 (holo. P).
Cynosorchis exilis Schltr. in Beih. Bot. Centralbl. 34 (2): 312 (1916), *non* (Frapp.) Schltr. (1915). TYPE: Madagascar, Anore, *H.Perrier* 106 (holo. P).

DISTRIBUTION: Madagascar: Toamasina.
HABITAT: humid, evergreen forest; rocky outcrops.
ALTITUDE: sea level – 800 m.
FLOWERING TIME: September.
LIFE FORM: terrestrial.
PHYTOGEOGRAPHICAL REGION: I.
DESCRIPTION: {896}.
NOTES: known only from the type. Flowers small, 11 mm long; lip 5-lobed, the basal lobes very small.

Cynorkis schmidtii *(Kraenzl.) Schltr.* in Oesterr. Bot. Zeit., 49 (1): 23 (1899). TYPE: Comoros, *Schmidt* 159 (holo. W; iso. K).
Holothrix schmidtii Kraenzl. in Bot. Jahrb. Syst. 17: 66 (1893).

DISTRIBUTION: Comoros.
FLOWERING TIME: June.
DESCRIPTION: {896}.
NOTES: flowers very small, c. 4 mm long, white; lip 5-lobed, spotted with pink; spur saccate.

Cynorkis sigmoidea *(K.Schum.) Kraenzl.*, Orch. Gen. Sp. 1: 490 (1901). TYPE: Comoros, Grande Comore, *Boivin* s.n. (holo. HBG; iso. P).
Herminium sigmoideum K.Schum. in Just's Bot. Jahresber. 26 (1): 336 (1900).

DISTRIBUTION: Comoros: Grande Comore.
ALTITUDE: 1100 – 1300 m.
FLOWERING TIME: March – May.
DESCRIPTION: {896}.
ILLUSTRATION: {896}.
NOTES: inflorescence laxly 10- to 20-flowered; lip finely papillose, 5-lobed; spur thick, 5 mm long, a little inflated in the upper third.

Cynorkis sororia *Schltr.* in Ann. Mus. Col. Marseille, sér. 3, 1: 154 (11), t.4 (1913). TYPE: Madagascar, Manongarivo massif, *H.Perrier* 1940 (holo. P; iso. K).

DISTRIBUTION: Madagascar: Antsiranana.
HABITAT: mossy forest, by side of streams in sand.

ALTITUDE: 800 – 1700 m.

FLOWERING TIME: May, September.

LIFE FORM: terrestrial or lithophyte.

PHYTOGEOGRAPHICAL REGION: III, IV.

DESCRIPTION: {896}.

ILLUSTRATION: {1205}, {1223}.

NOTES: flowers pink, 2 – 2.5 cm long; lip deeply 3-lobed, 9 – 15 mm long, the lateral lobes small, triangular and obtuse, the midlobe very big, the margins almost crenulate; spur apex distinctly inflated, 7 – 8 mm long. Related to *C. ridleyi* but differs in the shape of the rostellum.

Cynorkis souegesii *Bosser & Veyret* in Adansonia, sér. 2, 10 (2): 213 (1970). TYPE: Madagascar, Fort Dauphin, road to Manantenina, *Veyret* 1040 (holo. P).

DISTRIBUTION: Madagascar: Toliara.

HABITAT: marshes.

LIFE FORM: terrestrial.

FLOWERING TIME: April.

PHYTOGEOGRAPHICAL REGION: VI.

DESCRIPTION: {116}.

ILLUSTRATION: {116}.

NOTES: known only from the type. Plant lacking tubers, close to *C. rosellata* which has a longer spur and a different rostellum.

Cynorkis spatulata *H.Perrier* ex *Hermans* **sp. nov.** TYPE: Madagascar, nr. Fort-Dauphin, on rocks amongst the dunes, *Decary* 4070 (holo. P). See Appendix .

Cynosorchis spatulata H.Perrier in Fl. Mad. Orch. 1: 125, fig. XI, 1 – 6 (1939), *nom. nud.*

DISTRIBUTION: Madagascar: Toliara.

HABITAT: on sandstone rocks and wet sand, not far from the sea; grassland.

ALTITUDE: sea level – 100 m.

FLOWERING TIME: June.

LIFE FORM: terrestrial or lithophyte.

PHYTOGEOGRAPHICAL REGION: I.

DESCRIPTION: {896}.

ILLUSTRATION: {896}.

NOTES: herb terrestrial, with two oblong tubers; leaves two, radical, spreading on the soil, ovate or ovate-lanceolate, shortly acute at the tip, rounded or slightly narrowed at the base; spike with 2 sheaths with a narrow short blade; raceme 2- to 6-flowered; rachis short, at most as long as the ovary; floral bracts narrow, shorter than a quarter of the ovary; flowers large, pink, spotted with purple; dorsal sepal hood-shaped, ovate, acute; lateral sepals curved, very narrow towards the base; petals very narrow, falcate, narrowing almost from apex to the base; lip entire, narrow, the base rectangular, expanded and a little angular in thequarter, indented in the middle and a again little broadened at the rounded spathulate apex; spur cylindrical at the base, a little expanded towards the apical third; anther retuse.

Cynorkis squamosa *(Poir.) Lindl.*, Gen. Sp. Orchid. Pl.: 332 (1835). TYPE not identified.

Habenaria amphorchis Spreng., Syst. 3: 689 (1826).

Orchis squamosa Poir., Encyc. 4: 601 (1794).

Gymnadenia squamata (Poir.) A.Rich., Mem. Soc. Hist. Nat. Par. 4: 22 (1828).

Amphorchis squamosa (Poir.) Frapp., Fl. Ile Réunion: 231 (1895).

Amphorchis calcarata Thouars, Orch. Iles Afr. Tabl. Synopt. t.4 (1822). TYPE: Mauritius, *Thouars* (holo. P).

Barlaea calcarata (Thouars) Rchb.f., Linnaea, 41: 54 (1877).

Cynorkis calcarata (Thouars) T.Durand & Schinz, Consp. Fl. Afric. 5: 96 (1892 publ. 1893).

DISTRIBUTION: Comoros: Anjouan. Mascarenes: Mauritius, Réunion. There are unconfirmed reports from Madagascar.

HABITAT: amongst grasses; mossy rocks by streams.

ALTITUDE: 1500 – 2300 m.

FLOWERING TIME: June – August.
LIFE FORM: terrestrial or lithophyte.
DESCRIPTION: {206}.
ILLUSTRATION: {167}, {1028}.

Cynorkis stenoglossa *Kraenzl.* in Bot. Jahrb. Syst. 17: 63 (1893). TYPE: Madagascar, S Betsileo, Ankafina forest, *J.M. Hildebrandt* 4204 (holo. B?).

DISTRIBUTION: Madagascar: Antananarivo, Antsiranana, Fianarantsoa, Toamasina.
HABITAT: humid, evergreen forest.
ALTITUDE: 600 – 1500 m.
FLOWERING TIME: August – December.
PHYTOGEOGRAPHICAL REGION: I, III.
DESCRIPTION: {896}.
ILLUSTRATION: {896}.
NOTES: flowers medium-sized, violet or purple; lip with a narrow blade, 5 – 8 × 2.5 – 2.8 mm; spur 10 – 14 mm long, cylindrical. It is likely that the type has been destroyed, Perrier de la Bâthie indicated the following other collections: Madagascar, Maningory, *H.Perrier* 11377; between Moramanga & Anosibe, *H.Perrier* 18028; Mandraka basin, *Alleizette* 478; Manongarivo massif, *H.Perrier* 1941.

Cynorkis stenoglossa var. **pallens** *H.Perrier* ex *Hermans* **var. nov.** TYPE: Madagascar, Mt. Ambre, *H.Perrier* 17733 (holo. P). See Appendix .
Cynosorchis stenoglossa var. *pallens* H.Perrier in Fl. Mad. Orch. 1: 159 (1939), *nom. nud.*

DISTRIBUTION: Madagascar: Antsiranana.
HABITAT: humid, montane forest.
ALTITUDE: c. 1200 m.
LIFE FORM: epiphyte or terrestrial.
PHYTOGEOGRAPHICAL REGION: III.
DESCRIPTION: {896}.
NOTES: differs from the typical variety by the white or lilac-tinted flowers, the lip with 9 nerves, the 10 – 11 mm long spur being shorter than the 14 – 17 mm long ovary, the narrower viscidia, more acute at both ends, and the stigmatic processes that are fused together at the base and apex, with an elliptic gap in the middle.

Cynorkis stolonifera *(Schltr.) Schltr.* in Repert. Spec. Nov. Regni Veg. Beih. 33: 72 (1924). TYPE: Madagascar, nr. Antsirabe, *H.Perrier* 8126 (holo. P).
Bicornella stolonifera Schltr. in Beih. Bot. Centralbl. 34 (2): 304 (1916).

DISTRIBUTION: Madagascar: Antananarivo, Fianarantsoa.
HABITAT: marshes in peaty soil.
ALTITUDE: 700 – 1600 m.
FLOWERING TIME: January – March.
LIFE FORM: terrestrial.
PHYTOGEOGRAPHICAL REGION: III.
DESCRIPTION: {896}.
ILLUSTRATION: {896}, {1223}.
NOTES: leaves long and grass-like, lacking tubers but sending out stolons; flowers c. 10 mm long, reddish purple; lip ligulate; spur cylindrical, c. 1.5 mm long.

Cynorkis subtilis *Bosser* in Adansonia sér. 3, 26 (1): 56 (2004). TYPE: Toamasina, Maroantsetra, Masoala peninsula, Andranobe, *Labat, Andrianjafy & Poncy* 3426 (holo. P; iso. K, P, TAN).

DISTRIBUTION: Madagascar: Toamasina.
HABITAT: humid, lowland forest.
ALTITUDE: sea level – 10 m.
FLOWERING TIME: October.
LIFE FORM: terrestrial or amongst rock.
PHYTOGEOGRAPHICAL REGION: I.

Description: {145}.

Illustration: {145}.

Notes: similar to *C. gymnochiloides* but plant more slender with more leaves (3 – 5), a laxly-flowered inflorescence, smaller flowers, an oblong concave lip and a spur which is hook-shaped at the tip.

Cynorkis sylvatica *Bosser* in Adansonia, sér. 2, 9 (3): 352 (1969). TYPE: Madagascar, massif de Marivorahona, SW of Manambato (upper Mahavavy, Ambilobe district), *Humbert & Capuron* 25769 (holo. P; iso. K, TAN).

Distribution: Madagascar: Antsiranana.

Habitat: lichen-rich forest.

Altitude: 2000 – 2500 m.

Life form: terrestrial.

Phytogeographical region: IV.

Description: {113}.

Illustration: {113}.

Notes: the flowers resembles those of *C. tristis* but distinct by their 3-lobed, oblong lip and especially the rostellum.

Cynorkis tenella *Ridl.* in J. Linn. Soc., Bot. 22: 124 (1886). TYPES: Comoros, Grande Comore, Mohely, *Boivin* s.n. (syn. BM, P) & Madagascar, Ambatovory, *Fox* s.n. (syn. BM).

Distribution: Madagascar: Fianarantsoa. Comoros: Grande Comore.

Flowering time: May.

Phytogeographical region: III.

Description: {896}.

Notes: plant small, slender, few-flowered; lip 3-lobed, the lateral lobes curved and subacute; spur shortly expanded. The Comorean plants are somewhat different from those from Madagascar.

Cynorkis tenerrima *(Ridl.) Kraenzl.*, Orch. Gen. Sp. 1: 493 (1901). TYPE: Madagascar, Andrangaloaka, E Imerina, *Hildebrandt* 3732 (holo. BM; iso. K, P).
Habenaria tenerrima Ridl. in J. Linn. Soc., Bot. 21: 505 (1885).

Distribution: Madagascar: Antananarivo.

Habitat: marshes in peaty soil.

Altitude: 1400 – 2300 m.

Flowering time: November – December.

Life form: terrestrial.

Phytogeographical region: III, IV.

Description: {896}.

Illustration: {896}.

Notes: plant small, few-flowered; flowers small, white-pink or lilac; lip 3.5 mm long, broadly obcordate, the base narrow; spur cylindrical and straight, 2.2 – 2.8 mm long.

Cynorkis tenuicalcar *Schltr.* in Repert. Spec. Nov. Regni Veg. Beih. 33: 73 (1924). TYPE: Madagascar, Mt. Tsaratanana, *H.Perrier* 15269 (holo. P).

Distribution: Madagascar: Antsiranana.

Habitat: lichen-rich forest.

Altitude: 1700 – 2000 m.

Flowering time: October – January.

Life form: terrestrial.

Phytogeographical region: III, IV.

Description: {896}.

Notes: related to *C. lancilabia* but the lip which oblong and has a small callus at the base and the 2.7 cm long, filiform spur.

Cynorkis tenuicalcar subsp. **andasibeensis** *H.Perrier* ex *Hermans* **subsp. nov.** TYPE: Madagascar, Andasibe (Onive), *H.Perrier* 17417 (holo. P). See Appendix.
Cynosorchis tenuicalcar subsp. *andasibeensis* H.Perrier in Fl. Mad. Orch. 1: 153 (1939), *nom. nud.*

DISTRIBUTION: Madagascar: Antananarivo.
HABITAT: mossy forest.
ALTITUDE: c. 1000 m.
LIFE FORM: terrestrial.
PHYTOGEOGRAPHICAL REGION: III.
DESCRIPTION: {896}.
NOTES: flowers pale red; lateral sepals 7 – 9-veined; petals 3-veined; lip of the same shape as the typical subspecies, but perhaps with a lateral tooth on each side towards the base; viscidia very different, rounded at the tip, hollowed below and extended at one side into a broad transparent blade; middle lobe of the rostellum as high as the anther; stigmatic processes completely united to each other in a thick lobule and lacking a middle window; ovary and the floral bracts with a few glands.

Cynorkis tenuicalcar subsp. **onivensis** *H.Perrier* ex *Hermans* **subsp. nov.** TYPE: Madagascar, Forest of Andasibe (Onive), *H.Perrier* 17146 (holo. P). See Appendix .
Cynosorchis tenuicalar subsp. *onivensis* H.Perrier in Fl. Mad. Orch. 1: 153 (1939), *nom. nud.*

DISTRIBUTION: Madagascar: Antananarivo.
HABITAT: lichen-rich forest on moss- and lichen-covered trees.
ALTITUDE: c. 1300 m.
FLOWERING TIME: February.
LIFE FORM: epiphyte.
PHYTOGEOGRAPHICAL REGION: III.
DESCRIPTION: {896}.
NOTES: similar in habit to the typical subspecies; tubers very elongate; flowers pale red; spike, floral bracts, ovary and the outside of the perianth covered by scattered hairs; lateral sepals 5 – 6-nerved ; lip ovate-lanceolate or oblanceolate, wider in the middle or above and more or less elongate but always narrowed towards the base and lacking a callus at the base; viscidia forming a short spindle, acute at both ends; midlobe of the rostellum shorter than the anther and half the length of the rostellar arms.

Cynorkis tristis *Bosser* in Adansonia, sér. 2, 9 (3): 349 (1969). TYPE: Madagascar, Mt. N of Mangindrano, towards the Ambohimiravavy, *Humbert* 25222 (holo. P).

DISTRIBUTION: Madagascar: Mahajanga.
HABITAT: mossy forest.
FLOWERING TIME: January.
LIFE FORM: terrestrial.
PHYTOGEOGRAPHICAL REGION: IV.
ILLUSTRATION: {113}.
NOTES: known only from the type. Characterised by its short rostellum and the shape of the lip with a central callus.

Cynorkis tryphioides *Schltr.* in Ann. Mus. Col. Marseille, sér. 3, 1: 155, t.3 (1913). TYPE: Madagascar, Sambirano, *H.Perrier* 1865 (*Schlechter* 78) (holo. B†; iso. P).

DISTRIBUTION: Madagascar: Antsiranana, Mahajanga.
HABITAT: on limestone rocks in open forest; in grassland; dry forest.
ALTITUDE: 250 – 2000 m.
FLOWERING TIME: December – March.
LIFE FORM: terrestrial or lithophytic.
PHYTOGEOGRAPHICAL REGION: II, III, V.
DESCRIPTION: {896}.
ILLUSTRATION: {896}, {1205}, {1223}.
VERNACULAR NAME: Felantrandraka.
NOTES: plant very small; leaves spreading over the soil; flowers very small, c. 9 mm long; lip angular, 3-lobed; spur cylindrical, c. 3 mm long.

Cynorkis tryphioides var. **leandriana** *(H.Perrier) Bosser* in Adansonia, sér. 2, 9 (3): 358 (1969).
TYPE: Madagascar, Tsingy of Bemaraha, Res. Nat. 9, *Léandri* 951 (holo. P).
Benthamia leandriana H.Perrier in Notul. Syst. (Paris) 14: 139.

DISTRIBUTION: Madagascar: Mahajanga.
HABITAT: seasonally dry, deciduous forest or woodland on limestone.
ALTITUDE: sea level – 500 m.
FLOWERING TIME: December – January.
LIFE FORM: terrestrial or lithophyte.
PHYTOGEOGRAPHICAL REGION: V.
DESCRIPTION: {113}.
NOTES: differs from the typical variety by the shape of its short, 3-lobed lip, the shape of the rostellum, and its larger leaves.

Cynorkis uncinata *H.Perrier* ex *Hermans* **sp. nov.** TYPE: Madagascar, Moramanga area, *Decary* 7269 (holo. P). See Appendix .
Cynosorchis uncinata H.Perrier in Fl. Mad. Orch. 1: 137, fig. XII, 3 – 4 (1939), *nom. nud.*

DISTRIBUTION: Madagascar: Antananarivo, Toamasina.
HABITAT: epiphyte on *Pandanus* and occasionally other trees.
ALTITUDE: 600 – 1000 m.
FLOWERING TIME: February – March.
LIFE FORM: epiphyte.
PHYTOGEOGRAPHICAL REGION: III.
DESCRIPTION: {896}.
ILLUSTRATION: {59}, {470}, {479}, {482}, {611}, {834}, {862}, {896}, {1025} as *C. ridleyi*. The illustration in {281} is misidentified.
CULTIVATION: {700}.
NOTES: epiphytic herb, with 1 – 3 inflated, irregularly shaped fleshy roots; leaf 1, radical, large, ovate-oblong, acute, obtuse at the base, almost as long as the inflorescence; spike glabrous, with 1 – 2 wide, short, acuminate sheaths near the top, with thin recurved tips; raceme dense, subglobose, with 7 – 15-flowered; floral bracts almost as long as the flowers, broadly oval-lanceolate, very longly acuminate into a subfiliform point and curled-recurved; flowers pink, rather large, with a few glandular hairs on the ovary and the outside of the sepals; dorsal sepal ovate, obtuse, concave; lateral sepals arched, apiculate-acute near the apex at the upper edge; petals narrow, lanceolate-linear, obtuse; lip 4-lobed; lateral lobes subrectangular, sub-crenulate at the apex; middle lobe divided into 2 obovate-cuneiform lobules, crenulate at the apex; spur subcylindrical; anther large, shortly apiculate-acute, with a wide filament; caudicles rigid; viscidia small and oval; staminodes broadly rounded at the tip.
NOTE: The nomenclature of this species is in press, to be revised as follows: **Cynorkis calanthoides** *Kraenzl.* in Abh. Naturwiss. Vereine Bremen 7: 260 (1882). TYPE: Madagascar, Alarbi (E Imarina), *Rutenberg* s.n. (holo.†). TYPE: Madagascar, Hort. Kew, Dec.1900, ill. Bot. Mag. t.7852 (neo. K).
Syn. *Cynorkis uncinata* H.Perrier, Flore de Madagascar Orchidacées 1: 137 (1939), *nom. nud.* based on *Decary* 7062, 7173 & 7269 (all P).

Cynorkis uniflora *Lindl.*, Gen. Sp. Orch. Pl.: 331 (1835). TYPE: Madagascar, *Lyall* s.n. (holo. K).
Gymnadenia uniflora (Lindl.) Steud., Nom. Bot., ed. 2, 1: 712 (1840).
Cynosorchis grandiflora Ridl. in J. Linn. Soc., Bot. 20: 332 (1883). TYPE: Madagascar, *Deans Cowan* s.n. (holo. BM).
Cynosorchis grandiflora var. *albata* Ridl. in J. Linn. Soc., Bot. 20: 332 (1883). TYPE: Madagascar (holo. BM).
Cynosorchis grandiflora var. *purpurea* Ridl. in J. Linn. Soc., Bot. 20: 332 (1883). TYPE: Madagascar (holo. BM).
Cynosorchis grandiflora (Ridl.) T.Durand & Schinz, Consp. Fl. Afr. 5: 91 (1892).

DISTRIBUTION: Madagascar: Antananarivo, Fianarantsoa.
HABITAT: grassland; rocky outcrops, often on inselbergs.
ALTITUDE: 1200 – 1450 m.
FLOWERING TIME: December – March.
LIFE FORM: terrestrial or rarely epiphytic.

PHYTOGEOGRAPHICAL REGION: III.

DESCRIPTION: {552}, {896}.

HISTORY: {360}, {375}, {1068}, {1073}.

ILLUSTRATION: {246}, {281}, {360}, {552}, {841}, {896}, {906}, {959}, {1298}, {1468}.

NOTES: leaves green with brownish-red spots below especially towards the base; flowers 1 – 2, large; lip c.4 × 2.5 cm, 4-lobed, violet-red, with 2 paler spots on the disk.

Cynorkis verrucosa *Bosser* in Adansonia, sér. 2, 9 (3): 357 (1969). TYPE: Madagascar, Canton of Sahavondrona, Ambatondrazaka District, E of Manakambahiny, *Cons. Res. Nat.* 10490 (holo. P).

DISTRIBUTION: Madagascar: Toamasina.

HABITAT: humid, evergreen forest.

ALTITUDE: 1000 – 1500 m.

FLOWERING TIME: June.

LIFE FORM: terrestrial.

PHYTOGEOGRAPHICAL REGION: III.

DESCRIPTION: {113}.

ILLUSTRATION: {113}.

NOTES: flowers small; lip 3.5 – 5 × 4.5 – 5 mm, 5-lobed, with small warts on the surface.

Cynorkis villosa *Rolfe ex Hook.f.* in Curtis's Bot. Mag.: t.7845 (1902). TYPE: Madagascar, Tanambe, *Warpur* s.n. (holo. K).

DISTRIBUTION: Madagascar: Toliara.

HABITAT: in ravines; in humid, montane forest.

ALTITUDE: 1000 – 1200 m.

FLOWERING TIME: March – June.

LIFE FORM: epiphyte or terrestrial.

PHYTOGEOGRAPHICAL REGION: I.

DESCRIPTION: {555}, {896}.

HISTORY: {365}.

ILLUSTRATION: {281}, {470}, {480}, {555}, {906}, {1298}, {1457}.

CULTIVATION: {1457}.

NOTES: inflorescence hairy; flowers rose-purple, not opening much, very hairy; lip 3-lobed, pale.

Cynorkis violacea *Schltr.* in Ann. Mus. Col. Marseille, sér. 3, 1: 155, t.4 (1913). TYPE: Madagascar, along River Zangora (Manongarivo), *H.Perrier* 36 (holo. B†; iso. P).

DISTRIBUTION: Madagascar: Antsiranana, Fianarantsoa, Mahajanga.

HABITAT: rocky outcrops and in sand.

ALTITUDE: sea level – 1200 m.

FLOWERING TIME: February – March.

LIFE FORM: terrestrial.

PHYTOGEOGRAPHICAL REGION: II, III.

DESCRIPTION: {896}.

ILLUSTRATION: {896}, {1205}, {1223}.

NOTES: flowers violet; lip 10 – 12 mm long, 4-lobed, the lateral lobes large; spur cylindrical, 12 – 14 mm long.

Cynorkis zaratananae *Schltr.* in Repert. Spec. Nov. Regni Veg. Beih. 33: 75 (1924). TYPE: Madagascar, Mt. Tsaratanana, *H.Perrier* 15299 (holo. B†; iso. P).

DISTRIBUTION: Madagascar: Antsiranana.

HABITAT: mossy forest.

ALTITUDE: 1700 – 2200 m.

FLOWERING TIME: January.

LIFE FORM: epiphyte in moss.

PHYTOGEOGRAPHICAL REGION: III.

DESCRIPTION: {896}.

Notes: related to *C. stenoglossa*, but with smaller flowers and a longer pointed spur. Incorrectly changed by Perrier de la Bâthie {896} to *C. tsaratananae*.

Didymoplexis madagascariensis *(Schltr. ex H.Perrier) Summerh.* in Kew Bull. 8: 131 (1953). TYPE: Madagascar, Fandrarazana, *H.Perrier* 11349 (holo. P).
Gastrodia madagascariensis Schltr. ex H.Perrier in Repert. Spec. Nov. Regni Veg. Beih. 33: 123 (1924), *nom. nud.*; H.Perrier in Humbert ed., Fl. Mad. Orch. 1: 212 (1939).

Distribution: Madagascar: Toamasina.
Habitat: humid, evergreen forest; riverine forest, in humus.
Altitude: c. 200 m.
Flowering time: July, November.
Life form: terrestrial.
Phytogeographical region: I.
Life form: terrestrial.
Description: {896}.
Notes: plant 20 to 40 cm, leafless, saprophytic, lacking chlorophyll.

Disa andringitrana *Schltr.* in Repert. Spec. Nov. Regni Veg. Beih. 33: 98 (1924). TYPE: Madagascar, Andringitra massif, *H.Perrier* 14382 (holo. P).

Distribution: Madagascar: Fianarantsoa, Toliara.
Habitat: montane, ericaceous scrub.
Altitude: 2400 – 2600 m.
Flowering time: January.
Life form: terrestrial.
Phytogeographical region: III, IV.
Description: {896}.
Illustration: {896}, {1223}.
Notes: plant large, up to 60 cm tall; flowers 5 – 7 mm long, greenish-white, almost hidden by the floral bracts.

Disa buchenaviana *Kraenzl.* in Abh. Naturwiss. Vereine Bremen 7: 261 (1882). TYPE: Madagascar, N Betsileo, *Hildebrandt* 3845 (holo. BM; iso. K, P).
Disa rutenbergiana Kraenzl. in Abh. Naturwiss. Vereine Bremen 7: 262 (1882).
Satyrium calceatum Ridl. in J. Linn. Soc., Bot. 21: 520 (1885).

Distribution: Madagascar: Antananarivo, Fianarantsoa.
Habitat: wet grassland; montane ericaceous scrub; rocky outcrops.
Altitude: 1500 – 2400 m.
Flowering time: November – February.
Life form: terrestrial.
Phytogeographical region: III, IV.
Description: {896}.
Illustration: {472}, {896}.
Vernacular name: Vary mangatsiaka.
Notes: plant 35 – 70 cm tall; flowers c. 7 mm long, bright blue-mauve, not hidden by the floral bracts.

Disa caffra *Bolus* in J. Linn. Soc., Bot. 25: 171 (1889). TYPE: South Africa, Pondoland, Umkwani, *Tyson* 2611 (holo. G).
Disa compta Summerh. in Kew Bull. 17: 545 (1964). TYPE from Zambia.
Disa perrieri Schltr. in Repert. Spec. Nov. Regni Veg. Beih. 33: 100 (1924). TYPE: Madagascar, Mt. Ambohitrakidahy (Ambositra), *H.Perrier* 12434 (holo. B†; iso. P).

Distribution: Madagascar: Antananarivo, Fianarantsoa. Also in Eastern and southern Africa.
Habitat: marshes.
Altitude: sea level – 2000 m.
Flowering time: October – February.

LIFE FORM: terrestrial.
PHYTOGEOGRAPHICAL REGION: III, IV.
DESCRIPTION: {765}, {768}, {896}.
ILLUSTRATION: {104}, {768}, {896}.
NOTES: plant 25 – 50 cm tall; flowers 15 – 20 mm long, dull mauve; lip lanceolate-spathulate; spur conical.

Disa incarnata *Lindl.,* Gen. Sp. Orch. Pl.: 348 (1838). TYPE: Madagascar, without locality, *Lyall* 178 (holo. K).
Disa fallax Kraenzl. in Bot. Jahrb. Syst. 17: 64 (1893). TYPE: Madagascar, N Betsileo, *Hildebrandt* 3874 (holo. B†; iso. K?, P, W).

DISTRIBUTION: Madagascar: Antananarivo, Fianarantsoa.
HABITAT: grassland; marshes; near streams.
ALTITUDE: 1300 – 2000 m.
FLOWERING TIME: November – March.
LIFE FORM: terrestrial.
PHYTOGEOGRAPHICAL REGION: III, IV.
DESCRIPTION: {281}, {548}, {766}, {896}.
HISTORY: {357}.
ILLUSTRATION: {357}, {369}, {504}, {548}, {766}, {896}, {959}, {1298}.
NOTES: plant c. 50 cm tall; flowers 8 – 12 mm long, bright orange-red.

Disperis afzelii *Schltr.* in Repert. Spec. Nov. Regni Veg. Beih. 15: 326 (1918). TYPE: Madagascar, Andakambararata, *Afzelius* s.n. (holo. UPS or B†).

DISTRIBUTION: Madagascar.
ALTITUDE: 500 – 1000 m.
FLOWERING TIME: September.
LIFE FORM: terrestrial.
PHYTOGEOGRAPHICAL REGION: I.
DESCRIPTION: {896}, {686} .
NOTES: it is likely that this species is synonymous with *D. tripetaloides* but, as the type is lost, doubts persist.

Disperis ankarensis *H.Perrier* in Notul. Syst. (Paris) 8: 129 (1939). TYPE: Madagascar, Ankarana, N Ambilobe, *Humbert* 18825 (holo. P).

DISTRIBUTION: Madagascar: Antsiranana.
HABITAT: on limestone rocks in seasonally dry, deciduous forest or woodland; in semi-dry forest of the west, growing in humus amongst calcareous rocks.
ALTITUDE: c. 300 m.
FLOWERING TIME: January.
LIFE FORM: terrestrial.
PHYTOGEOGRAPHICAL REGION: V.
DESCRIPTION: {686}, {896}.
ILLUSTRATION: {686}.
NOTES: known only from the type. The flower shows affinities with *D. hildebrandtii* but it is distinguished readily by the nature of the hairs of the terminal appendage of the lip.

Disperis anthoceras var. **humbertii** *(H.Perrier) la Croix* in Adansonia sér. 3, 24 (1): 58 (2002). TYPE: Madagascar, Analavelona Forest, Fiherena, *Humbert* 14212 (holo. P).
D. humbertii H.Perrier in Not. Syst. 8: 35 (1939).

DISTRIBUTION: Madagascar: Toliara. The typical variety is widespread in tropical Africa.
HABITAT: in humus in bamboo forest in ravines; humid evergreen forest.
ALTITUDE: 950 – 1250 m.
FLOWERING TIME: March.
LIFE FORM: terrestrial.

PHYTOGEOGRAPHICAL REGION: III.

DESCRIPTION: {686, {896}.

ILLUSTRATION: {686}.

NOTES: known only from the type. Dorsal sepal and petals joined to form a distinct erect tapering spur; flowers of the variety are smaller than those of *D. anthoceros*.

Disperis bathiei *Bosser* in Adansonia sér. 3, 24 (1): 79 (2002). TYPE: Madagascar, Massif of Manongarivo, *H.Perrier* 1924bis (holo. P).

DISTRIBUTION: Madagascar: Antsiranana.

HABITAT: in the understorey of humid forest.

ALTITUDE: 1000 – 2000 m.

FLOWERING TIME: May.

LIFE FORM: terrestrial.

PHYTOGEOGRAPHICAL REGION: IV.

DESCRIPTION: {686}.

ILLUSTRATION: {686}.

NOTES: known only from the type. Lateral sepals 13 – 14 mm long; terminal lobe of the lip appendage tongue-shaped convex and glabrous on the upper surface, 5.5 mm long.

Disperis bosseri *P.J.Cribb & la Croix* in Adansonia sér. 3, 24 (1): 82 (2002). TYPE: Madagascar, Ankaratra, road from Ambatolampy to Faratsiho, *Bosser* 10869 (holo. P).

DISTRIBUTION: Madagascar: Antananarivo.

HABITAT: humid places in mist forest on steep slopes.

ALTITUDE: 2400 m.

FLOWERING TIME: February.

LIFE FORM: terrestrial.

PHYTOGEOGRAPHICAL REGION: IV.

DESCRIPTION: {686}.

ILLUSTRATION: {686}.

NOTES: known only from the type. Base of leaves clasp the stem; resembles *D. concinna* from South Africa and Zimbabwe, but different in the lip and appendage.

Disperis ciliata. *Bosser* in Adansonia sér. 3, 24 (1): 64 (2002). TYPE: Madagascar, mountains N of Mangindrano, 1951, *Humbert & Capuron* 25002 (holo. P; iso. K, MO, P).

DISTRIBUTION: Madagascar: Mahajanga.

HABITAT: Under trees in humid montane forest on laterite or gneiss, transition to lichen-rich forest.

ALTITUDE: 1800 – 2000 m.

FLOWERING TIME: January – February.

LIFE FORM: terrestrial.

PHYTOGEOGRAPHICAL REGION: V.

DESCRIPTION: {686}

ILLUSTRATION: {686}

NOTES: similar to *D. latigaleata*, distinguished by the ciliate petals and by the terminal appendage of its lip in which the mid-lobe is relatively short and covered by a short papillate pubescence.

Disperis concinna *Schltr*, in Bot. Jahrb. Syst. 20, Beible. 50: 43 (1895). TYPE: South Africa, Transvaal, near Wilge River, *Schlechter* (holo. B†).

DISTRIBUTION: Madagascar: Antananarivo. South Africa and Zimbabwe.

HABITAT: moist, submontane grassland (in mainland Africa).

ALTITUDE: 2400 m.

FLOWERING TIME: February.

LIFE FORM: terrestrial.

PHYTOGEOGRAPHICAL REGION: IV.

DESCRIPTION: {682}, {768}, {1156}, {1329}.
ILLUSTRATION: {768}, {1329}.
NOTES: found only once in Madagascar. The type is lost.

Disperis discifera *H.Perrier* in Notul. Syst. (Paris) 5: 222 (1936). TYPE: Madagascar, Mt. Tsaratanana, *H.Perrier* (lecto. P).

DISTRIBUTION: Madagascar: Antananarivo, Antsiranana, Fianarantsoa.
HABITAT: montane, ericaceous scrub; mossy forest.
ALTITUDE: 1000 – 2000 m.
FLOWERING TIME: January – June.
LIFE FORM: terrestrial.
PHYTOGEOGRAPHICAL REGION: III.
DESCRIPTION: {686} {896}.
ILLUSTRATION: {686}, {896}.
NOTES: lip with a simple, ovate appendage, covered in dense papillae.

Disperis discifera var. **borbonica** *Bosser* in Adansonia sér. 3, 24 (1): 65 (2002).
Disperis borbonica H.Perrier in Not. Syst. (Paris) 5 (3): 228 (1936). Type: Réunion, without exact loc., *Delteil* s.n (holo. P).

DISTRIBUTION: Madagascar: Antananarivo. Réunion.
HABITAT: under trees in humid forests and ericaceous scrub; also grows in pine plantations.
ALTITUDE: 1200 – 2350 m.
FLOWERING TIME: February – April, August.
LIFE FORM: terrestrial.
PHYTOGEOGRAPHICAL REGION: III, IV.
DESCRIPTION: {686}.
ILLUSTRATION: {686}.
NOTES: the variety is very similar in habit and lip morphology to the typical *D. discifera*, but its flowers are much smaller.

Disperis erucifera *H.Perrier* in Notul. Syst. (Paris) 5: 226 (1936). TYPE: Madagascar, Mt. de Français, nr. Antsiranana, *H.Perrier* 17509 (holo. P).

DISTRIBUTION: Madagascar: Antsiranana.
HABITAT: in leaf debris and humus between shaded and humid, calcareous rocks and limestone crevices in shade.
ALTITUDE: 30 – 250 m.
FLOWERING TIME: January – March.
LIFE FORM: terrestrial.
PHYTOGEOGRAPHICAL REGION: V.
DESCRIPTION: {686}, {896}.
ILLUSTRATION: {281}, {686}, {896}.
NOTES: apical free part of the lip 7 mm long, oblong, deeply concave at the base, narrowing above then dilated at the tip into a subrhombic lobe, densely white-tomentose on its upper side.

Disperis falcipetala *P.J.Cribb & la Croix* in Adansonia sér. 3, 24 (1): 85 (2002). TYPE: Madagascar, Antsiranana, Mangonarivo Special Reserve, Antsatrotro, E of summit, *Malcomber, Hutcheon, Razafimanantsoa & Zjhra* 1413 (holo. MO).

DISTRIBUTION: Madagascar: Antsiranana.
HABITAT: ridge-top, montane forest.
ALTITUDE: c. 1700 m.
FLOWERING TIME: April.
LIFE FORM: terrestrial.
PHYTOGEOGRAPHICAL REGION: V.
DESCRIPTION: {686}.
ILLUSTRATION: {686}.
NOTES: known from the type only. Base of leaves not clasping stem, lateral sepals fused in basal half.

Disperis hildebrandtii *Rchb.f.,* Otia Bot. Hamburg.: 73 (1881) & in Bot. Zeit. (Berlin) 39 (28): 449 (1881). TYPE: Madagascar, Lokobe, Nossi-Be, *Hildebrandt* 3158 (holo. W; iso. BM, K, P).

DISTRIBUTION: Madagascar: Antsiranana. Comoros: Mayotte.
HABITAT: humid evergreen forest; in dry woods and forests on laterite soils derived from gneiss or from basalt.
ALTITUDE: 90 – 1400 m.
FLOWERING TIME: September – February.
LIFE FORM: terrestrial.
PHYTOGEOGRAPHICAL REGION: II, III.
DESCRIPTION: {686}, {896}.
ILLUSTRATION: {686}, {896}.
NOTES: sepals all of similar length, 8 – 10 mm long; midlobe of lip appendage sessile; lip appendage bearing vesicular stalked hairs.

Disperis humblotii *Rchb.f.* in Flora 68: 377 (1885). TYPE: Comoros, *Humblot* s.n. (holo. W).
Disperis comorensis Schltr. in Bot. Jarhb. Syst. 24: 429 (1897). TYPE: Comoros, *Bang* s.n., in Herb. Schlechter (holo. B†).

DISTRIBUTION: Madagascar: Antananarivo, Antsiranana. Comoros: Anjouan, Grande Comore.
HABITAT: under trees in humid forest at intermediate elevations.
ALTITUDE: 800 – 2000 m.
FLOWERING TIME: July – September.
LIFE FORM: epiphyte, terrestrial or lithophyte.
PHYTOGEOGRAPHICAL REGION: III.
DESCRIPTION: {686}, {896}.
ILLUSTRATION: {686}, {896}.
NOTES: flower similar to *D. lanceana*, but the terminal appendage of the lip is larger and of a different shape.

Disperis lanceana *H.Perrier* in Notul. Syst. (Paris) 5: 223 (1936). TYPE: Madagascar, *Lance* s.n. (holo. P).

DISTRIBUTION: Madagascar.
ALTITUDE: 1600 – 1900 m.
PHYTOGEOGRAPHICAL REGION: I.
DESCRIPTION: {686}, {896}.
ILLUSTRATION: {686}, {896}.
NOTES: resembles *D. hildebrandtii* in habit, but distinguished by the terminal appendage of the lip which is stalked not sessile and by the indumentum of that appendage which lacks vesicular hairs.

Disperis lanceolata *Bosser* in Adansonia sér. 3, 24 (1): 81 (2002). TYPE: Madagascar, Ambohitantely, district of Ankazobe, *Bosser* 7951 (holo. P).

DISTRIBUTION: Madagascar: Antananarivo, Toamasina.
HABITAT: in humus in shade in moist forest; edge of forest in shade of small trees in deep leaf litter.
ALTITUDE: 1000 – 2000 m.
FLOWERING TIME: March – May.
LIFE FORM: terrestrial.
DESCRIPTION: {686}.
ILLUSTRATION: {686}.
NOTES: similar to *D. humblotii* and *D. saxicola* in its habit, distinguished by its much smaller and narrower, sessile leaves and by the terminal appendage of its lip in the form of a flat, elliptic or suborbicular plate having at its base a rounded protuberance.

Disperis latigaleata *H.Perrier* in Notul. Syst. (Paris) 8: 36 (1939). Type: Madagascar, Mt. Itrafanaomby, Ankazondrano, *Humbert* 13440 (lecto. P).

DISTRIBUTION: Madagascar: Fianarantsoa, Toliara.

HABITAT: seasonally dry, deciduous very shady forest or woodland; on moss-covered boulders; on river sides.

ALTITUDE: 600 – 2000 m.

FLOWERING TIME: December – February.

LIFE FORM: terrestrial, lithophyte or epiphyte.

PHYTOGEOGRAPHICAL REGION: III.

DESCRIPTION: {686}, {896}.

ILLUSTRATION: {686}

NOTES: close to *D. oppositifolia*, but differs in having a wider hood, in the lateral sepals being free to the base and lacking a spur, in the mid lobe of the lip appendage being as long as or slightly longer than the side lobes, and in the spathulate lobes of the rostellum.

Disperis majungensis *Schltr.* in Repert. Spec. Nov. Regni Veg. Beih. 33: 108 (1924). TYPE: Madagascar, nr. Mahajanga, *H.Perrier* 13038 (holo. P).

DISTRIBUTION: Madagascar: Mahajanga.

HABITAT: rocky forest; coastal forest.

ALTITUDE: sea level – 100 m.

FLOWERING TIME: January – February.

LIFE FORM: terrestrial in humus.

PHYTOGEOGRAPHICAL REGION: V.

DESCRIPTION: {686}, {896}.

ILLUSTRATION: {686}, {896}.

NOTES: this species is readily recognised by its very characteristic trident-like lip appendage.

Disperis masoalensis *P.J.Cribb & la Croix* in Adansonia sér. 3, 24 (1): 82 (2002). TYPE: Madagascar, Toamasina, Masoala Peninsula, *Schatz, van der Werff, Gray & Razafimandimbison* 3380 (holo. K; iso. MO, P).

DISTRIBUTION: Madagascar: Toamasina.

HABITAT: in moss on top of boulders in river-bed.

ALTITUDE: sea level – 25 m.

FLOWERING TIME: October.

LIFE FORM: terrestrial.

PHYTOGEOGRAPHICAL REGION: I.

DESCRIPTION: {686}.

ILLUSTRATION: {686}.

NOTES: allied to *D. bathiei* but differs from the former in its ovate, very shortly petiolate leaves, larger flowers in which the lateral sepals are fused almost to their mid-point and have the shallowest of spurs in the centre, broader undulate petals, and lip in which the mid-lobe is sessile and the side lobes are incurved and papillose rather than erect and pubescent.

Disperis oppositifolia *Sw.* in Rees, Cyclopaedia 11, no. 6 (1801). TYPE: Réunion, *J.E.Smith* s.n. (holo. LINN).

Dryopera oppositifolia (Sw.) Thouars, Hist. Orch.: t.1 (1822) as *Dryopeia*.

DISTRIBUTION: Madagascar: Antananarivo, Antsiranana, Fianarantsoa, Toamasina, Toliara {896}. Comoros: Grande Comore, Mayotte. Mascarenes: Réunion, Mauritius.

HABITAT: moisture-loving species of humid forests at low and intermediate elevations; terrestrial or on rocks by streams, sometimes in moss on the base of tree trunks; in riparian forest with *Pandanus* sp.; and sometimes in pine plantations.

ALTITUDE: sea level – 1500 m.

FLOWERING TIME: September – November.

LIFE FORM: epiphyte or terrestrial.

PHYTOGEOGRAPHICAL REGION: I, II, III.

DESCRIPTION: {686}, {896}.

ILLUSTRATION: {246}, {686}, {1398}.

NOTES: flowers of white, pale rose or lilac or white with sepals striped with red, magenta or mauve; terminal appendage of the lip is very variable, the mid-lobe often short and triangular, but it can be elongated into a point at the glabrous tip.

Disperis perrieri *Schltr.* in Ann. Mus. Col. Marseille, sér. 3, 1: 17, t.5 (1913). TYPE: Madagascar, Manongarivo, *H.Perrier* 39 (holo. P).

DISTRIBUTION: Madagascar: Antsiranana.
HABITAT: humid, evergreen forest; in humus and amongst lichens in damp woodland on gneiss.
ALTITUDE: 1000- 2000 m.
FLOWERING TIME: March, May.
LIFE FORM: epiphyte or terrestrial.
PHYTOGEOGRAPHICAL REGION: III.
DESCRIPTION: {686}, {896}.
ILLUSTRATION: {686}, {896}, {1205}.
NOTES: hood shallow, obovate, 12 – 15 mm tall, 16 – 20 mm wide at most but tapering to 2 mm wide at base; terminal lobe of the lip appendage reniform, lacking long hairs on the upper surface.

Disperis saxicola *Schltr.* in Repert. Spec. Nov. Regni Veg. Beih. 33: 110 (1924). TYPE: Madagascar, Itomampy Valley, *H.Perrier* 12650 (holo. P).

DISTRIBUTION: Madagascar: Fianarantsoa, Mahajanga, Toamasina.
HABITAT: on shaded rocks in humid, evergreen forest; seasonally dry, deciduous forest or woodland.
ALTITUDE: 400 – 1000 m.
FLOWERING TIME: March – August.
LIFE FORM: epiphyte or terrestrial.
PHYTOGEOGRAPHICAL REGION: I, III, V.
DESCRIPTION: {686}, {896}.
ILLUSTRATION: {686}, {896}.
NOTES: lateral sepals free to the base, 8–11 mm long; terminal lobe of the lip appendage reniform, 1.8 – 3 mm long, papillose on the upper surface.

Disperis similis *Schltr.* in Repert. Spec. Nov. Regni Veg. Beih. 33: 111 (1924). TYPE: Madagascar, Andringitra massif (lecto. P).

DISTRIBUTION: Madagascar: Antsiranana, Fianarantsoa.
HABITAT: mossy forest.
ALTITUDE: 1500 – 2000 m.
FLOWERING TIME: January – February.
LIFE FORM: terrestrial in humus.
PHYTOGEOGRAPHICAL REGION: III.
DESCRIPTION: {686}, {896}.
ILLUSTRATION: {686}.
NOTES: close to *D. oppositifolia* but the flower is much smaller and lateral sepals not strongly united at the base, and a tri-lobed appendage to the lip that has pubescent lateral lobes that are shorter than the glabrous mid-lobe.

Disperis trilineata *Schltr.* in Repert. Sp. Nov. Regni Veg. Beih. 33: 112 (1924). TYPE Madagascar, Sambirano valley, *H.Perrier* 15705 (holo. P).

DISTRIBUTION: Madagascar: Antsiranana. Comoros: Anjouan.
HABITAT: humid, evergreen forest.
ALTITUDE: c. 400 m.
FLOWERING TIME: January, March.
LIFE FORM: terrestrial in humus.
PHYTOGEOGRAPHICAL REGION: II.
DESCRIPTION: {686}, {896}.
ILLUSTRATION: {686}.
NOTES: leaf dull green on the upper side with three main rose-pink veins, appendage of lip 5-lobed; lateral sepals 3 – 4 mm long.

Disperis tripetaloides *(Thouars) Lindl.,* Gen. Sp. Orch. Pl.: 371 (1839). TYPE: Réunion, *Thouars* s.n. (holo. P).
Dryopera tripetaloidea Thouars, Hist. Orch.: t.3 (1822), as *Dryopeia.*

DISTRIBUTION: Madagascar: Antananarivo, Antsiranana, Toliara. Comoros: Mayotte. Mascarenes: Réunion, Mauritius, Rodrigues. Seychelles.

HABITAT: grows singly or in small colonies in rich soil and medium shade in the understorey of semi-dry forests or rarely in humid forest.

ALTITUDE: 20 – 400 m.

FLOWERING TIME: December – June.

LIFE FORM: terrestrial.

PHYTOGEOGRAPHICAL REGION: II.

DESCRIPTION: {686}, {896}.

ILLUSTRATION: {167}, {414}, {686}, {1398}.

NOTES: close to *D. discolor* in its habit and leaves which are violet on their underside. It can be distinguished readily by its free lateral sepals and the morphology of the terminal free part of the lip.

Eulophia angornensis *(Rchb.f.) Hermans* **comb. nov**. TYPE: Comoros, Anjouan (Angorna), *Peters* s.n. (holo. W.).
Galeandra angornensis Rchb.f. in Linnaea 20: 680 (1847).
Galeandra anjoanensis Rchb.f. in Walpers Ann. 6: 650 (1863), orth. var.
Lissochilus angornensis (Rchb.f.) Rchb.f. in Flora 48: 188 (1865), as *L. angoanensis.*
Eulophia anjoanensis (Rchb.f.) P.J.Cribb in Lindleyana 13, 3: 174 (1998).

DISTRIBUTION: Comoros: Anjouan.

DESCRIPTION: {896}.

NOTES: a large-flowered but poorly known species represented only by the type. Reichenbach first used the spelling '*angornensis*', referring to Angorna which became known as Anjouan. He then changed it to '*anjoanensis*' (1863), but, because this was not a typographical error, the original spelling should be used.

Eulophia beravensis *Rchb.f.,* Otia Bot. Hamburg.: 74 (1881) & in Bot. Zeit.: 449 (1881). TYPE: Madagascar, Beravi, *Hildebrandt* 3055 (holo. W; iso. BM, K, P).
Graphorchis beravensis Kuntze, Rev. Gen. Pl. 2: 662 (1891).
Lissochilus beravensis (Rchb.f.) H.Perrier in Humbert ed., Fl. Mad. Orch. 2: 42 (1941).

DISTRIBUTION: Madagascar: Fianarantsoa, Mahajanga, Toliara.

HABITAT: seasonally dry, deciduous forest and woodland, sand dunes.

ALTITUDE: sea level – 700 m.

FLOWERING TIME: May – November.

LIFE FORM: terrestrial, forming large clumps.

PHYTOGEOGRAPHICAL REGION: V, VI.

DESCRIPTION: {896}.

ILLUSTRATION: {896}.

VERNACULAR NAMES: Oron soro ala, Mafiravy.

NOTES: the plant has medicinal use. Plant with long stems; roots very fleshy; leaves narrowly linear, with a finely serrated edge; flowers c. 9 mm long.

Eulophia clitellifera *(Rchb.f.) Bolus.* in J. Linn. Soc, Bot. 25: 184 (1889). TYPE South Africa, Durban, *Gueinzius* (holo. W).
Lissochilus clitellifera Rchb.f. in Linnaea 20: 687 (1847) as *L. clitellifer.*
Lissochilus pulchellus Rendle in J. Bot. 33: 196 (1895).
Lissochilus rehmannii Rolfe in Dyer, Fl. Cap. 5, 3: 55 (1912).
Lissochilus flexuosus Schltr. in R. E. Fries Wiss. Ergebn. Schwed. Rhod.-Kongo-Exped. 1911 – 12, 1: 245 (1916), *non Eulophia flexuosa* Kraenzl. (1895).
Eulophia fractiflexa Summerh. in Kew Bull. 1953: 156 (1953).

DISTRIBUTION: Madagascar: Antananarivo, Fianarantsoa. Also throughout tropical and southern Africa.

HABITAT: grassy hills.

ALTITUDE: 1000 – 1600 m.

FLOWERING TIME: November – December.

LIFE FORM: terrestrial.

PHYTOGEOGRAPHICAL REGION: III.

DESCRIPTION: {768}, {1329}.

ILLUSTRATION: {768}, {1329}.

NOTES: plant slender, c. 20 – 35 cm tall; petals and lip white with reddish-purple lines on the inner surfaces, the callus crests on the lip bright yellow, fleshy ridges; spur conical.

Eulophia cucullata *(Sw.) Steud.* in Nom. Bot. 2: 205 (1840). TYPE: West Africa, without exact locality, *Afzelius* (holo. S).

Limodorum cucullatum Sw. in Svenska Vetensk.-Akad. Handl. 21: 243 (1800).

Lissochilus arenarius Lindl. in J. Linn. Soc., Bot. 6: 133 (1862).

Lissochilus dilectus Rchb.f., Otia Bot. Hamburg., 1: 62 (1878).

Lissochilus roscheri Rchb.f., Otia Bot. Hamburg., 1: 62 (1878).

Lissochilus stylites Rchb.f., Otia Bot. Hamburg., 1: 62 (1878).

Eulophia arenaria (Lindl.) Bolus in J. Linn. Soc. Bot. 25: 185 (1889).

Lissochilus monteiroi Rolfe in F.T.A. 7: 83 (1897).

Eulophia dilecta (Rchb.f.) Schltr., Westafr. Kautschuk-Exped. 391 (1900).

Lissochilus kassnerianus Kraenzl. in Bot. Jahrb. Syst. 51: 391 (1914). TYPE from Zambia.

Lissochilus amabilis Schltr. in Transvaal Mus. 10: 240 (1924).

Eulophia stylites (Rchb.f.) A.D.Hawkes in Orch. Rev. 72: 27 (1964).

Eulophia monteiroi (Rolfe) Butzin in Willdenowia 7: 589 (1975).

DISTRIBUTION: Comoros: Anjouan, Grande Comore. Also throughout tropical and southern Africa.

HABITAT: grassland; near forest; in *Eucalyptus* plantations.

ALTITUDE: 600 – 700 m.

FLOWERING TIME: November – March.

LIFE FORM: terrestrial.

DESCRIPTION: {221}, {768}.

ILLUSTRATION: {221}, {256b}, {768}, {1312}, {1345}.

NOTES: flowers showy; sepals green-purple or brownish; petals pale to deep pink; lip pale to deep pink, the throat white or yellowish with purple marking, base saccate; callus of two quadrate, erect flaps.

Eulophia ephippium *(Rchb.f.) Butzin* in Willdenowia, 7, 2: 588 (1975).

Lissochilus ephippium Rchb.f. in Flora 65: 533 (1882).

DISTRIBUTION: Madagascar ?

NOTES: said to be related to *Eulophia rosea* (Lindl.) A.D.Hawkes from Tropical Africa. No verifiable records have been found for Madagascar.

Eulophia filifolia *Bosser & Morat* in Adansonia sér. 3, 23 (1): 18 (2001). TYPE: Madagascar, Beraketa, *Morat* 1477 (holo. P).

DISTRIBUTION: Madagascar: Toliara.

HABITAT: sandy soil; limestone; in spiny forest.

FLOWERING TIME: February, July – October.

LIFE FORM: terrestrial {141}.

PHYTOGEOGRAPHICAL REGION: VI.

DESCRIPTION: {141}.

ILLUSTRATION: {141}.

NOTES: pseudobulbs ovoid; leaves linear; flowers close to those of *E. ramosa*, but a little larger and the lip has 3 well developed ridges on the upper surface.

Eulophia grandidieri *H.Perrier* in Bull. Soc. Bot. France 82: 149 (1935). TYPE: Madagascar, around Mahatsara, *Grandidier* s.n. (holo. P).

Lissochilus grandidieri (H.Perrier) H.Perrier in Humbert ed., Fl. Mad. Orch. 2: 8 (1941).

DISTRIBUTION: Madagascar.

FLOWERING TIME: April.

LIFE FORM: terrestrial.

DESCRIPTION: {896}.

ILLUSTRATION: {896}.

NOTES: similar in shape to *E. rutenbergiana*, but with more, smaller flowers; lip deeply 3-lobed, the side lobes broadly rounded, the midlobe oval-obtuse, the margins fimbriate.

Eulophia hians *Lindl.* var. **nutans** *(Sond.) S.Thomas* in la Croix & Cribb. Fl. Zamb. 11, 2: 476 (1998).

TYPES: South Africa, Katriviersberg, *Ecklon & Zeyher* & Uitenhage, *Ecklon & Zeyher* (both syn. S).

Eulophia nutans Sond. in Linnaea 19: 73 (1846).

Eulophia carunculifera Rchb.f. in Flora 64: 329 (1881). TYPE from S. Africa.

Eulophia madagascariensis Kraenzl. in Abh. Naturw. Ver. Bremen 7: 255 (1882). TYPE: Madagascar, Ankaratra, *Hildebrandt* 3864 (holo. HBG), *non Lissochilus madagascariensis* Kraenzl. (1882).

Eulophia vaginata Ridl. in J. Linn. Soc., Bot. 21: 467 (1885). TYPE: Madagascar, Ankaratra, *Hildebrandt* 3864 (syn. BM); Ankaratra, *Scott Elliot* 1967 (syn. BM).

Graphorchis madagascariensis (Kraenzl.) Kuntze, Rev. Gen. Pl. 2: 662 (1891).

Graphorkis nutans (Sond.) Kuntze, *loc.cit.*

Graphorchis vaginata (Sond.) Kuntze, *loc.cit.*

Eulophia aemula Schltr. in Bot. Jahrb. Syst. 20, Beibl. n. 50: 26 (1895). TYPE from S. Africa.

Eulophia galpinii Schltr., *loc.cit.*: 10 (1895). TYPE from S. Africa.

Eulophia laxiflora Schltr., *loc.cit.*: 4 (1895). TYPE from S. Africa.

Eulophia crinita Rolfe in Dyer, Fl. Trop. Afr. 7: 58 (1897), *non* G. Don (1839). TYPE from Malawi.

Eulophia flanaganii Bolus in Trans. S. African Philos. Soc. 16: 143 (1905). TYPE from S. Africa.

Eulophia gladioloides Rolfe in Kew Bull. 1910: 281 (1910). TYPE from S. Africa.

Eulophia purpurascens Rolfe, *loc. cit.* 1910: 281 (1910). TYPE from S. Africa.

Eulophia nelsonii Rolfe, *loc.c it.* 1910: 369 (1910). TYPE from S. Africa.

Eulophia ukingensis Schltr. in Bot. Jahrb. Syst. 53: 579 (1915). TYPE from Tanzania.

Eulophia triloba Rolfe in Kew Bull. 1910: 281 (1910). TYPE from S. Africa.

Eulophia amajubae Schltr. in Ann. Transvaal Mus. 10: 236 (1924). TYPE from S. Africa.

Eulophia decurva Schltr., *loc. cit.* 10: 235 (1924). TYPE from S. Africa.

Eulophia ernestii Schltr., *loc. cit.* 10: 236 (1924). TYPE from S. Africa.

Eulophia bathiei Schltr. in Repert. Spec. Nov. Regni Veg. Beih. 33: 260 (1925). TYPE: Madagascar, around Ambatofanyena, *H.Perrier* 12392 (holo. B†; iso. P).

Lissochilus vaginatus (Ridl.) H.Perrier in Humbert ed., Fl. Mad. Fam. 49, 2: 14 (1941).

Eulophia clavicornis Lindl. var. *nutans* (Sond.) A.V.Hall in J. S. Afr. Bot., Suppl. 5, 77 (1965).

Eulophia vleminckxiana Geerinck & Schaijes in Bull. Jard. Bot. Nation. Belg. 60, 1 – 2: 188 (1990). TYPE from Dem. Rep. Congo.

DISTRIBUTION: Madagascar: Antananarivo, Fianarantsoa. Also in East Africa, SC. Africa & South Africa.

HABITAT: grassland; dry deciduous scrubland; rocky outcrops.

ALTITUDE: 1500 – 2500 m.

FLOWERING TIME: December – March.

LIFE FORM: terrestrial.

PHYTOGEOGRAPHICAL REGION: III, IV.

DESCRIPTION: {768}, {896}.

ILLUSTRATION: {768}, {896}, {1223}, {1329}, {1472}.

VERNACULAR NAME: Tongolomboalavo.

NOTES: inflorescence up to 93 cm long; flowers with green to olivaceous sepals often flushed with purplish-brown, white to dull yellow-green petals, a white lip with a pale papillate callus, and a 2 – 4 mm long, incurved, unswollen spur. Medicinal use against boils.

Eulophia hologlossa *Schltr.* in Ann. Mus. Col. Marseille, sér. 3, 1: 171, t.9 (1913). TYPE: Madagascar, Matitana, *H.Perrier* 1918 (holo. P).

Lissochilus hologlossus (Schltr.) H.Perrier in Humbert ed., Fl. Mad. Orch. 2: 4 (1941).

DISTRIBUTION: Madagascar: Fianarantsoa, Toliara.

HABITAT: marshes and in sand dunes.

FLOWERING TIME: October – November.

LIFE FORM: terrestrial.
PHYTOGEOGRAPHICAL REGION: III.
DESCRIPTION: {896}.
ILLUSTRATION: {1205}, {1223}.
NOTES: plant saprophytic; inflorescence laxly 6- to 18-flowered; flower white with a pink lip; lip oblong, with 3 keels; spur 3.5 mm, conical-narrowed.

Eulophia ibityensis *Schltr.* in Repert. Spec. Nov. Regni Veg. Beih. 33: 262 (1925). TYPE: Madagascar, Mt. Ibity, *H.Perrier* 13977 (holo. P).
Lissochilus ibityensis (Schltr.) H.Perrier in Humbert ed. Fl. Mad. Orch. 2: 48 (1941).

DISTRIBUTION: Madagascar: Fianarantsoa.
HABITAT: rocky outcrops; sclerophyllous woodland.
ALTITUDE: 1000 – 2000 m.
FLOWERING TIME: April – November.
LIFE FORM: terrestrial.
PHYTOGEOGRAPHICAL REGION: III, IV.
DESCRIPTION: {896}.
ILLUSTRATION: {896}, {1223}.
NOTES: inflorescence with 15 – 20 rather laxly distributed flowers; flowers yellowish-green; lip 10 × 4 – 6 mm, 3-lobed, the mid-lobe almost rectangular, a little indented at the front, the margins pleated-fringed, the disc with 5 longitudinal ridges.

Eulophia livingstoniana *(Rchb.f.) Summerh.* in Kew Bull. 2: 132 (1947). TYPE: Malawi, Manganja Hills, *Meller* s.n. (holo. K).
Lissochilus fallax Rchb.f., Otia Bot. Hamb. 2: 115 (1881).
Lissochilus livingstonianus Rchb.f., *loc.cit.*
Lissochilus fallax Rchb.f., *loc.cit.* TYPE from Kenya.
Lissochilus malangensis Rchb.f., *loc.cit.* TYPE from Angola.
Lissochilus rutenbergianus Kraenzl. in Abh. Naturwiss. Ver. Bremen 7: 257 (1882). TYPE: Madagascar, Mahazamba, *Rutenberg* s.n. (holo. HBG).
Lissochilus affinis Rendle in J. Bot. 33: 193 (1895). TYPE from Burundi.
Lissochilus mediocris Rendle, *loc.cit.* TYPE from Kenya.
Lissochilus cornigerus Rendle, *loc.cit.* TYPE from Uganda.
Lissochilus cornigerus var. *minor* Rendle, *loc.cit.* TYPE from Uganda.
Lissochilus gracilior Rendle, *loc.cit.* TYPE from Burundi.
Lissochilus gracilior var. *angustata* Rendle in J. Bot. 33: 195 (1895). TYPES from Uganda and Tanzania.
Eulophia jumelleana Schltr. in Ann. Mus. Col. Marseille, ser.3, 1: 172, t.16 (1913). TYPE: Madagascar, Mahevarano, *H.Perrier* 74 (1902) (holo. B†; iso. P).
Eulophia robusta Schltr., *loc. cit.* (1913), *in obs. non* Rolfe (1910).
Lissochilus jumelleanus (Schltr.) Schltr. in Beih. Bot. Centralbl. 33 (2): 421 (1915).
Lissochilus laggiarae Schltr. in Repert. Spec. Nov. Regni Veg. Beih. 15: 332 (1918). TYPE: Madagascar, without locality, *Laggiara* s.n. (holo. B).
Lissochilus faradiensis De Wild. in Bull. Jard. Bot. État 6: 83 (1919), as "*faradjensis*". TYPE from Dem. Rep. Congo.
Eulophia gracilior (Rendle) Butzin in Willdenowia 7, 2: 588 (1975).

DISTRIBUTION: Madagascar: Antananarivo, Antsiranana, Fianarantsoa, Mahajanga, Toamasina, Toliara. Comoros: Grande Comore, Mayotte. Also in East and SC. Africa.
HABITAT: grassland; marshes.
ALTITUDE: sea level – 1300 m.
FLOWERING TIME: October – March.
LIFE FORM: terrestrial.
PHYTOGEOGRAPHICAL REGION: I, II, III, IV, V.
DESCRIPTION: {896}.
ILLUSTRATION: {221}, {256b}, {281}, {677}, {877}, {896}, {918}, {1205}, {1472}.
VERNACULAR NAME: Felatrandraka.
NOTES: a widespread species, sepals and petals pale to deep mauve-pink; lip purple with darker callus ridges and green or yellowish side lobes flushed and edges with pink, 3-lobed, with 5 ridges. The tubers are eaten.

Eulophia macra *Ridl.* in J. Linn. Soc., Bot. 22: 120 (1886). TYPE: Central Madagascar, *Baron* 3423 (holo. BM).

Eulophia ambositrana Schltr. in Beih. Bot. Centralbl. 34 (2): 331 (1916). TYPE: Madagascar, Ambatofitorana, Ambositra, *H.Perrier* 11483 (163) (holo. B†; iso. P).

Lissochilus macer H.Perrier in Humbert ed., Fl. Mad. Orch. 2: 43 (1941).

DISTRIBUTION: Madagascar: Fianarantsoa, Toliara.

HABITAT: open forest; exposed areas on rock.

ALTITUDE: sea level – 1500 m.

FLOWERING TIME: April – June.

LIFE FORM: terrestrial or lithophyte.

PHYTOGEOGRAPHICAL REGION: I, III, IV.

ILLUSTRATION: {257}.

DESCRIPTION: {896}.

NOTES: plant forming clumps with long narrowly linear leaves; inflorescence paniculate, laxly flowered; flowers 10 – 24 mm long; lip entire, obovate, fimbriate in the front, the disk with a few small keels; spur short, 3 mm long.

Eulophia mangenotiana *Bosser & Veyret* in Adansonia, ser.2, 10 (2): 216 (1970) . TYPE: Madagascar, Ingaro, nr. Ankaramena, *Y.Veyret* 1080 (holo. P).

DISTRIBUTION: Madagascar: Antsiranana.

HABITAT: humid, evergreen forest at low elevation.

LIFE FORM: terrestrial.

PHYTOGEOGRAPHICAL REGION: I.

DESCRIPTION: {116}.

ILLUSTRATION: {116}.

NOTES: known only from the type. Saprophyte, entirely brownish yellow; flowers terminal; lip 5.5 – 6 mm long, ovate-ligulate, obscurely 3-lobed, the basal lobes not prominent, the midlobe broadly ovate, with a central crest in the centre.

Eulophia megistophylla *Rchb.f.* in Gard. Chron. 23: 787 (1885). TYPE: Grande Comore, Combani forest, *Humblot* 1448 (448) (holo. W; iso. BM).

Eulophia pulchra (Thouars) Lindl. var. *divergens* Rchb.f. in Gard. Chron. n.s. 22: 102 (1884):

Lissochilus megistophyllus (Rchb.f.) H.Perrier in Humbert ed., Fl. Mad. Orch. 2: 42 (1941).

Eulophidium megistophyllum (Rchb.f.) Summerh. in Bull. Jard. Bot. Bruxelles 27: 398 (1957).

DISTRIBUTION: Records from Madagascar are unreliable. Comoros: Grande Comore.

LIFE FORM: terrestrial.

DESCRIPTION: {896}, {1007}.

ILLUSTRATION: {176}, {404}, {475}, {476}.

CULTIVATION: {176}.

NOTES: allied to *E. pulchra*, but much larger plant, with a paniculate inflorescence, and the sheaths of the peduncle broader and not approximate.

Eulophia nervosa *H.Perrier* in Bull. Soc. Bot. France 82: 150 (1935). TYPE: Madagascar, Antsirabe, *H.Perrier* 8116 (holo. P).

Lissochilus nervosus (H.Perrier) H.Perrier in Humbert ed., Fl. Mad. Orch. 2: 14 (1941).

DISTRIBUTION: Madagascar: Antananarivo.

HABITAT: peaty marshes.

ALTITUDE: 1000 – 1500 m.

FLOWERING TIME: January.

LIFE FORM: terrestrial.

PHYTOGEOGRAPHICAL REGION: III.

DESCRIPTION: {896}.

ILLUSTRATION: {896}.

NOTES: known only from the type. Leaves grass-like, up to 1 cm wide; flowers c. 12 mm long; lip obovate, 3-lobed, the lateral lobes obtuse, the midlobe 3-lobulate, the margins undulate-fimbriate, the disk with 2 calli.

Eulophia perrieri *Schltr.* in Ann. Mus. Col. Marseille, sér. 3, 1: 174, t.11 (1913). TYPE: Madagascar, Tsingy de Namoroka, Ambongo, *H.Perrier* 1548 (iso. P).

Lissochilus perrieri (Schltr.) H.Perrier in Humbert ed., Fl. Mad. Orch. 2: 46 (1941).

DISTRIBUTION: Madagascar: Mahajanga.
HABITAT: shaded and wet rocks.
ALTITUDE: sea level – 500 m.
FLOWERING TIME: July.
LIFE FORM: lithophyte.
PHYTOGEOGRAPHICAL REGION: V.
DESCRIPTION: {896}.
ILLUSTRATION: {1205}, {1223}.
NOTES: known only from the type. Pseudobulbs epigeous; leaves linear, acute, fleshy; flowers similar to those of *E. ramosa*; spur more inflated.

Eulophia pileata *Ridl.* in J. Linn. Soc., Bot. 21: 468 (1885).TYPE: Madagascar, Ankafana, Nutrongoa, *Deans Cowan* s.n. (holo. BM).

Graphorchis pileata (Ridl.) Kuntze, Rev. Gen. Pl. 2: 662 (1891).

Lissochilus pileatus (Ridl.) H.Perrier in Humbert ed., Fl. Mad. Orch. 2: 49 (1941).

DISTRIBUTION: Madagascar: Antananarivo. Comoros: unverified record.
HABITAT: montane forest.
ALTITUDE: 1000 – 1500 m.
FLOWERING TIME: July – November.
LIFE FORM: terrestrial.
PHYTOGEOGRAPHICAL REGION:III.
DESCRIPTION: {896}.
ILLUSTRATION: {246}, {896}, {281}.
NOTES: plant tall, slender; flowers orange-yellow, with reddish streaks; lip obovate, 10 × 8 mm, 3-lobed in the upper third, the disc with 2 very protruding keels towards the base.

Eulophia plantaginea *(Thouars) Rolfe ex Hochr.* in Ann. Conserv. Jard. Bot. Gen. 11: 56 (1908). TYPE: Madagascar, *Thouars* s.n. (holo. P).

Limodorum plantagineum Thouars, Hist. Orch.: t.41 – 42 (1822).

Cyrtopera plantaginea (Thouars) Lindl., Gen. Sp. Orch. Pl.: 189 (1833).

Cyrtopodium plantagineum (Thouars) Bentham in J. Linn. Soc. 18: 320 (1881), as *C. plantaginea*; Ridley in J. L. Soc., 21: 471 (1885); Cordem., Fl. Isle Réunion: 225 (1895).

Graphorchis plantaginea (Rolfe) Kuntze, Rev. Gen. Pl. 2: 662 (1891).

Cyrtopera bituberculata Rolfe in Gard. Chron., Ser.3, 18: 581 (1995).

Eulophia grandibracteata Kraenzl. in Mitt. Inst. Allg. Bot. Hamburg 5: 238 (1922). TYPE: Madagascar, Sahambendrana, *Majastre* (holo. HBG).

Lissochilus plantagineus (Thouars) H.Perrier in Humbert ed., Fl. Mad. Orch. 2: 10 (1941).

DISTRIBUTION: Madagascar: Antananarivo, Fianarantsoa, Toamasina, Toliara. Comoros: Grande Comore.
HABITAT: grassland; by ditches in arable land; marshes. Locally common.
ALTITUDE: sea level – 1300 m.
FLOWERING TIME: January – March.
LIFE FORM: terrestrial.
PHYTOGEOGRAPHICAL REGION:III.
DESCRIPTION: {896}.
ILLUSTRATION: {281}, {473}, {505}, {896}, {1398}.
VERNACULAR NAMES: Tenondahy, Tongolombato, Ovinakanga.
NOTES: inflorescence robust, 30 cm to 1 m tall; flowers c. 5 cm long, with yellowish green sepals, pure white petals and a white lip, spur and disk tinted with violet-red.

Eulophia pulchra *(Thouars) Lindl.,* Gen. Sp. Orch. Pl.: 182 (1830). TYPE: Réunion, *Thouars* s.n. (holo. P not found).

Limodorum pulchrum Thouars, Hist. Orchid.: t.43 – 44 (1822).

Eulophia macrostachya Lindl., Gen. Sp. Orch. Pl. 182 (1822). TYPE from Ceylon.

Eulophia emarginata Blume, Orch. Archip. Ind. & Jap. 180 (1885). TYPES from Sumatran.

Eulophia striata Rolfe in J. Linn. Soc., Bot. 29: 53 (1891). TYPE: Madagascar, nr. Fort Dauphin, *Scott Elliot* 2545 (holo. K).

Graphorkis pulchra (Thouars) Kuntze, Rev. Gen. Pl. 2: 662 (1891).

Eulophia versicolor Cordem., Fl. La Réunion: 223 (1895). TYPE: Réunion, Bassin du Diable, Mar. 1876, *Pottier* s.n. (lecto. MARS).

Eulophia coccifera Frapp., Cat. Orch. Réunion: 12 (1880), *nom. nud.*; Frapp. ex Cordem. Fl. La Réunion: 224 (1895). TYPE: Réunion, St. Benoit, without collector, lecto: MARS.

Eulophia papuana Bailey in Queensland Agric. J: 274 (1907), *non* (Ridley) J. J. Sm. (1909). TYPE from New Guinea.

Eulophia rouxii Kraenzl. in Sar. & Rouxs, Nova Caled. 1: 82 (1914). TYPE from New Caledonia.

Eulophia silvatica Schltr. in Bot. Jahrb. Syst. 53: 586 (1915). TYPE from Tanzania.

Lissochilus pulcher (Thouars) H.Perrier in Humbert ed., Fl. Mad. Orch. 2: 41 (1941).

Eulophidium pulchrum (Thouars) Summerh. in Bull. Jard. Bot. Bruxelles 27: 400 (1957).

Eulophidium silvaticum(Schltr.) Summerh., *loc.cit.*

Oeceoclades pulchra (Thouars) P.J.Cribb & M.A.Clem. in Austral. Orch. Res. 1: 99 (1989).

> DISTRIBUTION: Madagascar: Fianarantsoa, Toamasina, Toliara. Mascarenes: Réunion. Tropical Africa across to SW Pacific.
> HABITAT: humid, evergreen forest; open forest; littoral forest.
> ALTITUDE: sea level – 1000 m.
> FLOWERING TIME: December – April.
> LIFE FORM: terrestrial.
> PHYTOGEOGRAPHICAL REGION:I, III.
> DESCRIPTION: {896}.
> HISTORY: {8}, {373}, {1001}.
> ILLUSTRATION: {86}, {167}, {245}, {454}, {529}? {1054}, {1298}, {1394}, {1398}.
> CULTIVATION: {1054}, {1394}.
> NOTES: plant very variable in size and shape; inflorescence densely many-flowered; flowers yellow or pale green, marked with purple on the side lobes of lip and with an orange callus; lip 3-lobed but midlobe bilobulate, 7 mm × 18 mm; callus of 2 fleshy lobes at mouth of spur.

Eulophia ramosa *Ridl.* in J. Linn. Soc., Bot. 21: 470 (1885), *non* Hayata (1911). TYPE: Madagascar, Ankafana, *Deans Cowan* s.n. (holo. BM).

Graphorchis ramosa (Ridl.) Kuntze, Rev. Gen. Pl. 2: 662 (1891).

Eulophia leucorhiza Schltr. in Ann. Mus. Col. Marseille, sér. 3, 1: 29, t.9 (1913). TYPE: Madagascar, between Mahajiba & the Manambolo (Menabe), *H.Perrier* 1926 (holo. B†; iso. P).

Eulophia pseudoramosa Schltr. in Ann. Mus. Col. Marseille, sér. 3, 1: 175 (32), t.10 (1913). TYPE: Madagascar, Afiafitatra, nr. Mt. Tsitondraina, Ambongo, *H.Perrier* 1109 (holo. B†; iso. P).

Lissochilus ramosus (Ridl.) H.Perrier in Humbert ed., Fl. Mad. Orch. 2: 44 (1941).

> DISTRIBUTION: Madagascar: Antananarivo, Fianarantsoa, Mahajanga, Toamasina, Toliara.
> HABITAT: open ground; dry forest; limestone outcrops.
> ALTITUDE: sea level – 1000 m.
> FLOWERING TIME: February, July – November.
> LIFE FORM: terrestrial or lithophyte.
> PHYTOGEOGRAPHICAL REGION: I, III, IV, VI.
> DESCRIPTION: {896}.
> ILLUSTRATION: {1205}, {1223}.
> NOTES: widespread and somewhat variable species. Inflorescence paniculate, very laxly several-flowered; flower 1 cm long, yellowish and lined with red; lip oblong, 3-lobed, the lateral lobes small and obtuse, the midlobe bigger and crispate-undulate at the front, the disk with 5 verrucose keels.

Eulophia reticulata *Ridl.* in J. Linn. Soc., Bot. 21: 470 (1885). TYPE: Madagascar, Imani, *Hilsenberg* s.n. (holo. BM).

Lissochilus madagascariensis Kraenzl. in Abh. Naturwiss. Vereine Bremen 7: 256 (1882). TYPES: Madagascar, Vohemar, *Rutenberg* s.n.; Vondruzona, *Rutenberg* s.n.; Antananarivo, *Rutenberg* s.n. (all syn. B).

Graphorchis reticulata (Ridl.) Kuntze, Rev. Gen. Pl. 2: 662 (1891).

Eulophia camporum Schltr. in Ann. Mus. Col. Marseille, ser.3, 1: 170 (27), t.16 (1913). TYPE: Madagascar, Haut-Isandrano, *H.Perrier* 1497 (holo. B†; iso. P).

DISTRIBUTION: Madagascar: Antananarivo, Antsiranana, Fianarantsoa.
HABITAT: grassland; amongst rocks.
ALTITUDE: 500 – 2000 m.
FLOWERING TIME: October – December.
LIFE FORM: terrestrial.
PHYTOGEOGRAPHICAL REGION: III.
DESCRIPTION: {896}.
ILLUSTRATION: {453}, {1205}.
VERNACULAR NAMES: Kamasina, Tandrokondrylahy, Kitandrokondrilahy.
NOTES: leaves and capsules are eaten and the roots used as an aphrodisiac. Rhizome underground; lip 3-lobed, the lateral lobes small and obtuse, the midlobe obcuneate and obtuse, margins crispate-undulate, the disk with a yellow callus.

Eulophia rutenbergiana *Kraenzl.* in Abh. Naturwiss. Vereine Bremen 7: 255 (1882). TYPE: Madagascar, Antananarivo, *Rutenberg* s.n. (holo. B†); Imerina, *J.H.Hildebrandt* 3842 (lecto. K; isolecto. BM, K, W, all selected here).
Graphorchis rutenbergiana (Kraenzl.) Kuntze, Rev. Gen. Pl. 2: 662 (1891).
Lissochilus kranzlini H.Perrier in Humbert ed., Fl. Mad. Orch. 2: 13 (1941).

DISTRIBUTION: Madagascar: Antananarivo, Antsiranana, Fianarantsoa, Toamasina.
HABITAT: marshes; secondary grassland; open ground; rocky slopes.
ALTITUDE: 500 – 2000 m.
FLOWERING TIME: December – May.
LIFE FORM: terrestrial.
PHYTOGEOGRAPHICAL REGION: I, III.
DESCRIPTION: {896}.
ILLUSTRATION: {281}, {896}.
NOTES: rhizome underground; inflorescence 40 – 60 cm tall; flower c. 2 cm long, sulphur-yellow; lip golden-yellow with the 2 lateral lobes red-brown.

Eulophia wendlandiana *Kraenzl.* in Gard. Chron. sér. 3, 22: 262 (1897). TYPE: Madagascar, without locality, *Braun* s.n. (holo.?).
Lissochilus wendlandianus (Kraenzl.) H.Perrier in Humbert ed., Fl. Mad. Orch. 2: 40 (1941).

DISTRIBUTION: Madagascar.
LIFE FORM: terrestrial.
DESCRIPTION: {896}.
NOTES: described from a cultivated plant, the type has not been found but is reputedly related to *E. pulchra*.

Eulophiella capuroniana *Bosser & Morat* in Adansonia, sér. 2, 12 (1): 73 (1972). TYPE: Madagascar, east coast, Jard. Bot. Tan. 1405 (holo. P; iso. TAN).

DISTRIBUTION: Madagascar: Toamasina.
HABITAT: humid, lowland forest.
FLOWERING TIME: October.
LIFE FORM: epiphyte.
PHYTOGEOGRAPHICAL REGION: I.
DESCRIPTION: {121}.
ILLUSTRATION: {121}.
NOTES: plant similar to *E. elisabethae*, but it is more slender, with narrower leaves; flowers smaller, yellow and with the lip bearing crests of a different shape.

Eulophiella elisabethae *Linden & Rolfe* in Lindenia 7: 77, t.325 (1891). TYPE: Madagascar, *Sallerin* s.n. (holo. K).
Eulophiella perrieri Schltr. in Orchis 14 (2): 27 (1920). TYPE: Madagascar, *Heimat* s.n. (holo. B†).

DISTRIBUTION: Madagascar: Antsiranana, Toamasina.

HABITAT: coastal forest on *Dypsis fibrosa.*

ALTITUDE: sea level – 200 m.

FLOWERING TIME: November – December.

LIFE FORM: epiphyte.

PHYTOGEOGRAPHICAL REGION: I.

DESCRIPTION: {896}, {1388}.

ILLUSTRATION: {50}, {199}, {231}, {247}, {281}, {380}, {429}, {462}, {467}, {529}, {550}, {630}, {694}, {733}, {760}, {761}, {810}, {905}, {1060}, {1087}, {1160}, {1223}, {1285}, {1298}, {1388}, {1442}, {1458}.

HISTORY: {13}, {22}, {23}, {24}, {43}, {231}, {300}, {301}, {329}, {377}, {380}, {429}, {430}, {462}, {467}, {498}, {550}, {630}, {694}, {753}, {757}, {758}, {759}, {761}, {928}, {1069}, {1070}, {1071}, {1072}, {1074}, {1098}, {1149}, {1299}, {1347}, {1456}, {1496}, {1504}.

CULTIVATION: {329}, {441}, {462}, {630}, {800}, {1055}, {1076}, {1402}, {1508}.

NOTES: plant large, with narrowly lanceolate plicate leaves; flowers c. 4 cm across; sepals and petals pale pinkish-white; lip white with a yellow callus and central area and red spotting around the callus.

Eulophiella ericophila *Bosser* in Adansonia, sér. 2, 14 (2): 215 (1974). TYPE: Madagascar, Betsomanga massif, Antongondriha, *Humbert & Capuron* 24368 (holo. P).

DISTRIBUTION: Madagascar: Antsiranana.

HABITAT: montane, ericaceous scrub.

ALTITUDE: 1300 – 1500 m.

FLOWERING TIME: November.

LIFE FORM: terrestrial.

PHYTOGEOGRAPHICAL REGION: IV.

DESCRIPTION: {122}.

ILLUSTRATION: {122}.

NOTES: distinct from the other members of the genus by the more slender plant, with smaller leaves and flowers, and the shallowly two-crested lip.

Eulophiella galbana (*Ridl.*) *Bosser & Morat* in Adansonia sér. 3, 23 (1): 22 (2001). TYPE: Madagascar, Ankafana, Fianarantsoa, *Deans Cowan* s.n. (holo. BM).

Eulophia galbana Ridl. in J. Linn. Soc., Bot. 21: 469 (1885).

Graphorchis galbana (Ridl.) Kuntze, Rev. Gen. Pl. 2: 662 (1891).

Lissochilus galbanus (Ridl.) H.Perrier in Humbert ed., Fl. Mad. Orch. 2: 36 (1941).

DISTRIBUTION: Madagascar: Antananarivo, Fianarantsoa.

HABITAT: humid, evergreen forest.

FLOWERING TIME: January.

LIFE FORM: epiphyte.

PHYTOGEOGRAPHICAL REGION: III.

DESCRIPTION: {896}.

ILLUSTRATION: {141}, {246}.

NOTES: plant small, with shiny ovoid pseudobulbs; flowers greenish yellow, not opening widely, lacking a spur.

Eulophiella roempleriana (*Rchb.f.*) *Schltr.* in Orchis 9: 109 (1915). TYPE: Madagascar, *Roempler* s.n. (holo. W).

Grammatophyllum roemplerianum Rchb.f. in Gard. Chron. n.s., 7: 240 (1877).

Eulophiella peetersiana Kraenzl. in Gard. Chron. sér. 3, 21: 182 (1897). TYPE: Madagascar, *J. Braun* s.n. (holo. B).

Eulophiella hamelini Baill. ex Rolfe in Orch. Rev. 8: 197 (1900).

Eulophiella peetersiana Kraenzl. forma *pallida* Hochr. in Ann. Conserv. Jard. Bot. Genève 11/12 (2): 54 (1908). TYPE: Madagascar, Vatomandry district, *Guillot*, 15 (holo. P).

Eulophiella peetersiana Kraenzl. forma *rubra* Hochr. in Ann. Conserv. Jard. Bot. Genève 11/12 (2): 54 (1908). TYPE: Madagascar, Vatomandry district, *Guillot*, 16 (holo. P).

DISTRIBUTION: Madagascar: Antananarivo, Fianarantsoa, Toamasina.

HABITAT: on *Pandanus* sp. in coastal forest and montane forest.

ALTITUDE: sea level – 1300 m.

FLOWERING TIME: December – January.

LIFE FORM: epiphyte.

PHYTOGEOGRAPHICAL REGION: I.

DESCRIPTION: {896}.

ILLUSTRATION: {135}, {208}, {199}, {222}, {266}, {281}, {331}, {363}, {371}, {451}, {494}, {498}, {501}, {529}, {553}, {592}, {704}, {707}, {709}, {710}, {724}, {904}, {905}, {933}, {947}, {948}, {957}, {1038}, {1149}, {1188}, {1228}, {1264}, {1266}, {1298}, {1325}, {1497}.

HISTORY: {13}, {200}, {201}, {203}, {289}, {371}, {409}, {498}, {553}, {571}, {704}, {707}, {845}, {948}, {1098}, {1149}, {1481}, {1497}.

CULTIVATION: {3}, {159}, {290}, {451}, {498}, {700}, {707}, {708}, {933}, {947}.

NOTES: plant very large; rhizome stout, creeping; leaves large, lanceolate, plicate; inflorescence up to 120 cm tall; flowers c. 9 cm, rose-purple with a white disc and yellow-tipped callus ridges.

Eulophiella saboureaui *Ursch & Toill.-Gen.* in Nat. Malg. 5 (2): 149 (1953), *nom. nud.* TYPE: Madagascar, Masoala, Santaha, near the sea, *Duran & Jean de Dieu Rakoto* 275.

DISTRIBUTION: Madagascar: Toamasina.

FLOWERING TIME: May.

LIFE FORM: epiphyte.

PHYTOGEOGRAPHICAL REGION: I.

DESCRIPTION: {1415}.

NOTES: *E. saboureaui* was not validly published, it seems likely that it is a natural hybrid between *E. roempleriana* and *E. elisabethae*.

Galeola humblotii *Rchb.f.* in Flora 68: 378 (1885). TYPES: Grande Comore, *Humblot* 1425 (425), 1429 (429) & 1431 (431) (all syn. W; isosyn. P).

DISTRIBUTION: Madagascar: Toamasina, Toliara. Comoros: Grande Comore.

HABITAT: eastern forest.

FLOWERING TIME: October – February.

LIFE FORM: epiphyte.

PHYTOGEOGRAPHICAL REGION: I.

DESCRIPTION: {896}.

NOTES: Related to *Galeola hydra* from mainland Asia, but the lip is different.

Gastrorchis françoisii *Schltr.* in Repert. Spec. Nov. Regni Veg. Beih. 33: 168 (1924). TYPE: Madagascar, Mandraka Gorge, *H.Perrier* 13492 (holo. P).

Phaius françoisii (Schltr.) Summerh. in Kew Bull. 17 (3): 558 (1964).

Gastrorchis françoisii var. *pauliani* Ursch & Toilliez-Genoud in Nat. Malg. 2 (2): 151 (1950), *nom. nud.*

DISTRIBUTION: Madagascar: Antananarivo, Antsiranana, Fianarantsoa, Toamasina.

HABITAT: humid, evergreen forest on plateau.

ALTITUDE: 1200 – 1800m.

FLOWERING TIME: January – March.

LIFE FORM: epiphyte or terrestrial.

PHYTOGEOGRAPHICAL REGION: III.

DESCRIPTION: {214}, {896}.

ILLUSTRATION: {214}, {281}, {305}, {308}, {492}, {529}, {573}, {896}, {1268}, {1413}.

CULTIVATION: {214}, {573}.

NOTES: somewhat similar *G. humblotii* but flower colour and shape of the lip are very different. *Gastrorchis françoisii* var. *orientalis* Ursch & Toilliez-Genoud (*nom. nud.* in Naturaliste Malagache 2 (2): 151, 1950) is likely to be a hybrid of unknown origin.

Gastrorchis geffrayi (*Bosser) Senghas* in Die Orchideen, 3rd ed. 1A: 887 (1984). TYPE: Madagascar, Mt. Ambre, *J. P. Peyrot* 50 (holo. P).

Phaius geffrayi Bosser in Adansonia, ser.2, 11 (3): 538 (1971).

DISTRIBUTION: Madagascar: Antsiranana.

ALTITUDE: c. 1000 m.

HABITAT: humid, evergreen forest.

LIFE FORM: terrestrial.

PHYTOGEOGRAPHICAL REGION: III.

DESCRIPTION: {120}.

ILLUSTRATION: {120}.

CULTIVATION: {33}.

NOTES: this is likely to be a peloric form of *Gastrorchis lutea*. Flower with fleshy, yellowish green petals and sepals, yellowing with age, and very long, c. 6 cm long floral bracts.

Gastrorchis humblotii *(Rchb.f.) Schltr.* in Repert. Spec. Nov. Regni Veg. Beih. 33: 169 (1924). TYPE: Madagascar, *Humblot* s.n. (holo. W).

Phaius humblotii Rchb.f. in Gard. Chron. n.s., 14: 365, 812 (1880).

Phaius humblotii var. *albiflora* W.Watson & H.J.Chapman, Orch. Cult. Managm.: 417 (1903).

Gastrorchis schlechteri var. *milotii* Ursch & Genoud in Nat. Malg. 2 (2): 156 (1950), *nom. nud.* Based upon: Madagascar, Mt. Ambre, Jard. Bot. Tan. 301 (P).

DISTRIBUTION: Madagascar: Antananarivo, Antsiranana, Fianarantsoa, Toamasina.

HABITAT: humid, evergreen forest on plateau; on the edge of sphagnum bog.

ALTITUDE: 850 – 1500 m.

FLOWERING TIME: October – March.

LIFE FORM: epiphyte or terrestrial.

PHYTOGEOGRAPHICAL REGION: I, III.

DESCRIPTION: {120}, {896}.

ILLUSTRATION: {3}, {10}, {50}, {120}, {208}, {246}, {252}, {253}, {261}, {281}, {354}, {378}, {482}, {505}, {529}, {618}, {724}, {750}, {896}, {920}, {958}, {1018}, {1193}, {1257}, {1268}, {1338}, {1427}, {1483}, {1491}.

HISTORY: {187}, {237}, {349}, {571}, {750}, {1106}, {1442}, {1483}.

CULTIVATION: {10}, {26}, {33}, {237}, {920}, {1193}, {1427}, {1442}, {1453}, {1467}, {1483}, {1491}.

NOTES: *Phaius humblotii* var. *albus* was described and illustrated (in The Gardening World, n.s. 21: 359 & 361, 1904). The species has a characteristic 'V'-shaped callus at the base of the lip.

Gastrorchis humblotii var. **rubra** *(Bosser) Bosser & P.J.Cribb* in DuPuy *et al.*, Orch. Madag.: 145 (1999). TYPE: Madagascar, La Mandraka, *Bosser* 20410 (holo. P).

Phaius humblotii var. *ruber* Bosser in Adansonia, sér. 2, 11 (3): 526 (1971).

DISTRIBUTION: Madagascar: Antananarivo.

HABITAT: humid, mossy, evergreen forest.

LIFE FORM: terrestrial.

PHYTOGEOGRAPHICAL REGION: III.

FLOWERING TIME: March.

DESCRIPTION: {120}, {134}.

ILLUSTRATION: {134}.

CULTIVATION: {134}.

NOTES: sepals and petals deep violet-red and differ from the typical variety in the colour and shape of the lip.

Gastrorchis humblotii var. **schlechteri** *(H.Perrier) Senghas ex Bosser & P.J.Cribb* in DuPuy *et al.*, Orch. Madag.: 145 (1999).

Gastrorchis schlechteri François in Revue Horticole, 100: 304 (1928) & H.Perrier in Bull. Acad. Malg., n.s. 12: 33 (1930). TYPE: Madagascar, Mt. Tsaratanana, *H.Perrier* 18355 (holo. P).

Phaius humblotii var. *schlechteri* (H.Perrier) Bosser in Adansonia, sér. 2, 11 (3): 526 (1971).

Phaius schlechteri (H.Perrier) Summerh. in Kew Bull. 17, 3: 560 (1964).

DISTRIBUTION: Madagascar: Antsiranana.

HABITAT: mossy forest.

ALTITUDE: 1200 – 2000m.

FLOWERING TIME: January – April.

PHYTOGEOGRAPHICAL REGION: III.

DESCRIPTION: {896}.

ILLUSTRATION: {72}, {305}, {529}, {850}, {876}, {896}, {1257}, {1264}, {1268}.
NOTES: sepals and petals white instead of pink; floral bracts white instead of green.

Gastrorchis lutea (*Ursch & Genoud ex Bosser*) *Sengh.* in Die Orchideen, 3rd ed. 1A: 887 (1984).
TYPE: Madagascar, Andranoditra marsh, Perinet, *E. Ursch.* in Jard. Bot. Tan. 27 (holo. P).
Phaius luteus Ursch & Genoud ex Bosser in Adansonia, sér.2, 11 (3): 534 (1971).
Gastrorchis luteus Ursch & Genoud in Nat. Malg. 2 (1): 159, (1950), *nom. nud.*

DISTRIBUTION: Madagascar: Toamasina.
HABITAT: humid, evergreen forest; marshes.
ALTITUDE: 500 – 1100 m.
FLOWERING TIME: October.
LIFE FORM: epiphyte or terrestrial.
PHYTOGEOGRAPHICAL REGION: I, III.
DESCRIPTION: {120}.
ILLUSTRATION: {281}, {484}, {529}, {1413}.
NOTES: This species is distinguished easily from the other Madagascan species by its fleshy yellowish-green tepals and yellow-pink lip and its very long floral bracts.

Gastrorchis peyrotii (*Bosser*) *Senghas* in Die Orchideen, 3rd. ed. 1A: 887 (1984). TYPE: Madagascar between Moramanga and Anosibe, *J.P.Peyrot* s.n. (holo. P).
Phaius peyrotii Bosser in Adansonia, ser.2, 11 (3): 540 (1971).

DISTRIBUTION: Madagascar: Toamasina.
HABITAT: humid, evergreen forest on plateau and mountains.
ALTITUDE: 500 – 1500 m.
FLOWERING TIME: December – January.
LIFE FORM: terrestrial.
PHYTOGEOGRAPHICAL REGION: III.
DESCRIPTION: {120}.
ILLUSTRATION: {120}, {281}.
NOTES: characterised by the leaves which are relatively small, undulate & crispate at the edges and greyish-green in colour.

Gastrorchis pulchra *Humbert & H.Perrier* in Mém. Inst. Scient. Madagascar, sér. B, Biol. Veg. 6: 259 (1955). TYPE: Madagascar, Marojejy massif, *Humbert* 22537 (holo. P; iso. TAN).
Phaius pulcher (Humbert & H.Perrier) Summerh. in Kew Bull. 17, 3: 560 (1964); Bosser (1971) in Adansonia, sér. 2, 11 (3): 528 .

DISTRIBUTION: Madagascar: Antsiranana, Toamasina.
HABITAT: humid, mossy, evergreen forest, in shade and near humid rocks.
ALTITUDE: 400- 2000 m.
FLOWERING TIME: November – December.
LIFE FORM: terrestrial.
PHYTOGEOGRAPHICAL REGION: III, IV.
ILLUSTRATION: {83}, {281}, {707}, {902}, {1257}, {1264}, {1268}.
CULTIVATION: {33}.
NOTES: flower with pure white tepals; lip pinkish-red, trapezoid and widely oblong with undulating margins and with a central raised mound covered in yellow hairs.

Gastrorchis pulchra var. **perrieri** (*Bosser*) *Bosser & P.J.Cribb* in Du Puy *et al.*, Orch. Mad.: 146 (1999).
Phaius pulcher var. *perrieri* Bosser in Adansonia, ser.2, 11 (3): 530 (1971). TYPE: Madagascar, Perinet, *E. Ursch*, cult. Jard. Bot. Tan. 42 (holo. P).
Gastrorchis tuberculosa var. *perrieri* Ursch & Genoud in Nat. Malg. 2 (2): 156, (1950), *nom. nud.*

DISTRIBUTION: Madagascar: Toamasina.
HABITAT: humid, mossy, evergreen forest; marshes.
ALTITUDE: 500 – 1000 m.
FLOWERING TIME: October – November.

Life form: terrestrial.

Description: {120}.

Illustration: {120}, {481}, {1413}.

Notes: characterised by its narrow leaves and the shape of the flower, especially by the shape of the lip.

Gastrorchis simulans *(Rolfe) Schltr.* in Repert. Spec. Nov. Regni Veg. Beih. 33: 170 (1924). TYPE: Madagascar, East Coast, *Warpur* s.n. (holo. K).

Phaius simulans Rolfe in Orch. Rev. 9: 43 (1901).

Phaius fragrans Grignan in Jardin 15: 359 (1901). TYPE: lecto.: the illustration in Grignan, Le Jardin, 15: 359 (1901).

Distribution: Madagascar: Toamasina.

Habitat: montane forest.

Life form: epiphyte.

Phytogeographical region: I.

Description: {120}, {413}, {896}, {1109}.

Illustration: {379}, {1117}, {1455}.

Cultivation: {1104}, {1152}, {1157}.

Note: there has been debate over the hybrid status of this taxon {1}, {25}, {26}, {186}, {364}, {1061}, {1106}, {1117}, {1119}, {1151}, {1153}, {1165}.

Note: close to *G. tuberculosa*, the main difference being that it is an epiphyte with a relatively slender, elongated rhizome.

Gastrorchis steinhardtiana *Senghas* in J. Orchideenfr. 4: 133, fig. (1997). TYPE: Madagascar, cult. *Heidelburg B. G.* 882 (holo. HEID).

Distribution: Madagascar.

Habitat: montane forest.

Life form: terrestrial {1268}.

Description: {1268}.

Illustration: {481}, {1268}.

Notes: plant tall; flowers 6 cm across; sepals and petals white; lip white, speckled with pink; callus slightly raised, bearing yellow hairs.

Gastrorchis tuberculosa *(Thouars) Schltr.* in Repert. Spec. Nov. Regni Veg. Beih. 33: 169 (1924). TYPE: Madagascar, *Thouars* s.n. (holo. P).

Limodorum tuberculosum Thouars, Hist. Orch.: t.31 (1822).

Bletia tuberculosa (Thouars) Spreng., Syst. Veg. 3: 744 (1826).

Phaius tuberculosus (Thouars) Blume, Mus. Lugd. Batav. 2: 181 (1856).

Phaius tuberculatus Blume, Coll. Orch. Arch. Ind. 1: t.11 (1858).

Phaius warpuri Weathers in Gard. Chron. sér. 3, 29: 82 (1901). TYPE: Madagascar, *Warpur* s.n. (holo. not located).

Gastrorchis humbertii Ursch & Genoud in Nat. Malg. 2 (2): 154 (1950), *nom. nud.* TYPE: Madagascar, Perinet, cult. Jard. Bot. Tan. 640 (holo. P).

Distribution: Madagascar: Antananarivo, Toamasina.

Habitat: humid, evergreen forest.

Altitude: sea level – 1500 m.

Flowering time: September – February.

Life form: epiphyte or terrestrial.

Phytogeographical region: I.

Description: {50}, {120}, {1061}, {1442}.

Illustration: {1}, {50}, {94}, {120}, {208}, {226}, {253}, {322}, {349}, {364}, {444}, {473}, {501}, {607}, {673}, {724}, {850}, {896}, {949}, {1018}, {1061}, {1104}, {1153}, {1268}, {1298}, {1398} {1413}, {1442}, {1448}, {1455}, {1467}, {1468}.

Cultivation: {1}, {35}, {160}, {252}, {322}, {607}, {673}, {761}, {925}, {949}, {1061}, {1083}, {1489}.

History: {1}, {25}, {150}, {186}, {226}, {364}, {571}, {994}, {1061}, {1104}, {1153}, {1165}, {1257}.

Notes: tepals white; lip 3-lobed, the side lobes yellow covered with red spots, the midlobe white with lilac- or violet-spotted margins, the disc bearing 3 yellow, verrucose ridges.

Goodyera afzelii *Schltr.* in Repert. Spec. Nov. Regni Veg. Beih. 15: 327 (1918). TYPE: Madagascar, nr. Moramanga, *Afzelius* s.n. (holo. B†; iso. S).

DISTRIBUTION: Madagascar: Antsiranana, Fianarantsoa, Toamasina. Also Mozambique.
HABITAT: humid, mid-elevation forest.
ALTITUDE: 700 – 1800 m.
FLOWERING TIME: October.
LIFE FORM: terrestrial.
PHYTOGEOGRAPHICAL REGION: I, III.
ILLUSTRATION: {884}, {1223}.
DESCRIPTION: {224}, {896}.
NOTES: plant with an elongate, creeping rhizome; lip 4.5 × 4 mm, bipartite, the hypochile globose, deeply saccate, bearing numerous papillae on inner surface, the epichile recurved, ovate-ligulate, obtuse, 2 mm long.

Goodyera flaccida *Schltr.* in Repert. Spec. Nov. Regni Veg. Beih. 33: 124 (1924). TYPE: Madagascar, Anove Basin, *H.Perrier* 11395 (holo. P).

DISTRIBUTION: Madagascar: Toamasina.
HABITAT: humid, evergreen forest.
ALTITUDE: c. 200 m.
FLOWERING TIME: September.
LIFE FORM: epiphyte or terrestrial.
PHYTOGEOGRAPHICAL REGION: I.
DESCRIPTION: {896}.
ILLUSTRATION: {884}, {896}.
NOTES: plant erect, on an ascending rhizome; lip 4 × 3 mm, 3-lobed above the middle, the lateral lobes very small, the midlobe expanded at the tip into a transversal blade, 1 × 1.8 mm, the disc with 2 diverging keels.

Goodyera goudotii *Ormd. & Cavestro.* in Taiwania. 51 (3): 154 (2006). TYPE: Madagascar, sin. loc, *Gaudot* s.n. (holo. G).
Related to *Goodyera rosea* but differs in its smaller plant size, shorter inflorescence and labellum with an obcuneate-subquadrate, rather than ovate, epichile.

Goodyera humicola *(Schltr.) Schltr.* in Repert. Spec. Nov. Regni Veg. Beih. 33: 124 (1924). TYPE: Madagascar, Maningory, *H.Perrier* 11383 (116) (holo. P).
Platylepis humicola Schltr. in Beih. Bot. Centralbl. 34 (2): 319 (1916).

DISTRIBUTION: Madagascar: Toamasina.
HABITAT: humid, evergreen forest, in humus.
ALTITUDE: 500 – 1200 m.
FLOWERING TIME: September – October.
LIFE FORM: terrestrial.
PHYTOGEOGRAPHICAL REGION: I.
DESCRIPTION: {896}.
ILLUSTRATION: {672}, {896}, {1223}.
NOTES: stem rambling, with c. 8 petiolate leaves; lip very concave, with 2 verrucose keels.

Goodyera perrieri *(Schltr.) Schltr.* in Repert. Spec. Nov. Regni Veg. Beih. 33: 125 (1924). TYPE: Madagascar, Mt. Vatovavy, *H.Perrier* 11846 (holo. B†; iso. P).
Platylepis perrieri Schltr. in Ann. Mus. Col. Marseille, ser.3, 1: 161 (18), t.6 (1913).

DISTRIBUTION: Madagascar: Fianarantsoa.
HABITAT: evergreen forest.
ALTITUDE: sea level – 500 m.
FLOWERING TIME: October – December.
LIFE FORM: epiphyte or terrestrial in humus.
PHYTOGEOGRAPHICAL REGION: I.
DESCRIPTION: {896}.

ILLUSTRATION: {1205}, {1223}.

NOTES: raceme less than 10 cm long; sepals 4 mm at most; petals oblanceolate-linear and obtuse; lip verrucose on the outside, with protruding calli at the base.

Goodyera rosea (*H.Perrier*) *Ormd. & Cavestro* in Taiwania 51, 3: 158 (2006). TYPE: Madagascar, Mt. Maromizaha, nr. Analamazaotra, *H.Perrier* 15964 (holo. P).
Bathiorchis rosea (H.Perrier) Bosser & P.J.Cribb in Adansonia sér. 3, 25 (2): 229 (2003).
Gymnochilus roseum H.Perrier in Bull. Soc. Bot. France 83: 24 (1936).

DISTRIBUTION: Madagascar: Toamasina.
HABITAT: humid, evergreen forest on plateau.
ALTITUDE: c. 1000 m.
FLOWERING TIME: February.
LIFE FORM: terrestrial.
PHYTOGEOGRAPHICAL REGION: III.
DESCRIPTION: {896}.
ILLUSTRATION: {896}, {672}, {144}.

Grammangis ellisii (*Lindl.*) *Rchb.f.* in Hamb. Gart. Blumenz. 16: 520 (1860). TYPE: Madagascar, *Ellis* s.n. (holo. K; iso. W).
Grammatophyllum ellisii Lindl. in Hooker in Bot. Mag. 86: t.5179 (1860).
Grammangis ellisii var. *dayanum* Rchb.f. in Gard. Chron. 14: 326 (1880). TYPE: Madagascar, *Humblot* s.n. (holo. W).
Grammangis fallax Schltr. in Orchis 9: 120 (1915). TYPE: Madagascar, cult. Dahlem B.G., *Ferko* s.n. (holo. B).

DISTRIBUTION: Madagascar: Antananarivo, Antsiranana, Fianarantsoa, Toamasina.
HABITAT: coastal forest; humid, evergreen forest; branches overhanging rivers; on *Raphia farinifera* and on *Pandanus*.
ALTITUDE: sea level – 1300 m.
FLOWERING TIME: November – January.
LIFE FORM: epiphyte.
PHYTOGEOGRAPHICAL REGION: I, III.
DESCRIPTION: {133}, {695}, {896}, {1236}.
ILLUSTRATION: {3}, {45}, {50}, {112}, {133}, {187}, {208}, {222}, {226}, {253}, {266}, {281}, {397}, {434}, {436}, {444}, {484}, {529}, {564}, {593}, {615}, {629}, {724}, {733}, {850}, {862}, {896}, {905}, {940}, {944}, {1025}, {1037}, {1066}, {1158}, {1236}, {1264}, {1265}, {1321}, {1421}, {1442}, {1467}, {1468}, {1507}.
HISTORY: {37}, {45}, {226}, {236}, {337}, {392}, {429}, {564}, {571}, {984}, {988}, {1037}, {1066}, {1158}, {1265}, {1490}.
CULTIVATION: {236}, {392}, {443}, {754}, {761}, {938}, {940}, {1025}, {1265}, {1421}, {1321}, {1427}, {1442}.
NOTES: plant robust with large tetragonal pseudobulbs; flowers large, golden-brown, glossy; lip 3-lobed, with pronounced white keels.

Grammangis spectabilis *Bosser & Morat* in Adansonia, sér. 2, 9 (2): 303 (1969). TYPE: Madagascar, near Sakaraha, *Lauffenburger* 1395 (holo. P; iso. TAN).

DISTRIBUTION: Madagascar: Toliara.
HABITAT: seasonally dry, deciduous forest or woodland on *Ficus*.
FLOWERING TIME: November.
LIFE FORM: epiphyte.
PHYTOGEOGRAPHICAL REGION: VI.
ILLUSTRATION: {112}, {281}, {479}, {482}.
NOTES: overall more slender than *G. ellisii* and is easily distinguished by floral characteristics, particularly thorn-like appendages on the lip surface.

Graphorkis concolor (*Thouars*) *Kuntze* var. **alphabetica** *F.N.Rasm.* in Bot. Notul. 132 (3): 387 (1979). TYPE: Réunion or Madagascar, *Thouars* s.n. (holo. P).
Epidendrum scriptum Thouars, Cahier de six planches: t.3 (1819*), auct. non L.* (1753).

Eulophia scripta Pfitzer in Engler & Prantl, Pflanzenfam. 2 (6): 183 (1889).
Graphorkis aiolographis Thouars, Cahier de six planches: t.3, (1819), *nom. illeg.*
Graphorkis scripta var. *scripta sensu* Senghas in Die Orchidee 15: 65 (1964).
Limodorum scriptum Thouars, Hist. Orch.: t.46 & 47 (1822).
Lissochilus scriptus sensu H.Perrier in Humbert ed., Fl. Mad. Orch. 2: 37 (1941).

DISTRIBUTION: Madagascar: Antsiranana, Toamasina, Toliara. Comoros. Mascarenes: Réunion. Seychelles.

HABITAT: humid, lowland forest; littoral forest, on raffia palms, on mango trees, on *Tiphonodorum lindleyanum.*

ALTITUDE: sea level – 100 m.

FLOWERING TIME: September – October.

LIFE FORM: epiphyte.

PHYTOGEOGRAPHICAL REGION: I.

DESCRIPTION: {896}

ILLUSTRATION: {167}, {222}, {281}, {470}, {482}, {1240}, {1398}.

NOTES: root-mass extensive, developing around the base of the plant, the erect roots sharply pointed; older pseudobulbs carrying teat at the apex; flowers 15 – 25 mm, yellow spotted with red, the petals and sepals greenish yellow. The typical variety is found in Réunion.

Graphorkis ecalcarata *(Schltr.)* Summerh. in Kew Bull. 8 (1): 161 (1953). TYPE: Madagascar, Mt. Vatovavy, *H.Perrier* 11482 (holo. B†; iso. P).
Eulophiopsis ecalcarata Schltr. in Beih. Bot. Centralbl. 34 (2): 332 (1916).
Lissochilus ecalcaratus (Schltr.) H.Perrier in Humbert ed., Fl. Mad. Orch. 2: 36 (1941).
Eulophia ecalcarata (Schltr.) M. Lecoufle in Orchid Review 94: 313 (1986).

DISTRIBUTION: Madagascar: Toamasina.

HABITAT: humid, lowland forest.

ALTITUDE: sea level – 1000 m.

FLOWERING TIME: October – January.

LIFE FORM: epiphyte.

PHYTOGEOGRAPHICAL REGION: I.

DESCRIPTION: {896}, {1039}.

ILLUSTRATION: {896}, {1039}, {1045}, {1223}.

NOTES: pseudobulbs ovoid-elliptical; flowers c. 2 cm long; lip 7 mm long, without a spur.

Graphorkis medemiae *(Schltr.)* Summerh. in Kew Bull. 8 (1): 161 (1953). TYPE: Madagascar, Soalala, Ambongo, *H.Perrier* 1582 (holo. P).
Eulophia medemiae Schltr. in Ann. Mus. Col. Marseille, sér. 3, 1: 173 (30), t.12 (1913).
Eulophiopsis medemiae (Schltr.) Schltr. in Beih. Bot. Centralbl. 33, 2: 422 (1915).
Lissochilus medemiae (Schltr.) H.Perrier in Humbert ed., Fl. Mad. Orch. 2: 38 (1941).

DISTRIBUTION: Madagascar: Mahajanga.

HABITAT: on trunks of *Medemia nobilis.*

ALTITUDE: sea level – 500 m.

FLOWERING TIME: August- September.

LIFE FORM: epiphyte.

PHYTOGEOGRAPHICAL REGION: V.

DESCRIPTION: {896}.

ILLUSTRATION: {1205}, {1223}.

NOTES: known only from the type. Inflorescence slender, c. 25 cm long; flowers 2.5 cm long, brown with a yellow lip.

Habenaria acuticalcar *H.Perrier* in Notul. Syst. (Paris) 14: 141 (1951). TYPES: Madagascar, summit E of Marojejy massif, *Humbert* 23776 (lecto. P, selected here); W of Marojejy, *Humbert* 22450 & Mt. Ambatosoratra (Lokoho Valley), *Humbert & Cours* 22842 (both syn. P).

DISTRIBUTION: Madagascar: Antsiranana.

HABITAT: humid, evergreen forest.

ALTITUDE: 500 – 2500 m.

FLOWERING TIME: December – January, March – April.
LIFE FORM: terrestrial.
PHYTOGEOGRAPHICAL REGION: IV.
DESCRIPTION: {901}.
NOTES: related to *H. praealta* but plant different, raceme denser, flowers larger, and spur 12 – 15 mm long, inflated, acutely narrowed.

Habenaria alta *Ridl.* in J. Linn. Soc., Bot. 21: 509 (1885). TYPE: Madagascar, Ankafana, *Deans Cowan* (holo. BM).

DISTRIBUTION: Madagascar: Antananarivo, Fianarantsoa.
HABITAT: humid, mossy, evergreen forest; shaded and humid rocks.
ALTITUDE: 1500 – 2000 m.
FLOWERING TIME: February – April.
LIFE FORM: terrestrial.
PHYTOGEOGRAPHICAL REGION: III, IV.
DESCRIPTION: {896}.
ILLUSTRATION: {504}, {896}.
NOTES: leaves ovate; lip as long as the lateral sepals; spur thickened towards the apex.

Habenaria ambositrana *Schltr.* in Beih. Bot. Centralbl. 34 (2): 315 (1916). TYPE: Madagascar: between Ambatomainty & Itremo, *H.Perrier* 11368 (131) (iso. P).

DISTRIBUTION: Madagascar: Antananarivo, Fianarantsoa.
HABITAT: river margins; marshes; amongst rocks.
ALTITUDE: 1500 – 2000 m.
FLOWERING TIME: February – June.
LIFE FORM: terrestrial.
PHYTOGEOGRAPHICAL REGION: III.
DESCRIPTION: {896}.
ILLUSTRATION: {896}, {1223}.
NOTES: leaves narrowly linear; flowers not approximate to the rachis, yellow; petals obtuse; spur c. 3 cm; anther not apiculate.

Habenaria arachnoides *Thouars,* Hist. Orch.: t.18 & 19 (1822). TYPE: Madagascar, *Thouars* s.n. (holo. P; iso. W).

DISTRIBUTION: Madagascar: Antananarivo, Fianarantsoa, Toamasina. Mascarenes: Réunion.
HABITAT: humid, evergreen and montane forest.
ALTITUDE: 750 – 2000 m.
FLOWERING TIME: March, July, October.
LIFE FORM: terrestrial.
PHYTOGEOGRAPHICAL REGION: III, IV.
DESCRIPTION: {896}.
ILLUSTRATION: {1398}.
NOTES: plant with 5 – 6 leaves, rounded at the base, laxly-flowered; lip papillose, divided into 3 narrow lobes; spur scarcely inflated.

Habenaria bathiei *Schltr.* in Repert. Spec. Nov. Regni Veg. Beih. 33: 89 (1924). TYPE: Madagascar, Ipatina, Ifasina, on the Imorona, Mania Valley, *H.Perrier* 12448. (holo. P).

DISTRIBUTION: Madagascar: Fianarantsoa, Toliara.
HABITAT: humid, evergreen forest; highland savannah.
ALTITUDE: 800 – 1400 m.
FLOWERING TIME: January – February.
LIFE FORM: terrestrial.
PHYTOGEOGRAPHICAL REGION: I, III.
DESCRIPTION: {896}.
ILLUSTRATION: {1223}.

NOTES: related to *H. monadenioides*, but differs in lip structure, gymnostemium and rostellum; spur filiform, 5 cm long.

Habenaria beharensis *Bosser* in Adansonia, sér. 2, 9 (2): 293 (1969). TYPE: Madagascar, Ambatomika, Behara, *Bosser* 14791 (holo. P).

DISTRIBUTION: Madagascar: Toliara.
HABITAT: dry, deciduous scrubland.
ALTITUDE: sea level – 500 m.
FLOWERING TIME: March- April.
LIFE FORM: terrestrial.
PHYTOGEOGRAPHICAL REGION: VI.
DESCRIPTION: {111}.
ILLUSTRATION: {111}.
NOTES: sepals with short hairs on the exterior; lip with two small, obtuse swellings at the base; spur very long.

Habenaria boiviniana *Kraenzl.*, Orch. Gen. Sp. 1: 238 (1897). TYPE: Comoros, Mohely, *Boivin* s.n., & Grande Comore, *Boivin* s.n. (both syn. P; isosyn. HBG?).
Habenaria nigricans Schltr. in Ann. Mus. Col. Marseille, sér. 3, 1: 157, t.1 (1913); H.Perrier in Notul. Syst. (Paris) 14: 142 (1951). TYPE: Madagascar, Manongarivo, Ambongo, *H.Perrier* 939 (holo. P).
Habenaria perrieri Schltr. in Ann. Mus. Col. Marseille, sér. 3, 1: 158, t.1 (1913) cfr. H.Perrier in Notul. Syst. (Paris) 14: 142 (1951). TYPE: Madagascar, Sambirano Valley, *H.Perrier* 1928 (29) (holo. P).

DISTRIBUTION: Madagascar: Antsiranana, Fianarantsoa, Mahajanga. Comoros: Grande Comore, Mayotte, Moheli. Also in East Africa.
HABITAT: semi-deciduous forest; humid, evergreen forest; on rocks; in woods.
ALTITUDE: sea level – 1000 m.
FLOWERING TIME: March – April, August.
LIFE FORM: terrestrial.
PHYTOGEOGRAPHICAL REGION: III, V.
DESCRIPTION: {529}, {896}, {901}, {1379}.
ILLUSTRATION: {1205}, {1223}.
NOTES: lip with 1.5 mm long, basal teeth obvious; spur filiform, c. 2 cm long; stigmatic processes broadened and flattened at the front.

Habenaria cirrhata *(Lindl.) Rchb.f.* in Flora 48: 180 (1865). TYPE: Madagascar, Imerina, *Lyall* (holo. K).
Bonatea cirrhata Lindl., Gen. Sp. Orch. Pl.: 327 (1835).
Habenaria schweinfurthii Rchb.f., Otia. Bot. Hamburg.: 58 (1878). TYPE from Sudan.
Habenaria zenkerana Kraenzl. in Bot. Jahrb. Syst. 19: 247 (1894). TYPE from Cameroon.
Habenaria longistigma Rolfe in Fl. Trop. Afr. 7: 248 (1898). TYPE from Tanganyika.
Habenaria dawei Rolfe in Kew Bull. 1912: 134 (1912). TYPE from Uganda.
Habenaria megistosolen Schlechter in Bot. Jahrb. Syst. 53: 512 (1915). TYPE from Tanganyika.

DISTRIBUTION: Madagascar: Antananarivo, Fianarantsoa, Mahajanga. Comoros: Moheli. Also in tropical Africa.
HABITAT: dry meadows and grassland.
ALTITUDE: 100 – 1200 m.
FLOWERING TIME: October – March.
PHYTOGEOGRAPHICAL REGION: III, V.
DESCRIPTION: {896}.
ILLUSTRATION: {246}, {256b}, {1472}.
NOTES: flowers large; sepals up to 2 cm long; spur 7 – 8 cm long.

Habenaria clareae *Hermans* nom. nov.
Habenaria elliotii Rolfe in J. Linn. Soc., Bot. 29: 57 (1891), *non* Beck, Bot. N. Middle States, 348 (1833). TYPE: Madagascar, Fianarantsoa, *Scott Elliot* 2037 (holo. K).

DISTRIBUTION: Madagascar: Antananarivo, Antsiranana, Fianarantsoa, Mahajanga, Toamasina. Also in East Africa.

HABITAT: coastal forest; humid, evergreen forest; marshes; rocky outcrops.

ALTITUDE: sea level – 1800 m.

FLOWERING TIME: February.

LIFE FORM: terrestrial.

PHYTOGEOGRAPHICAL REGION: I, II, III, V.

DESCRIPTION: {896}.

ILLUSTRATION: {677}, {1062}.

VERNACULAR NAME: Sinananga.

NOTES: plant tall, laxly-flowered; spur 4–5 cm, a little expanded-fusiform towards the apex. Owing to the earlier *Habenaria elliotii* Beck, based on a North American collection, this name cannot be used for the Madagascan plant described by Rolfe. I therefore propose the epithet *clareae*, for the Madagascan species.

Habenaria cochleicalcar *Bosser* in Adansonia, sér. 2, 9 (2): 294 (1969). TYPE: Madagascar, canton de Behara, Amboasary district, *Rakotoson* 9420 (holo. P; iso. TAN).

DISTRIBUTION: Madagascar: Toliara.

HABITAT: dry, deciduous scrub.

FLOWERING TIME: March.

LIFE FORM: terrestrial.

PHYTOGEOGRAPHICAL REGION: VI.

DESCRIPTION: {111}.

ILLUSTRATION: {111}.

NOTES: spur cylindrical, 15 mm long, twisted into a spiral, globulose and a little flattened at the tip.

Habenaria comorensis *H.Perrier* in Notul. Syst. (Paris) 14: 142 (1951). TYPE: Comoros, *Humblot* s.n. (holo. P).

DISTRIBUTION: Comoros.

LIFE FORM: terrestrial.

DESCRIPTION: {901}.

NOTES: known only from the type. Plant leafless; lip 3-lobed.

Habenaria conopodes *Ridl.* in J. Linn. Soc., Bot. 22: 124 (1886). TYPE: Madagascar, Ambatovory, *Fox* 5 (holo. K; iso. BM).

DISTRIBUTION: Madagascar: Antananarivo.

HABITAT: savannah; on rocks in the highlands.

ALTITUDE: c. 1000 m.

FLOWERING TIME: January – March.

LIFE FORM: terrestrial.

PHYTOGEOGRAPHICAL REGION: III.

DESCRIPTION: {896}.

NOTES: raceme laxly flowered; spur recurved toward the top, inflated at the apex.

Habenaria deanscowaniana *Hermans* **nom. nov.**

Habenaria stricta Ridl. in J. Linn. Soc., Bot. 21: 510 (1885), *non* A.Rich. & Gal. in Ann. Sci. Nat. (Paris) sér. 3. 3: 29 (1845), *nec* (Lindl.) Rijdb. in Bull. Torrey Bot. Club: 24: 4, 189 (1897). TYPE: Madagascar, Imerina, *Deans Cowan* s.n. (holo. BM 32245).

DISTRIBUTION: Madagascar: Antananarivo.

PHYTOGEOGRAPHICAL REGION: III.

DESCRIPTION: {896}.

NOTES: differs from *H. hilsenbergii* by its numerous strict narrow leaves and many-flowered raceme. *Habenaria stricta* Rich. & Gal. (1845) was based on a Mexican species and thus the name cannot be used for the Madagascan plant described by Ridley in 1885. I propose the epithet *deanscowaniana* in honour of the Rev. Deans Cowan who first collected this species.

Habenaria decaryana *H.Perrier* in Bull. Soc. Bot. France 83: 582 (1936). TYPE: Madagascar, Ikongo massif, *Decary* 5528 (holo. P).

DISTRIBUTION: Madagascar: Antananarivo, Fianarantsoa.
HABITAT: open woodland and open humid forests.
ALTITUDE: 1000 – 1200 m.
FLOWERING TIME: October, November.
LIFE FORM: terrestrial.
PHYTOGEOGRAPHICAL REGION: III.
DESCRIPTION: {896}.
ILLUSTRATION: {20}, {836}, {1416}.
NOTES: plant very tall; inflorescence unevenly laxly-flowered; spur a little expanded.

Habenaria demissa *Schltr.* in Repert. Spec. Nov. Regni Veg. Beih. 33: 90 (1924). TYPE: Madagascar, Mt. Tsiafajavona, *H.Perrier* 13526 (holo. P).

DISTRIBUTION: Madagascar: Antananarivo, Toliara.
HABITAT: mossy forest.
ALTITUDE: 1200 – 2400 m.
FLOWERING TIME: December, March.
LIFE FORM: terrestrial.
PHYTOGEOGRAPHICAL REGION: III, IV.
DESCRIPTION: {896}.
ILLUSTRATION: {1223}.
NOTES: related to *H. arachnoides* and *H. alta* but differs from the former by the wider, differently shaped leaves, and from the latter by the slender, smaller leaves, shape of petals and lip.

Habenaria ferkoana *Schltr.* in Repert. Spec. Nov. Regni Veg. Beih. 15: 325 (1918). TYPE: Madagascar, *Laggiara* (holo. B†), neo.: the illustration in Repert. Spec. Nov. Reg. Veg., 68: t.26, 101 (1932), selected here.

DISTRIBUTION: Madagascar.
LIFE FORM: terrestrial.
PHYTOGEOGRAPHICAL REGION: I.
DESCRIPTION: {896}.
ILLUSTRATION: {1223}.
NOTES: the type of this species was destroyed, but was said to be close to *H. arachnoides* but with bigger and more leaves, slightly larger flowers and more pointed petal segments, the lip and a more pointed spur.

Habenaria foxii *Ridl.* in J. Linn. Soc., Bot. 22: 124 (1886). TYPE: Madagascar, Ambatovory, *Fox* (holo. K).

DISTRIBUTION: Madagascar: Antananarivo, Fianarantsoa.
HABITAT: in forest.
FLOWERING TIME: March, June.
LIFE FORM: terrestrial.
PHYTOGEOGRAPHICAL REGION: III.
DESCRIPTION: {896}.
NOTES: leaves narrow; flowers small; spur swollen, c. 12 mm long.

Habenaria hilsenbergii *Ridl.* in J. Linn. Soc., Bot. 21: 509 (1885). TYPE: Madagascar, Imerina, *Hilsenberg & Bojer* s.n. (holo. BM).
Habenaria atra Schltr. in Beih. Bot. Centralbl. 34 (2): 314 (1916). TYPE: Madagascar, near Hutrirala, *H.Perrier* 8095 (holo. P).
Habenaria ankaratrana Schltr. in Repert. Spec. Nov. Regni Veg. Beih. 33: 87 (1924). TYPE: Madagascar, Ankaratra, *H.Perrier* 13568 (holo. P).

DISTRIBUTION: Madagascar: Antananarivo, Fianarantsoa, Toamasina.
HABITAT: grassland, rocky outcrops.

ALTITUDE: 1500 – 2000 m.

FLOWERING TIME: February – March, June.

LIFE FORM: terrestrial.

PHYTOGEOGRAPHICAL REGION: III.

DESCRIPTION: {896}.

ILLUSTRATION: {896}, {1223}.

NOTES: a variable species allied to *H. truncata*, distinguished by its shorter and blunter petals and lobes of the lip, and its longer filiform spur.

Habenaria incarnata *(Lindl.) Rchb.f.* in Flora 48: 180 (1865). TYPE: Madagascar, Imerina, *Lyall* s.n. (holo. K).

Bonatea incarnata Lindl., Gen. Sp. Orch. Pl.: 327 (1835).

Habenaria rutenbergiana Kraenzl. in Abh. Naturwiss. Vereine Bremen 7: 258 (1882). TYPE: Madagascar, Antananarivo, *Rutenberg* s.n. (holo. HBG).

Habenaria humblotii Rchb.f. in Flora 68: 535 (1885). TYPE: Comoros, *Humblot* 426 (1426) (holo. W; iso. BM, P).

Habenaria diptera Schltr. in Beih. Bot. Centralbl. 39 (2): 316 (1916). TYPE: Madagascar, near Inanatona, W of Betafo, Andratsay basin, *H.Perrier* 8123 (XXXIV) (holo. P).

DISTRIBUTION: Madagascar: Antananarivo, Antsiranana, Fianarantsoa, Toliara. Comoros: Grande Comore, Mayotte, Moheli.

HABITAT: grassland; forest and woodland margins.

ALTITUDE: sea level – 1400 m.

FLOWERING TIME: January – June.

LIFE FORM: terrestrial.

PHYTOGEOGRAPHICAL REGION: I, III.

DESCRIPTION: {896}.

ILLUSTRATION: {896}.

NOTES: lateral sepals almost 2 times longer than the reflexed dorsal sepal; spur c. 25 mm long.

Habenaria johannae *Kraenzl.* in Bot. Jahrb. Syst. 16: 77 (1892). TYPE: Comoros, Anjouan, *Hildebrandt* 1883 (holo. W).

DISTRIBUTION: Comoros: Anjouan.

ALTITUDE: sea level – 500 m.

DESCRIPTION: {896}.

NOTES: flowers and floral bracts sometimes a little pubescent on the exterior.

Habenaria lastelleana *Kraenzl.*, Orch. Gen. Sp. 1: 357 (1898). TYPE: Madagascar, Maroantsetra, *de Lastelle* 1841 (holo. HBG?; iso. P).

DISTRIBUTION: Madagascar: Toamasina.

HABITAT: semi-deciduous, lowland forest.

ALTITUDE: 400 – 600 m.

FLOWERING TIME: April,

PHYTOGEOGRAPHICAL REGION: I.

DESCRIPTION: {896}.

NOTES: flowers small; raceme few- to c. 10-flowered, lax, the flowers c. 15 mm apart.

Habenaria leandriana *Bosser* in Adansonia, sér. 2, 9 (2): 297 (1969). TYPE: Madagascar, Zombitsy forest, Sakaraha, *Bosser* 19278 (holo. P; iso. TAN).

DISTRIBUTION: Madagascar: Toliara.

HABITAT: seasonally dry, deciduous forest, woodland or dry deciduous scrub.

ALTITUDE: sea level – 500 m.

FLOWERING TIME: March.

LIFE FORM: terrestrial.

PHYTOGEOGRAPHICAL REGION: VI.

DESCRIPTION: {111}.

ILLUSTRATION: {111}.

NOTES: related to *H. clarae* but differs by the more numerous and larger flowers, its longer stigmatic processes and by its more attenuate, pseudo-petiolate leaves.

Habenaria monadenioides *Schltr.* in Beih. Bot. Centralbl. 34 (2): 315 (1916). TYPE: Madagascar, Mt. Ibity, S of Antsirabe, *H.Perrier* 8083 (XIII) (holo. P).

DISTRIBUTION: Madagascar: Antananarivo, Fianarantsoa.

HABITAT: rocky outcrops.

ALTITUDE: 1500 – 2200 m.

FLOWERING TIME: February – March.

LIFE FORM: terrestrial.

PHYTOGEOGRAPHICAL REGION: III, IV.

DESCRIPTION: {896}.

ILLUSTRATION: {896}, {1223}.

NOTES: plant up to 45 cm tall; raceme uniformly densely flowered; flowers white; spur cylindrical.

Habenaria nautiloides *H.Perrier* in Notul. Syst. (Paris) 14: 142 (1951). TYPE: Madagascar, Mt. Vohimavo, Manampanihy, *Humbert* 20690 (holo. P).

DISTRIBUTION: Madagascar.

HABITAT: humid,montane forest.

ALTITUDE: c. 600 m.

FLOWERING TIME: March.

LIFE FORM: terrestrial.

PHYTOGEOGRAPHICAL REGION: I.

DESCRIPTION: {901}.

NOTES: known only from the type. Differs from *H. hilsenbergii* by its narrowed leaves not enveloping the stem, the sepals and the lip of a different shape, and the hairs on the keels of the sepals.

Habenaria praealta *(Thouars) Spreng.*, Syst. Veg. 3: 691 (1826). TYPE: Réunion, *Thouars* s.n. (holo. P). *Satyrium praealtum* Thouars, Hist. Orch.: t.2, 11 & t.12, f.6 (1822).

DISTRIBUTION: Madagascar. Mascarenes: Réunion, Mauritius ?

HABITAT: humid, evergreen forest; riverine forest.

ALTITUDE: c. 1300 m.

FLOWERING TIME: November – December.

LIFE FORM: terrestrial.

PHYTOGEOGRAPHICAL REGION: III.

DESCRIPTION: {896}.

ILLUSTRATION: {149b}, {1398}.

NOTES: robust rhizomatous plant; leaves somewhat grass-like; petals with a lateral tooth; mid-lobe of the lip much shorter than the laterals; spur 5 – 6 mm long.

Habenaria quartzicola *Schltr.* in Repert. Spec. Nov. Regni Veg. Beih. 33: 93 (1924). TYPE: Madagascar, between Mania and Ivato, *H.Perrier* 12383 (holo. P).

DISTRIBUTION: Madagascar: Fianarantsoa.

HABITAT: rocky quartzite outcrops and wet flushes.

ALTITUDE: 800 – 1600 m.

FLOWERING TIME: February.

LIFE FORM: terrestrial.

PHYTOGEOGRAPHICAL REGION: III.

DESCRIPTION: {896}.

ILLUSTRATION: {1223}.

NOTES: plant slender; flowers white; spur filiform, narrowed-acute, 3 cm long.

Habenaria saprophytica *Bosser & P.J.Cribb* in Adansonia sér. 4, 18, 3 – 4: 335 (1997). TYPE: Madagascar, Mt. Ambre, *Malcolm & Rapanarivo* 1214 (holo. P; iso. K., MO, TAN).

DISTRIBUTION: Madagascar: Antsiranana.
HABITAT: montane forest.
ALTITUDE: 610m.
LIFE FORM: saprophyte.
PHYTOGEOGRAPHICAL REGION: I.
DESCRIPTION: {136}.
ILLUSTRATION: {136}.
NOTES: plant small, achlorophyllous, saprophytic, aphyllous, glabrous; lip ligulate.

Habenaria simplex *Kraenzl.* in Abh. Naturwiss. Vereine Bremen 7: 260 (1882). TYPE: Madagascar, Efitra, *Rutenberg* s.n. (type not located).
Habenaria ichneumoniformis Ridl. in J. Linn. Soc., Bot. 22: 125 (1886). TYPE: Central Madagascar, without locality, *Baron* 3879 (holo. BM).

DISTRIBUTION: Madagascar: Antananarivo, Fianarantsoa.
HABITAT: grass amongst rocks.
FLOWERING TIME: October – January.
PHYTOGEOGRAPHICAL REGION: III.
DESCRIPTION: {896}.
ILLUSTRATION: {281}, {896}.
NOTES: leaves short, lowermost approximate to the rachis; petals acute; anther apiculate.

Habenaria tomentella *Rchb.f.* in Flora 68: 536 (1885). TYPE: Comoros, *Humblot* s.n. (holo. W).

DISTRIBUTION: Comoros.
DESCRIPTION: {896}.
NOTES: known only from the type, said to be different from *H. alta* by the smaller flowers and shorter anther canals.

Habenaria tropophila *H.Perrier* in Notul. Syst. (Paris) 14: 142 (1951). TYPE: Madagascar, gorge of Manombo, *Humbert* 19979 (holo. P; iso. K, TAN).

DISTRIBUTION: Madagascar: Toliara.
HABITAT: humid, evergreen forest; rocky outcrops.
ALTITUDE: 100 – 300 m.
FLOWERING TIME: January – February.
LIFE FORM: terrestrial .
PHYTOGEOGRAPHICAL REGION: VI.
DESCRIPTION: {901}.
NOTES: leaves fleshy, broadly oval; lip 3-lobed, the lobes filiform; spur c. 12 mm long, slightly thickened.

Habenaria truncata *Lindl.,* Gen. Sp. Orch. Pl.: 311 (1938). TYPE: Madagascar, Imerina, *Lyall* (holo. K).

DISTRIBUTION: Madagascar: Antananarivo, Fianarantsoa, Toamasina.
HABITAT: grassland; rocky outcrops.
ALTITUDE: 1200 – 2000 m.
FLOWERING TIME: January – April.
LIFE FORM: terrestrial .
PHYTOGEOGRAPHICAL REGION: III.
DESCRIPTION: {896}.
ILLUSTRATION: {246}.
NOTES: a variable species; leaves broad; raceme densely-flowered; lip consists of 3 narrowly linear lobes.

Habenaria tsaratananensis *H.Perrier* in Bull. Soc. Bot. France 83: 583 (1936). TYPE: Madagascar, Mt. Tsaratanana, *H.Perrier* 16105 (holo. P).

DISTRIBUTION: Madagascar: Antsiranana.
HABITAT: mossy montane forest.
ALTITUDE: 1200 – 1500 m.
FLOWERING TIME: April.
LIFE FORM: terrestrial.
PHYTOGEOGRAPHICAL REGION: III.
DESCRIPTION: {896}.
NOTES: known only from the type. Leaves, sheaths, floral bracts, stem, ovary and sepals pubescent-scaly; spur 9 mm long, cylindrical, a little expanded.

Hetaeria vaginalis *Rchb.f.* in Flora 68: 537 (1885). TYPE: Comoros, *Humblot* s.n. (holo. W).

DISTRIBUTION: Comoros.
NOTES: an enigmatic species, imprecisely described and the origin is dubious.

Imerinaea madagascarica *Schltr.* in Repert. Spec. Nov. Regni Veg. Beih. 33: 153 (1924). TYPE: Madagascar, La Mandraka, *H.Perrier* 14631 (holo. P).
Phajus gibbosulus H.Perrier in Humbert, Mém. Inst. Sci. Mad., ser. , 6: 262 (1955), *nom. nud.* Based upon Madagascar, Lakoho valley, N of Maroambihy, *Humbert* 23552 (P).

DISTRIBUTION: Madagascar: Antananarivo, Antsiranana, Toamasina.
HABITAT: mossy and lichen-rich montane forest; shaded and humid rocks.
ALTITUDE: 1000 – 1500 m.
FLOWERING TIME: December – May.
LIFE FORM: terrestrial.
PHYTOGEOGRAPHICAL REGION: III.
DESCRIPTION: {896}, {109}.
ILLUSTRATION: {109}, {672}, {896}, {902}.
NOTES: plant small; lip uppermost, rolled and carrying a distinct callus.

Jumellea ambrensis *H.Perrier* in Notul. Syst. (Paris) 7: 61 (1938). TYPE: Madagascar, Mt. Ambre, Ambohitra, *H.Perrier* 19015 (holo. P).

DISTRIBUTION: Madagascar: Antsiranana.
HABITAT: mossy, montane forest.
ALTITUDE: c. 1000 m.
FLOWERING TIME: February.
LIFE FORM: epiphyte.
PHYTOGEOGRAPHICAL REGION: III.
DESCRIPTION: {896}, {1259}.
ILLUSTRATION: {1259}.
NOTES: leaves fairly little rigid, thickened, semi-cylindrical, narrowly linear, 15 – 23 cm × 5 mm.

Jumellea amplifolia *Schltr.* in Repert. Spec. Nov. Regni Veg. Beih. 33: 288 (1925). TYPE: Madagascar, Mt. Tsaratanana, *H.Perrier* 15325 (holo. P).

DISTRIBUTION: Madagascar: Antsiranana.
HABITAT: mossy, montane forest.
ALTITUDE: c. 2000 m.
FLOWERING TIME: January.
LIFE FORM: epiphyte.
PHYTOGEOGRAPHICAL REGION: III, IV.
DESCRIPTION: {896}.
NOTES: related to *J. maxillarioides* which is smaller and has a different lip.

Jumellea angustifolia *H.Perrier* in Notul. Syst. (Paris) 7: 53 (1938). TYPE: Madagascar, Mt. Tsaratanana, *H.Perrier* 16482 (holo. P).

DISTRIBUTION: Madagascar: Antsiranana.
HABITAT: lichen-rich forest.
ALTITUDE: 1400 – 2000 m.
FLOWERING TIME: April.
LIFE FORM: terrestrial.
PHYTOGEOGRAPHICAL REGION: III, IV.
DESCRIPTION: {896}.
NOTES: similar to *J. dendrobioides*; lip very concave, 15 mm long, almost geniculate above the narrow base.

Jumellea anjouanensis *(Finet) H.Perrier* in *Humbert* ed., Fl. Mad. Orch. 2: 170 (1941). TYPES: Comoros, Anjouan, *Boivin* s.n. & *Lavanchie* 35 (both syn. P).
Angraecum anjouanense Finet in Bull. Soc. Bot. France 54, Mém. 9: 11, t.2 (1907).

DISTRIBUTION: Comoros: Anjouan, Grande Comore.
HABITAT: humid montane forest.
ALTITUDE: 950 – 1800 m.
FLOWERING TIME: February, May, July.
LIFE FORM: epiphyte.
DESCRIPTION: {896}.
ILLUSTRATION: {292}, {529}, {1027}.
NOTES: sheaths and bracts covering about a third of the pedicellate ovary; lip lanceolate; spur c. 10 mm long, thickened.

Jumellea arachnantha *(Rchb.f.) Schltr.* in Beih. Bot. Centralbl. 33 (2): 428 (1915). TYPE: Comoros, Combani, Grande Comore, *Humblot* 1423 (423) (holo. W; iso. BM, P).
Aeranthes arachnanthus Rchb.f. in Flora 68: 539 (1885).

DISTRIBUTION: Comoros: Grande Comore.
LIFE FORM: epiphyte.
DESCRIPTION: {449}, {896}.
ILLUSTRATION: {3}, {449}, {529}.
CULTIVATION: {449}.
NOTES: plant large, short-stemmed; lip broadly lanceolate; spur filiform, 4 – 4.5 cm long.

Jumellea arborescens *H.Perrier* in Notul. Syst. (Paris) 7: 58 (1938). TYPE: Madagascar, Tsinjoarivo, *H.Perrier* 16968 (holo. P).

DISTRIBUTION: Madagascar: Antananarivo, Toamasina.
HABITAT: mossy montane forest, on trunks.
ALTITUDE: 1100 – 1400 m.
FLOWERING TIME: October, February.
LIFE FORM: epiphyte or lithophyte.
PHYTOGEOGRAPHICAL REGION: III.
DESCRIPTION: {872}, {896}.
ILLUSTRATION: {529}, {872}, {1027}.
NOTES: plant very tall; flowers fleshy; lip expanded towards the middle; spur 11 – 12 cm long.

Jumellea bathiei *Schltr.* in Repert. Spec. Nov. Regni Veg. Beih. 33: 290 (1925). TYPE: Madagascar, Mt. Tsiafajavona, *H.Perrier* 13509 (holo. P).

DISTRIBUTION: Madagascar: Antananarivo.
HABITAT: mossy, upper montane forest on moss- and lichen-covered trees.
ALTITUDE: c. 2400 m.
FLOWERING TIME: March.
LIFE FORM: epiphyte.

PHYTOGEOGRAPHICAL REGION: IV.
DESCRIPTION: {896}.
ILLUSTRATION: {1223}.
NOTES: characterised by its long stem and short, 35 – 45 mm long spur.

Jumellea brachycentra *Schltr.* in Repert. Spec. Nov. Regni Veg. Beih. 33: 291 (1925). TYPES: Madagascar, Mt. Tsiafajavona, *H.Perrier* 13538 (lecto. P selected here); Andringitra massif, *H.Perrier* 14576 (syn. P).

Jumellea floribunda Schltr. in Repert. Spec. Nov. Regni Veg. Beih. 33: 294 (1925) TYPE: Madagascar, Mt. Tsaratanana, *H.Perrier* 15321 (holo. P).

DISTRIBUTION: Madagascar: Antsiranana, Fianarantsoa.
HABITAT: moss- and lichen-rich forest on moss- and lichen-covered trees.
ALTITUDE: 1800 – 2400 m.
FLOWERING TIME: January – March.
LIFE FORM: epiphyte.
PHYTOGEOGRAPHICAL REGION: III, IV.
DESCRIPTION: {896}.
ILLUSTRATION: {896}, {1223}.
NOTES: plant large, stemless; lip oblong, acute, 25 mm long, a little contracted between the third and the centre; spur 12 mm long.

Jumellea brevifolia *H.Perrier* in Notul. Syst. (Paris) 8: 44 (1939). TYPE: Madagascar, Mt. Kalambatitra, *Humbert* 11792 (holo. B; iso. P, TAN).

DISTRIBUTION: Madagascar: Toliara.
HABITAT: rocky outcrops of granite and gneiss.
ALTITUDE: 1500 – 1650 m.
FLOWERING TIME: November.
LIFE FORM: lithophyte.
PHYTOGEOGRAPHICAL REGION: III.
DESCRIPTION: {896}.
NOTES: plant robust; leaves leathery, broad; lip 22 mm long, abruptly expanded into a wide; spur filiform, 11 cm long.

Jumellea comorensis *(Rchb.f.) Schltr.* in Beih. Bot. Centralbl. 33 (2): 428 (1915). TYPE: Comoros, *Humblot* 1247 (247) (holo. W).

Aeranthes comorensis Rchb.f. in Flora 68: 540 (1885), as *Aeranthus*.
Mystacidium comorense (Rchb.f.) T.Durand & Schinz, Consp. Fl. Afric. 5: 52 (1892).
Angraecum comorense (Rchb.f.) Finet in Bull. Soc. Bot. France 54, Mém. 9: 13 (1907), *non* Kraenzl.

DISTRIBUTION: Comoros: Grande Comore.
HABITAT: humid, low-elevation forest.
ALTITUDE: c. 600 m.
FLOWERING TIME: August – November.
LIFE FORM: epiphyte.
DESCRIPTION: {896}.
ILLUSTRATION: {3}, {292}, {529}, {1027}, {1318}.
NOTES: plant densely leaved; lip 2 cm long, a little wider above the middle; spur 9 – 11 cm long, filiform.

Jumellea confusa *(Schltr.) Schltr.* in Beih. Bot. Centralbl. 33 (2): 429 (1915). TYPE: Madagascar, Manongarivo massif, *H.Perrier* 1937 (holo. B†; iso. P).

Angraecum confusum Schltr. in Ann. Mus. Col. Marseille, sér. 3, 1: 190 (47), t.14 (1913).
Jumellea ankaratrana Schltr. in Repert. Spec. Nov. Regni Veg. Beih. 33: 289 (1925). TYPE: Madagascar, Ankaratra, *H.Perrier* 13403 (holo. P).

DISTRIBUTION: Madagascar: Antananarivo, Antsiranana, Toamasina.
HABITAT: humid, evergreen, montane forest.

ALTITUDE: 600 – 1500 m.

FLOWERING TIME: September, December – May.

LIFE FORM: epiphyte or lithophyte.

PHYTOGEOGRAPHICAL REGION: I, II, III.

DESCRIPTION: {95}, {896}.

ILLUSTRATION: {95}, {529}, {1027}, {1205}, {1223}, {1412}.

CULTIVATION: {95}.

VERNACULAR NAME: Fontilahyjanahary madiniky.

NOTES: a variable species; plant up to 60 cm tall; lip c. 6 mm long, wide in the middle; spur filiform, 12 – 13 cm long.

Jumellea cowanii *(Ridl.) Garay* in Bot. Mus. Leafl. 23: 183 (1972). TYPE: Madagascar, Imerina, *Deans Cowan* s.n. (holo. BM).

Angraecum cowanii Ridl. in J. Linn. Soc., Bot. 21: 484 (1885).

Angorchis cowanii (Ridl.) Kuntze, Rev. Gen. Pl. 2: 651 (1891).

DISTRIBUTION: Madagascar: Antananarivo.

LIFE FORM: epiphyte {44}.

PHYTOGEOGRAPHICAL REGION: III.

DESCRIPTION: {1031}

NOTES: a poorly known species, known only from the type.

Jumellea cyrtoceras *Schltr.* in Repert. Spec. Nov. Regni Veg. Beih. 15: 335 (1918). TYPE: Madagascar, *Laggiara* s.n. (holo. B†); neo.: the illustration in Repert. Spec. Nov. Reg. Veg., 68 t.84, 333 (1932), selected here.

DISTRIBUTION: Madagascar.

HABITAT: in forest.

PHYTOGEOGRAPHICAL REGION: I.

DESCRIPTION: {896}.

ILLUSTRATION: {1223}.

NOTES: the type was destroyed; plant c. 15 cm tall, densely leaved; lip ligulate; spur short.

Jumellea dendrobioides *Schltr.* in Repert. Spec. Nov. Regni Veg. Beih. 33: 292 (1925). TYPE: Madagascar, Mt. Tsaratanana, *H.Perrier* 15702 (holo. P).

DISTRIBUTION: Madagascar: Antsiranana.

HABITAT: lichen-rich, montane forest on moss- and lichen-covered trees.

ALTITUDE: c. 2000 m.

FLOWERING TIME: January.

LIFE FORM: epiphyte.

PHYTOGEOGRAPHICAL REGION: III, IV.

DESCRIPTION: {896}.

NOTES: known only from the type. Plant tall, densely leaved; lip 16 × 8 mm; spur c. 12 mm long.

Jumellea densefoliata *Senghas* in Adansonia sér. 2, 4: 308 (1964). TYPE: Madagascar, Antsirabe, *Rauh* 7288/1961 (holo. HEID).

DISTRIBUTION: Madagascar: Antananarivo, Fianarantsoa.

HABITAT: humid, evergreen forest; shaded and humid rocks; in *Uapaca* woodland.

ALTITUDE: 1200 – 1600 m.

FLOWERING TIME: September – November.

LIFE FORM: epiphyte or lithophyte.

PHYTOGEOGRAPHICAL REGION: III.

DESCRIPTION: {1239}.

ILLUSTRATION: {529}, {638}, {1239}, {1264}.

NOTES: plant small, succulent; lip c. 16 × 4.5 mm unguiculate; spur c. 12 cm long.

Jumellea flavescens *H.Perrier* in Notul. Syst. (Paris) 7: 63 (1938). TYPE: Madagascar, Ankeramadinika, *H.Perrier* 18470 (lecto. P, selected here); *E.François* in *H.Perrier* 17921 (syn. P).

DISTRIBUTION: Madagascar: Antananarivo, Antsiranana.
HABITAT: mossy, montane forest on moss- and lichen-covered trees.
ALTITUDE: c. 1200 m.
FLOWERING TIME: January – February.
LIFE FORM: epiphyte.
PHYTOGEOGRAPHICAL REGION: III.
DESCRIPTION: {896}.
NOTES: plant small, branching; lip 2 cm, oblanceolate, acute, 7 – 8 mm wide above the middle; spur 10 – 12 cm long.

Jumellea françoisii *Schltr.* in Repert. Spec. Nov. Regni Veg. Beih. 33: 294 (1925). TYPES: Madagascar, Mandraka, *E.François* in *H.Perrier* 14903 & Ambositra, *H.Perrier* 11370 (both syn. P).

DISTRIBUTION: Madagascar: Antananarivo, Fianarantsoa.
HABITAT: humid, mossy, evergreen forest.
ALTITUDE: 1200 – 1500 m.
FLOWERING TIME: January- May.
LIFE FORM: epiphyte.
PHYTOGEOGRAPHICAL REGION: III.
DESCRIPTION: {896}.
ILLUSTRATION: The illustration in {281} is *J. spathulata*.
NOTES: plant tall with branching stem; lip 17 × 7 mm; spur c. 12 mm long.

Jumellea gladiator *(Rchb.f.) Schltr.* in Beih. Bot. Centralbl. 33 (2): 429 (1915). TYPE: Comoros, Combani, Grande Comore, *Humblot* 1415 (415) (holo. W; iso. BM, P).
Aeranthes gladiator Rchb.f. in Flora 68: 539 (1885).
Mystacidium gladiator (Rchb.f.) T.Durand & Schinz, Consp. Fl. Afric. 5: 53 (1892).

DISTRIBUTION: Comoros: Anjouan, Grande Comore.
HABITAT: humid, lowland forest.
ALTITUDE: c. 700 m.
FLOWERING TIME: April, October.
LIFE FORM: epiphyte.
DESCRIPTION: {896}.
ILLUSTRATION: {501}, {724}.
NOTES: plant very large, stemless; lip lanceolate, acute, c. 22 mm long; spur 25 – 35 cm long.

Jumellea gracilipes *Schltr.* in Repert. Spec. Nov. Regni Veg. Beih. 18: 324 (1922). TYPE: Madagascar, Ambatoloana, *Viguier & Humbert* 1216 (holo. P).
Jumellea ambongensis Schltr. in Repert. Spec. Nov. Regni Veg. Beih. 33: 288 (1925) TYPE: Madagascar, Manongarivo, *H.Perrier* 1903 (holo. P).
Jumellea exilipes Schltr., *loc.cit.* 293 (1925). TYPE: Madagascar, Tsaratanana massif, *H.Perrier* 15320 (holo. P).
Jumellea imerinensis Schltr., *loc.cit.* 296 (1925). TYPE: Madagascar, Mandraka, *H.Perrier* 14005 (holo. P).
Jumellea unguicularis Schltr., *loc.cit.* 304 (1925). TYPE: Madagascar, Tsaratanana massif, *H.Perrier* 15279 (holo. P).

DISTRIBUTION: Madagascar: Antananarivo, Antsiranana, Fianarantsoa, Toamasina.
HABITAT: mossy, montane forest.
ALTITUDE: 1000 – 2000 m.
FLOWERING TIME: November – May.
LIFE FORM: epiphyte.
PHYTOGEOGRAPHICAL REGION: III, IV.
DESCRIPTION: {890}, {896}.
ILLUSTRATION: {281}, {529}, {1025}, {1281}, {1392} this may be *J. flavescens*, {1425}.
NOTES: plant stemless; lip 22 × 7 mm; spur 11 – 15 mm long.

Jumellea gregariiflora *H.Perrier* in Notul. Syst. (Paris) 8: 45 (1939). TYPES: Madagascar, Mt. Amboahangy, Mandrare, *Humbert* 6845 (lecto. P; isolecto. TAN, both selected here), Sakamalio valley, Mandrare, *Humbert* 13373 (syn. P).

DISTRIBUTION: Madagascar: Toliara.
HABITAT: rocky outcrops of granite and gneiss.
ALTITUDE: 1000 – 1150 m.
FLOWERING TIME: November.
LIFE FORM: lithophyte.
PHYTOGEOGRAPHICAL REGION: III.
DESCRIPTION: {896}.
NOTES: plant branching; flowers several, grouped at the inter-nodes; lip c. 21 mm long; spur c. 8 cm long.

Jumellea hyalina *H.Perrier* in Notul. Syst. (Paris) 7: 54 (1938). TYPE: Madagascar, Ankeramadinika, *E.François* 10bis in *H.Perrier* s.n. (holo. P).

DISTRIBUTION: Madagascar: Antananarivo.
HABITAT: mossy, montane forest on moss- and lichen-covered trees.
ALTITUDE: c. 1500 m.
FLOWERING TIME: January.
LIFE FORM: epiphyte.
PHYTOGEOGRAPHICAL REGION: III.
DESCRIPTION: {896}.
NOTES: plant small; lip 10 × 8 mm, with a small callus at the base; spur c. 11 mm long.

Jumellea ibityana *Schltr.* in Repert. Spec. Nov. Regni Veg. Beih. 33: 296 (1925). TYPE: Madagascar, Mt. Ibity, *H.Perrier* 13583 (holo. P).

DISTRIBUTION: Madagascar: Antananarivo.
HABITAT: rocky outcrops.
ALTITUDE: 2000 – 2100 m.
FLOWERING TIME: February – March.
LIFE FORM: lithophyte.
PHYTOGEOGRAPHICAL REGION: IV.
DESCRIPTION: {896}.
ILLUSTRATION: {470}, {529}, {896}, {1223}.
NOTES: plant up to 30 cm tall; leaves rigid and leathery; lip lanceolate, acute, 10 × 6 mm; spur c. 3.5 mm long.

Jumellea intricata *H.Perrier* in Notul. Syst. (Paris) 7: 62 (1938). TYPE: Madagascar, Mt. Tsaratanana, *H.Perrier* 15697 (holo. P).

DISTRIBUTION: Madagascar: Antsiranana, Fianarantsoa, Toamasina.
HABITAT: lichen-rich montane forest on moss- and lichen-covered trees.
ALTITUDE: 1600 – 2000 m.
FLOWERING TIME: January.
LIFE FORM: epiphyte.
PHYTOGEOGRAPHICAL REGION: III, IV.
DESCRIPTION: {896}.
ILLUSTRATION: {529}, {896}.
NOTES: plant spreading, tangled; lip ligulate, 20 × 6 mm; spur 10 cm long.

Jumellea jumelleana *(Schltr.) Summerh.* in Kew Bull. 7 (3): 472 (1951). TYPE: Madagascar, Analamazaotra, *H.Perrier* 1939 (holo. B†; iso. P).
Angraecum jumelleanum Schltr. in Ann. Mus. Col. Marseille, sér. 3, 1: 51, t.24 (1913).
Jumellea henryi Schltr. in Beih. Bot. Centralbl. 33 (2): 429 (1915).

DISTRIBUTION: Madagascar: Antananarivo, Toamasina.
HABITAT: on shrubs and small trees in humid, evergreen forest.

ALTITUDE: 800 – 1500 m.

FLOWERING TIME: September – November, February.

LIFE FORM: epiphyte.

PHYTOGEOGRAPHICAL REGION: III.

DESCRIPTION: {896},

ILLUSTRATION: {1205}.

NOTES: plant small; lip obovate, elliptical, 13 × 7 mm, subobtuse; spur c. 11 cm long.

Jumellea lignosa *(Schltr.) Schltr.* in Beih. Bot. Centralbl. 33 (2): 429 (1915). TYPE: Madagascar, Ambositra, *H.Perrier* 1855 (holo. B†; iso. P).

Angraecum lignosum Schltr. in Ann. Mus. Col. Marseille, sér. 3, 1: 52, t.24 (1913).

Jumellea ferkoana Schltr. in Repert. Spec. Nov. Regni Veg. Beih. 15: 335 (1918).

Jumellea lignosa subsp. *typica* H.Perrier in Notul. Syst. (Paris) 7: 59 (1938).

Jumellea lignosa subsp. *ferkoana* (Schltr.) H.Perrier, *loc.cit.* 60 (1938), **syn. nov.** TYPE: Madagascar, *Laggiara* s.n. (holo. B†; neo. illustration in Repert. Spec. Nov. Reg. Veg., 68 t.84, 334, 1932; selected here).

DISTRIBUTION: Madagascar: Antananarivo, Fianarantsoa.

HABITAT: lichen-rich forest; rocky outcrops of granite and gneiss.

ALTITUDE: 700 – 2000 m.

FLOWERING TIME: December – January, March.

LIFE FORM: epiphyte or lithophyte.

PHYTOGEOGRAPHICAL REGION: I, III, IV.

DESCRIPTION: {896}.

ILLUSTRATION: {1205}, {1223}, {1392}.

NOTES: plant very tall with a woody stem; lip c. 30 mm long, widest (c.11 mm) in the middle; spur c. 11 cm long. *Jumellea lignosa* subsp. *ferkoana* (Schltr.) H.Perrier is a large form of this species.

Jumellea lignosa subsp. **acutissima** *H.Perrier ex Hermans* **subsp. nov.** TYPE: Madagascar, Ankeramadinika, *E.François* in *H.Perrier* 18518 (holo. P). See Appendix .

Jumellea lignosa var. *acutissima* H.Perrier in Notul. Syst. (Paris) 7: 59 (1938), *nom. nud.*

DISTRIBUTION: Madagascar: Antananarivo.

HABITAT: mossy forest on moss- and lichen -covered trees.

ALTITUDE: c. 1400 m.

FLOWERING TIME: March.

LIFE FORM: epiphyte.

PHYTOGEOGRAPHICAL REGION: III.

DESCRIPTION: {896}.

NOTES: differs from all other subspecies by the wings which are longly acute and by its much shorter spur, 6 cm at most; anther broadly indented at the front, its two protrusions being shallower and its retinaculum much narrower and elongate, 3 × 1.5 mm near the tip; central tooth of rostellum 0.6 mm long, very acute and shorter than the wings.

Jumellea lignosa subsp. **latilabia** *H.Perrier* ex *Hermans* **subsp. nov.** TYPE: Madagascar, Andasibe, Onive, *H.Perrier* 17129 (holo. P). See Appendix .

Jumellea lignosa var. *latilabia* H.Perrier in Notul. Syst. (Paris) 7: 60 (1938), *nom. nud.*

DISTRIBUTION: Madagascar: Antananarivo.

HABITAT: humid, evergreen forest.

ALTITUDE: c. 1000 m.

FLOWERING TIME: February.

LIFE FORM: epiphyte.

PHYTOGEOGRAPHICAL REGION: III.

DESCRIPTION: {896}.

NOTES: plants up to 80 cm tall with 12 – 14 cm long, 2.2 – 2.6 cm wide leaves. It differs from the other subspecies in its broadly oval, acute lip with 17 obvious veins and a central keel all along the length of the lip, 8 cm long spur, and column in which the central tooth of the rostellum is almost as long as the wings that are obtuse at the back.

Jumellea lignosa subsp. **tenuibracteata** *H.Perrier* ex *Hermans* **subsp. nov.** TYPE: Madagascar, Mt. Tsaratanana, *H.Perrier* 15331 (holo. P).

Jumellea lignosa var. *tenuibracteata* H.Perrier in Notul. Syst. (Paris) 7: 59 (1938), *nom. nud.*

> DISTRIBUTION: Madagascar: Antsiranana.
> HABITAT: mossy montane forest on moss- and lichen-covered trees.
> ALTITUDE: c. 2000 m.
> FLOWERING TIME: January.
> LIFE FORM: epiphyte.
> PHYTOGEOGRAPHICAL REGION: III, IV.
> DESCRIPTION: {896}.
> NOTES: differs from the typical subspecies by its longer leaves, up to 20 cm long, the sheaths on the peduncle and its thin, tubular floral bracts that are shorter and neither keeled nor flattened.

Jumellea linearipetala *H.Perrier* in Notul. Syst. (Paris) 7: 56 (1938). TYPES: Madagascar, Bezofo, nr. Maromandia, *Decary* 1448 (lecto. P, selected here); Katsory, nr. Maromandia, *Decary* 1457; Manongarivo massif, *H.Perrier* 8049 (both syn. P).

> DISTRIBUTION: Madagascar: Antsiranana, Mahajanga.
> HABITAT: humid forest.
> FLOWERING TIME: February.
> LIFE FORM: epiphyte on trunks.
> PHYTOGEOGRAPHICAL REGION: II.
> DESCRIPTION: {896}.
> ILLUSTRATION: {529}.
> NOTES: characterised by its wide, short leaves and narrow petals; spur 10 – 12 mm long.

Jumellea longivaginans *H.Perrier* in Notul. Syst. (Paris) 7: 56 (1938). TYPE: Madagascar, Mt. Tsaratanana, *H.Perrier* 15701 (holo. P)

Jumellea longivaginans var. *grandis* H.Perrier in Notul. Syst. (Paris) 7: 57 (1938), *nom. nud.* Based upon: Madagascar, Analamazaotra, *Viguier & Humbert* 980 (P).

> DISTRIBUTION: Madagascar: Antsiranana, Toamasina.
> HABITAT: lichen-rich forest and humid, evergreen forest, on moss- and lichen-covered trees.
> ALTITUDE: 900 – 2000 m.
> FLOWERING TIME: January, October.
> LIFE FORM: epiphyte.
> PHYTOGEOGRAPHICAL REGION: II, III, IV.
> DESCRIPTION: {896}.
> NOTES: plant medium-sized; pedicellate ovary 8 cm long; lip lanceolate; spur 9 – 10 cm long. *J. longivaginans* var. *grandis* is a larger form of the species and has a short ridge at the apex of the throat.

Jumellea majalis *(Schltr.) Schltr.* in Beih. Bot. Centralbl. 33 (2): 429 (1915). TYPE: Madagascar, Manongarivo massif, *Schlechter* 88 (*H.Perrier* 1857) (holo. B†; iso. P).

Angraecum majale Schltr. in Ann. Mus. Col. Marseille, sér. 3, 1: 196 (53), t.24 (1913).

> DISTRIBUTION: Madagascar: Antsiranana.
> HABITAT: lichen-rich forest, on moss- and lichen-covered trees.
> ALTITUDE: 1500 – 2000 m.
> FLOWERING TIME: January – May.
> LIFE FORM: epiphyte.
> PHYTOGEOGRAPHICAL REGION: III.
> DESCRIPTION: {896}.
> ILLUSTRATION: {1205}, {1223}.
> NOTES: stem long and pendant; lip pandurate-ligulate, 16 × 5 mm; spur c. 15 cm long.

Jumellea major *Schltr.* in Repert. Spec. Nov. Regni Veg. Beih. 33: 298 (1925). TYPE: Madagascar, Manongarivo massif, *H.Perrier* 8130 (holo. P).

> DISTRIBUTION: Madagascar: Antsiranana.
> HABITAT: mossy, montane forest and river margins on large trees.
> ALTITUDE: c. 1500 m.
> FLOWERING TIME: March, May.
> LIFE FORM: epiphyte.
> PHYTOGEOGRAPHICAL REGION: III.
> DESCRIPTION: {896}.
> ILLUSTRATION: {529}, {1392}.
> NOTES: plant very large; stem long and thick; flowers large; lip 3.5 mm long; spur c. 7 mm long.

Jumellea marojejiensis *H.Perrier* in Notul. Syst. (Paris) 14: 161 (1951). TYPES: Madagascar, summit E of Marojejy massif, *Humbert & Cours* 23645bis (lecto. P, selected here); W of Marojejy, *Humbert* 23645; Mt. Beandroka, Lokoho valley, *Humbert* 23593 (both syn. P).

> DISTRIBUTION: Madagascar: Antsiranana.
> HABITAT: mossy, montane forest.
> ALTITUDE: 1400 – 2000 m.
> FLOWERING TIME: March – April.
> LIFE FORM: epiphyte.
> PHYTOGEOGRAPHICAL REGION: III, IV.
> DESCRIPTION: {901}.
> NOTES: related to *J. cyrtoceras* but differs by the shorter leaves, the more acute, narrower sepals, petals and lip, and filiform spur, slightly shorter than the sepals and lip.

Jumellea maxillarioides *(Ridl.) Schltr.* in Repert. Spec. Nov. Regni Veg. Beih. 33: 299 (1925). TYPES: Madagascar, Ankafana, S Betsileo, *Deans Cowan* s.n. & Imerina, *Parker* s.n. (both syn. BM).
Angraecum maxillarioides Ridl. in J. Linn. Soc., Bot. 21: 479 (1885).
Angorchis maxillarioides (Ridl.) Kuntze in Rev. Gen. Pl. 2: 651 (1891).

> DISTRIBUTION: Madagascar: Antananarivo, Fianarantsoa.
> HABITAT: humid, evergreen forest.
> ALTITUDE: 1200 – 2000 m.
> FLOWERING TIME: January – March.
> LIFE FORM: epiphyte or lithophyte.
> PHYTOGEOGRAPHICAL REGION: III.
> DESCRIPTION: {896}.
> ILLUSTRATION: {246}, {896}, {1324}.
> CULTIVATION: {1333}.
> NOTES: plant robust, stemless or short-stemmed; stem covered by fibrous, disintegrating sheaths; flowers large and waxy; spur c. 4 cm long.

Jumellea ophioplectron *(Rchb.f.) Schltr.* in Beih. Bot. Centralbl. 33 (2): 430 (1915). TYPE: Madagascar, cult. *Low & Co.* s.n. (holo. W).
Aeranthes ophioplectron Rchb.f. in Gard. Chron. 2: 91 (1888).
Mystacidium ophioplectron (Rchb.f.) T.Durand & Schinz, Consp. Fl. Afric. 5: 54 (1892).
Rhaphidorhynchus ophioplectron (Rchb.f.) Poiss., Recherch. Fl. Merid. Madagascar: 185 (1912).

> DISTRIBUTION: Madagascar.
> DESCRIPTION: {896}.
> NOTES: not satisfactorily known, it may well be an *Aerangis* {890}.

Jumellea pachyceras *Schltr.* in Repert. Spec. Nov. Regni Veg. Beih. 33: 299 (1925). TYPES: Madagascar, Andringitra massif, *H.Perrier* 14573 (lecto. P selected here); Mt. Tsiafajavona, *H.Perrier* 13405 & *H.Perrier* 14375 (both syn. P).

> DISTRIBUTION: Madagascar: Antananarivo, Fianarantsoa.

HABITAT: mossy, montane forest on moss- and lichen-rich-covered trees.
ALTITUDE: c. 2000 m.
FLOWERING TIME: December – February.
LIFE FORM: epiphyte.
PHYTOGEOGRAPHICAL REGION: III, IV.
DESCRIPTION: {896}.
ILLUSTRATION: {896}, {1223}.
NOTES: close to *J. hyalina*, but taller, with bigger leaves, longer floral bracts larger flowers, and a
15 mm long spur.

Jumellea pachyra *(Kraenzl.)* H.Perrier in Humbert ed., Fl. Mad. Orch. 2: 177 (1941). TYPE:
Madagascar, Ankafina, S Betsileo, *Hildebrandt* 3988 (holo. W, Reichenbach Herbarium 14895).
Angraecum pachyrum Kraenzl. in Ann. Naturh. Mus. Wein 44: 323 (1930).

DISTRIBUTION: Madagascar: Fianarantsoa.
PHYTOGEOGRAPHICAL REGION: III.
DESCRIPTION: {896}.
NOTES: known only from the type. Close to *J. spathulata*,

Jumellea pandurata *Schltr.* in Beih. Bot. Centralbl. 34 (2): 334 (1916). TYPE: Madagascar, Betafo,
W of Antsirabe, *H.Perrier* 8105 (holo. P).

DISTRIBUTION: Madagascar: Antananarivo.
HABITAT: rocky outcrops.
ALTITUDE: c. 1200 m.
FLOWERING TIME: October.
LIFE FORM: epiphyte or lithophyte.
PHYTOGEOGRAPHICAL REGION: III.
DESCRIPTION: {896}.
ILLUSTRATION: {1223}.
NOTES: plant 20 cm tall; leaves ligulate; lip pandurate-lanceolate; spur c. 12 cm long.

Jumellea papangensis *H.Perrier* in Notul. Syst. (Paris) 7: 57 (1938). TYPE: Madagascar, Mt.
Papanga, Befotaka, *Humbert* 6934 (holo. P).

DISTRIBUTION: Madagascar: Fianarantsoa.
HABITAT: ericaceous scrub.
ALTITUDE: 1300 – 1500 m.
FLOWERING TIME: December.
LIFE FORM: epiphyte.
PHYTOGEOGRAPHICAL REGION: III.
DESCRIPTION: {896}.
ILLUSTRATION: {71}.
NOTES: plant 15 – 25 cm tall; leaves ligulate; lip c. 1.5 cm long; spur 13 cm long.

Jumellea peyrotii *Bosser* in Adansonia, sér. 2, 10 (1): 95 (1970). TYPE: Madagascar, S of
Moramanga, *Bosser & J.P.Peyrot* 16779 (holo. P).

DISTRIBUTION: Madagascar: Toamasina.
HABITAT: humid, mossy, evergreen forest, on moss- and lichen-covered trees.
ALTITUDE: 500 – 1000 m.
FLOWERING TIME: February – March.
LIFE FORM: epiphyte.
PHYTOGEOGRAPHICAL REGION: III.
DESCRIPTION: {115}.
ILLUSTRATION: {115}.
NOTES: vegetatively similar to *J. teretifolia*, but tepals linear, lip longly acuminate, and spur c. 12
cm long.

Jumellea phalaenophora *(Rchb.f.) Schltr.* in Beih. Bot. Centralbl. 33 (2): 430 (1915). TYPE: Comoros, *Humblot* s.n. (holo. W).

Aeranthes phalaenophorus Rchb.f. in Flora 68: 539 (1885).

Mystacidium phalaenophorum (Rchb.f.) T.Durand & Schinz, Consp. Fl. Afric. 5: 54 (1892).

DISTRIBUTION: Comoros: Anjouan.
DESCRIPTION: {896}.
NOTES: poorly known species, recorded only from the type.

Jumellea porrigens *Schltr.* in Repert. Spec. Nov. Regni Veg. Beih. 33: 300 (1925). TYPE: Madagascar, Mt. Tsaratanana, *H.Perrier* 15691 (holo. P).

DISTRIBUTION: Madagascar: Antsiranana.
HABITAT: lichen-rich forest on moss- and lichen-covered trees.
ALTITUDE: c. 2000 m.
FLOWERING TIME: January.
LIFE FORM: epiphyte.
PHYTOGEOGRAPHICAL REGION: III, IV.
DESCRIPTION: {896}.
NOTES: known only from the type. Lip oblanceolate-spathulate; spur c. 11 cm long.

Jumellea punctata *H.Perrier* in Notul. Syst. (Paris) 7: 64 (1938). TYPE: Madagascar, between Moramanga & Anosibe, *H.Perrier* 18289 (holo. P).

DISTRIBUTION: Madagascar: Toamasina.
HABITAT: humid, evergreen forest on trunks.
ALTITUDE: 600 – 1100 m.
FLOWERING TIME: September – October.
LIFE FORM: epiphyte.
PHYTOGEOGRAPHICAL REGION: I, III.
DESCRIPTION: {896}.
NOTES: plant tall; stem sheaths finely flecked with black dots; leaves narrowly linear; lip 3 cm long, oval-lanceolate; spur c. 11 cm long.

Jumellea rigida *Schltr.* in Repert. Spec. Nov. Regni Veg. Beih. 33: 301 (1925). TYPE: Madagascar, Mt. Angavokely, E Imerina, *H.Perrier* 12941 (holo. P).

DISTRIBUTION: Madagascar: Antananarivo.
HABITAT: rocky outcrops.
ALTITUDE: 1500 – 1800 m.
FLOWERING TIME: December – March.
LIFE FORM: epiphyte or lithophyte.
PHYTOGEOGRAPHICAL REGION: III.
DESCRIPTION: {896}.
ILLUSTRATION: {1223}, {1259}, {1027}.
NOTES: similar to *J. porrigens* but with shorter, wider leaves, bigger flowers a different lip, and a 9 – 10 cm spur.

Jumellea rigida var. **altigena** *Schltr.* in Repert. Spec. Nov. Regni Veg. Beih. 33: 301 (1925). TYPE: Madagascar, Andringitra massif, *H.Perrier* 14359 (holo. P).

DISTRIBUTION: Madagascar: Fianarantsoa.
HABITAT: rocky outcrops.
ALTITUDE: c. 2400 m.
FLOWERING TIME: January.
LIFE FORM: lithophyte.
PHYTOGEOGRAPHICAL REGION: IV.
DESCRIPTION: {896}.
NOTES: plant more compact, inflorescence shorter and flowers fleshy.

Jumellea sagittata *H.Perrier* in Notul. Syst. (Paris) 7: 52 (1938). TYPE: Madagascar, Ankeramadinika, *H.Perrier* 18423 (holo. P).

Angraecum gracilipes Rolfe in Curtis Bot. Mag.: t.8758 (1918). TYPE: Madagascar, cult. Messrs. Charlesworth (holo. K).

DISTRIBUTION: Madagascar: Antananarivo.
HABITAT: mossy, montane forest on moss- and lichen-covered trees.
ALTITUDE: c. 1400 m.
FLOWERING TIME: January – February.
LIFE FORM: epiphyte.
PHYTOGEOGRAPHICAL REGION: III.
DESCRIPTION: {896}.
ILLUSTRATION: {3}, {50}, {64}, {204}, {529}, {872}, {896}, {944}, {1027}, {1167}, {1168}, {1259}, {1298}, {1501}.
HISTORY: {1501}.
CULTIVATION: {529}, {1333}.
NOTES: plant stemless; lip lanceolate, 4 × 2 cm; spur 5 – 6 cm long.

Jumellea similis *Schltr.* in Repert. Spec. Nov. Regni Veg. Beih. 33: 302 (1925). TYPE: Madagascar, Manankazo, NE of Ankazobe, *H.Perrier* 11319 (holo. P).

DISTRIBUTION: Madagascar: Antananarivo.
HABITAT: humid, evergreen forest.
ALTITUDE: c. 1500 m.
FLOWERING TIME: November.
LIFE FORM: epiphyte.
PHYTOGEOGRAPHICAL REGION: III.
DESCRIPTION: {896}.
NOTES: similar to *J. confusa* but the leaves narrower and the flowers smaller.

Jumellea spathulata *(Ridl.) Schltr.* in Repert. Spec. Nov. Regni Veg. Beih. 33: 329 (1925). TYPE: Madagascar, Ankafana, S Betsileo, *Hildebrandt* 3988 (holo. BM).

Angraecum spathulatum Ridl. in J. Linn. Soc., Bot. 21: 478 (1885).
Angorchis spathulata (Ridl.) Kuntze, Rev. Gen. Pl. 2: 652 (1891).

DISTRIBUTION: Madagascar: Fianarantsoa.
HABITAT: rocky outcrops.
FLOWERING TIME: March.
LIFE FORM: epiphyte.
PHYTOGEOGRAPHICAL REGION: III.
DESCRIPTION: {281} as *J. francoisii*, {896}.
NOTES: plant up to 30 cm tall; leaves ligulate; lip oval-spathulate, c. 20 mm long; spur 25 mm long.

Jumellea stenoglossa *H.Perrier* in Notul. Syst. (Paris) 14: 162 (1951). TYPE: Madagascar, Manampanihy Valley, Ampasimena, *Humbert* 20566 (holo. P).

DISTRIBUTION: Madagascar: Toliara.
HABITAT: coastal forest.
ALTITUDE: 20 – 100 m.
FLOWERING TIME: March.
LIFE FORM: epiphyte.
PHYTOGEOGRAPHICAL REGION: I.
DESCRIPTION: {901}.
NOTES: plant small; lip 3 cm long, broadened at the base; spur c. 12 mm long.

Jumellea teretifolia *Schltr.* in Repert. Spec. Nov. Regni Veg. Beih. 33: 303 (1925). TYPES: Madagascar, Manankazo, NE of Ankazobe, *H.Perrier* 11317 (lecto. P, selected here); Mt. Vohitrilongo, N Imerina, *H.Perrier* 14944 (syn. P).

DISTRIBUTION: Madagascar: Antananarivo, Fianarantsoa, Toamasina.

HABITAT: humid, evergreen forest.

ALTITUDE: 1100 – 1500 m.

FLOWERING TIME: September – December.

LIFE FORM: epiphyte.

PHYTOGEOGRAPHICAL REGION: III.

DESCRIPTION: {896}.

ILLUSTRATION: {529}, {896}, {1223}.

NOTES: leaves cylindrical; lip narrow and sub-unguiculate at the base, c. 30 mm long; spur c. 13 cm long.

Jumellea zaratananae *Schltr.* in Repert. Spec. Nov. Regni Veg. Beih. 33: 305 (1925). TYPES: Madagascar, Mt. Tsaratanana, *H.Perrier* 15703 (lecto. P, selected here); *H.Perrier* 15320bis (syn. P).

DISTRIBUTION: Madagascar: Antsiranana.

HABITAT: mossy forest on moss- and lichen-covered trees.

ALTITUDE: 600 – 1700 m.

FLOWERING TIME: December – January.

LIFE FORM: epiphyte.

PHYTOGEOGRAPHICAL REGION: II, III.

DESCRIPTION: {896}.

ILLUSTRATION: {3}.

NOTES: plant robust, stemless; lip 9 × 4 mm; spur 9 cm long. The specific name was erroneously changed to *J. tsaratananae* in {877}, {896}.

Lemurella culicifera *(Rchb.f.)* *H.Perrier* in Humbert ed., Fl. Mad. Orch. 2: 334 (1941). TYPE: Comoros, Combani Forest, Mayotte, *Humblot* 378 (1378) (holo. W).

Angraecum culiciferum Rchb.f. in Flora 68: 538 (1885).

Oeonia culicifera (Rchb.f.) Finet in Bull. Soc. Bot. France 54, Mém. 9: 60, t.12 (1907).

Angraecum ambongense Schltr. in Ann. Mus. Col. Marseille, sér. 3, 1: 188 (45), t.21 (1913); Bosser in Adansonia, sér. 2, 10 (3): 368 (1970). TYPE: Madagascar, Manongarivo, Ambongo, *H.Perrier* 967a (holo. B†; iso. P).

Lemurella ambongensis (Schltr.) Schltr. in Repert. Sp. Nov. Regni Veg. Beih. 33: 367 (1925).

Beclardia humbertii H.Perrier in Notul. Syst. (Paris) 14: 160 (1951); & in Adansonia, sér. 2, 10 (3): 368 (1970). TYPE: Madagascar, Analamarina forest (Teheza basin), *Humbert* 19673 (holo. P).

DISTRIBUTION: Madagascar: Mahajanga, Toliara. Comoros: Anjouan, Grande Comore, Mayotte.

HABITAT: seasonally dry, deciduous forest and woodland, on tree trunks.

ALTITUDE: sea level – 700 m.

FLOWERING TIME: October – November.

LIFE FORM: epiphyte.

PHYTOGEOGRAPHICAL REGION: V.

DESCRIPTION: {117}, {896}.

ILLUSTRATION: {281}, {292}, {896}, {1205}, {1259}.

NOTES: the plant looses its leaves during the dry season; lip with the terminal lobe truncate and notched at the tip, with a short apicule in the sinus.

Lemurella pallidiflora *Bosser* in Adansonia, sér. 2, 10 (3): 370 (1970). TYPE: Madagascar, Ambavaniasy, E Perinet, *Bosser* 19803 (holo. P).

DISTRIBUTION: Madagascar: Antsiranana, Toamasina.

HABITAT: mossy, montane forest.

ALTITUDE: 800 – 1200 m.

FLOWERING TIME: January – March.

LIFE FORM: epiphyte.

PHYTOGEOGRAPHICAL REGION: III.

DESCRIPTION: {117}.

ILLUSTRATION: {117}.

NOTES: differs from *L. culicifera* and *L. virescens* in its habit, the larger flowers, its single- to 2-flowered inflorescences and the short peduncle.

Lemurella papillosa *Bosser* in Adansonia, sér. 2, 10 (3): 372 (1970). TYPE: Madagascar, nr. Maroantsetra, *Bosser* 14188 (holo. P).

DISTRIBUTION: Madagascar: Antsiranana, Toamasina.
HABITAT: humid, evergreen forest.
ALTITUDE: 800 – 1000 m.
FLOWERING TIME: January – May.
LIFE FORM: epiphyte.
PHYTOGEOGRAPHICAL REGION: I.
DESCRIPTION: {117}.
ILLUSTRATION: {117}.
NOTES: distinct in its papillose pubescent stems, flowers spaced along the stem, and the slender spur.

Lemurella virescens *H.Perrier* in Notul. Syst. (Paris) 7: 135 (1938). TYPE: Madagascar, Mt. Maromizaha, *H.Perrier* 16085 (holo. P).

DISTRIBUTION: Madagascar: Toamasina.
HABITAT: forest.
ALTITUDE: 1000 – 1200 m.
FLOWERING TIME: February, June.
LIFE FORM: epiphyte on moss- and lichen-rich-covered tree trunks.
PHYTOGEOGRAPHICAL REGION: III.
DESCRIPTION: {896}.
NOTES: inflorescence several-flowered, on a c.3 cm long peduncle; flowers small; spur less than 1 cm long.

Lemurorchis madagascariensis *Kraenzl.* in Bot. Jahrb. Syst. 17: 58 (1893). TYPE: Madagascar, Ankafana, S Betsileo, *Hildebrandt* 4212 (holo. not located).

DISTRIBUTION: Madagascar: Antananarivo, Fianarantsoa.
HABITAT: mossy, montane forest on moss and lichen-covered trees.
ALTITUDE: c. 2000 m.
FLOWERING TIME: February – March.
LIFE FORM: epiphyte.
PHYTOGEOGRAPHICAL REGION: III, IV.
DESCRIPTION: {896}.
ILLUSTRATION: {281}, {645}, {896}, {1259}.
NOTES: plant large, stemless; inflorescence pendant, with up to 100 yellow flowers.

Liparis ambohimangana *Hermans* **nom. nov.**
Liparis monophylla H.Perrier in Notul. Syst. (Paris) 5: 248 (1936), *non* Ames, Orchid. 6: 294 (1920). TYPE: Madagascar, Imerina, Ambohimanga, *Decary* 6222 (holo. P!).

DISTRIBUTION: Madagascar: Antananarivo.
HABITAT: in shady areas by rocks, in clay.
ALTITUDE: 1200 – 1600m.
FLOWERING TIME: March – May.
LIFE FORM: terrestrial.
PHYTOGEOGRAPHICAL REGION: III.
DESCRIPTION: {896}.
ILLUSTRATION: {896}.
NOTES: plant small with a single leaf per pseudobulb with slender stolons; lip broadly oval or suborbicular. The earlier *L. monophylla* Ames (1920), based on a plant from the Philippines, precludes the use of this name for the Madagascan plant described by Perrier. I therefore propose for it the epithet *ambohimangana*, referring to the type locality and once home of Madagascan Royalty.

Liparis andringitrana *Schltr.* in Repert. Sp. Nov. Regni Veg. Beih. 33: 135 (1924). TYPE: Madagascar, Andringitra massif, *H.Perrier* 14396 (holo. P).

DISTRIBUTION: Madagascar: Fianarantsoa.
HABITAT: shaded and humid rocks and rocky outcrops.
ALTITUDE: c. 1800 m.
FLOWERING TIME: February.
LIFE FORM: lithophyte.
PHYTOGEOGRAPHICAL REGION: III.
DESCRIPTION: {896}.
ILLUSTRATION: {1223}.
NOTES: known only from the type. Pseudobulbs compressed; lip obovate-cuneate, with a transversal callus.

Liparis anthericoides *H.Perrier* in Notul. Syst. (Paris) 5: 243 (1936). TYPES: Madagascar, Andasibe, Onive, *H.Perrier* 17104 (lecto. P, selected here); *H.Perrier* 17103 (syn. P).

DISTRIBUTION: Madagascar: Antananarivo, Toamasina.
HABITAT: mossy, montane forest.
ALTITUDE: 900 – 1300 m.
FLOWERING TIME: February – August.
LIFE FORM: epiphyte or terrestrial.
PHYTOGEOGRAPHICAL REGION: III.
DESCRIPTION: {896}.
ILLUSTRATION: {896}.
NOTES: pseudobulbs ovoid; leaves 2, narrowly lanceolate; lip indented in the middle; anther with a straight, acute beak.

Liparis bathiei *Schltr.* in Repert. Sp. Nov. Regni Veg. Beih. 33: 135 (1924). TYPE: Madagascar, Ambatolampy, *H.Perrier* 13548 (holo. P).

DISTRIBUTION: Madagascar: Antananarivo.
HABITAT: in shade of *Acacia*.
ALTITUDE: 1500 – 1600 m.
FLOWERING TIME: February – March.
LIFE FORM: terrestrial.
PHYTOGEOGRAPHICAL REGION: III.
DESCRIPTION: {896}.
ILLUSTRATION: {1223}.
NOTES: pseudobulbs short and sub-globular; flowers small, c. 6 mm across; raceme densely flowered; lip angular at the front; anther with a narrow beak, 0.6 mm long.

Liparis bicornis *Ridl.* in J. Linn. Soc., Bot. 21: 458 (1885). TYPE: Madagascar, Imerina, *Hildebrandt* 3849 (holo. BM).
Leptorkis bicornis (Ridl.) Kuntze, Rev. Gen. Pl. 2: 671 (1891), as *Leptorchis*.

DISTRIBUTION: Madagascar: Antananarivo.
HABITAT: marshes; rocky outcrops.
ALTITUDE: c. 1400 m.
FLOWERING TIME: January – March.
LIFE FORM: lithophyte or terrestrial.
PHYTOGEOGRAPHICAL REGION: III.
DESCRIPTION: {896}.
ILLUSTRATION: {896}.
NOTES: pseudobulbs elongate; leaves lanceolate, acute; disk of the lip with a small 2-horned callus at the base.

Liparis bulbophylloides *H.Perrier* in Notul. Syst. (Paris) 5: 242 (1936). TYPE: Madagascar, confl. Onive & Mangoro, *H.Perrier* 16998 (holo. P).

DISTRIBUTION: Madagascar: Toamasina.
HABITAT: humid, evergreen forest.
ALTITUDE: 500 – 1000 m.
FLOWERING TIME: February.
LIFE FORM: epiphyte.
PHYTOGEOGRAPHICAL REGION: I.
DESCRIPTION: {896}.
ILLUSTRATION: {896}.
NOTES: plant similar to a small *Bulbophyllum*; leaves broadly oval, small; lip tri-sinuate; anther with a wide beak.

Liparis caespitosa *(Lam.) Lindl.* in Bot. Reg.: sub t.882 (1824); Rasmussen in Bot. Notul. 132: 385 – 392 (1979). TYPE: Réunion, Herb. *Jussieu* s.n. (holo. P- LA).
Epidendrum cespitosum Lam., Encycl. Meth. Bot. 1: 187 (1783).
Stichorkis cestichis Thouars, Cahier de six planches t.r1 (1819).
Malaxis angustifolia Blume, Bijdr.: 393 (1825). TYPES from Java.
Liparis angustifolia (Blume) Lindl., Gen. Sp. Orch. Pl: 31 (1830).
Liparis auriculata Rchb.f. in Flora 55: 277 (1872), *non* Miq., *nom. illegit.* TYPE from India.
Liparis pusilla Ridl. in J. Linn. Soc. 22: 294 (1886). TYPE as *L. auriculata.*
Leptorkis caespitosa (Lam.) Kuntze, Rev. Gen. Pl. 2: 671 (1891), as *Leptorchis.*
*Liparis comos*a Ridl. in J. Linn. Soc. 32: 229 (1896). TYPE from Malaya.
Cestichis caespitosa (Lam.) Ames, Orch., 2: 132 (1908).

DISTRIBUTION: Madagascar: Toamasina. Comoros: Grande Comore. Mascarenes: Réunion, Mauritius. Also in East Africa, Sri Lanka, NE India to the Philippines, New Guinea, Solomon Is. and Fiji.
HABITAT: mossy, evergreen forest.
ALTITUDE: 400 – 1400 m.
FLOWERING TIME: February.
LIFE FORM: epiphyte or terrestrial.
PHYTOGEOGRAPHICAL REGION: III.
DESCRIPTION: {896}.
ILLUSTRATION: {676}, {1398}.
NOTES: plant less than 9 cm tall; pseudobulbs single-leaved; leaf erect; flower small; sepals less than 6 mm long; lip lanceolate and recurved.

Liparis cladophylax *Schltr.* in Beih. Bot. Centralbl. 34 (2): 321 (1916). TYPE: Madagascar, high Sambirano Valley, *H.Perrier* 11360 (139) (holo. B†; iso. P).

DISTRIBUTION: Madagascar: Antsiranana.
HABITAT: humid, evergreen forest.
ALTITUDE: c. 800 m.
FLOWERING TIME: December.
LIFE FORM: epiphyte.
PHYTOGEOGRAPHICAL REGION: II.
DESCRIPTION: {896}.
NOTES: known only from the type. Pseudobulbs often with 3 leaves; lip with a transversal callus at the base; and with 2 lateral keels higher; anther clearly truncate at the front.

Liparis clareae *Hermans* nom. nov.
Liparis cardiophylla H.Perrier in Notul. Syst. (Paris) 5: 244 (1936), *non* Ames, Orchid. 3: 92 (1908).
  TYPE: Madagascar, Mandraka Gorge, *E.François* in *H.Perrier* 17797 (holo. P).

DISTRIBUTION: Madagascar: Antananarivo, Fianarantsoa, Mahajanga, Toamasina.
HABITAT: mossy, montane forest.
ALTITUDE: 800 – 1500 m.

FLOWERING TIME: February – May.
LIFE FORM: epiphyte or terrestrial.
PHYTOGEOGRAPHICAL REGION: III.
DESCRIPTION: {896}.
NOTES: leaves almost heart-shaped; lip 7 × 4 mm, auriculate at the base, the front margin 3-toothed in the middle, lacking a callus. The earlier use of *Liparis cardiophylla* Ames (1908), based on a plant from Jamaica, precludes the use of this name for the Madagascan plant described by Perrier. I therefore propose the epithet *clareae*, in honour of my wife who has contributed so much to our study of the Orchid Flora of Madagascar, for the Madagascan species. *Liparis cardiophylla* H.Perrier var. *angustifolia* H.Perrier was described after 1st January 1935 and, lacking a Latin diagnosis, is therefore invalid under the ICBN. See Appendix.

Liparis clareae var. **angustifolia** *H.Perrier ex Hermans* **var. nov.** TYPE: Madagascar, area of the confluence of the Mangoro & Onive, Eastern forest, c. 700 m, Feb. 1925 *H.Perrier* 17018 (holo. P). See Appendix .

Liparis cardiophylla var. *angustifolia* H.Perrier in Notul. Syst. (Paris) 5: 244 (1936), *nom. nud.*

DISTRIBUTION: Madagascar: Toamasina.
HABITAT: humid, evergreen forest on moss- and lichen-covered trees.
ALTITUDE: c. 700 m.
FLOWERING TIME: February.
LIFE FORM: epiphyte or terrestrial.
PHYTOGEOGRAPHICAL REGION: I.
SUBSTRATE: epiphyte.
DESCRIPTION: {896}.
NOTES: it differs from the typical variety in having narrower leaves, more numerous flowers (3 – 5), longer sepals, petals with 2 veins, a larger lip, with a rectangular blade and a disk with a pronounced callus at the base, a taller 5.5 mm high column and a smaller anther, the beak being wider and more obtuse.

Liparis danguyana *H.Perrier* in Notul. Syst. (Paris) 5: 248 (1936). TYPE: Madagascar, Mt. Angavokely, E of Antananarivo, *H.Perrier* 16064 (holo. P).

DISTRIBUTION: Madagascar: Antananarivo.
HABITAT: mossy, montane forest.
ALTITUDE: 1500 – 1800 m.
FLOWERING TIME: March.
LIFE FORM: epiphyte or terrestrial.
PHYTOGEOGRAPHICAL REGION: III.
DESCRIPTION: {896}.
NOTES: related to *L. listeroides* but with wider leaves, a rounded lip, apiculate at the front and with a different callus and an anther without a beak.

Liparis densa *Schltr.* in Repert. Sp. Nov. Regni Veg. Beih. 33: 137 (1924). TYPE: Madagascar, Andringitra massif, *H.Perrier* 14358 (holo. P).

DISTRIBUTION: Madagascar: Fianarantsoa, Toliara.
HABITAT: shaded and humid rocks.
ALTITUDE: 1500 – 2400 m.
FLOWERING TIME: January.
LIFE FORM: lithophyte or terrestrial.
PHYTOGEOGRAPHICAL REGION: IV.
DESCRIPTION: {896}.
ILLUSTRATION: {896}, {1223}.
NOTES: pseudobulbs oblong, compressed-ancipitate; raceme much longer than the peduncle; lip without a callus.

Liparis disticha *(Thouars) Lindl.* in Bot. Reg.: sub t.882 (1824). TYPE: Réunion, Thouars 17 (holo. P).
Malaxis disticha Thouars, Hist. Orch.: t.89 (1822).
Liparis gregaria Lindl., Gen. Sp. Orchid. Pl.: 33 (1830).
Cestichis disticha (Thouars) Pfitzer in Engler & Prantl, Nat. Pflanzenfam. 2(6): 131 (1888).
Leptorkis disticha (Thouars) Kuntze, Revis. Gen. Pl. 2: 671 (1891).
Stichorkis disticha (Thouars) Pfitzer in Engler & Prantl., Nat. Pflanzenfam, Nachtr. 1: 103 (1897).
Stelis micrantha Sieb. in Herb. Maur. 168 (iso. C).

DISTRIBUTION: Mascarenes: Réunion, Mauritius. Comoros: Grande Comore.
HABITAT: humid forest, in scrub.
ALTITUDE: 500 – 830 m.
FLOWERING TIME: November – July.
LIFE FORM: epiphyte.
DESCRIPTION: {826}.
ILLUSTRATION: {167}, {1398}.
NOTES: vegetatively similar to the Asian species of this group but the lip is distinct as is the shape of the column.

Liparis dryadum *Schltr.* in Repert. Sp. Nov. Regni Veg. Beih. 33: 138 (1924). TYPE: Madagascar, Mt. Tsaratanana, *H.Perrier* 15245 (holo. P).

DISTRIBUTION: Madagascar: Antsiranana.
HABITAT: lichen-rich forest.
ALTITUDE: 1200 – 1600 m.
FLOWERING TIME: January.
LIFE FORM: epiphyte or lithophyte.
PHYTOGEOGRAPHICAL REGION: III.
DESCRIPTION: {896}.
ILLUSTRATION: {896}.
NOTES: known only from the type. Plant with 3 rather big leaves, up to 30 cm long; inflorescence shorter than the leaves; disk with a short and bilobed callus near the base.

Liparis flavescens *(Thouars) Lindl.* in Bot. Reg.: sub t.882 (1825). TYPE: Mascarene Isles, *Thouars* s.n. (holo. P).
Malaxis flavescens Thouars, Hist. Orch.: t.25 (1822).
Leptorkis flavescens (Thouars) Kuntze, Rev. Gen. Pl. 2: 671 (1891), as *Leptorchis*.

DISTRIBUTION: Madagascar: Antananarivo. Comoros: Anjouan. Mascarenes: Réunion, Mauritius. Seychelles.
HABITAT: forest.
FLOWERING TIME: June.
PHYTOGEOGRAPHICAL REGION: III.
DESCRIPTION: {896}.
ILLUSTRATION: {167}, {1398}.
NOTES: leaves plicate, suberect; inflorescence longer than the leaves; lip rounded, crenulate and cuspidate.

Liparis gracilipes *Schltr.* in Repert. Sp. Nov. Regni Veg. Beih. 33: 138 (1924). TYPE: Madagascar, Mt. Tsaratanana, *H.Perrier* 15747 (holo. P).

DISTRIBUTION: Madagascar: Antsiranana.
HABITAT: mossy, montane forest; river margins; and shaded and humid rocks.
ALTITUDE: c. 1800 m.
FLOWERING TIME: January.
LIFE FORM: lithophyte or terrestrial.
PHYTOGEOGRAPHICAL REGION: III.
DESCRIPTION: {896}.
ILLUSTRATION: {896}.
NOTES: known from the type only. Pseudobulbs stem-like, less than 13 cm long; leaves c. 7 cm long; lip 6.5 × 7.5 mm, wider than high, the disk with a 3-toothed callus near the base.

Liparis henricii *Schltr.* in Repert. Sp. Nov. Regni Veg. Beih. 33: 139 (1924). TYPE: Madagascar, Andringitra massif, *H.Perrier* 14395 (holo. P).

Liparis latilabris Schltr. in Repert. Sp. Nov. Regni Veg. Beih. 33: 142 (1924). TYPE: Madagascar, Andringitra massif, *H.Perrier* 14394 (holo. P).

Liparis verecunda Schltr. in Repert. Sp. Nov. Regni Veg. Beih. 33: 150 (1924). TYPE: Madagascar, Ambatofianandrano, *H.Perrier* 12420 (holo. P).

DISTRIBUTION: Madagascar: Antananarivo, Fianarantsoa.
HABITAT: shaded and humid rocks.
ALTITUDE: 800 – 1800 m.
FLOWERING TIME: January – March.
LIFE FORM: lithophyte or terrestrial.
PHYTOGEOGRAPHICAL REGION: III.
DESCRIPTION: {887}, {896}.
ILLUSTRATION: {1223}.
NOTES: pseudobulbs elongate, 2 – 4 cm long; lip 5 × 4 mm, rounded at the front; anther with a short triangular beak.

Liparis imerinensis *Schltr.* in Repert. Sp. Nov. Regni Veg. Beih. 33: 141 (1924). TYPE: Madagascar, nr. Manjakandriana, E Imerina, *H.Perrier* 12923 (holo. P).

DISTRIBUTION: Madagascar: Antananarivo.
HABITAT: mossy, montane forest.
ALTITUDE: 700 – 1500 m.
FLOWERING TIME: December – March.
LIFE FORM: epiphyte, lithophyte or terrestrial.
PHYTOGEOGRAPHICAL REGION: III.
DESCRIPTION: {896}.
VERNACULAR NAME: Famany.
NOTES: pseudobulbs elongate, 4 – 6 cm long; lip oval-rhombic, acute at the tip.

Liparis jumelleana *Schltr.* in Beih. Bot. Centralbl. 34 (2): 320 (1916). TYPE: Madagascar, Sambirano valley, *H.Perrier* 11366 (133) (holo. B†; iso. P).

DISTRIBUTION: Madagascar: Antsiranana.
HABITAT: humid, evergreen forest.
ALTITUDE: c. 800 m.
FLOWERING TIME: December.
LIFE FORM: epiphyte or terrestrial.
PHYTOGEOGRAPHICAL REGION: II.
DESCRIPTION: {896}.
ILLUSTRATION: {896}, {1223}.
NOTES: pseudobulbs at an angle to one another; lip much wider than long, 7 – 8 × 10 – 11 mm, shortly acuminate at the front; callus very thick and very obtuse.

Liparis listeroides *Schltr.* in Repert. Sp. Nov. Regni Veg. Beih. 33: 143 (1924). TYPE: Madagascar, Andringitra massif, *H.Perrier* 14440 (holo. P).

DISTRIBUTION: Madagascar: Antananarivo, Antsiranana, Fianarantsoa.
HABITAT: mossy, montane forest.
ALTITUDE: 1250 – 1500 m.
FLOWERING TIME: January- April.
LIFE FORM: epiphyte or terrestrial.
PHYTOGEOGRAPHICAL REGION: III.
DESCRIPTION: {896}.
ILLUSTRATION: {1223}.
NOTES: a variable species; leaves oval-lanceolate; lip calli elongate, horn-shaped, oblique at the front; anther with a short, obtuse beak.

Liparis longicaulis *Ridl.* in J. Linn. Soc., Bot. 21: 461 (1885). TYPES: Madagascar, Ankafana, *Hildebrandt* 3979 (lecto. BM, selected here); *Deans Cowan* s.n. (syn. BM, P).
Leptorkis longicaulis (Ridl.) Kuntze in Rev. Gen. Pl. 2: 671 (1891).

> DISTRIBUTION: Madagascar: Antananarivo, Antsiranana, Toamasina, Toliara.
> HABITAT: mossy, montane forest.
> ALTITUDE: 700 – 1800 m.
> FLOWERING TIME: January – April.
> LIFE FORM: epiphyte or terrestrial.
> PHYTOGEOGRAPHICAL REGION: III.
> DESCRIPTION: {896}.
> ILLUSTRATION: {246}, {896}.
> NOTES: pseudobulbs stem-like, up to 60 cm long, angular; lip almost 3-lobed; base of the lip with a big transversal callus on top, indented at the front.

Liparis longipetala *Ridl.* in J. Linn. Soc., Bot. 21: 459 (1885). TYPE: Madagascar, Ankafana, S Betsileo, *Hildebrandt* 3980 (holo. BM; iso. K).
Leptorkis longipetala (Ridl.) Kuntze, Rev. Gen. Pl. 2: 671 (1891).

> DISTRIBUTION: Madagascar: Fianarantsoa.
> HABITAT: humid, evergreen forest.
> ALTITUDE: c. 1500 m.
> FLOWERING TIME: March.
> LIFE FORM: terrestrial.
> PHYTOGEOGRAPHICAL REGION: III.
> DESCRIPTION: {896}.
> NOTES: leaves 6 – 10 × 1.2 cm; lip oblong, very shortly apiculate, the disk with a small obtuse callus; anther with an acute beak.

Liparis lutea *Ridl.* in J. Linn. Soc., Bot. 21: 458 (1885). TYPE: Madagascar, Ankafana, *Deans Cowan* s.n. (holo. BM).
Leptorkis lutea (Ridl.) Kuntze, Rev. Gen. Pl. 2: 671 (1891).

> DISTRIBUTION: Madagascar: Fianarantsoa.
> HABITAT: marshes.
> ALTITUDE: c. 1500 m.
> LIFE FORM: terrestrial.
> PHYTOGEOGRAPHICAL REGION: III.
> DESCRIPTION: {896}.
> ILLUSTRATION: {246}.
> NOTES: known from the type only. Leaves 5 × 1.2 cm; flowers small; lip oval-cordate, obtuse.

Liparis microcharis *Schltr.* in Repert. Sp. Nov. Regni Veg. Beih. 33: 144 (1924). TYPE: Madagascar, Mt. Vohitrakidahy, W Betsileo, *H.Perrier* 12423 (holo. P).

> DISTRIBUTION: Madagascar: Fianarantsoa.
> HABITAT: on humid rocky piles.
> ALTITUDE: c. 2000 m.
> FLOWERING TIME: February.
> LIFE FORM: lithophyte or terrestrial.
> PHYTOGEOGRAPHICAL REGION: III, IV.
> DESCRIPTION: {896}.
> NOTES: known from the type only. Small plant, inflorescence at least 2 times longer than the leaves; callus on the disk crescent-shaped.

Liparis nephrocardia *Schltr.* in Repert. Sp. Nov. Regni Veg. Beih. 33: 145 (1924). TYPE: Madagascar, Mt. Tsaratanana, *H.Perrier* 15746 (holo. P).

> DISTRIBUTION: Madagascar: Antsiranana, Toamasina.
> HABITAT: humid, mossy, evergreen forest.
> ALTITUDE: 600 – 1600m.

FLOWERING TIME: December – March.
LIFE FORM: epiphyte or terrestrial.
PHYTOGEOGRAPHICAL REGION: I, II, III.
DESCRIPTION: {896}.
NOTES: pseudobulbs conical; lip expanded into a wide, subcordate blade and rounded at the front; callus tri-sinuate.

Liparis ochracea *Ridl.* in J. Linn. Soc., Bot. 21: 461 (1885). TYPE: Madagascar, Ankafana, *Deans Cowan* s.n. (holo. BM).
Liparis connata Ridl. in J. Linn. Soc., Bot. 21: 462 (1885). TYPE: Madagascar, Imerina, *Deans Cowan* s.n. (holo. BM).
Leptorkis ochracea (Ridl.) Kuntze, Rev. Gen. Pl. 2: 671 (1891), as *Leptorchis*.
Leptorkis connata (Ridl.) Kuntze, *loc.cit.*, as *Leptorchis*.
Liparis hildebrandtiana Schltr. in Repert. Sp. Nov. Regni Veg. Beih. 33: 140 (1924). TYPE: Madagascar, Mt. Tsiafajavona, *H.Perrier* 13511 (holo. P).

DISTRIBUTION: Madagascar: Antananarivo, Antsiranana, Fianarantsoa, Toamasina.
HABITAT: humid, evergreen forest; lichen-rich forest.
ALTITUDE: 700 – 2000 m.
FLOWERING TIME: February – March.
LIFE FORM: epiphyte or terrestrial.
PHYTOGEOGRAPHICAL REGION: I, III, IV.
DESCRIPTION: {896}.
ILLUSTRATION: {246}, {1223}.
NOTES: a very variable species; pseudobulbs elongate, up to 7 cm; leaves petiolate; lip elongate, 12 – 15 mm long, the disk with 2 very obvious calli.

Liparis ornithorrhynchos *Ridl.* in J. Linn. Soc., Bot. 21: 460 (1885). TYPES: Madagascar, Ankafana (S Betsileo), *Hildebrandt* 4049 (lecto. P; isolecto. K, both selected here); *Deans Cowan* s.n. (syn. BM).
Leptorchis ornithorrhynchos (Ridl.) Kuntze, Rev. Gen. Pl. 2: 671 (1891), as *Leptorchis*.

DISTRIBUTION: Madagascar: Fianarantsoa.
HABITAT: humid areas near evergreen forest, at the base of or low down on trees .
FLOWERING TIME: January – February.
LIFE FORM: terrestrial or epiphyte.
PHYTOGEOGRAPHICAL REGION: III.
DESCRIPTION: {896}.
ILLUSTRATION: {246}.
NOTES: plant up to 30 cm tall; lip 7 – 10 mm without a callus, broadly cordate, the base narrow; anther with a long beak.

Liparis panduriformis *H.Perrier* in Notul. Syst. (Paris) 5: 253 (1936). TYPE: Madagascar, Manongarivo massif, *H.Perrier* 1946 (holo. P).

DISTRIBUTION: Madagascar: Antsiranana.
HABITAT: shaded and humid rocks.
ALTITUDE: 500 – 800 m.
FLOWERING TIME: February – May.
LIFE FORM: lithophyte.
PHYTOGEOGRAPHICAL REGION: II.
DESCRIPTION: {896}.
NOTES: plant up to 30 cm tall; lip broadly 3-lobed, pandurate, the lateral lobes small, angular-obtuse and recurved towards the base, the front edge crenulate-dentate; callus large, bilobed and thick.

Liparis parva *(Kuntze) Ridl.* in J. Linn. Soc., Bot. 21: 462 (1885). TYPE: Madagascar, Ankafana, *Deans Cowan* s.n. (holo. BM).
Leptorkis parva Kuntze, Rev. Gen. Pl. 2: 671 (1891).

DISTRIBUTION: Madagascar: Antananarivo, Fianarantsoa.

HABITAT: mossy, montane forest on moss- and lichen-covered trees.
ALTITUDE: c. 1500 m.
FLOWERING TIME: February.
LIFE FORM: epiphyte.
PHYTOGEOGRAPHICAL REGION: III.
DESCRIPTION: {896}.
ILLUSTRATION: {246}, {896}.
NOTES: plant small, 8 – 10 cm tall; disc of the lip with a big subrectangular callus, erect, thick, near the base; anther obtuse at the front.

Liparis perrieri *Schltr.* in Ann. Mus. Col. Marseille, sér. 3, 1: 164 (21), t.7 (1913). TYPE: Madagascar, Firingalava, Boina, *H.Perrier* 499 (iso. P).

DISTRIBUTION: Madagascar: Antsiranana, Mahajanga, Toliara.
HABITAT: humid, evergreen forest and shaded and humid rock.
ALTITUDE: sea level – 600 m.
FLOWERING TIME: September, December – March.
LIFE FORM: lithophyte.
PHYTOGEOGRAPHICAL REGION: II, V.
DESCRIPTION: {896}.
ILLUSTRATION: {1205}, {1223}.
NOTES: plant up to 40 cm tall; raceme carrying many small flowers; lip reddish, with 2 small calli at the base.

Liparis perrieri var. **trinervia** *H.Perrier* ex *Hermans* **var. nov.** TYPE: Madagascar, confl. Onive & Mangoro, *H.Perrier* 17136 (holo. P). See Appendix .
Liparis perrieri var. *trinervia* H.Perrier in Notul. Syst. (Paris) 5: 251 (1936), *nom. nud.*

DISTRIBUTION: Madagascar: Toamasina.
HABITAT: in the Savoka, humid, evergreen forest.
ALTITUDE: c. 700 m.
FLOWERING TIME: February.
LIFE FORM: epiphyte or terrestrial.
PHYTOGEOGRAPHICAL REGION: I.
DESCRIPTION: {896}.
NOTES: differs from the typical variety in having lateral sepals with 5 veins, a 3-veined lip that is shortly apiculate in the middle of the front indentation, the middle vein being unbranched, in having a distinct, very pronounced, conical callus, and an anther with a subrectangular front lobe.

Liparis puncticulata *Ridl.* in J. Linn. Soc., Bot. 22: 119 (1886). TYPE: Madagascar, *Baron* 4334 (holo. BM; iso. K, P).
Leptorkis puncticulata (Ridl.) Kuntze, Rev. Gen. Pl. 2: 671 (1891).

DISTRIBUTION: Madagascar: Antananarivo, Antsiranana, Toamasina.
HABITAT: mossy, montane forest.
ALTITUDE: 800 – 2000m.
FLOWERING TIME: January – May.
LIFE FORM: epiphyte or terrestrial.
PHYTOGEOGRAPHICAL REGION: III.
DESCRIPTION: {896}.
ILLUSTRATION: {896}.
NOTES: pseudobulbs stem-like; leaves oblong-lanceolate; sheaths finely punctuate; anther with an obtuse beak.

Liparis rectangularis *H.Perrier* in Notul. Syst. (Paris) 5: 254 (1936). TYPE: Madagascar, Mt. Maromizaha, nr. Analamazaotra, *H.Perrier* 16043 (holo. P).

DISTRIBUTION: Madagascar: Toamasina.

HABITAT: mossy, montane forest on moss- and lichen-covered trees.
ALTITUDE: c. 1000 m.
FLOWERING TIME: February.
LIFE FORM: epiphyte.
PHYTOGEOGRAPHICAL REGION: III.
DESCRIPTION: {896}.
NOTES: known from the type only. Pseudobulbs rounded and short; lip abruptly expanded towards the upper third in a subrectangular transverse blade, subcordate at theedge, the disk with 2 small, obsolete swelling at the base.

Liparis rivalis *Schltr.* in Repert. Sp. Nov. Regni Veg. Beih. 33: 146 (1924). TYPE: Madagascar, Mt. Tsaratanana, *H.Perrier* 15244 (holo. P).

DISTRIBUTION: Madagascar: Antsiranana.
HABITAT: mossy, montane forest; river margins.
ALTITUDE: 1200 – 1600 m.
FLOWERING TIME: December – January.
LIFE FORM: terrestrial.
PHYTOGEOGRAPHICAL REGION: III, IV.
DESCRIPTION: {896}.
ILLUSTRATION: {896}.
NOTES: pseudobulbs stem-like, 20 cm long, raceme with at least 25 flowers, lip broadly obovate, 5.5 × 4.5 mm, very obtuse at the front, at the margins a little crenulate; with a small shallow callus near the base.

Liparis salassia *(Pers.) Summerh.* in Kew Bull. 8, 133 (1953). TYPE: without known provenance, *Jussieu* s.n. (holo. P not located).
Epipactis salassia Pers. Syn. Pl. 2, 513 (1807).
Serapias salassia (Pers.) Steud., Nomencl. Bot. ed. 1, 767 (1821).
Malaxis purpurascens Thouars, Hist. Orchid.: tt. 26 & 27 (1822).
Liparis purpurascens (Thouars) Lindl. in Edwards' Bot. Reg. 11, sub. t.882 (1825).
Neottia salassia (Pers.) Steud. Nomencl. Bot. ed. 2. 2, 189 (1841).
Leptorkis purpurascens (Thouars) Kuntze, Rev. Gen. 2: 670 (1891).
Ophrys salassia Herb. Commers.; A.Rich., Monogr. Orch. Iles France & Bourbon, 53 (1828).

DISTRIBUTION: Madagascar: Antsiranana, Mahajanga. Mascarenes: Réunion, Mauritius.
HABITAT: humid, evergreen forest.
ALTITUDE: 1000 – 1200 m.
FLOWERING TIME: December – April, September.
LIFE FORM: epiphyte or terrestrial.
PHYTOGEOGRAPHICAL REGION: III, IV.
DESCRIPTION: {896}.
ILLUSTRATION: {281}, {1398}.
NOTES: plant reddish or violet; raceme short, subcorymbose; flowers purple-violet.

Liparis sambiranoensis *Schltr.* in Repert. Sp. Nov. Regni Veg. Beih. 33: 147 (1924). TYPE: Madagascar, Sambirano Valley, *H.Perrier* 15194 (holo. P).

DISTRIBUTION: Madagascar: Antsiranana. Comoros: Mayotte.
HABITAT: coastal forest; humid, evergreen forest.
ALTITUDE: 200 – 500 m.
FLOWERING TIME: October, December – February.
LIFE FORM: epiphyte or terrestrial.
PHYTOGEOGRAPHICAL REGION: II.
DESCRIPTION: {896}.
NOTES: pseudobulbs narrow and elongate, 8 – 10 cm long; lip abruptly widened in the middle, the front part broadly obovate-suborbicular; callus big, broadened and bi-gibbose at the front.

Liparis stenophylla *Schltr.* in Repert. Sp. Nov. Regni Veg. Beih. 33: 148 (1924). TYPES: Madagascar, Manongarivo massif, *H.Perrier* 8048 (lecto. P, selected here); *H.Perrier* 8045 (syn. P).

DISTRIBUTION: Madagascar: Antsiranana.
HABITAT: humid, evergreen forest.
ALTITUDE: 400 – 800 m.
FLOWERING TIME: April.
LIFE FORM: epiphyte or terrestrial.
PHYTOGEOGRAPHICAL REGION: II, III.
DESCRIPTION: {896}.
NOTES: pseudobulbs 2-leaved; leaves linear, 12 – 16 cm × 8 – 9 mm; lip 5 × 2.5 mm, obovate-suborbicular, with 2 small calli above the base.

Liparis trulliformis *Schltr.* in Repert. Sp. Nov. Regni Veg. Beih. 33: 149 (1924). TYPE: Madagascar, Mt. Tsaratanana, *H.Perrier* 15744 (holo. P).

DISTRIBUTION: Madagascar: Antsiranana.
HABITAT: mossy, montane forest on moss- and lichen-covered trees.
ALTITUDE: c. 1800 m.
FLOWERING TIME: January.
LIFE FORM: epiphyte.
PHYTOGEOGRAPHICAL REGION: III.
DESCRIPTION: {896}.
NOTES: known from the type only. Plant small, up to 12 cm tall; pseudobulbs short, up to 2 cm long; lip trullate and thick.

Liparis warpurii *Rolfe* in Kew Bull. 1908: 69 (1908). TYPE: Madagascar, *Warpur* s.n. (holo. K).

DISTRIBUTION: Madagascar.
DESCRIPTION: {896}.
NOTES: described from plant cultivated at Kew; related to *L. parva*, but flowers bigger and lip crenulate.

Liparis xanthina *Ridl.* in J. Linn. Soc., Bot. 22: 275 (1886). TYPE: Madagascar, Ankafana (S Betsileo), *Deans Cowan* s.n. (holo. BM).
Leptorkis xanthina (Ridl.) Kuntze, Rev. Gen. Pl. 2: 671 (1891).

DISTRIBUTION: Madagascar: Fianarantsoa.
HABITAT: humid, evergreen forest.
ALTITUDE: c. 1500 m.
LIFE FORM: epiphyte.
PHYTOGEOGRAPHICAL REGION: III.
DESCRIPTION: {896}.
ILLUSTRATION: {246}.
NOTES: known only from the type. Pseudobulbs oval-conical; flowers small; lip oblong and obtuse.

Liparis zaratananae *Schltr.* in Repert. Sp. Nov. Regni Veg. Beih. 33: 151 (1924). TYPE: Madagascar, Mt. Tsaratanana, *H.Perrier* 15743 (holo. P).

DISTRIBUTION: Madagascar: Antsiranana.
HABITAT: mossy, montane forest.
ALTITUDE: 1500 – 2000 m.
FLOWERING TIME: April, October, December.
LIFE FORM: epiphyte.
PHYTOGEOGRAPHICAL REGION: III.
DESCRIPTION: {896}.
ILLUSTRATION: {896}.
NOTES: plant up to 20 cm tall; leaves broadly oval-acute, rounded at the base and petiolate; lip

lacking a callus; anther extended by a 1 mm long, narrow beak. Perrier de la Bâthie {896} changed the species name to *L. tsaratananae* believing Schlechter to have made a typographic error, but there is no basis for this.

Malaxis atrorubra *(H.Perrier) Summerh.* in Kew Bull. 8: 578 (1953). TYPE: Madagascar, Manerinerina, Tampoketsa, *H.Perrier* 16828 (holo. P).
Microstylis atro-ruber H.Perrier in Notul. Syst. (Paris) 5: 233 (1936).

DISTRIBUTION: Madagascar: Antananarivo.
HABITAT: humid, evergreen forest on plateau.
ALTITUDE: c. 1500 m.
FLOWERING TIME: October.
LIFE FORM: epiphyte or terrestrial.
PHYTOGEOGRAPHICAL REGION: III.
DESCRIPTION: {896}.
ILLUSTRATION: {896}.
NOTES: known only from the type. Raceme with about 8 flowers; flowers dark purple; disk of the lip carrying 2 calli.

Malaxis cardiophylla *(Rchb.f.) Kuntze*, Rev. Gen. Pl. 2: 673 (1891). TYPE: Comoros, without locality, *Humblot* 437 (holo. W; iso. BM, P).
Microstylis cardiophylla Rchb.f. in Flora 68: 543 (1885).

DISTRIBUTION: Comoros: Grande Comore.
HABITAT: low-elevation, humid, evergreen forest.
FLOWERING TIME: February.
LIFE FORM: lithophyte.
DESCRIPTION: {896}.
NOTES: leaves almost heart-shaped; peduncle small; rostellum flanked by 2 ascending, triangular, acute lobules.

Malaxis françoisii *(H.Perrier) Summerh.* in Kew Bull., 8: 578 (1953). TYPE: Madagascar, Mandraka, E Imerina, *E.François* in *H.Perrier* 17585 (holo. P).
Microstylis françoisii H.Perrier in Notul. Syst. (Paris) 5: 234 (1936).

DISTRIBUTION: Madagascar: Antananarivo.
HABITAT: humid, evergreen forest on plateau.
ALTITUDE: c. 1200 m.
FLOWERING TIME: February.
LIFE FORM: epiphyte or terrestrial.
PHYTOGEOGRAPHICAL REGION: III.
DESCRIPTION: {896}.
NOTES: known from the type only. Raceme with at least 15 flowers; lip broadly orbicular, carrying 2 small dot-like calli near the base.

Malaxis madagascariensis *(Klinge) Summerh.* in Kew Bull. 8: 578 (1953). TYPE: Madagascar, *Hamelin* s.n. (holo. LE).
Microstylis madagascariensis Klinge in Acta Hort. Petrop. 17 (1): 140, t.2 (1899).

DISTRIBUTION: Madagascar.
DESCRIPTION: {896}.
ILLUSTRATION: {633}.
NOTES: plant 20 – 30 cm tall; flowers violet; lip clearly 3-lobed, the lobes obtuse, the mid-lobe a little bigger.

Malaxis physuroides *(Schltr.) Summerh.* in Kew Bull., 8: 578 (1953). TYPE: Madagascar, Manongarivo, *H.Perrier* 87 (8010) (holo. B†; iso. P).
Microstylis physuroides Schltr. in Ann. Mus. Col. Marseille, sér. 3, 1: 164 (21), t.7 (1913).

DISTRIBUTION: Madagascar: Antsiranana, Toamasina.

HABITAT: humid, evergreen forest; shaded and humid rocks.

ALTITUDE: 400 – 700 m.

FLOWERING TIME: February – April.

LIFE FORM: epiphyte or lithophyte.

PHYTOGEOGRAPHICAL REGION: I, II.

DESCRIPTION: {896}.

ILLUSTRATION: {1205}, {1223}.

NOTES: raceme densely more than 50-flowered; flowers red; lip papillose, with a single callus.

Megalorchis regalis *(Schltr.) H.Perrier* in Bull. Soc. Bot. France 83: 580 (1936). TYPE: Madagascar, Mt. Tsaratanana, *H.Perrier* 15263 (holo. P).

Habenaria regalis Schltr. in Repert. Sp. Nov. Regni Veg. Beih. 33: 94 (1924).

DISTRIBUTION: Madagascar: Antsiranana, Mahajanga.

HABITAT: mossy, montane forest.

ALTITUDE: 1800 – 2000 m.

FLOWERING TIME: October – November.

LIFE FORM: terrestrial.

PHYTOGEOGRAPHICAL REGION: IV.

DESCRIPTION: {896}.

ILLUSTRATION: {896}, {945}.

NOTES: plant robust, up to 1.5 m tall; flowers large, white; lip 3 cm long.

Microcoelia aphylla *(Thouars) Summerh.* in Bull. Misc. Inform. Kew. 1936: 233 (1936). TYPE: *Thouars* 22 (holo. P).

Angraecum aphyllum Thouars in Hist. Orch; t.73 (1822).

Gussonea aphylla A.Rich. in Mém. Soc. Hist. Nat. 4: 68 (1828).

Saccolabium aphyllum Lindl., Gen. Sp. Orch. Pl. 223 (1833).

Epidorchis aphylla (Thouars) Kuntze, Rev. Gen. Pl. 2: 660 (1891).

Gastrochilus aphyllus (Thouars) Kuntze, *loc. cit.* 661 (1891).

Mystacidium aphyllum T.Durand & Schinz, Consp. Fl. Afr. 5: 51 (1892).

Rhaphidorhynchus aphyllus Finet in Mém. Soc. Bot. France. 9: 35 (1907).

Angraecum defoliatum Schltr. in Ann. Mus. Col. Marseille ser. 3, 1: 48, t.20 (1913). TYPE: Madagascar, dry forest of Manongarivo, *H.Perrier* 22 (holo. B†).

Gussonea defoliata Schltr. in Beih. Bot. Centralbl. 33, 2: 425 (1915).

Gussonea aphylla var. *typica* H.Perrier in Humbert, Fl. Mad. Orch. 2: 78 (1941), *nom. inval.*

Gussonea aphylla var. *defoliata* (Schltr.) H.Perrier in Humbert, Fl. Mad. Orch. 2: 78 (1941), **syn. nov.**

Gussonea aphylla var. *orientalis* H.Perrier in Humbert, Fl. Mad. Orch. 2: 78 (1941), *nom. nud.*, TYPE: Madagascar, Mahanoro, Nov. 1921, *H.Perrier* 14198 (holo. P).

Solenangis aphylla Summerh. in Bot. Mus. Leafl. 11, 5: 159 (1943).

DISTRIBUTION: Madagascar: Antsiranana, Toamasina, Toliara. Comoros: Mayotte. Mascarenes: Réunion, Mauritius. Also in Kenya, Mozambique.

HABITAT: humid, lowland forest.

ALTITUDE: sea level – 30 m.

FLOWERING TIME: July - December.

LIFE FORM: epiphyte.

PHYTOGEOGRAPHICAL REGION: I, II.

DESCRIPTION: {768}, {896}.

ILLUSTRATION: {34}, {167}, {182}, {281}, {292}, {470}, {685}, {768}, {1223}, {1259}, {1345}, {1398}.

NOTES: leafless species, stem elongate; flowers white, marked with pink; lip obscurely 3-lobed; spur c. 5 mm long, incurved. The varieties described by Perrier de la Bâthie {896} are merely minor variants of this widespread and polymorphous species.

Microcoelia bispiculata *L.Jonsson* in Symb. Bot. Upsal. 23 (4): 97 (1981). TYPE: Madagascar, Kalalao Forest, St Marie, SE Lonkintsy, *Jonsson* 1035 (holo. UPS; iso. AMES, K, P, S, TAN).

DISTRIBUTION: Madagascar: Antsiranana, Toamasina.

HABITAT: coastal, humid, evergreen forest.

ALTITUDE: sea level – 100 m.

FLOWERING TIME: February – April.

LIFE FORM: epiphyte.

PHYTOGEOGRAPHICAL REGION: I.

DESCRIPTION: {596}.

ILLUSTRATION: {596}.

NOTES: distinct by the speculum-shaped thickenings at the mouth of the spur.

Microcoelia cornuta (*Ridley*) *Carlsward* in Amer. J. Bot. 93, 5: 779 (2006). TYPE: Grande Comore, Combani forest, *Humblot* 238 (1238) (holo. BM; iso. P, W).

Gussonea cornuta Ridley in J. Bot., London 23: 310 (1885).

Angraecum cornutum Rchb.f. in Flora 68: 538 (1885).

Angraecum cyclochilum Schltr. in Bot. Jahrb. Syst. 38: 160 (1906). TYPE: Madagascar, Marofandiha, near Morandava, *H.Perrier* 1841 (holo. B†; iso. P).

Rhaphidorhynchus cornutus Finet in Bull. Soc. Bot. Fr. 54, mem. 9, 34(1907).

Gussonea cyclochila Schltr. in Beih. Bot. Centralbl. 36, 2: 92 (1918).

Solenangis cornuta Summerh. in Bot. Mus. Leafl. 11, 5: 160 (1943).

DISTRIBUTION: Madagascar: Antsiranana, Mahajanga. Comoros: Grande Comore, Mayotte.

HABITAT: dry, semi-deciduous forest on rocks.

ALTITUDE: sea level – 300 m.

LIFE FORM: epiphyte.

PHYTOGEOGRAPHICAL REGION: I, II, V.

DESCRIPTION: {1008}, {1259}.

ILLUSTRATION: {1259}.

NOTES: differs from *M. aphylla* in the fewer-flowered inflorescences, the more or less orbicular lip lamina, retuse at the apex, the spur tapering towards the apex, and the more acute rostellum lobes. Rice makes a new combination and separates *Solenangis cyclochila* (Schltr.) R.Rice in Oasis 3, edition 1; 4 (2006) from *Microcoelia cornuta* by the distinct lip, petals and column. Further work is needed to confirm this.

Microcoelia decaryana *L.Jonsson* in Symb. Bot. Upsal. 23 (4): 95 (1981). TYPE: Madagascar, Majunga, surroundings of Bekodoka, *Decary* 2271 (holo. P).

DISTRIBUTION: Madagascar: Mahajanga.

HABITAT: seasonally dry, deciduous forest, woodland, and scrub, on *Dalbergia*, *Commiphora* and *Hildegardia*.

ALTITUDE: c. 300 m.

FLOWERING TIME: October – December.

LIFE FORM: epiphyte.

PHYTOGEOGRAPHICAL REGION: V.

DESCRIPTION: {596}.

ILLUSTRATION: {596}.

NOTES: lip large, 3-lobed, the mid-lobe cup-shaped, with a distinct ridge-like thickening at each side of the spur opening, c. 3 × 2.5 mm; spur cylindrical, c. 4 mm long.

Microcoelia dolichorhiza (*Schltr.*) *Summerh.* in Bot. Mus. Leafl. 11 (5): 153 (1943), (as *M. dolichorrhiza*). TYPE: Madagascar, Diego-Suarez, Manongarivo, *H.Perrier* 1919 (holo. B†; iso. P).

Angraecum dolichorhizum Schltr. in Ann. Inst. Bot.-Geol. Colon. Marseille 1: 192 (1913).

Gussonea dolichorhiza (Schltr.) Schltr. in Beih. Bot. Centralbl. 33 (2): 425 (1915).

DISTRIBUTION: Madagascar: Antsiranana, Mahajanga.

HABITAT: dense rain-forest.

ALTITUDE: c. 1000 m.

FLOWERING TIME: October – December.

LIFE FORM: epiphyte.

PHYTOGEOGRAPHICAL REGION: II.

DESCRIPTION: {596}.

ILLUSTRATION: {596}, {1205}, {1223}.

NOTES: known only from the type. Distinct by the two, plate-like, transverse lamellae, one on each side at the mouth of the spur. There has been confusion over the spelling of this name, Schlechter {1205}, described *Angraecum dolichorhizum* (the correct spelling) but spelt it as *A. dolichorrizum* in the accompanying plate.

Microcoelia elliotii *(Finet) Summerh.* in Bot. Mus. Leafl. 11 (5): 153 (1943). TYPE: Madagascar, Tulear, nr. Fort Dauphin, *Scott Elliot* 2653 (holo. P).
Listrostachys elliotii Finet in Bull. Soc. Bot. France 54, Mém. 9: 50, Pl. X (1907).
Gussonea elliotii Schltr. in Beih. Bot. Centralbl. 33 (2): 425 (1915).

DISTRIBUTION: Madagascar: Antsiranana, Mahajanga, Toliara.
HABITAT: dense rainforest.
ALTITUDE: sea level – 1300 m.
FLOWERING TIME: January – April.
LIFE FORM: epiphyte.
PHYTOGEOGRAPHICAL REGION: I, II.
DESCRIPTION: {596}.
ILLUSTRATION: {292}, {596}, {1271}.
NOTES: flowers orange, like *M. gilpinae*, but distinguished by the more or less folded, broadly elliptical lip with a row of tubercles at the spur opening, and by the long, curved spur.

Microcoelia exilis *Lindl.*, Gen. Sp. Orchid. Pl.: 61 (1830). TYPE: Madagascar, *J.Forbes* s.n. (holo. K-Lindl.).
Angraecum chiloschistae Rchb.f. in Linnaea 20: 678 (1847). TYPE from South Africa.
Gussonea exilis (Lindl.) Ridley in J. Linn. Soc. Bot. 21: 493 (1885).
Epidorkis exilis (Lindl.) Kuntze, Rev. Gen. Pl. 2: 660 (1891).
Mystacidium exile (Lindl.) T.Durand & Schinz, Consp. Fl. Afr. 5: 52 (1892).
Rhaphidorhynchus chiloschistae (Rchb.f.) Finet in Bull. Soc. Bot. France 54, Mém. 9: 35 (1907).
Gussonea chiloschistae (Rchb.f.) Schltr. in Beih. Bot. Centralbl. 36 (2): 91 (1918), as *Gussonea chilochistae*.

DISTRIBUTION: Madagascar: Toliara. Also in tropical and South Africa.
HABITAT: coastal forest; seasonally dry, deciduous woodland and scrub, on small branches of understorey trees and bushes.
ALTITUDE: sea level – 2000 m.
FLOWERING TIME: January – December.
LIFE FORM: epiphyte.
PHYTOGEOGRAPHICAL REGION: VI.
DESCRIPTION: {34}, {596}, {768}.
ILLUSTRATION: {34}, {292}, {596}, {676}, {768}, {1329} {1505}.
NOTES: distinguished by the highly branched root system, the very small flowers and almost globular spur.

Microcoelia gilpinae *(Rchb.f. & S.Moore) Summerh.* in Bot. Mus. Leafl. 11 (5): 153 (1943). TYPE: Madagascar, surroundings of Antananarivo, *H. Gilpin* (holo. K).
Angraecum gilpinae Rchb.f. & S.Moore in J. Linn. Soc., Bot. 16: 206 (1877).
Gussonea gilpinae (Rchb.f. & S.Moore) Ridl. in J. Linn. Soc., Bot. 21: 491 (1885).
Epidorkis gilpinae (Rchb.f. & S.Moore) Kuntze, Rev. Gen. Pl. 2: 660 (1891).
Mystacidium gilpinae (Rchb.f. & S.Moore) T.Durand & Schinz, Consp. Fl. Afric. 5: 52 (1892).
Rhaphidorhynchus gilpinae (Rchb.f. & S.Moore) Finet in Bull. Soc. Bot. France 54, Mém. 9: 34 (1907).
Gussonea gilpinae var. *minor* Schltr. in Beih. Bot. Centralbl. 34 (2): 333 (1916). TYPE: Madagascar, Tamatave, Mananara River, *H.Perrier* 11486 (holo. B†; iso. P).
Gussonea melinantha Schltr. in Repert. Spec. Nov. Regni. Veg. Beih. 33: 377 (1925). TYPE: Madagascar, Diego-Suarez, Tsaratanana massif, *H.Perrier* 15693 (holo. P).
Microcoelia melinantha (Schltr.) Summerh. in Bot. Mus. Leafl. 11 (5): 155 (1943).

DISTRIBUTION: Madagascar: Antananarivo, Antsiranana, Fianarantsoa, Mahajanga, Toamasina.
HABITAT: dense, humid, evergreen forest and in secondary and disturbed vegetation on various understory shrubs and trees, seems to prefer shaded habitats.
ALTITUDE: 200 – 1800 m.

FLOWERING TIME: throughout the year.

LIFE FORM: epiphyte.

PHYTOGEOGRAPHICAL REGION: I, II, III.

DESCRIPTION: {596}.

ILLUSTRATION: {246}, {281}, {292}, {461}, {896}, {957}, {635}, {470}, {687}, {959}.

NOTES: distinct from *M. elliotii* by the sigmoid spur parallel to the ovary and the slipper-shaped lip. The exact identity of *M. aurantiaca* (Schltr) Summerh. (1943) (*Gussonea aurantiaca* Schlechter 1918) needs verifying.

Microcoelia macrantha *(H.Perrier) Summerh.* in Bot. Mus. Leafl. 11 (5): 155 (1943). TYPE: Madagascar, SSW Anosibe, confluence Mangoro R. & Onive R., *H.Perrier* 17028 (holo. P).

Gussonea macrantha H.Perrier in Notul. Syst. (Paris) 7: 29 (1938).

DISTRIBUTION: Madagascar: Antsiranana, Fianarantsoa, Toamasina, Toliara.

HABITAT: evergreen forest, mostly on smaller branches and twigs, often near water courses.

ALTITUDE: sea level – 1000 m.

FLOWERING TIME: January – April.

LIFE FORM: epiphyte.

PHYTOGEOGRAPHICAL REGION: I.

DESCRIPTION: {596}, {798}.

ILLUSTRATION: {596}, {798}, {281}, {479}, {505}, {687}.

NOTES: flower large, c. 13 mm long; lip prominent; spur slightly incurved.

Microcoelia perrieri *(Finet) Summerh.* in Bot. Mus. Leafl. 11 (5): 147 (1943). TYPE: Madagascar, Majunga, Ambato-Boeni, nr. Maevatanana, *H.Perrier* 58 (holo. P).

Rhaphidorhynchus perrieri Finet in Notul. Syst. (Paris) 1: 89 (1909).

Angraecum perrieri (Finet) Schltr. in Ann. Mus. Col. Marseille, sér. 3, 1: 198 (1913).

Gussonea perrieri (Finet) Schltr. in Beih. Bot. Centralbl. 33 (2): 425 (1915).

DISTRIBUTION: Madagascar: Antsiranana, Mahajanga, Toliara.

HABITAT: seasonally dry, deciduous woods & scrub and open woodland, on *Didiereaceae*.

ALTITUDE: sea level – 500 m.

FLOWERING TIME: April – October.

LIFE FORM: epiphyte.

PHYTOGEOGRAPHICAL REGION: II, V.

DESCRIPTION: {596}.

ILLUSTRATION: {281}, {292}, {459}, {596}, {687}, {1259}.

NOTES: characterised by the stout stem, long slender rhachis and hair-like pedicels and the almost cylindrical spur about three times as long as the lip lamina.

Microcoelia physophora *(Rchb.f.) Summerh.* in Bot. Mus. Leafl. 11 (5): 148 (1943). TYPE: Madagascar, *Hildebrandt* 3255 (holo. W; iso. K, P).

Angraecum physophorum Rchb.f., Otia Bot. Hamb. 2: 77 (1881) & in Bot. Zeit. 39: 449 (1881).

Gussonea physophora (Rchb.f.) Ridl. in J. Linn. Soc., Bot. 21: 492 (1885).

Angorkis physophora (Rchb.f.) Kuntze, Rev. Gen. Pl. 2: 651 (1891).

Epidorkis physophora (Rchb.f.) Kuntze, *loc.cit.* 660.

Mystacidium physophorum (Rchb.f.) T.Durand & Schinz, Consp. Fl. Afric. 5: 54 (1892).

DISTRIBUTION: Madagascar: Antsiranana, Mahajanga. Also in East Africa.

HABITAT: seasonally dry, deciduous woods and open woodland.

ALTITUDE: sea level – 500 m.

FLOWERING TIME: September – December.

LIFE FORM: epiphyte.

PHYTOGEOGRAPHICAL REGION: V.

DESCRIPTION: {596}.

ILLUSTRATION: {596}, {716}, {958}, {959}, {1259}.

NOTES: recognised by the flattened roots, the long spur thickened at the apex and the narrow ligulate, erect and concave lip lamina.

Microterangis boutonii *(Rchb.f.) Senghas* in Orchidee 36 (1): 22 (1985). TYPE: Comoros, *Bouton* 1837 (holo. W).
Angraecum boutoni Rchb.f., Otia Bot. Hamburg. 2: 117 (1881).
Angorchis boutoni (Rchb.f.) Kuntze, Rev. Gen. Pl. 2: 651 (1891).
Angraecopsis boutoni (Rchb.f.) H.Perrier in Humbert ed., Fl. Mad. Orch. 2: 82 (1941).
Chamaeangis boutoni (Rchb.f.) Garay in Bot. Mus. Leafl. 23 (4): 165 (1972).

DISTRIBUTION: Comoros.
LIFE FORM: epiphyte {896}.
DESCRIPTION: {896}.
NOTES: close to *M. hildebrandtii*; lip angular, then expanded-tridentate; spur filiform.

Microterangis coursiana *(H.Perrier) Senghas* in Die Orchidee 36 (1): 22 (1985). TYPE: Madagascar, Mountains near lake Alaotra, Menaloha river, Ambatodrazaja district, *Cours* 744 (holo. P).
Chamaeangis coursiana H.Perrier in Notul. Syst. (Paris) 14: 160 (1951).

DISTRIBUTION: Madagascar: Toamasina.
HABITAT: humid, highland forest.
ALTITUDE: c. 900 m.
FLOWERING TIME: September.
LIFE FORM: epiphyte.
DESCRIPTION: {901}.
NOTES: known only from the type. Differs from *M. hariotiana* by the smaller leaves, shorter inflorescences with racemes in a series, and broadly rhombic lip.

Microterangis divitiflora *(Schltr.) Senghas* in Orchidee 36 (1): 22 (1985). TYPE: Madagascar, Tampoketsa (Ambongo), *H.Perrier* 1774 (holo. B†; iso. P).
Angraecum divitiflorum Schltr. in Ann. Mus. Col. Marseille, sér. 3, 1: 192 (49), t.22 (1913).
Chamaeangis divitiflora (Schltr.) Schltr. in Beih. Bot. Centralbl. 33 (2): 426 (1915).

DISTRIBUTION: Madagascar: Antananarivo, Mahajanga.
HABITAT: humid, highland forest.
FLOWERING TIME: July – September.
LIFE FORM: epiphyte.
PHYTOGEOGRAPHICAL REGION: V.
DESCRIPTION: {896}.
ILLUSTRATION: {896}, {1205}.
NOTES: distinct by its dense inflorescence and small c. 3 mm long flowers; spur cylindrical-obtuse.

Microterangis hariotiana *(Kraenzl.) Senghas* in Orchidee 36 (1): 22 (1985). TYPES: Grande Comore, *Humblot* 1292 (292) (lecto. P; isolecto. W, both selected here); Anjouan, *Lavanchie* s.n., (syn. P); Comoros, *Boivin* 3081 (syn. P).
Mystacidium hariotianum Kraenzl. in J. Bot. (Morot) 11: 153 (1897).
Saccolabium hariotianum (Kraenzl.) Finet in Bull. Soc. Bot. France 54, Mém. 9: 32 (1907).
Chamaeangis hariotiana (Kraenzl.) Schltr. in Beih. Bot. Centralbl. 33 (2): 426 (1915).

DISTRIBUTION: Comoros: Anjouan, Grande Comore, Mayotte.
HABITAT: in humid forest.
ALTITUDE: sea level – 500 m.
FLOWERING TIME: August – November.
LIFE FORM: epiphyte.
DESCRIPTION: {896}, {1258}, {1411}.
ILLUSTRATION: {3}, {281}, {485}, {529}, {896}, {1258}, {1259}, {1411}.
NOTES: plant with many inflorescences and small orange flowers; lip entire; spur inflated.

Microterangis hildebrandtii *(Rchb.f.) Senghas* in Orchidee 36 (1): 22 (1985). TYPE: Madagascar, *Hildebrandt* s.n. (holo. W).
Angraecum hildebrandtii Rchb.f. in Gard. Chron. n.s. 9: 725 (1878).
Chamaeangis hildebrandtii (Rchb.f.) Garay in Bot. Mus. Leafl. 23 (4): 165 (1972).

DISTRIBUTION: Comoros.

HABITAT: in humid forest.

DESCRIPTION: {982}.

NOTES: a poorly known species, further investigation on the differences between it and *M. hariotiana* is needed.

Microterangis humblotii *(Rchb.f.) Senghas* in Orchidee 36 (1): 22 (1985). TYPE: Comoros, *Humblot* s.n. (holo. W).

Saccolabium humblotii Rchb.f. in Flora 68: 537 (1885).

Angraecum saccolabioides H.Perrier in Humbert ed., Fl. Mad. Orch. 2: 209 (1941).

Chamaeangis humblotii (Rchb.f.) Garay in Bot. Mus. Leafl. 23 (4): 165 (1972).

DISTRIBUTION: Comoros {896}.

DESCRIPTION: {896}.

NOTES: known only from the type. Flowers small, yellow; lip oblong; spur cylindrical, obtuse.

Microterangis oligantha *(Schltr.) Senghas* in Orchidee 36 (1): 22 (1985). TYPE: Madagascar, Manongarivo, *H.Perrier* 1862 (holo. B†; iso. P).

Angraecum oliganthum Schltr. in Ann. Mus. Col. Marseille, sér. 3, 1: 197 (54), t.22 (1913).

Chamaeangis oligantha (Schltr.) Schltr. in Beih. Bot. Centralbl. 33: 426 (1915).

DISTRIBUTION: Madagascar: Antsiranana.

HABITAT: dry, semi-deciduous western forest.

FLOWERING TIME: June.

LIFE FORM: epiphyte.

PHYTOGEOGRAPHICAL REGION: II.

DESCRIPTION: {896}.

ILLUSTRATION: {1205}, {1223}.

NOTES: plant small; flowers relatively large; lip oval-lanceolate, acute, 2.4 × 1.5 mm; spur obtuse, 1.25 mm long.

Neobathiea grandidieriana *(Rchb.f.) Garay* in Bot. Mus. Leafl. 23: 188 (1972). TYPE: Madagascar, *Humblot* s.n. (holo. W).

Aeranthus grandidierianus Rchb.f. in Flora 68: 381 (1885).

Angraecum grandidierianum (Rchb.f.) Carrière in Rev. Hort. 59: 42, t.9 (1887).

Mystacidium grandidierianum (Rchb.f.) T.Durand & Schinz, Consp. Fl. Afr. 5: 53 (1892).

Neobathiea filicornu Schltr. in Repert. Sp. Nov. Regni Veg. Beih. 33: 369 (1925). TYPE: Madagascar, Manakazo, NE Ankazobe, *H.Perrier* 11320 (holo. P).

DISTRIBUTION: Madagascar: Antananarivo, Fianarantsoa, Toliara. Comoros: Anjouan, Grande Comore.

HABITAT: humid, evergreen forest; lichen-rich forest.

ALTITUDE: 1000 – 1650 m.

FLOWERING TIME: September – December.

LIFE FORM: epiphyte on branches.

PHYTOGEOGRAPHICAL REGION: III.

DESCRIPTION: {896}.

ILLUSTRATION: {3}, {20}, {50}, {88}, {175}, {529}, {838}, {896}, {1223}, {1319}.

CULTIVATION: {538}, {581}.

NOTES: lip entire; sepals and petals lanceolate, acute; lip acute at the tip.

Neobathiea hirtula *H.Perrier* in Notul. Syst. (Paris) 7 (2): 49 (1938). TYPE: Madagascar, Besafotra, Menavava, Boina, *H.Perrier* 179 (holo. P).

DISTRIBUTION: Madagascar: Mahajanga.

HABITAT: humid forest and semi-deciduous, western forest, on branches.

FLOWERING TIME: December – February.

LIFE FORM: epiphyte.

PHYTOGEOGRAPHICAL REGION: V.

DESCRIPTION: {896}.

ILLUSTRATION: {20}, {896}, {1259}.

NOTES: inflorescences long; lip 3-lobed, the mid-lobe wider than the laterals, orbicular or sub-rectangular.

Neobathiea hirtula var. **floribunda** *H.Perrier* ex *Hermans* **var. nov.** TYPE: Madagascar, Ankarafantsika (Boina), *Ursch* 20 (holo. P). See Appendix .
Neobathiea hirtula var. *floribunda* H.Perrier in Notul. Syst. (Paris) 7: 50 (1938), *nom. nud.*

DISTRIBUTION: Madagascar: Mahajanga.

HABITAT: semi-deciduous western forest.

FLOWERING TIME: April.

LIFE FORM: epiphyte.

PHYTOGEOGRAPHICAL REGION: V.

DESCRIPTION: {896}.

NOTES: differs from the typical variety in its larger size, longer stem, larger leaves, up to 11 × 2 cm, 6 – 10 inflorescences per stem each 30 – 50 cm long and carrying 6 to 10 flowers, and flowers that are slightly larger with an 18 mm long dorsal sepal and 21 – 22 mm long lateral sepals; the mid-lobule of the rostellum of the column does not exceed the wings. This likely to be just a more vigorous and mature plant, further research is needed.

Neobathiea keraudrenae *Toill.-Gen. & Bosser* in Nat. Malg. 13: 25 (1962). TYPE: Madagascar, Lake Alaotra, *M. Keraudren* s.n. (holo. P).

DISTRIBUTION: Madagascar: Toamasina.

HABITAT: humid, mossy, evergreen forest on plateau.

FLOWERING TIME: May.

LIFE FORM: epiphyte.

DESCRIPTION: {1408}.

ILLUSTRATION: {3}, {1408}.

NOTES: lip entire; sepals and petals spathulate at the tip, obtuse; lip truncate, broadened at the front. This may be a local variant of *N. grandidieriana.*

Neobathiea perrieri *(Schltr.) Schltr.* in Repert. Sp. Nov. Regni Veg. Beih. 33: 371 (1925). TYPE: Madagascar, Besafotra, Menavava, Boina, *H.Perrier* 180 (holo. B†; iso. P).
Aeranthes perrieri Schltr. in Ann. Mus. Col. Marseille, sér. 3, 1: 187, t.19 (1913).
Bathiea perrieri (Schltr.) Schltr. in Beih. Bot. Centralbl. 33 (2): 440 (1915).

DISTRIBUTION: Madagascar: Antsiranana, Mahajanga.

HABITAT: evergreen forest and semi-deciduous western forest, on shrubs.

ALTITUDE: sea level – 350 m.

FLOWERING TIME: January – June.

LIFE FORM: epiphyte.

PHYTOGEOGRAPHICAL REGION: V.

DESCRIPTION: {896}.

ILLUSTRATION: {50}, {520}, {529}, {896}, {1205}, {1223}, {281}.

NOTES: lip with three indistinct lobes, the mid-lobe acute at the tip; spur 10 – 12 cm long.

Neobathiea spatulata *H.Perrier* in Notul. Syst. (Paris) 7: 50 (1938).TYPES: Madagascar, Anosiravo, Mt. des Francais, nr. Diego-Suarez, *Decary* s.n. & *Poisson* 44 (syn. P).

DISTRIBUTION: Madagascar: Antsiranana.

HABITAT: calcareous rocks and mid-elevation, deciduous, western forest.

ALTITUDE: sea level – 300 m.

FLOWERING TIME: December – January.

LIFE FORM: lithophyte or epiphyte.

PHYTOGEOGRAPHICAL REGION: V.

DESCRIPTION: {896}.

ILLUSTRATION: {728}, {896}.

NOTES: lip 3-lobed; sepals and petals spathulate, rounded or obtuse at the end.

Nervilia affinis *Schltr.* in Repert. Sp. Nov. Regni Veg. Beih. 33: 116 (1924). TYPE: Madagascar, Mt. Tsaratanana, *H.Perrier* 15206 (holo. P).

Nervilia perrieri Schltr. in Repert. Sp. Nov. Regni Veg. Beih. 33: 117 (1924). TYPE: Madagascar, Mt. Tsaratanana, *H.Perrier* 11335 (holo. P).

Nervilia pilosa Schltr. & H.Perrier in Repert. Sp. Nov. Regni Veg. Beih. 33: 118 (1924). TYPE: Madagascar: Tsaratanana, *H.Perrier* 15716 (holo. P).

DISTRIBUTION: Madagascar, Antananarivo, Antsiranana, Fianarantsoa, Toamasina. Mascarenes: Réunion.

HABITAT: rather dry to very wet, humid, evergreen forest.

ALTITUDE: 700 – 1500 m.

FLOWERING TIME: October – January.

LIFE FORM: terrestrial.

PHYTOGEOGRAPHICAL REGION: I, II, III.

DESCRIPTION: {896}, {910}.

ILLUSTRATION: {896}, {910}, {1223}.

NOTES: leaf c. 3 × 6 cm; lip c. 16 × 15 mm, 3-lobed, pubescent.

Nervilia bicarinata *(Blume) Schltr.* in Bot. Jahrb. Syst. 45: 405 (1911). TYPE: Madagascar, Sambirano, Nosy-Be, *Pervillé* 383 (583) (lecto. P).

Begonia monophylla Pour. in de Candolle., Prodr. Reg. Veg. 15, 1: 402 (1864), *nom. nud.* Based upon: Réunion, *Barber* s.n. (G).

Pogonia bicarinata Blume, Coll. Orch. Arch. Id. 1: 152, pl. 60/1 A – D (1858). TYPE: Madagascar, Sambirano, Nosy-Be, Nossi-Keili (= Tany Kely), *Pervillé* 383 (583) (lecto. P).

Pogonia umbra Rchb.f. in Flora 50: 102 (1867). TYPE from Sao Tome.

Pogona viridiflava Rchb.f. in Flora 65: 532 (1882). TYPE from Angola.

Pogonia chariensis Chev., Expl. Fl. Afr. Centr. 1: 297 (1913), *nom. nud.* TYPE from Central African Republic (P).

Pogonia djalonensis Chev., Expl. Fl. Afr. Centr. 1: 620 (1913), *nom. nud.* TYPE from Guinea ?

Pogonia commersonii Blume, Coll. Orch. Arch. Ind.: 152 (1858). TYPE: Madagascar Nossi-Keili, *Commerson* s.n. (holo. L).

Pogonia barklayana Rchb.f. in Flora 68: 378 (1885) & in Gard. Chron. n.s. 23: 726 (1885). TYPE: Comoros, Grande Comore, Combani forest, *Humblot* 1416 (416) (holo. W).

Nervilia viridiflava (Blume) Schltr. in Bot. Jahrb. Syst. 45: 404 (1911).

Nervilia barklyana (Rchb.f.) Schltr. in Bot. Jahrb. Syst. 45: 405 (1911).

Pogonia ghindana Fiori in Nuov. Giorn. Bot. Ital. (n.s.) 19: 439 (1912). TYPE from Ethiopia.

Nervilia commersonii (Blume) Schltr. in Bot. Jahrb. Syst. 45: 405 (1911).

Pogonia renschiana sensu Moreau in Rev. Gen. Bot. 24: 98, fig. 10 & 10bis (1912), *non* Rchb.f.

Nervilia renschiana sensu H.Perrier in Humbert ed., Fl. Mad. Orch. 1: 203 (1939), p.p. excl. *Hildebrandt* 3303 & 3383 in Humbert, Mém. Inst. Sci. Mad. sér. B, 6: 256 (1955), *non* (Rchb.f.) Schltr.

Nervilia ghindiana (Fiori) Cufodontis in Bull. Jard. Bot. Nat. Belg. 42, Suppl.: 1611 (1972).

DISTRIBUTION: Madagascar: Antsiranana, Mahajanga, Toamasina. Comoros: Grande Comore, Mayotte. Mascarenes: Réunion, Mauritius. Also in East, South & Western Africa.

HABITAT: coastal forest; riverine forest often in *Syzygium* thicket; deciduous western forest; and highland forest.

ALTITUDE: sea level – 1500 m.

FLOWERING TIME: October – December.

LIFE FORM: terrestrial.

PHYTOGEOGRAPHICAL REGION: I, II, III, V.

DESCRIPTION: {768}, {910}.

ILLUSTRATION: {94}, {256b}, {444}, {677}, {768}, {827}, {910}, {1430}.

VERNACULAR NAME: Agoagoala.

NOTES: leaf very large, heart-shaped, c. 22 × 26 mm, on a long stalk; lip ovate, 20 – 31 × 17 – 25 mm, obscurely 3-lobed, with 2 parallel fleshy pubescent ridges running from the base of the lip to base of mid-lobe.

Nervilia hirsuta *(Blume) Schltr.* in Bot. Jahrb. Syst. 45: 405 (1911). TYPE: Comoros: Mayotte, *Boivin* 3087 (lecto. L; isolecto. P).

Pogonia hirsuta Blume, Coll. Orch. (1859).

DISTRIBUTION: Comoro: Mayotte.

HABITAT: on denuded cliffs.

LIFE FORM: terrestrial.

DESCRIPTION: {768}.

NOTES: *Nervilia hirsita* is probably conspecific with *N. stolziana*, but there are no flowers with the type material.

Nervilia kotschyi *(Rchb.f.) Schltr.* in Bot. Jahrb. Syst. 45: 404 (1911). TYPE: Sudan, An Nil al Azraq, Blue Nile Prov., *Cienkowski* 236 (holo. W).

Pogonia kotschyi Rchb.f. in Kotschy, Oesterr. Bot. Z. 14: 338 (1864).

Nervilia sakoae Jum. & H.Perrier in Ann. Fac. Sci. Marseille 21 (2): 194 (1912). TYPE: Madagascar, Maevrono (Mahevarano), Mahajanga, *H.Perrier* 1892 (holo. P).

Pogonia sakoae (Jum. & H.Perrier) Moreau in Rev. Gen. Bot. 24: 98, fig 11 (1912), *nom. nud.*

Pogonia abyssinica Chiov., in Ann. Bot. 9: 136 (1911). TYPE from Ethiopia.

Nervilia similis Schltr. in Bot. Jahrb. Syst. 53: 554 (1915). TYPE from Tanzania.

Nervilia abyssinica (Chiova.) Schltr. in Repert. Sp. Nov. Regni. Veg. 16: 356 (1920).

Nervilia diantha Schltr. in Bot. Jahrb. Syst. 53: 553 (1915). TYPE from Tanzania.

Pogonia thouarsii Chev. in Expl. Bot. Afr. Occ. Fr. 1: 620 (1920), *sensu non* Blume.

Nervilia kotschyi var. *kotschyi* Pettersson in Nord. Journ. Bot. 9; 489 (1990).

DISTRIBUTION: Madagascar: Antsiranana, Mahajanga. Comoros: Mayotte. Also in tropical Africa.

HABITAT: grassland; plantations; dry savannah; and semi-deciduous, western forest.

ALTITUDE: sea level – 2300 m.

FLOWERING TIME: September – January.

LIFE FORM: terrestrial.

PHYTOGEOGRAPHICAL REGION: V.

DESCRIPTION: {910}.

ILLUSTRATION: {219}, {256b}, {682}, {828}, {896}, {1345}.

NOTES: a variable species; leaf prostrate, av.12 cm × 14 cm; flowers one or two; lip elliptical, 10 – 19 × 7 – 12 mm, obscurely 3-lobed, bearing 2 parallel fleshy pubescent ridges running from base of the lip to base of mid-lobe.

Nervilia kotschyi var. **purpurata** *(Rchb.f. & Sond.) Börge Pett.* in Nord. Journ. Bot. 9: 489 (1990). TYPE: South Africa, Magaliesberg, Transvaal, *Zeyher* 1584 (holo. W).

Pogonia purpurata Rchb.f. & Sond. in Rchb.f., Flora 48: 184 (1865).

Nervilia purpurata (Rchb.f. & Sond.) Schltr. in Warburg, Kunene-Samb.-Exped. Baum; 210 (1903).

Nervilia dalbergiae Jum. & H.Perrier in Ann. Fac. Sci. Marseille 21 (2): 196 (1912). TYPE: Madagascar, Ankara plateau, Mahajanga, *H.Perrier* 1506 (holo. P).

Pogonia lanceolata Moreau in Rev. Gen. Bot. 24: 98, fig 12 (1912), *nom. nud.*

DISTRIBUTION: Madagascar: Mahajanga. Also in East and southern Africa.

HABITAT: grassland and marshes, under *Dalbergia*.

ALTITUDE: 475 – 2300 m.

FLOWERING TIME: November – January.

LIFE FORM: terrestrial.

PHYTOGEOGRAPHICAL REGION: V.

DESCRIPTION: {896}, {910}.

ILLUSTRATION: {256b}, {896}, {910}.

NOTES: leaf glabrous, suberect, stalked, less than 4 cm wide; flowers like those of var. *kotschyi*; fruit purple.

Nervilia leguminosarum *Jum. & H.Perrier* in Ann. Fac. Sci. Marseille 21 (2): 195 (1912). TYPE: Madagascar, Ankirihitra, Ankaladina, *H.Perrier* 1377 (lecto. P).

Pogonia leguminosarum Moreau in Rev. Gen. Bot. 24: 98, fig. 2 – 8 (1912), *nom. nud.*

DISTRIBUTION: Madagascar: Mahajanga.

HABITAT: in dry sandy forest, amongst *Tamarindus*.

ALTITUDE: sea level – 200 m.

FLOWERING TIME: December – January.

LIFE FORM: terrestrial.
PHYTOGEOGRAPHICAL REGION: V.
DESCRIPTION: {896}, {910}.
ILLUSTRATION: {896}.
NOTES: leaf prostrate, av. 19 × 20 cm; lip 3-lobed, covered by short, clavate hairs.

Nervilia lilacea *Jum. & H.Perrier* in Ann. Fac. Sci. Marseille 21 (2): 197 (1912). TYPE: Madagascar, Manongarivo massif, *H.Perrier* 1873 (holo. P).

DISTRIBUTION: Madagascar: Antsiranana.
HABITAT: humid, evergreen forest.
ALTITUDE: c. 1000 m.
LIFE FORM: terrestrial.
PHYTOGEOGRAPHICAL REGION: III.
DESCRIPTION: {896}, {910}.
ILLUSTRATION: {910}.
NOTES: leaf erect, angular-cordate, 1.5 × 2 cm, stalked, glabrous; flowers pinkish lilac; lip 3-lobed, the side lobes acute, shorter than the lanceolate mid-lobe, front margin finely toothed.

Nervilia petraea *(Afzel. ex Sw.) Summerh.* in Bot. Mus. Leafl. 11: 249 (1945). TYPE: Sierra Leone, *Afzelius* s.n. (holo. UPS).
Arethusa petraea Afzel. ex Sw. in Kon. Sven. Vet. Ak. Hand. 21: 230 (1800) & in Schrad. Neu. J. Bot. 1, 1: 62 (1805).
Nervilia afzelii Schltr., Bot. Jahrb. Syst. 45: 402 (1911), *nom. superfl.*

DISTRIBUTION: Madagascar: Mahajanga. Also tropical Africa.
HABITAT: grassland; coastal forest; seasonally dry, deciduous woods; plantations.
ALTITUDE: sea level – 1500 m.
FLOWERING TIME: October – January.
LIFE FORM: terrestrial.
PHYTOGEOGRAPHICAL REGION: V.
DESCRIPTION: {910}.
ILLUSTRATION: {219}, {910}, {1398}.
NOTES: the nomenclature of this species is discussed in {145}. Leaf prostrate, av. 10 × 18 cm; lip obovate-cuneate, 3-lobed, the lateral lobes oblong-triangular, the mid-lobe fimbriate, suborbicular, with subulate appendages.

Nervilia renschiana *(Rchb.f.) Schltr.* in Bot. Jahrb. Syst. 45: 404 (1911). TYPE: Madagascar, Sambirano, Nossi-Be, Lokobe forest, *Hildebrandt* 3303 (holo. W).
Pogonia renschiana Rchb.f. in Bot. Zeit. (Berlin) 39 (28): 449 (1881).
Nervilia insolata Jum. & H.Perrier in Ann. Fac. Sci. Marseille 21 (2): 192 (1912). TYPE: Madagascar, Antsirabe, Ambongo, *H.Perrier* 1763 (holo. P).
Nervilia grandiflora Schltr. in Ann. Transv. Mus. 10: 241 (1924). TYPE from South Africa.
Nervilia natalensis Schelpe in Orch. Rev. 74: 394 (1966). TYPE from South Africa.

DISTRIBUTION: Madagascar: Antananarivo, Antsiranana, Mahajanga. Also in East Africa and southern Africa.
HABITAT: seasonally dry deciduous woods; riverine forest; rocky outcrops.
ALTITUDE: sea level – 1760 m.
FLOWERING TIME: September – December.
LIFE FORM: terrestrial.
PHYTOGEOGRAPHICAL REGION: II, V.
DESCRIPTION: {768}, {896}, {910}, {1203}
ILLUSTRATION: {768}, {910}.
NOTES: leaf prostrate, reniform, av. 15 × 16 cm; flowers large; lip ovate, obscurely 3-lobed, with 2 parallel fleshy pubescent ridges from base to base of mid-lobe.

Nervilia simplex *(Thouars) Schltr.* in Bot. Jahrb. Syst. 45: 401 (1911). TYPE: Mauritius, the illustration in Thouars, Orch. Iles Austr. Afr.: t.g. 24 (1822) (lecto.).

Arethusa simplex Thouars, Hist. Orchid.: t.24 (1822).
Stellorkis aplostellis Thouars, Hist. Orchid.: t.24 (1822), *nom. rejic.*
Epidendrum simplex (Thouars) Spreng., Syst. Veg. 3: 736 (1826).
Aplostellis ambigua A.Rich., Monogr. Orch. Iles France Bourbon: 41 (1828), *nom. illeg.* Based on
    Arethusa simplex.
Haplostellis truncata Lindl. Gen. Sp. Orch. Pl. (1840) 411.
Bolborchis crociformis Zoll. & Mor. in Moritzi, Syst. Verz.: 89 (1846) .
Pogonia crispata Blume, Mus. Bot. Lugd.-Bat. 32 (1849). TYPE from Indonesia, painting by Keultjes
    of a no longer extant specimen collected by van Hasselt in Indonesia (lecto. W.).
Pogonia thouarsii Blume, Coll. Orch.: 152, pl.59, fig. 1, A – B (1859). TYPE: Mauritius, Thouars s.n.
    (lecto. L; isolecto. P).
Pogonia simplex (Thouars) Rchb.f. in Xenia Orch. 2: 92 (1874).
Pogonia bollei Rchb.f., Xenia Orch. 2: 88 (1874). TYPE from the Cape Verde Islands.
Pogonia prainiana King & Pant., Journ. A. Soc. Beng. 65: 129 (1896). TYPE from India.
Nervilia crispata (Blume) Schltr. ex Kraenzl. in Schumann & Lauterbach, Fl. Deutsch. Schutzg.
    Südsee (1901) 82.
Nervilia bollei (Rchb.f.) Schltr., in Bot. Jahrb. Syst. 45: 405 (1911).
Nervilia humilis Schltr., in Bot. Jahrb. Syst. 53: 551 (1915). TYPE from Tanzania.
Nervilia reniformis Schltr. in Bot. Jahrb. Syst. 53: 551 (1915), *non N. reniformis* var. *afzelii auct.* TYPE
    from Tanzania.
Pogonia fineti Chev., Expl. Bot. Afr. Occ. Fr. 1: 620 (1920), *nom. nud.*
Nervilia françoisii H.Perrier ex François in Rev. Hort. n.s. 21: 304 (1928), *nom. nud.*
Nervilia afzelii var.*grandiflora* Summerh. in Kew Bull. 222 (1936). TYPE from Guinea.
Nervilia bathiei Senghas in Adansonia sér. 2, 4: 303 (1964). TYPE: Madagascar, between Antsirabe &
    Ambositra, *Rauh* 1622 (holo. HEID).
Nervilia erosa P.J.Cribb in Kew Bull. 32: 155 (1977). TYPE from Zambia.
Nervilia crociformis (Zoll.& Mor.) Seidenf. in Dansk Bot. Arkiv. 32, 2: 151 (1978). TYPE from Java.
Nervilia prainiana (King & Pantl.) Seidenf. in Dansk. Bot. Ark. 32, 2 (1978) 149, fig. 91.

> DISTRIBUTION: Madagascar: Antananarivo, Antsiranana, Mahajanga. Mascarenes: Réunion,
>   Mauritius. Also in tropical and southern Africa, India, Nepal, Thailand, Indo-China, Malaysia,
>   Indonesia, Philippines, New Guinea, Australia.
> HABITAT: grassland; humid evergreen forest; riverine forest; plantations; dry *Uapaca* forest.
> ALTITUDE: 75 – 2000 m.
> FLOWERING TIME: October – January.
> LIFE FORM: terrestrial.
> PHYTOGEOGRAPHICAL REGION: III, V.
> DESCRIPTION: {768}, {910}.
> ILLUSTRATION: {256b}, {768}, {910}, {1234}, {1239}, {1242}, {1243}, {1270}.
> CULTIVATION: {1239}.
> NOTES: the nomenclature of this species is discussed in {145}. Leaf variable, reniform, av. 8 × 10
>   cm, slightly pubescent; lip oblong-cuneate, 12 – 18 × 9 – 11 mm, 3-lobed, inner surface more
>   or less covered with thin hairs and a few thicker formations.

Oberonia disticha *(Lam.) Schltr.* in Repert. Sp. Nov. Regni Veg. Beih. 33: 132 (1924). TYPE:
    Réunion, *Commerson* s.n. (lecto. P-JU 3871 A).
Epidendrum distichum Lam., Encycl. Meth. Bot. 1: 189 (1783).
Epidendrum equitans G.Forst. in Florulae Ins. Austr. Prodr.: 60 (1786). TYPE: Tahiti, *Forster* 170,
    Cook 2nd voyage (lecto. BM; isolecto. P).
Cymbidium equitans (G.Forst.) Swartz, Nov. Act. 6: 72 (1799).
Pleurothallis disticha (Lam.) A.Rich., Monogr. Orch. Iles France Bourbon: 55 (1828).
Oberonia brevifolia Lindl., Gen. Sp. Orchid. Pl.: 16 (1830). TYPE: Mauritius, *Bouton* (holo. K).
Malaxis brevifolia (Lindl.) Rchb.f. in Walpers Ann. Bot. Syst. 6: 215 (1861).
Iridorchis equitans (G. Forst.) Kuntze, Rev. Gen. Pl. 2: 669 (1891).
Oberonia equitans (G. Forst.) Schltr. in Beih. Bot. Centralbl. 33 (2): 411 (1915).

> DISTRIBUTION: Madagascar: Antananarivo, Antsiranana, Fianarantosa, Toamasina, Toliara.
>   Comoros: Anjouan, Grande Comore. Mascarenes: Réunion, Mauritius, Rodrigues. Also in
>   East and southern Africa.

HABITAT: humid, evergreen forest.

ALTITUDE: sea level – 2000 m.

FLOWERING TIME: September – May.

LIFE FORM: epiphyte.

PHYTOGEOGRAPHICAL REGION: I, II, IV, III.

DESCRIPTION: {219}, {768}, {896}.

ILLUSTRATION: {34}, {165}, {167}, {219}, {246}, {256b}, {281}, {468}, {478}, {504}, {676}, {677}, {681}, {682}, {684}, {768}, {896}, {1028}, {1345}, {1398}.

VERNACULAR NAMES: Fontsilahinjanahary, Ahipisabato.

NOTES: leaves distichous, bilaterally flattened; inflorescence terminal, racemose, densely many-flowered; flowers tiny, c. 1 mm in diameter, yellow.

Oeceoclades alismatophylla *(Rchb.f.) Garay & Taylor* in Bot. Mus. Leafl. 24: 258 (1976). TYPE: Madagascar, Ankaye forest, *Humblot* s.n. (holo. W).

Eulophia alismatophylla Rchb.f. in Flora 68: 543 (1885).

Eulophidium alismatophyllum (Rchb.f.) Summerh. in Bull. Jard. Bot. Bruxelles 27: 394 (1957).

DISTRIBUTION: Madagascar? Comoros?

DESCRIPTION: {1008}.

NOTES: a poorly known species, said to be related to *O. sclerophylla*; leaves long-stalked, lanceolate; flowers yellow.

Oeceoclades ambongensis *(Schltr.) Garay & P.Taylor* in Bot. Mus. Leafl. 24: 258 (1976). TYPE: Madagascar, Manongarivo, Ambongo, *H.Perrier* 1684 (holo. B†; iso. P).

Eulophidium ambongense Schltr. in Ann. Mus. Col. Marseille, sér. 3, 1: 182 (39), t.17 (1913).

Eulophia schlechteri H.Perrier in Bull. Soc. Bot. France 82: 154 (1935).

Lissochilus schlechteri (H.Perrier) H.Perrier in Humbert ed., Fl. Mad. Orch. 2: 27 (1941).

DISTRIBUTION: Madagascar: Mahajanga.

HABITAT: seasonally dry, deciduous forest and woodland.

FLOWERING TIME: January.

LIFE FORM: terrestrial.

PHYTOGEOGRAPHICAL REGION: V.

DESCRIPTION: {896}.

ILLUSTRATION: {1205}.

NOTES: known only from the type. Closely related to *O. maculata* but flowers larger; spur subglobose; callus of the lip strongly bilobed.

Oeceoclades ambrensis *(H.Perrier) Bosser & Morat* in Adansonia sér. 3, 23 (1): 11 (2001). TYPE: Madagascar, Mt. d'Ambre, *Humbert* 3964 (holo. P).

Lissochilus ambrensis H.Perrier in Notul. Syst. (Paris) 14 (2): 159 (1951).

Eulophia ambrensis (H.Perrier) Butzin in Willdenowia 7 (3): 587 (1975).

DISTRIBUTION: Madagascar: Antsiranana.

HABITAT: humid forest at medium-elevation.

ALTITUDE: 1000 – 1500 m.

FLOWERING TIME: December – January.

PHYTOGEOGRAPHICAL REGION: I.

LIFE FORM: terrestrial.

DESCRIPTION: {901}.

ILLUSTRATION: {141}.

NOTES: vegetatively similar to *O. pulchra*, distinct by the lip with more developed, rounded terminal lobes.

Oeceoclades analamerensis *(H.Perrier) Garay & P.Taylor* in Bot. Mus. Leafl. 24: 259 (1976). TYPE: Madagascar, Rodo, Analabe River, Diego-Suarez, *Humbert* 19247 (holo. P).

Lissochilus analamerensis H.Perrier in Notul. Syst. (Paris) 8: 42 (1939).

Eulophidium analamerense (H.Perrier) Summerh. in Bull. Jard. Bot. Bruxelles 27: 394 (1957).

DISTRIBUTION: Madagascar: Antsiranana.

HABITAT: shaded and wet rocks.

ALTITUDE: sea level – 500 m.

FLOWERING TIME: January.

LIFE FORM: terrestrial.

PHYTOGEOGRAPHICAL REGION: V.

LIFE FORM: terrestrial on chalk.

DESCRIPTION: {896}.

NOTES: vegetatively similar to *O. alismatophylla* but with two small lamellae at the base of the lip and a hirsute disc of the lip.

Oeceoclades analavelensis *(H.Perrier) Garay & P.Taylor* in Bot. Mus. Leafl. 24: 259 (1976). TYPE: Madagascar, Analavelona forest, N of Fiherenana, *Humbert* 14218 (holo. B; iso. P, K, TAN).

Eulophidium analavelense (H.Perrier) Summerh. in Bull. Jard. Bot. Bruxelles 27: 395 (1957).

Lissochilus analavelensis H.Perrier in Notul. Syst. (Paris) 8: 41 (1939).

DISTRIBUTION: Madagascar: Toliara.

HABITAT: humid, medium-elevation forest.

ALTITUDE: 500 – 1500 m.

FLOWERING TIME: March.

LIFE FORM: terrestrial.

PHYTOGEOGRAPHICAL REGION: V.

DESCRIPTION: {896}.

NOTES: pseudobulbs narrow, 2 – 3 cm long, normally with a longly petiolate single leaf; lip with three thickened ridges in front of the callus.

Oeceoclades angustifolia *(Senghas) Garay & P.Taylor* in Bot. Mus. Leafl. 24: 258 (1976). TYPE: Madagascar, nr. Diego Suarez, *Rauh & Buchloch* 7987 (holo. HEID).

Eulophidium angustifolium Senghas in Adansonia, sér. 2, 6: 557 (1967).

Eulophidium angustifolium subsp. *diphyllum* Senghas in Adansonia, sér. 2, 6: 561 (1967). TYPE: Madagascar, nr. Sakaraha, River Fiherenana, *Rauh* 10423 (holo. HEID).

DISTRIBUTION: Madagascar: Antsiranana, Toliara.

FLOWERING TIME: July, November.

LIFE FORM: terrestrial.

PHYTOGEOGRAPHICAL REGION: V.

DESCRIPTION: {50}, {1244}.

ILLUSTRATION: {50}, {1244}, {1344}.

NOTES: leaves shortly petiolate, variegated on tip, paler underneath; lip 4-lobed, av. 11 mm × 15 mm, the lobes distinctly convex, the lateral lobes semi-orbicular, with a 2-lobed callus at the base; spur short, bilobed.

Oeceoclades antsingyensis *G.Gerlach* in J. Orchideenfreund 2 (2): 62 – 65 (1995). TYPE: Madagascar, Tsingy of Antsalova, *Bogner* 2145 (holo. M).

DISTRIBUTION: Madagascar: Mahajanga.

HABITAT: humus pockets in limestone 'Tsingy'.

ALTITUDE: 60 – 100 m.

FLOWERING TIME: February.

PHYTOGEOGRAPHICAL REGION: V.

DESCRIPTION: {386}.

ILLUSTRATION: {386}, {281}, {484}.

NOTES: related to *O. rauhii* but differs by the broader, oval leaves, densely flowered inflorescence, and larger flowers with rounded petals and sepals and acuminate calli.

Oeceoclades aurea *Loubr.* in L'Orchidee 12: 105 (1994). TYPE: Northern Madagascar, *Garreau de Loubresse* s.n. (holo. herb. Loubresse).

DISTRIBUTION: Madagascar {383}.

DESCRIPTION: {383}.

ILLUSTRATION: {383}.

CULTIVATION: {383}.

NOTES: known only from the type. Flowers golden-yellow with a prominent double callus.

Oeceoclades boinensis *(Schltr.) Garay & P.Taylor* in Bot. Mus. Leafl. 24: 260 (1976). TYPE: Madagascar, Andranofasy, Boina, *H.Perrier* 1384 (holo. B†; iso. P).

Eulophidium boinense Schltr. in Ann. Mus. Col. Marseille, sér. 3, 1: 39, t.17 (1913).

Lissochilus boinensis (Schltr.) H.Perrier in Humbert ed., Fl. Mad. Orch. 2: 26 (1941).

DISTRIBUTION: Madagascar: Mahajanga.

HABITAT: humid, evergreen, lowland forest.

ALTITUDE: sea level – 500 m.

FLOWERING TIME: January – March.

LIFE FORM: terrestrial.

PHYTOGEOGRAPHICAL REGION: I, V.

DESCRIPTION: {896}, {1205} .

ILLUSTRATION: {896}, {1205}, {1223}.

NOTES: base to the leaves more or less cordate; flowers 16 – 26 mm long, yellow with the lip with dark red markings; lip 3-lobed, almost orbicular.

Oeceoclades calcarata *(Schltr.) Garay & P.Taylor* in Bot. Mus. Leafl. 24: 261 (1976). TYPE: Madagascar, Manongarivo, Ambongo, *H.Perrier* 1681 (iso. P).

Eulophia paniculata Rolfe in Gard. Chron. sér. 3, 29: 197 (1905), *non Oeceoclades paniculata* Lindl. TYPE: Madagascar, cult. Paris, *Bleu* s.n. (holo. K).

Cymbidium calcaratum Schltr. in Ann. Mus. Col. Marseille, sér. 3, 1: 181 (38), t.16 (1913).

Eulophia calcarata (Schltr.) Schltr. in Repert Sp. Nov. Regn. Veg. Beih. 33: 262 (1925).

Lissochilus paniculatus (Rolfe) H.Perrier in Humbert ed., Fl. Mad. Orch. 2: 29 (1941).

Eulophidium paniculatum (Rolfe) Summerh. in Bull. Jard. Bot. Bruxelles 27: 399 (1957).

DISTRIBUTION: Madagascar: Antsiranana, Fianarantsoa, Mahajanga, Toliara.

HABITAT: rocks in highlands; semi-deciduous, western forest.

ALTITUDE: sea level – 1100 m.

FLOWERING TIME: November – April.

LIFE FORM: terrestrial.

PHYTOGEOGRAPHICAL REGION: V, VI.

DESCRIPTION: {896}.

HISTORY: {1131}.

ILLUSTRATION: {281}, {934}, {1205}, {1223}.

VERNACULAR NAMES: Bekapiaky, Tsikapiaka.

NOTES: plant large, with conical pseudobulbs; inflorescence up to 1.5 m tall, paniculate; lip with 3 keels.

Oeceoclades cordylinophylla *(Rchb.f.) Garay & P.Taylor* in Bot. Mus. Leafl. 24: 261 (1976). TYPE: Comoros, *Humblot* s.n. (holo. W).

Eulophia cordylinophylla Rchb.f. in Flora 68: 541 (1885).

Eulophia lokobensis H.Perrier in Bull. Soc. Bot. France 82: 153 (1935). TYPE: Madagascar, Lokobe forest, Nossi Be, *H.Perrier* 19013 (holo. P).

Lissochilus cordylinophyllus (Rchb.f.) H.Perrier in Humbert ed., Fl. Mad. Orch. 2: 20 (1941).

Lissochilus lokobensis (H.Perrier) H.Perrier in Humbert ed., Fl. Mad. Orch. 2: 22 (1941).

Eulophidium cordylinophyllum (Rchb.f.) Summerh. in Bull. Jard. Bot. Bruxelles 27: 395 (1957).

Eulophidium lokobense (H.Perrier) Summerh. in Bull. Jard. Bot. Bruxelles 27: 396 (1957).

DISTRIBUTION: Madagascar: Antsiranana. Comoros: Mayotte.

HABITAT: shaded and humid rocks.

ALTITUDE: sea level – 300 m.

FLOWERING TIME: February.

LIFE FORM: terrestrial.

PHYTOGEOGRAPHICAL REGION: II.

DESCRIPTION: {50}, {896}.

ILLUSTRATION: {808}, {896}.

CULTIVATION: {382}, {808}.

NOTES: plant large, up to 80 cm tall; flowers small; lip 4-lobed, villous near the base and with 2 calli, spur conical, 6 mm long.

Oeceoclades decaryana *(H.Perrier) Garay & P.Taylor* in Bot. Mus. Leafl. 24: 262 (1976). TYPE: Madagascar, probably southern Madagascar, cult. Paris, *Decary* K467, (holo. P).

Eulophia decaryana H.Perrier ex Guillaumin & Manguin in Bull. Soc. Bot. France sér. 2 6: 521 (1934), *nom. nud.*

Eulophia decaryana H.Perrier in Bull. Soc. Bot. France 82: 154 (1935).

Lissochilus decaryanus (H.Perrier) H.Perrier in Humbert ed., Fl. Mad. Orch. 2: 32 (1941).

Eulophidium decaryanum (H.Perrier) Summerh. in Bull. Jard. Bot. Bruxelles 27: 395 (1957).

DISTRIBUTION: Madagascar: Mahajanga, Toliara. Also in Kenya, Zimbabwe, Mozambique, South Africa, Australia ?

HABITAT: scrubland, amongst xerophytic vegetation.

ALTITUDE: sea level – 300 m.

FLOWERING TIME: January – April.

LIFE FORM: terrestrial.

PHYTOGEOGRAPHICAL REGION: V, VI.

DESCRIPTION: {896}, {1345}, {768}.

ILLUSTRATION: {281}, {470}, {477}, {768}, {896}, {913}, {959}, {1270}, {1296}, {1345}.

VERNACULAR NAME: Tsikapiaky.

NOTES: pseudobulbs obscurely five-angled; leaves dark green or brownish red with varying quantities of grey or pale brown mottling or marbling; callus at the base of the lip with two raised oblong lobes, creating a pronounced V-shape; spur cylindrical and short.

Oeceoclades flavescens *Bosser & Morat* in Adansonia sér. 3, 23 (1): 15 (2001). TYPE: Madagascar, Maroantsetra, *Bosser & Descoings* 210 (holo. P).

DISTRIBUTION: Madagascar: Toamasina.

HABITAT: coastal forest on sand.

FLOWERING TIME: November.

LIFE FORM: terrestrial.

PHYTOGEOGRAPHICAL REGION: I.

DESCRIPTION: {141}.

ILLUSTRATION: {141}.

NOTES: known from the type only. Leaves thin; lip 4-lobed, with a keeled disc.

Oeceoclades furcata *Bosser & Morat* in Adansonia sér. 3, 23 (1): 14 (2001). TYPE: Madagascar, Soalala, *R.Decary* 18905 (holo. P).

DISTRIBUTION: Madagascar: Mahajanga.

HABITAT: sand.

FLOWERING TIME: April.

LIFE FORM: terrestrial.

PHYTOGEOGRAPHICAL REGION: V.

DESCRIPTION: {141}.

ILLUSTRATION: {141}.

NOTES: known from the type only. Raceme simple; flowers small; lip with 2 small triangular calli; spur which is forked at the tip.

Oeceoclades gracillima *(Schltr.) Garay & P.Taylor* in Bot. Mus. Leafl. 24: 262 (1976). TYPE: Madagascar, Besafotra, affluent de Menavava, *H.Perrier* 1059 (holo. B†; iso. P).

Eulophia gracillima Schltr. in Ann. Mus. Col. Marseille, sér. 3, 1: 170 (27), t.14 (1913).

Lissochilus gracillimus (Schltr.) H.Perrier in Humbert ed., Fl. Mad. Orch. 2: 28 (1941).

Eulophidium gracillimum (Schltr.) Schltr. in Repert. Sp. Nov. Regni Veg. Beih. 33: 255 (1925).

Eulophidium roseovariegatum Senghas in Adansonia, sér. 2, 6: 561 (1967), **syn. nov.**

Oeceoclades roseovariegata (Senghas) Garay & Taylor in Bot. Mus. Leafl. 24: 270 (1976), **syn. nov.**
   TYPE: Madagascar, nr. Diego-Suarez, Mt. des Français, *Rauh & Buchloch* 7985 (holo. HEID).

   DISTRIBUTION: Madagascar: Antsiranana, Mahajanga.
   HABITAT: seasonally dry, deciduous forest and woodland on limestone.
   FLOWERING TIME: March – June, November.
   LIFE FORM: terrestrial.
   PHYTOGEOGRAPHICAL REGION: V.
   DESCRIPTION: {50}, {896}, {1244}.
   ILLUSTRATION: {470}, {1205}, {1223}, {1244}, {1296}.
   NOTES: leaves oval-oblong, 3 – 7.5 × 1.7 – 3 cm, intricately patterned, margins corrugate; flowers small, 1 – 1.2 cm long; lip 3-lobed. *Oeceoclades roseovariegata* is said to differ by the shape of the plant, colour of foliage and shape of lip and spur. This is a vegetatively variable species, the shape of lip and spur are very close, in *O. gracillima* the spur is a little narrower towards the tip.

Oeceoclades hebdingiana *(Guillaumin) Garay & P.Taylor* in Bot. Mus. Leafl. 24: 263 (1976). TYPE: Madagascar, Anipanihy, *Montagnac*, cult. Jard. Bot. Les Cèdres (holo. P).
Lissochilus hebdingianus Guillaumin in Bull. Mus. Nat. Hist. Nat., sér. 2, 35 (2): 521 (1963).

   DISTRIBUTION: Madagascar: Mahajanga.
   HABITAT: scrubland and forest undergrowth.
   ALTITUDE: 80 – 100 m.
   FLOWERING TIME: January.
   LIFE FORM: terrestrial.
   PHYTOGEOGRAPHICAL REGION: V.
   DESCRIPTION: {422}.
   NOTES: leaf linear, up to 100 cm long, 2 cm wide; inflorescence up to 140 cm long; flower similar to those of *O. calcarata*.

Oeceoclades humbertii *(H.Perrier) Bosser & Morat* in Adansonia sér. 3, 23 (1): 10 (2001). TYPE: Madagascar, Manambolo valley, Mandrare basin, *Humbert* 12843 (holo. B; iso. P).
Lissochilus humbertii H.Perrier in Notul. Syst. (Paris) 8: 40 (1939).
Eulophia humbertii (H.Perrier) Butzin in Willdenowia 7 (3): 588 (1975).

   DISTRIBUTION: Madagascar: Toliara.
   HABITAT: xerophytic forest.
   ALTITUDE: sea level – 1000 m.
   FLOWERING TIME: December – January.
   LIFE FORM: terrestrial.
   PHYTOGEOGRAPHICAL REGION: V.
   DESCRIPTION: {896}.
   ILLUSTRATION: {141}.
   NOTES: pseudobulbs small, conical; leaves develop after flowering; flowers completely green.

Oeceoclades lanceata *(H.Perrier) Garay & Taylor* in Bot. Mus. Leafl. 24: 263 (1976). TYPE: Madagascar, Manerinerina, Tampoketsa, *H.Perrier* 16843 (holo. P).
Eulophia lanceata H.Perrier in Bull. Soc. Bot. France 82: 156 (1935).

   DISTRIBUTION: Madagascar: Antananarivo.
   HABITAT: open forest.
   ALTITUDE: 1500 m.
   FLOWERING TIME: December.
   LIFE FORM: terrestrial.
   PHYTOGEOGRAPHICAL REGION: III.
   DESCRIPTION: {882}.
   NOTES: this may be the same as *O. pandurata*, but the lip is reputedly distinct.

Oeceoclades lonchophylla *(Rchb.f.) Garay & P.Taylor* in Bot. Mus. Leafl. 24: 264 (1976). TYPE: Mozambique, Lourenço Marques, Lebombo Mts., *Daintree* s.n. (lecto. W).

Eulophia lonchophylla Rchb.f. in Flora 68: 542 (1885).

Eulophia tainoides Schltr. in Bot. Jahrb. Syst. 26: 339 (1899). TYPE from Mozambique.

Eulophidium lonchophyllum (Rchb.f.) Schltr. in Repert. Sp. Nov. Regni Veg. Beih. 33: 256 (1925).

Lissochilus lonchophyllus (Rchb.f.) H.Perrier in Humbert ed., Fl. Mad. Orch. 2: 26 (1941).

Eulophidium tainiodes (Schltr.) Summerh. in Bull. Jard. Bot. État 27: 403 (1957).

Eulophia dissimilis Dyer in Fl. Pl. Afr. 27: pl. 1066 (1949). TYPE from Mozambique.

Eulophidium dissimile Dyer in Fl. Pl. Afr. 27: pl. 1066 (1949), *nom. alt.* in obs.

> DISTRIBUTION: Comoros: Grande Comore, Mayotte. Also in East Africa.
> HABITAT: rocky places and bright, lowland forest; coastal forest.
> ALTITUDE: sea level – 200 m.
> FLOWERING TIME: January.
> LIFE FORM: terrestrial.
> DESCRIPTION: {768}, {896}.
> ILLUSTRATION: {768}, {896}, {1328},
> NOTES: leaf petiolate; flowers small, yellowish green, tinged purple; lip 3-lobed, with 2 – 3 keels; spur 4 – 5 mm long.

Oeceoclades longibracteata *Bosser & Morat* in Adansonia sér. 3, 23 (1): 15 (2001). TYPE: Madagascar, Tsaramasao, S of Sakaraha, *Bosser* 19909 (holo. P).

> DISTRIBUTION: Madagascar: Toliara.
> HABITAT: dry forest undergrowth.
> ALTITUDE: 500 – 800 m.
> FLOWERING TIME: February – March.
> LIFE FORM: terrestrial.
> PHYTOGEOGRAPHICAL REGION: V.
> DESCRIPTION: {141}.
> ILLUSTRATION: {141}.
> NOTES: close to *O. decaryana*, but distinct by the very narrow, linear, erect leaf and by the floral bracts and stem sheaths which are long and linear.

Oeceoclades maculata *(Lindl.) Lindl.*, Gen. Sp. Orch. Pl.: 237 (1833) TYPE: Brazil, without precise locality, cultivated *Loddiges* (holo. BM).

Angraecum maculatum Lindl., Coll. Bot., t.15 (May 1821).

Limodorum maculatum Lodd., Bot. Cab. 5, t.496 (June 1821).

Geodorum pictum Link & Otto, Ic. Pl. Sel. 3, t.14 (July 1821) lecto. t.14.

Angraecum monophyllum A.Rich., Monogr. Orch. Iles France Bourbon: 58 (66) t.9 (1828), **syn. nov.** TYPE: Mauritius, *Commerson* s.n. (holo. P?).

Eulophia maculata (Lindl.) Rchb.f. in Walp. Ann. 6: 647 (1863).

Aerobion maculatum (Lindl.) Sprengel, Syst. Veg. 3: 718 (1863).

Eulophia monophylla (A.Rich.) S.Moore in Baker, Fl. Mauritius: 336 (1877), **syn. nov.**

Eulophidium maculatum (Lindl.) Pfitz., Entw. Nat. Anordn. Orch. 88 (1887).

Eulophia ledienii N.E. Br. in Kew Bull. 1889: 90 (1889) TYPE from Zaire (WRSL). But also see Stein 1888.

Graphorchis maculata (Lindl.) Kuntze., Rev. Gen. Pl. 2: 662 (1891).

Eulophia mackeni Rolfe ex Hemsl. in Gard. Chron. ser.3, 12: 583 (1892); in F.C. 5, 3: 22 (1912). Type from South Africa.

Eulophidium warneckianum Kraenzl. in Bot. Jahrb. Syst. 33: 70 (1901) TYPE from Togo.

Eulophidium ledienii (N.E. Br.) De Wild. in Ann. Mus. Congo, Bot. sér.5, 1: 115 (1904).

Eulophidium monophyllum (Richard) *Schltr.* in Ann. Mus. Col. Marseille, sér. 3, 1: (40) (1913), **syn. nov.**

Eulophidium nyassanum Schltr. in Bot. Jahrb. Syst. 53: 593 (1915). TYPE from Tanzania.

Oeceoclades mackenii (Rolfe ex. Hemsl.) Garay & P.Taylor in Bot. Mus. Leafl. 24: 265 (1976).

Oeceoclades monophylla (A.Rich.) Garay & P.Taylor in Bot. Mus. Leafl. 24: 267 (1976), **syn. nov.**

> DISTRIBUTION: Madagascar: Mahajanga, Toliara. Mascarenes: Réunion, Mauritius. Also Tropical Africa, South & Central America.
> HABITAT: lowland and coastal, seasonally dry forest.
> ALTITUDE: sea level – 200 m.

FLOWERING TIME: February – April.
LIFE FORM: terrestrial.
DESCRIPTION: {769}, {770}
PHYTOGEOGRAPHICAL REGION: I, V.
ILLUSTRATION: {182}, {194}, {770}, {1345}.
CULTIVATION: {684}
NOTES: a cosmopolitan species. Leaf oblong-oblanceolate or oblong-elliptic, grey-green mottled with dark green; lip white with 2 purple-red blotches at the base of the mid-lobe, the disk with 2 approximate, divergent, glabrous basal keels. *O. monophylla* (A.Rich.) Garay & Taylor from the Mascarenes is a local variant with a larger, articulated leaf and a darker colouration of leaf and flowers; both these characteristics are not constant in other populations or in seed-grown plants.

Oeceoclades pandurata *(Rolfe) Garay & P. Taylor* in Bot. Mus. Leafl. 24: 268 (1976). TYPE: Madagascar, nr. Fort Dauphin, *Scott Elliot* 2546 (holo. K).
Eulophia pandurata Rolfe in J. Linn. Soc. 29: 52 (1891).
Lissochilus panduratus (Rolfe) H.Perrier in Humbert ed., Fl. Mad. Orch. 2: 29 (1941).
Eulophidium panduratum (Rolfe) Summerh. in Bull. Jard. Bot. Bruxelles 27: 399 (1957).

DISTRIBUTION: Madagascar: Antananarivo, Toliara.
HABITAT: seasonally dry, deciduous forest.
ALTITUDE: sea level – 1500 m.
FLOWERING TIME: December, May.
LIFE FORM: terrestrial.
PHYTOGEOGRAPHICAL REGION: I.
DESCRIPTION: {896}.
NOTES: lateral lobes of the lip free, the lateral veins on the disc are papillose-ciliolate.

Oeceoclades perrieri *(Schltr.) Garay & P.Taylor* in Bot. Mus. Leafl. 24: 268 (1976). TYPE: Madagascar, Manongarivo, Ambongo, *H.Perrier* 1654 (holo. P).
Eulophia ambongensis Schltr. in Ann. Mus. Col. Marseille, sér. 3, 1: 169 (26), t.13 (1913).
Eulophidium perrieri Schltr. in Repert. Sp. Nov. Regni Veg. Beih. 33: 256 (1925).
Lissochilus ambongensis (Schltr.) H.Perrier in Humbert ed., Fl. Mad. Orch. 2: 19 (1941).

DISTRIBUTION: Madagascar: Mahajanga.
HABITAT: seasonally dry, deciduous woodland on sandstone and sand.
FLOWERING TIME: December.
LIFE FORM: terrestrial.
PHYTOGEOGRAPHICAL REGION: V.
DESCRIPTION: {896}.
ILLUSTRATION: {896}, {1205}.
NOTES: pseudobulbs acutely quadrangular with a single linear leaf; inflorescence paniculate.

Oeceoclades petiolata *(Schltr.) Garay & Taylor* in Bot. Mus. Leafl. 24: 269 (1976). TYPES: Madagascar, Manongarivo, Ambongo, *H.Perrier* 478 (holo. B†; iso. P).
Eulophia petiolata Schltr. in Ann. Mus. Col. Marseille, sér. 3, 1: 175 (32), t.13 (1913).
Eulophidium petiolatum (Schltr.) Schltr. in Repert. Sp. Nov. Regni. Veg. Beih. 33: 256 (1925).
Lissochilus petiolatus (Schltr.) H.Perrier in Humbert ed., Fl. Mad. Orch. 2: 25 (1941).

DISTRIBUTION: Madagascar: Antsiranana, Mahajanga, Toliara. Comoros: Mayotte.
HABITAT: seasonally dry, deciduous woodland and xerophytic scrub on sandstone and sand.
ALTITUDE: sea level – 600 m.
FLOWERING TIME: December – April.
LIFE FORM: terrestrial.
PHYTOGEOGRAPHICAL REGION: V.
DESCRIPTION: {896}.
ILLUSTRATION: {281}, {470}, {896}, {1205}.
NOTES: close to *Oeceoclades lonchophylla*. Pseudobulbs conical; leaf longly petiolate, 5 – 10 cm long; lip indistinctly 4-lobed, with 2 ridges at the base.

Oeceoclades peyrotii *Bosser & Morat* in Adansonia sér. 3, 23 (1): 11 (2001). TYPE: Madagascar, Bona forest, Ankazoaba, *Bosser & Morat* in Jard. Bot. 1479 (holo. P).

DISTRIBUTION: Madagascar: Toliara.
HABITAT: forest undergrowth in evergreen forest.
LIFE FORM: terrestrial.
FLOWERING TIME: December – February.
PHYTOGEOGRAPHICAL REGION: V.
DESCRIPTION: {141}.
ILLUSTRATION: {141}.
NOTES: leaves oval-acute; petiole c. 10 mm long; lip 4-lobed, the disk with a bi-lamellate callus; spur cylindrical-conical, 3 – 4 mm long.

Oeceoclades quadriloba *(Schltr.) Garay & P.Taylor* in Bot. Mus. Leafl. 24: 269 (1976). TYPE: Madagascar, Manongarivo, Ambongo, *H.Perrier* 1696 (holo. B†; iso. P).
Eulophia quadriloba Schltr. in Ann. Mus. Col. Marseille, sér. 3, 1: 176 (33), t.12 (1913).
Eulophidium quadrilobum (Schltr.) Schltr. in Repert. Sp. Nov. Regni. Veg. Beih. 33: 256 (1925).
Lissochilus quadrilobus (Schltr.) H.Perrier in Humbert ed., Fl. Mad. Orch. 2: 30 (1941).

DISTRIBUTION: Madagascar.
HABITAT: coastal forest; semi-deciduous, western forest.
LIFE FORM: terrestrial, sandstone and sand.
FLOWERING TIME: January.
PHYTOGEOGRAPHICAL REGION: V.
DESCRIPTION: {896}.
ILLUSTRATION: {896}.
NOTES: known from the type only. Leaves oblanceolate-linear; flowers small; lip 4-lobed with 2 short triangular calli at the base; spur c. 5 mm long.

Oeceoclades rauhii *(Senghas) Garay & P.Taylor* in Bot. Mus. Leafl. 24: 270 (1976). TYPE: Madagascar, S of Anivorano, *Rauh & Senghas* 22865 (holo. HEID).
Eulophidium rauhii Senghas in Orchidee 24: 61 (1973).

DISTRIBUTION: Madagascar: Antsiranana.
FLOWERING TIME: July.
LIFE FORM: terrestrial.
PHYTOGEOGRAPHICAL REGION: V.
DESCRIPTION: {1250}.
ILLUSTRATION: {1250}.
NOTES: close to *O. boinensis* but sepals and petals linear-lanceolate, lip 4-lobed, and spur 6 – 7 mm long.

Oeceoclades sclerophylla *(Rchb.f.) Garay & P.Taylor* in Bot. Mus. Leafl. 24: 271 (1976). TYPE: Madagascar, Ankaye forest, *Humblot* s.n. (holo. W).
Eulophia sclerophylla Rchb.f. in Flora 68: 542 (1885).
Eulophia elliotii Rolfe in J. Linn. Soc. 29: 52 (1891). TYPE: Madagascar, around Fort Dauphin, *Scott Elliot* 2424 (holo. K).
Lissochilus elliotii (Rolfe) H.Perrier in Humbert ed., Fl. Mad. Orch. 2: 47 (1941).
Eulophidium sclerophyllum (Rchb.f.) Summerh. in Bull. Jard. Bot. Bruxelles 27: 402 (1957).

DISTRIBUTION: Madagascar: Toliara. Comoros ?
HABITAT: seasonally dry, deciduous coastal forest on sand.
FLOWERING TIME: April.
LIFE FORM: terrestrial.
PHYTOGEOGRAPHICAL REGION: I.
DESCRIPTION: {896}.
ILLUSTRATION: {896}.
NOTES: leaves graminiform; lip 8 × 4 mm, the front lobe deeply bifid; spur 4 mm long, a little wider at the obtuse tip than at the base.

Oeceoclades spathulifera *(H.Perrier) Garay & P.Taylor* in Bot. Mus. Leafl. 24: 272 (1976). TYPE: Madagascar, Ambongo-Boina, *H.Perrier* 15930 (holo. P).
Eulophia spathulifera H.Perrier in Bull. Soc. Bot. France 82: 157 (1935).
Lissochilus spathulifer (H.Perrier) H.Perrier in Humbert ed., Fl. Mad. Orch. 2: 33 (1941).
Eulophidium spathuliferum (H.Perrier) Summerh. in Bull. Jard. Bot. Bruxelles 27: 403 (1957).

DISTRIBUTION: Madagascar: Antsaranana, Mahajanga, Toliara.
HABITAT: seasonally dry, deciduous woodland on sand and coastal dunes.
FLOWERING TIME: November – January.
LIFE FORM: terrestrial.
PHYTOGEOGRAPHICAL REGION: V.
DESCRIPTION: {896}.
ILLUSTRATION: {257}, {281}, {470}, {479}, {896}, {1246}, {1296}.
NOTES: leaves elliptic-linear, marked with green-blackish spots; sepals and petals distinctly spathulate; spur globose.

Oeonia brauniana *H.Wendl. & Kraenzl.,* Xenia Orch. 3: 172, t.300 (1900). TYPE: Madagascar, without locality, *I.Braun* s.n. (holo. not located).

DISTRIBUTION: Madagascar: Toamasina.
HABITAT: humid, low-elevation forest.
FLOWERING TIME: October.
PHYTOGEOGRAPHICAL REGION: I.
DESCRIPTION: {129}, {896}.
ILLUSTRATION: {129}.
NOTES: sepals and lip 12 – 15 mm long; lip 6-lobed, the terminal lobe deeply incised.

Oeonia brauniana var. **sarcanthoides** *(Schltr.) Bosser* in Adansonia 11 (2): 162 (1989) & in Adansonia 11 (4): 381 (1989). TYPE: Madagascar, Sambirano valley, *H.Perrier* 15700 (holo. P).
Oeoniella sarcanthoides Schltr. in Repert. Sp. Nov. Regni. Veg. Beih. 33: 364 (1925).
Oeonia subacaulis H.Perrier in Notul. Syst. (Paris) 7: 138 (1938). TYPE: Madagascar, confl. Mangoro & Onive, *H.Perrier* 17075 (holo. P).
Lemurella tricalcariformis H.Perrier in Notul. Syst. (Paris) 14: 164 (1951). TYPE: Madagascar, Betampona, Reserve Nat. 1, *Decary* 16908 (holo. P).
Lemurella sarcanthoides (Schltr.) Senghas in Schltr., Die Orchideen, ed. 3: 1018 (1986).

DISTRIBUTION: Madagascar: Antsiranana, Fianarantsoa, Toamasina.
HABITAT: humid, evergreen forest.
ALTITUDE: 500 – 1000 m.
FLOWERING TIME: January – February.
LIFE FORM: epiphyte.
PHYTOGEOGRAPHICAL REGION: I, II.
DESCRIPTION: {129}, {281}, {896}, {1221}.
ILLUSTRATION: {80}, {129}, {454}, {468}, {896}.
NOTES: sepals and petals 7 – 9 mm long; lip 5-lobed, the terminal lobe retuse or only slightly indented.

Oeonia curvata *Bosser* in Adansonia 22 (2): 233 (2000). TYPE: Madagascar, Ranomafana, *Malcomber, Hemingway & Randriamantera* 2429 (holo. P).

DISTRIBUTION: Madagascar: Fianarantsoa.
HABITAT: humid forest.
ALTITUDE: 1200 – 1400 m.
FLOWERING TIME: April.
LIFE FORM: epiphyte.
PHYTOGEOGRAPHICAL REGION: III.
DESCRIPTION: {140}.
ILLUSTRATION: {140}.
NOTES: known from the type only. Similar to *O. madagascariensis*, but distinct by the 2.5 × 2.5 cm lip, which is 4-lobed, the 2 lateral lobes erect, broad, rounded, the edges undulate, and the terminal lobes smaller and narrower.

Oeonia madagascariensis *(Schltr.) Bosser* in Adansonia 6 (3): 370 (1984). TYPE: Madagascar, Mt. Tsaratanana, *H.Perrier* 15334 (holo. P).
Perrieriella madagascariensis Schltr. in Repert. Sp. Nov. Regn. Veg., Beih. 33: 366 (1925).

DISTRIBUTION: Madagascar: Antsiranana.
HABITAT: humid evergreen forest on plateau; lichen-rich forest.
ALTITUDE: 1500 – 2000 m.
FLOWERING TIME: January – March.
LIFE FORM: epiphyte.
PHYTOGEOGRAPHICAL REGION: III.
DESCRIPTION: {125},
ILLUSTRATION: {125}, {896}, {1259}.
NOTES: inflorescence single-flowered; spur cylindrical, up to 1.5 cm long; anther with a 2-lobed, elongated lobule at the front.

Oeonia rosea *Ridl.* in J. Linn. Soc., Bot. 21: 496 (1885). TYPE: Madagascar, Ankafana, *Deans Cowan* s.n. (holo. BM).
Oeonia oncidiiflora Kraenzl. in Bot. Jahrb. Syst. 17: 56 (1893). TYPE: Madagascar, S Betsileo, Ankafana, *Hildebrandt* 4205 (holo. W).
Oeonia volucris auct. non (Thouars) Spreng. in Fl. Réunion: 217 (1895).
Oeonia forsythiana Kraenzl. in Bot. Jahrb. Syst. 28: 191 (1900). TYPE: Madagascar, Irobimanitra, *Forsyth Major* 18 (holo. BM; iso. G).

DISTRIBUTION: Madagascar: Antananarivo, Fianarantsoa, Toamasina, Toliara. Mascarenes: Réunion.
HABITAT: humid, mossy, evergreen forest, on branches.
ALTITUDE: 500 – 2000 m.
FLOWERING TIME: September – May.
LIFE FORM: epiphyte.
PHYTOGEOGRAPHICAL REGION: I, III.
DESCRIPTION: {129}, {896}.
ILLUSTRATION: {3}, {50}, {78}, {167}, {246}, {281}, {470}, {468}, {478}, {482}, {520}, {529}, {615}, {958}, {1440}.
CULTIVATION: {455}, {1333}.
NOTES: lip white with a red throat, 4-lobed, the terminal lobes large, spathulate, divergent; spur 5 – 6 mm long, slightly inflated.

Oeonia volucris *(Thouars) Spreng.*, Syst. Veg. 3: 727 (1826). TYPE: Madagascar, *Thouars* s.n. (holo. P).
Epidendrum volucre Thouars, Hist. Orchid.: t.81 (1822).
Oeonia auberti Lindl. in Bot. Reg. 10: sub t.817 (1824), *nom. illeg.*
Aeranthes volucris (Thouars) Rchb.f. in Walpers Ann. Bot. Syst. 6: 900 (1864).
Epidorchis volucris (Thouars) Kuntze, Rev. Gen. Pl. 2: 659 (1891).
Oeonia elliotii Rolfe in J. Linn. Soc., Bot. 29: 55, t.11 (1891). TYPES: Madagascar, Fort Dauphin, *Scott Elliot* 2842 & *Scott Elliot* 2843 (both syn. K).
Oeonia volucris (Thouars) T.Durand & Schinz, Consp. Fl. Afric. 5: 51 (1892), *nom. superfl.*
Oeonia humblotii Kraenzl. in Bot. Jahrb. Syst. 43: 397 (1909). TYPE: Madagascar, Ivondro, *Humblot* 667 (holo. K not located).

DISTRIBUTION: Madagascar: Antsiranana, Fianarantsoa, Toamasina, Toliara. Mascarenes: Réunion, Mauritius.
HABITAT: coastal forest; humid, evergreen forest, on branches.
ALTITUDE: sea level – 1500 m.
FLOWERING TIME: March – July.
LIFE FORM: epiphyte.
PHYTOGEOGRAPHICAL REGION: I, III.
DESCRIPTION: {129}, {896}.
ILLUSTRATION: {129}, {281}, {292}, {468}, {529}, {702}, {896}, {1062}, {1398}.
NOTES: flower completely white; lip 3-lobed, the mid-lobe more or less indented; spur cylindrical and slender, 3.5 – 4 mm long.

Oeoniella polystachys *(Thouars) Schltr.* in Beih. Bot. Centralbl. 33 (2): 439 (1915). TYPE: Réunion, *Thouars* s.n. (holo. P).
Epidendrum polystachys Thouars, Hist. Orchid.: t.82 (1822).
Angraecum polystachyum (Thouars) A.Rich., Monogr. Orch. Iles France Bourbon: 74, t.10 (1828).
Listrostachys polystachys (Thouars) Rchb.f. in Walpers Ann. Bot. Syst. 6: 909 (1861).
Beclardia polystachya (Thouars) Frapp., Orch. Réunion, Cat.: 12 (1880).
Oeonia polystachya (Thouars) Benth., Gen. Pl. 3: 584 (1883).
Angraecum kimballianum Hort. in Kew Bull. App. ii: 38 (1890).
Angorchis polystachya (Thouars) Kuntze, Rev. Gen. Pl.: 651 (1891).
Monixus polystachys (Thouars) Finet in Bull. Soc. Bot. France 54, Mém. 9: 18, t.3, f. 1 – 2 (1907).

DISTRIBUTION: Madagascar: Antsiranana, Fianarantsoa, Toamasina, Toliara. Comoros: Grande Comore. Mascarenes: Réunion, Mauritius. Seychelles.
HABITAT: coastal, lowland andmontane forest, on trunks.
ALTITUDE: sea level – 100 m.
FLOWERING TIME: August – May.
LIFE FORM: epiphyte.
PHYTOGEOGRAPHICAL REGION: I, II.
DESCRIPTION: {896}, {1238}.
ILLUSTRATION: {3}, {50}, {135}, {167}, {168}, {281}, {292}, {444}, {505}, {529}, {532}, {716}, {722}, {821}, {896}, {1028}, {1238}, {1259}, {1265}, {1278}, {1281}, {1319}, {1324}, {1398}, {1425}.
CULTIVATION: {182}, {410}, {462}, {615}, {821}, {1278}, {1319}, {1333}.
VERNACULAR NAME: Foutsilang zanahar.
NOTES: a variable and locally common species; flowers 12 – 16 mm, white; lip trumpet-shaped, apiculate. Bentham in Gen. Pl. 3: 584 (1883) mentions *Oeonia polystachae* var. *longifoliae* Rchb.f.

Paralophia epiphytica *(P.J.Cribb, DuPuy & Bosser)* P.J.Cribb in Curtis's Bot. Mag. 22, 1: t.519 (2005). TYPE: Madagascar, Toliara Province, 7 km N of Taolanaro, Ampasinahampoana, cult. Kew, *DuPuy, Cribb, Andriantiana & Ranaivijaona* M857 (holo. K; iso. P, TAN).
Eulophia epiphytica P.J.Cribb, DuPuy & Bosser in Adansonia sér. 3, 24 (2): 169 – 172 (2002).

DISTRIBUTION: Madagascar: Fianarantsoa, Toliara.
HABITAT: on *Elaeis guineensis, Raphia farinifera* and probably also *Dypsis* palm trees, climbing amongst leaf bases on trunk of palm trees.
ALTITUDE: sea level – 600 m.
FLOWERING TIME: December – February.
LIFE FORM: epiphyte.
PHYTOGEOGRAPHICAL REGION: I.
DESCRIPTION: {225}.
ILLUSTRATION: {225} as *Eulophia palmicola*, {281}.
NOTES: plant large, epiphytic; rhizomes elongate, creeping; stems trailing, non-pseudobulbous; leaves thin-textured; flowers c. 3.5 cm long; sepals and petals yellow-green; lip 3-lobed, white with 3 purple-marked callus ridges.

Paralophia palmicola *(H.Perrier)* P.J.Cribb & Hermans in Curtis's Bot. Mag. 22, 1: 51 (2005). TYPE: Madagascar, between Menarandra & the Manambovo, *H.Perrier* 18893 (holo. P).
Eulophia palmicola H.Perrier in Bull. Soc. Bot. France 82: 158 (1935).
Lissochilus palmicolus (H.Perrier) H.Perrier in Humbert ed., Fl. Mad. Orch. 2: 47 (1941).

DISTRIBUTION: Madagascar: Toliara.
HABITAT: on the palm *Ravenia xerophila* in *Didiereaceae* forest.
FLOWERING TIME: October.
LIFE FORM: epiphyte.
PHYTOGEOGRAPHICAL REGION: VI.
DESCRIPTION: {480}, {896},
ILLUSTRATION: {706}, {896}.
NOTES: plant with rhizomes; pseudobulbs three- or four-leaved; inflorescence longer than the leaves, erect; flowers with apiculate sepals and petals; lip with a 2-ridged callus at the base, a shortly apiculate mid-lobe and a 4 mm long spur.

Phaius pulchellus *Kraenzl.* in Abh. Naturwiss. Vereine Bremen 7: 254 (1882). TYPE: Madagascar, Ambaravambato, *Rutenberg* s.n. (holo. not located).

DISTRIBUTION: Madagascar: Antananarivo, Antsiranana, Toamasina, Toliara. Mascarenes: Mauritius.
HABITAT: humid, evergreen forest.
ALTITUDE: 700 – 1700 m.
FLOWERING TIME: December – March.
LIFE FORM: epiphyte or terrestrial.
PHYTOGEOGRAPHICAL REGION: I, II, III.
DESCRIPTION: {120}, {896}.
ILLUSTRATION: {3}, {82}, {120}, {135}, {167}, {281}, {896}, {1268}, {1281}, {1391}.
CULTIVATION: {33}.
VERNACULAR NAME: Tadidiala.
NOTES: lip with the lateral lobes longitudinally fused to the column and surrounding the column, with a small spur.

Phaius pulchellus var. **ambrensis** *Bosser* in Adansonia, ser.2, 11 (3): 534 (1971). TYPE: Madagascar, Mt. Ambre, *Bosser* 20785 (holo. P).

DISTRIBUTION: Madagascar: Antsiranana.
HABITAT: humid, evergreen forest on plateau.
ALTITUDE: 1000 – 1600 m.
LIFE FORM: terrestrial.
PHYTOGEOGRAPHICAL REGION: III.
DESCRIPTION: {120}.
NOTES: sepals and petals white, more than 3 cm long, the callus ridges verrucose.

Phaius pulchellus var. **andrambovatensis** *Bosser* in Adansonia, sér. 2, 11 (3): 533 (1971). TYPE: Madagascar, Andrambovato, nr. Fianarantsoa, *Humbert* in *Jard. Bot. Tan.* 1079 (holo. P).

DISTRIBUTION: Madagascar: Fianarantsoa.
HABITAT: humid, evergreen forest.
FLOWERING TIME: October – November.
LIFE FORM: terrestrial.
PHYTOGEOGRAPHICAL REGION: I.
DESCRIPTION: {120}.
ILLUSTRATION: {120}.
NOTES: leaves with very undulating edges; sepals and petals pinkish-purple, reflexed at the tip.

Phaius pulchellus var. **sandrangatensis** *Bosser* in Adansonia, ser.2, 11 (3): 534 (1971). TYPE: Madagascar, Sandrangato, Moramanga, *Bosser & J.P.Peyrot* 18971 (holo. P).

DISTRIBUTION: Madagascar: Toamasina.
HABITAT: humid, evergreen forest.
ALTITUDE: 500 – 1000 m.
FLOWERING TIME: February.
LIFE FORM: terrestrial.
PHYTOGEOGRAPHICAL REGION: I, III.
DESCRIPTION: {120}.
ILLUSTRATION: {120}, {281}.
NOTES: sepals and petals light green; lip completely dark purple.

Physoceras australis *Boiteau.* in Bull. Acad. Malgache n.s. 24: 88 (1942). TYPES: Madagascar, Ambondrombe, S Betsileo, *Boiteau* in Jard. Bot. Tananarive 4921 (holo. P).

DISTRIBUTION: Madagascar: Fianarantsoa.
HABITAT: mossy, montane forest.
ALTITUDE: 1800 m.
LIFE FORM: epiphyte or terrestrial.
PHYTOGEOGRAPHICAL REGION: III.

DESCRIPTION: {100}.

NOTES: plant up to 15 cm tall; flowers c. 10 mm long, pink; lip very narrowly 4-lobed; spur c. 1 mm long.

Physoceras bellum *Schltr.* in Repert. Sp. Nov. Regni Veg. Beih. 33: 79 (1924). TYPES: Madagascar, Mt. Tsaratanana, *H.Perrier* 15267 (lecto. P, selected here); *H.Perrier* 15709 (syn. P).

DISTRIBUTION: Madagascar: Antsiranana.

HABITAT: forest, lichen-rich, montane forest.

ALTITUDE: 1500 – 2000 m.

FLOWERING TIME: October – January.

LIFE FORM: lithophyte or terrestrial.

PHYTOGEOGRAPHICAL REGION: III, IV.

DESCRIPTION: {896}.

ILLUSTRATION: {896}, {945}.

NOTES: leaf 4.5 – 10 × 2 – 4.6 cm., slightly cordate at the base; flowers 1.5 cm long; lip angular at the base, deeply 3-lobed; spur 7 mm long.

Physoceras betsomangense *Bosser* in Adansonia, sér. 2, 20 (2): 258 (1980). TYPE: Madagascar, Androranga valley, Bemarivo, *Humbert & Capuron* 24370 (holo. P).

DISTRIBUTION: Madagascar: Antsiranana.

HABITAT: montane, ericaceous scrub.

ALTITUDE: 1000 – 1500 m.

FLOWERING TIME: November.

LIFE FORM: terrestrial.

PHYTOGEOGRAPHICAL REGION: III.

DESCRIPTION: {124}.

ILLUSTRATION: {124}.

NOTES: distinguished by its 1- to 2-flowered inflorescence, and large flowers with a bottle-shaped spur, conical at the tip.

Physoceras bifurcum *H.Perrier* in Humbert, Mém. Inst. Sci. Mad. sér. B, 6: 254 (1955). TYPE: Madagascar, Eastern summit of Marojejy, W of high Manantenina, *Humbert & Cours* 23878 (holo. P).

DISTRIBUTION: Madagascar.

HABITAT: peaty hollows at high elevation.

ALTITUDE: 2000 – 2500 m.

FLOWERING TIME: March – April.

LIFE FORM: terrestrial.

PHYTOGEOGRAPHICAL REGION: IV.

DESCRIPTION: {902}.

ILLUSTRATION: {902}.

NOTES: lip 4-lobed; spur 6 mm long, terete, apex expanded, flattened and forked.

Physoceras epiphyticum *Schltr.* in Repert. Sp. Nov. Regni Veg. Beih. 33: 80 (1924). TYPE: Madagascar, Mt. Tsaratanana, *H.Perrier* 15268 (holo. P).

DISTRIBUTION: Madagascar: Antsiranana.

HABITAT: mossy, montane forest; river margins on moss- and lichen-covered trees.

ALTITUDE: 1600 – 1700 m.

FLOWERING TIME: April, October.

LIFE FORM: epiphyte.

PHYTOGEOGRAPHICAL REGION: III.

DESCRIPTION: {896}.

NOTES: differs from *P. bellum* by the narrower leaves, larger flowers, spread-out lip of a different shape, and distinct anther.

Physoceras lageniferum *H.Perrier* in Humbert, Mém. Inst. Sci. Mad. sér. B, 6: 254 (1955). TYPE: Madagascar, Mt. Beondroka, *Humbert* 23603 (holo. P).

DISTRIBUTION: Madagascar.
HABITAT: humid, mid-elevation forest and rocks.
ALTITUDE: 1000 – 1500 m.
LIFE FORM: terrestrial or lithophyte.
PHYTOGEOGRAPHICAL REGION: III, IV.
DESCRIPTION: {902}.
ILLUSTRATION: {902}.
NOTES: perianth c. 2 cm long, distinct by the absence of rostellar arms and fused anther.

Physoceras mesophyllum *(Schltr.) Schltr.* in Repert. Sp. Nov. Regni Veg. Beih. 33: 82 (1924). TYPE: Madagascar, Mt. Tsaratanana, *H.Perrier* 11355 (holo. P).
Cynosorchis mesophylla Schltr. in Beih. Bot. Centralbl. 34 (2): 308 (1916).

DISTRIBUTION: Madagascar: Antsiranana.
HABITAT: mossy and lichen-rich forest.
ALTITUDE: c. 2000 m.
FLOWERING TIME: October, December.
LIFE FORM: terrestrial.
PHYTOGEOGRAPHICAL REGION: III, IV.
DESCRIPTION: {896}.
ILLUSTRATION: {1223}.
NOTES: known only from the type. Plant c. 15 cm tall; leaf c. 5 × 1.5 cm; spur obovate, 3 mm long, slightly inflated at the tip.

Physoceras perrieri *Schltr.* in Repert. Sp. Nov. Regni Veg. Beih. 33: 82 (1924). TYPE: Madagascar, foot of Mt. Tsaratanana, *H.Perrier* 15710 (holo. P).

DISTRIBUTION: Madagascar: Antsiranana.
HABITAT: dry forest; humid, evergreen forest on plateau.
ALTITUDE: 1000 – 1800 m.
FLOWERING TIME: January.
LIFE FORM: terrestrial.
PHYTOGEOGRAPHICAL REGION: II, III.
DESCRIPTION: {896}.
NOTES: plant up to 45 cm tall; leaves c. 10 × 2.5 cm; spur 6 mm long, inflated into a sphere at the tip.

Physoceras rotundifolium *H.Perrier* in Notul. Syst. (Paris) 14: 144 (1951). TYPE: Madagascar, Mts. between Vohemar & Ambilobe, *Decary* 14701 (holo. P).

DISTRIBUTION: Madagascar: Antsiranana.
HABITAT: humid, evergreen forest on plateau.
FLOWERING TIME: July.
LIFE FORM: epiphyte.
PHYTOGEOGRAPHICAL REGION: III, V.
DESCRIPTION: {901}.
NOTES: known only from the type. Leaf broadly acute-oval, av. 3 × 2.5 cm, the base rounded-subtruncate; lip 3-lobed in front; spur obtuse, claviform, 8 mm long.

Physoceras violaceum *Schltr.* in Repert. Sp. Nov. Regni Veg. Beih. 33: 83 (1924). TYPE: Madagascar, Masoala, *H.Perrier* 11332 (holo. P).

DISTRIBUTION: Madagascar: Antsiranana, Toamasina.
HABITAT: humid, evergreen forest.
ALTITUDE: c. 500 m.
FLOWERING TIME: August.

LIFE FORM: epiphyte.
PHYTOGEOGRAPHICAL REGION: I.
DESCRIPTION: {896}.
ILLUSTRATION: {281}, {896}, {945}.
NOTES: leaves broad; spur 7.5 mm long, in a long sybcylindric mass, contracted towards the base.

Platycoryne pervillei *Rchb.f.* in Bonplandia 3: 212 (1855). TYPE: Madagascar, Ambongo, *Pervillé* 593 (holo. W; iso. BR, K, P).
Habenaria depauperata Kraenzl. in Abh. Naturwiss. Vereine Bremen 7: 259 (1882). TYPE: Madagascar, Efitra, *Rutenberg* s.n. (holo. B).
Habenaria pervillei (Rchb.f.) Kraenzl. in Bot. Jahrb. Syst. 16: 209 (1892).
Habenaria tenuicaulis Rendle in J. Linn. Soc., Bot. 30: 396 (1895). TYPE from Tanzania.
Habenaria buchwaldiana Kraenzl. in Bot. Jahrb. Syst. 24: 503 (1898). TYPE from Tanzania.
Platycoryne tenuicaulis (Rendle) Rolfe in Fl. Trop. Afr. 7: 257 (1898).
Platycoryne buchananiana sensu Rolfe, *loc. cit.* 7: 257 (1898), partim., *non* (Kraenzl.) Rolfe.
Habenaria flammea Kraenzl. in Bot. Jahrb. Syst. 28: 173 (1900). TYPE from Tanzania.

DISTRIBUTION: Madagascar: Mahajanga. Also in East Africa.
HABITAT: grassland; marshes; humid woods.
FLOWERING TIME: January – April.
PHYTOGEOGRAPHICAL REGION: V.
LIFE FORM: terrestrial.
DESCRIPTION: {896}, {1374}.
ILLUSTRATION: {896}, {1345}.
NOTES: plant slender; leaves linear-lanceolate; flowers several in a head, bright orange; dorsal sepal hooded; lip ligulate, spurred at base.

Platylepis bigibbosa *H.Perrier* in Bull. Soc. Bot. France 83: 26 (1936). TYPES: Madagascar, Analamahitso Forest, *H.Perrier* 7972 & foot of Manongarivo massif, *H.Perrier* 1949 (both syn. P).

DISTRIBUTION: Madagascar: Antsiranana, Mahajanga.
HABITAT: humid, highland forest.
ALTITUDE: 600 – 800 m.
PHYTOGEOGRAPHICAL REGION: II, III.
LIFE FORM: terrestrial.
DESCRIPTION: {896}.
ILLUSTRATION: {505}, {896}.
NOTES: floral bracts more than 3 times longer than wide; lip strongly gibbose at the base, with 6 calli at the base of the lip, the front keels of the lip protruding and clearly ridged.

Platylepis densiflora *Rolfe* in Kew Bull. 378 (1906). TYPES: Madagascar, centre, without locality, *Baron* and/or *Warpur*? 6550 & 6753 (holo. K?).
? *Erporkis cryptorpis* Thouars, Bull. Herb. Boiss. ii. (1894) 462 2: 462 (1894).

DISTRIBUTION: Madagascar. Mascarenes: Mauritius.
DESCRIPTION: {884}, {896}.
NOTES: similar to *P. margaritifera*, further research is needed. The type could not be found at K and at BM, the type indicated at K is a cultivated plant (orig. *Warpur*?) the leaves are those of *Platylepis*, the flowers of *Liparis* (Paul Ormerod pers. comm).

Platylepis margaritifera *Schltr.* in Repert. Sp. Nov. Regni. Veg. Beih. 15: 328 (1918). TYPE: Madagascar, *Laggiara* s.n. (holo. B†). NEOTYPE: the illustration in Repert. Spec. Nov. Reg. Veg., 68: t.50, 200 (1932).

DISTRIBUTION: Madagascar: Toamasina.
LIFE FORM: terrestrial.
PHYTOGEOGRAPHICAL REGION: I.
DESCRIPTION: {896}.
ILLUSTRATION: {1223}.

NOTES: the type was destroyed, and only Schlechter's description and sketch survive, it is likely that this is the same species as *P. densiflora*.

Platylepis occulta *(Thouars) Rchb.f.* in Linnaea 41: 62 (1877). TYPE: *Thouars* s.n. (holo. P).
Goodyera occulta Thouars, Hist. Orch.: t.28 (1822).
Platylepis goodyeroides A.Rich., Monogr. Orchid. Iles France Bourbon: 39 (1828). TYPES: Mauritius, *Commerson* s.n. & Réunion, *Thouars* s.n. (both syn. P).
Hetaeria occulta (Thouars) Lindl. in Bot. Reg. 24: 94 (1838).
Aetheria occulta (Thouars) Lindl., Gen. Sp. Orch. Pl. 491 (1840).
Notiophrys occulta (Thouars) Lindl. in J. Linn. Soc. 1: 189 (1857).
Erporkis bracteata Kuntze, Rev. Gen. Pl. 2: 660 (1891), *nom. illeg.*
Orchiodes occultum (Thouars) Kuntze, Rev. Gen. Pl. 2: 675 (1891).

> DISTRIBUTION: Madagascar: Antsiranana, Toamasina. Comoros: Grande Comore. Mascarenes: Réunion, Mauritius. Seychelles.
> HABITAT: montane forest.
> ALTITUDE: 1000 – 1600 m.
> FLOWERING TIME: January, July.
> LIFE FORM: terrestrial.
> PHYTOGEOGRAPHICAL REGION: III.
> DESCRIPTION: {896}.
> ILLUSTRATION: {167}, {884}, {896}, {1028}, {1398}.
> NOTES: similar in shape to *P. bigibbosa* but the inflorescence is almost glabrous, the floral bracts 3 times longer than wide, and the lip and the rostellum distinct.

Platylepis polyadenia *Rchb.f.* in Flora 68: 537 (1885). TYPE: Comoros, Grande Comore, *Humblot* 1427 (holo. W; iso. P).

> DISTRIBUTION: Madagascar: Antananarivo, Mahajanga, Toamasina, Toliara. Comoros: Grande Comore.
> HABITAT: seasonally dry, deciduous woods; humid, evergreen forest; in marshes.
> ALTITUDE: sea level – 1200 m.
> FLOWERING TIME: November – February.
> LIFE FORM: terrestrial.
> PHYTOGEOGRAPHICAL REGION: I, III, V.
> DESCRIPTION: {896}.
> ILLUSTRATION: {505}, {672}.
> NOTES: floral bracts more than 3 times longer than wide; lip thickened in the upper half, with 2 big calli at the base.

Polystachya anceps *Ridl.* in J. Linn. Soc., Bot. 21: 473 (1885). TYPES: Madagascar, Ankafana, *Hildebrandt* 4222 (lecto. BM; isolecto. K, both selected here); Imerina, *Deans Cowan* s.n.; *Baron* 192 (both syn. BM).
Dendrorkis anceps (Ridl.) Kuntze in Rev. Gen. Pl. 2: 658 (1891).
Polystachya hildebrandtii Schltr. in Repert. Sp. Nov. Regni Veg. Beih. 33: 155 (1924), *non* Kraenzl.
Polystachya mauritiana var. *anceps* (Ridl.) H.Perrier, Fl. Mad. Orch. 1: 251 (1939).

> DISTRIBUTION: Madagascar: Antananarivo, Fianarantsoa, Toamasina. Comoros: Anjouan, Grande Comore.
> HABITAT: humid, highland forest.
> ALTITUDE: 800 – 1700 m.
> FLOWERING TIME: December – March.
> LIFE FORM: epiphyte or lithophyte.
> PHYTOGEOGRAPHICAL REGION: III.
> DESCRIPTION: {896}.
> ILLUSTRATION: {246}, {1331}.
> NOTES: a variable and poorly understood species; sheaths of the scape large and reaching the base of the panicle.

Polystachya aurantiaca Schltr. in Ann. Mus. Col. Marseille, sér. 3, 1: 165, t.8 (1913). TYPE: Madagascar, Belambo, nr. Mevatavana, left bank of the Ikopa, *H.Perrier* 1058 (holo. P).

DISTRIBUTION: Madagascar: Mahajanga.
HABITAT: on shaded rocks; semi-deciduous forest.
ALTITUDE: 150 – 700 m.
FLOWERING TIME: February – April.
LIFE FORM: epiphyte, in forest.
PHYTOGEOGRAPHICAL REGION: V.
DESCRIPTION: {896}, {1205}.
ILLUSTRATION: {492}, {491}, {1205}.
NOTES: flowers small and russet – orange; mentum short.

Polystachya clareae *Hermans*, in Orch. Rev. 111, 1254: 354 – 357 (2003). TYPE: Madagascar, Antananarivo province, Manjakandriana area, fl. in cult., *Hermans* 3523 (holo. K).

DISTRIBUTION: Madagascar: Antananarivo.
HABITAT: in mid-elevation, humid, evergreen forest.
ALTITUDE: 700 – 950m.
LIFE FORM: epiphyte or terrestrial.
PHYTOGEOGRAPHICAL REGION: III.
DESCRIPTION: {491}.
ILLUSTRATION: {491}, {492}
NOTES: flowers close together, large, bright orange-yellow.

Polystachya concreta *(Jacq.) Garay & H. R. Sweet* in Orquideologia 9 (3): 206 (1974). TYPE from West Indies.
Epidendrum concretum Jacq., Enum. Syst. Pl.: 30 (1760).
Epidendrum minutum Aubl., Hist. Pl. Guin. Fr. 2: 824 (1775), based on *Helleborine ramosa, floribus minimus, luteis*, Plum. Cat. 9, 1703 & Icon Burm. t.185, f.1 (1759), *non Polystachya minuta* Rich. & Gal. (1845).
Cranichis luteola Sw., Fl. Ind. Occ. 3: 1433 (1804), *nom. illeg.*
Dendrobium polystachyum Thouars, Hist. Orchid.: t.85 (1822).
Onychium flavescens Blume, Bijd. Fl. Nederl. Indie 7: 325 (1825).
Polystachya luteola Hook., Exot. Fl. 2: t.103 (1825), *nom. illeg.*
Polystachya mauritiana Spreng., Syst. Veg. 3: 742 (1826) Based upon: *Dendrobium polystachys* Thouars.
Dendrobium flavescens Lindl., Gen. & Sp. Orch. Pl.: 85 (1830).
Polystachya zeylanica Lindl. in Bot. Reg. 24: Misc. p.78 (1838).
Polystachya estrellensis Rchb.f. in Linnaea 25: 231 (1852).
Polystachya zollingeri Rchb.f. in Bonpl. 5: 39 (1857).
Polystachya tessellata Lindl. in J. Linn. Soc. 6: 130 (1862).
Polystachya extinctora Rchb.f. in Walp. Ann. Bot. Syst. 6: 638 (1863).
Polystachya wightii Rchb.f., *loc.cit.* 640.
Polystachya jussieuana Rchb.f., *loc.cit.* 641. TYPE: Madagascar, ex herb. Jussieu (holo. W).
Polystachya modesta Rchb.f., Flora 50: 114 (1867).
Polystachya nitidula Rchb.f., *loc.cit.*
Polystachya rigidula Rchb.f., *loc.cit.* 117.
Polystachya tricruris Rchb.f., *loc.cit.* 118.
Polystachya rufinula Rchb.f. in Gard. Chron. n.s. 11: 41 (1879).
Polystachya hypocrita Rchb.f. in Gard. Chron. n.s. 16: 685, (1881).
Polystachya similis Rchb.f. in Otia Bot. Hamb. 2: 82 (1881).
Polystachya shirensis Rchb.f., *loc.cit.*
Callista flavescens Kuntze, Rev. Gen. Pl.2: 654 (1891).
Dendrorchis minuta Kuntze, *loc.cit.* 658.
Dendrorchis estrellensis Kuntze, Rev. Gen. Pl. pt.2: 659 (1891).
Dendrorchis extinctora Kuntze, *loc.cit.*
Dendrorkis jussieuana (Rchb.f.) Kuntze, *loc.cit.* 658.
Dendrorchis rufinula Kuntze, *loc.cit.* 659.
Dendrorchis shirensis Kuntze, *loc.cit.* 658.

Dendrorchis similis Kuntze, *loc.cit.* 659.

Dendrorchis tesselata Kuntze, *loc.cit.*

Dendrorchis wightii Kuntze, *loc.cit.*

Dendrorchis zollingeri Kuntze, *loc.cit.*

Polystachya buchanani Rolfe in Kew Bull. 335 (1893).

Epidendrum resupinatum Sessé & Moc., Fl. Mex. ed.2: 202 (1894).

Polystachya minuta Frapp. in Cat. Orch. Reun. 14 (1880), *nom. nud.;* Cordem. ex Frapp. in Fl. Réunion: 190 (1895), *non* Rich. & Gal. (1845).

Polystachya siamensis Ridley in J. Linn. Soc. Bot. 32: 343 (1896).

Polystachya zanguebarica Rolfe in in Fl. Trop. Afr. 7: 115 (1897).

Polystachya gracilis De Wild., Not. Pl. Util. Congo 1: 136 (1903).

Polystachya latifolia De Wild., *loc.cit.* 138.

Polystachya mukandaensis De Wild., *loc.cit.* 139.

Polystachya huyghei De Wild., *loc.cit.* 314 (1904).

Polystachya praealta Kraenzl. in Bot. Jahrb. Syst. 36, 118 (1905).

Polystachya plehniana Schltr.in Bot. Jahrb. Syst. 38: 8 (1905).

Polystachya flavescens (Blume) J.J.Sm., Fl. Buitenz. 6: 284 (1905).

Polystachya dorotheae Rendle in Cat. Talb. Niger. Pl. 103 (1913).

Polystachya colombiana Schltr. in Repert. Spec. Nov. Regni Veg. Beih. 7: 156 (1920).

Polystachya caquetana Schltr. in Repert. Spec. Nov. Regni Veg. Beih. 27: 79 (1924).

Polystachya lepidantha Kraenzl. in Kew Bull. 290 (1926).

Polystachya lettowiana Kraenzl. in Repert. Spec. Nov. Regni Veg. Beih. 39: 127 (1926).

Polystachya reichenbachiana Kraenzl., *loc.cit.*

Polystachya gabonensis Summerh. in Kew. Bull. 1929: 496 (1939).

Polystachya kraenzliniana Pabst in Rodriguesia 30 – 31: 34 (1957).

> DISTRIBUTION: Madagascar: Antananarivo, Antsiranana, Fianarantsoa, Mahajanga, Toamasina, Toliara. Comoros: Anjouan, Grande Comore, Mayotte. Mascarenes: Réunion, Mauritius. Seychelles. Also Florida, Guyana, Surinam, Brazil, W. Indies, tropical Africa & Asia.
>
> HABITAT: seasonally dry, deciduous woods; humid evergreen forest; shaded/humid rocks.
>
> ALTITUDE: sea level – 1400 m.
>
> FLOWERING TIME: December – August.
>
> LIFE FORM: epiphyte or lithophyte.
>
> PHYTOGEOGRAPHICAL REGION: I, II, III.
>
> DESCRIPTION: {896}, {1375}, {768}.
>
> ILLUSTRATION: {167}, {168}, {414}, {504}, {557}, {768}, {1205}, {1223}, {1263}, {1398}.
>
> CULTIVATION: {1263}.
>
> NOTES: a very variable and widespread species; plant with many broad, oblanceolate leaves; flowers small; lip oblong, descending between the triangular petals and fused with them on both sides. For type citations see Garay & Sweet in Orquideologia 9: 206 (1974).

Polystachya cornigera *Schltr.* in Repert. Sp. Nov. Regni Veg. Beih. 33: 155 (1924). TYPE: Madagascar, Mt. Tsaratanana, *H.Perrier* 15751 (holo. P).

> DISTRIBUTION: Madagascar: Antananarivo, Antsiranana, Fianarantsoa, Toamasina.
>
> HABITAT: lichen-rich forest on moss- and lichen-covered trees.
>
> ALTITUDE: 1000 – 2000 m.
>
> FLOWERING TIME: January – March.
>
> LIFE FORM: epiphyte.
>
> PHYTOGEOGRAPHICAL REGION: III.
>
> DESCRIPTION: {896}.
>
> NOTES: plant and flowers small; lip 3-lobed, with a distinct horn at the base, margins slightly undulate.

Polystachya cornigera var. **integrilabia** *Senghas* in Adansonia sér. 2, 4: 307 (1964). TYPE: Madagascar, Ranomafana, *Rauh* M 654 (holo. HEID).

> DISTRIBUTION: Madagascar: Fianarantsoa.
>
> HABITAT: humid, evergreen forest on plateau.

ALTITUDE: c. 1200 m.
FLOWERING TIME: October.
PHYTOGEOGRAPHICAL REGION: III.
DESCRIPTION: {1239}.
NOTES: differs from the typical variety by its entire lip.

Polystachya cultriformis *(Thouars) Spreng.*, Syst. Veg. 3: 742 (1826). TYPE: *Thouars* s.n. (holo. P).
Dendrobium cultriforme Thouars, Hist. Orchid.: t.87 (1822).
Polystachya cultrata Lindl. in Bot. Reg.: sub t.851 (1824), *nom. superfl.*
Polystachya cultriformis var. *humblotii* Rchb.f. in Flora 68: 379 (1885).
Polystachya cultriformis var. *nana* S.Moore, Fl. Maur.: 362 (1877).
Polystachya gerrardii Harv., Thes. Cap. ii. 49. 2: 49 (1863). TYPE from South Africa.
Polystachya lujae Wildem., Pl. Nov. Hort. Then. t.18.: t.18 (1904). TYPE from Dem. Rep. Congo.
Polystachya cultriformis var. *humblotii* Rchb.f. in Flora 68: 379 (1885).
Polystachya kirkii Rolfe in Kew Bull. 1895: 283 (1895) TYPE from Tanzania.
Dendrorkis appendiculata (Kraenzl.) Kuntze, Deutsche Bot. Monatsschr. 21: 173 (1903).
Polystachya appendiculata Kraenzl. in Notizbl. Bot. Gart. Berl. 3: 238 (1903).
Polystachya cultriformis var. *africana* Schltr. in Notizbl. Bot. Gart. Berlin 9: 20 (1924).
Polystachya cultriformis var. *autogama* Schltr. in Notizbl. Bot. Gart. Berlin 9: 20 (1924).
Polystachya cultriformis var. *occidentalis* Kraenzl. in Repert. Sp. Nov. Regni Veg. Beih. 39: 119 (1926).

DISTRIBUTION: Madagascar: Antananarivo, Antsiranana, Fianarantsoa, Toamasina, Toliara, Mahajanga. Comoros: Grande Comore, Mayotte. Mascarenes: Réunion, Mauritius. Seychelles. Also in tropical and southern Africa.
HABITAT: coastal forest, seasonally dry deciduous woods, humid evergreen forest.
ALTITUDE: sea level – 2500 m.
FLOWERING TIME: January – December.
LIFE FORM: epiphyte, lithophyte, terrestrial.
PHYTOGEOGRAPHICAL REGION: I, II, III, IV.
DESCRIPTION: {219}, {896}, {768}.
ILLUSTRATION: {3}, {50}, {167}, {256b}, {414}, {503}, {676}, {768}, {1223}, {1281}, {1329}, {1345}, {1398}, {1472}.
VERNACULAR NAME: Sonjomboae.
NOTES: a very common and widespread species, distinct by its single leaf, branching inflorescence and large, whitish-yellow flowers. Cordemoy lists 2 varieties (*alba* and *erubescens*), {206}, {310}.

Polystachya fusiformis *(Thouars) Lindl.* in Bot. Reg.: sub t.851 (1824). TYPE: Réunion, *Thouars* s.n. (holo. P).
Dendrobium fusiforme Thouars, Hist. Orchid.: t.86 (1822).
Polystachya minutiflora Ridl. in J. Linn. Soc., Bot. 20: 330 (1883). TYPE: Madagascar, Ankafana, *Deans Cowan* s.n. (holo. BM).
Polystachya multiflora Ridl. in J. Linn. Soc., Bot. 21: 475 (1885).
Dendrorkis minutiflora (Ridl.) Kuntze, Rev. Gen. Pl. 2: 658 (1891) .
Polystachya composita Kraenzlin in Repert. Spec. Nov. Regni Veg. Beih. 39: 103 (1926).

DISTRIBUTION: Madagascar: Antananarivo, Antsiranana, Fianarantsoa, Toamasina, Toliara. Mascarenes: Réunion, Mauritius. Seychelles? Also in tropical and southern Africa.
HABITAT: dry forest; humid, evergreen forest on plateau.
ALTITUDE: 600 – 2000 m.
FLOWERING TIME: December – April.
LIFE FORM: epiphyte or lithophyte.
PHYTOGEOGRAPHICAL REGION: III, IV.
DESCRIPTION: {896}, {768}.
ILLUSTRATION: {34}, {246}, {384}, {473}, {676}, {684}, {768}, {1329}, {1345}, {1398}, {1472}.
NOTES: a common and widespread species; stems cylindrical-fusiform, superposed and pendulous; inflorescence branched, pubescent or hirsute; flowers tiny. Cordemoy recognises two varieties in Réunion, *virescens* and *purpurascens* {206}.

Polystachya heckeliana *Schltr.* in Ann. Mus. Col. Marseille, sér. 3, 1: 167, t.8 (1913). TYPE: Madagascar, Manongarivo, *H.Perrier* 1936 (21) (holo. P).

DISTRIBUTION: Madagascar: Antsiranana.
HABITAT: lichen-rich forest; dry forest.
ALTITUDE: 1000 – 1700 m.
FLOWERING TIME: May.
LIFE FORM: terrestrial.
PHYTOGEOGRAPHICAL REGION: III.
DESCRIPTION: {896}.
ILLUSTRATION: {1205}, {1223}.
NOTES: inflorescence not branched; flowers large; lip sulphur yellow, other parts whitish pink.

Polystachya henrici *Schltr.* in Repert. Sp. Nov. Regni Veg. Beih. 33: 158 (1924). TYPE: Madagascar, nr. Ambatofangena, *H.Perrier* 12391 (holo. P).

DISTRIBUTION: Madagascar: Fianarantsoa.
HABITAT: grassy and xerophytic scrub; rocky outcrops.
ALTITUDE: 1600 – 1700 m.
FLOWERING TIME: February.
LIFE FORM: epiphyte or lithophyte.
PHYTOGEOGRAPHICAL REGION: III.
DESCRIPTION: {896}.
ILLUSTRATION: {1223}.
NOTES: related to *P. rosea*, but leaves broader, plant less than 25 cm tall, inflorescence short, and flowers yellowish.

Polystachya humbertii *H.Perrier* in Bull. Soc. Bot. France 83: 33 (1936). TYPE: Madagascar, Andringitra massif, *Humbert* 3716 (lecto. B; iso. P); Mt. Itrafanaomby (Ankazondrano) and foothills SW (Haut-Mandrare), *Humbert* 13495 (syn. K, P, TAN).

DISTRIBUTION: Madagascar: Fianarantsoa, Toliara, Antananarivo, Toamasina.
HABITAT: ericaceous scrub and humid montane forest.
ALTITUDE: 1000 – 2000 m.
FLOWERING TIME: November – December.
LIFE FORM: epiphyte.
PHYTOGEOGRAPHICAL REGION: III, IV.
DESCRIPTION: {896}.
ILLUSTRATION: {896}.
NOTES: plant small, under 8 cm tall; leaves broad; flowers c. 10 mm long, white with a yellow lip, spotted with purple in the centre; lip with a large obtuse callus at its base.

Polystachya monophylla *Schltr.* in Beih. Bot. Centralbl. 34 (2): 322 (1916). TYPE: Madagascar, Mt. Ibity, S of Antsirabe, *H.Perrier* 8099 (XXIII) (holo. P).

DISTRIBUTION: Madagascar: Antananarivo, Antsiranana.
HABITAT: quartz rocks.
ALTITUDE: 1800 – 2300 m.
FLOWERING TIME: February.
PHYTOGEOGRAPHICAL REGION: IV.
DESCRIPTION: {896}, {1221}.
ILLUSTRATION: {1223}
NOTES: plant with 1 or 2 leaves; flowers c. 11 mm long, yellow; lip 3-lobed with 3 fleshy crests.

Polystachya oreocharis *Schltr.* in Repert. Sp. Nov. Regni Veg. Beih. 33: 160 (1924). TYPE: Madagascar, Mt. Tsaratanana, *H.Perrier* 15752 (holo. P).

DISTRIBUTION: Madagascar: Antsiranana, Fianarantsoa, Antananarivo, Toamasina.
HABITAT: lichen-rich forest on moss- and lichen-covered trees.
ALTITUDE: 950 – 2000 m.

FLOWERING TIME: January – April.
LIFE FORM: epiphyte.
PHYTOGEOGRAPHICAL REGION: III, IV.
DESCRIPTION: {896}.
ILLUSTRATION: {896}, {281}.
NOTES: pseudobulbs sympodial and ascending, very rarely branched, inflorescence simple, flowers reddish or greenish, small, lip 4 × 2 mm, with an incurved callus.

Polystachya pergibbosa *H.Perrier* in Notul. Syst. (Paris) 14: 149 (1951). TYPES: Madagascar, Mt. Beondroka, Lokoho Valley, *Humbert* 23390 (lecto. P, selected here); E summit of Mt. Marojejy, *Humbert* & Cours 23869; Mt. Beondroka, Lokoho valley, *Humbert* 23598 (both syn. P).

DISTRIBUTION: Madagascar: Antsiranana.
HABITAT: humid, mossy, evergreen forest on plateau.
ALTITUDE: 500 – 2000 m.
FLOWERING TIME: January – March.
LIFE FORM: epiphyte.
PHYTOGEOGRAPHICAL REGION: III, IV.
DESCRIPTION: {901}.
NOTES: plant and pseudobulbs small; lip 3-lobed, the mid-lobe apiculate.

Polystachya perrieri *Schltr.* in Beih. Bot. Centralbl. 34 (2): 322 (1916). TYPE: Madagascar, Mt. Ibity, S of Antsirabe, *H.Perrier* 8077 (holo. P).

DISTRIBUTION: Madagascar: Antananarivo.
HABITAT: rocky outcrops.
ALTITUDE: c. 1900 m.
FLOWERING TIME: February.
LIFE FORM: epiphyte or lithophyte.
PHYTOGEOGRAPHICAL REGION: III.
DESCRIPTION: {896}.
ILLUSTRATION: {1223}.
NOTES: a poorly known species, possibly the same as *P. anceps*.

Polystachya rhodochila *Schltr.* in Beih. Bot. Centralbl. 34 (2): 323 (1916). TYPE: Madagascar, nr. Tsaratanana, *H.Perrier* 136 (holo. P).

DISTRIBUTION: Madagascar: Antsiranana, Mahajanga.
HABITAT: lichen-rich forest.
ALTITUDE: 1400 – 2000 m.
FLOWERING TIME: October – March.
LIFE FORM: terrestrial or rarely epiphytic.
PHYTOGEOGRAPHICAL REGION: III.
DESCRIPTION: {896}.
ILLUSTRATION: {281}, {672}, {896}, {1223}.
NOTES: plant up to 50 cm tall; flowers white; lip 9 × 10 mm, 3-lobed, red.

Polystachya rosea *Ridl.* in J. Linn. Soc., Bot. 21: 474 (1885). TYPE: Madagascar, Imerina, *Deans Cowan* s.n. (holo. BM).
Dendrorkis rosea (Ridl.) Kuntze in Rev. Gen. Pl. 2: 658 (1891).
Polystachya bicolor Rolfe in Kew Bull.: 114 (1906). TYPE: Seychelles, *Thomasset* 58 (holo. K).
Polystachya hildebrandtii Kraenzl. in Ann. Naturh. Mus. Wien 36: 6 (1923). TYPE: Madagascar, S Betsileo, Ankafana, *Hildebrandt* 4222 (holo. W).

DISTRIBUTION: Madagascar: Antananarivo, Fianarantsoa, Mahajanga, Antsiranana, Toamasina. Seychelles ?
HABITAT: shaded, humid, rocky outcrops; lichen-rich forests.
ALTITUDE: 800 – 1500 m.
FLOWERING TIME: January- May.

LIFE FORM: lithophyte or terrestrial.
PHYTOGEOGRAPHICAL REGION: III, IV.
DESCRIPTION: {896}.
ILLUSTRATION: {246}, {504}, {896}, {911}.
NOTES: a very variable species both in colour and shape. Plant slender with narrow leaves; mentum long; lip 3-lobed without an obvious callus.

Polystachya rosellata *Ridl.* in J. Linn. Soc., Bot. 20: 330 (1884). TYPE: Madagascar, Ankafana, *Deans Cowan* s.n. (holo. BM).
Dendrorkis rosellata (Ridl.) Kuntze, Rev. Gen. Pl. 2: 658 (1891).

DISTRIBUTION: Madagascar: Antananarivo, Antsiranana.
HABITAT: montane, ericaceous scrub and humid forest, lichen-rich forest.
ALTITUDE: 1000 – 1800 m.
FLOWERING TIME: January – March.
LIFE FORM: epiphyte.
PHYTOGEOGRAPHICAL REGION: III, IV.
DESCRIPTION: {896}.
ILLUSTRATION: {246}, {896}.
NOTES: leaves c. 6 mm wide; bracts less than a quarter of the ovary; flowers small, c. 6 mm long, pink; lip with a callus.

Polystachya tsaratananae *H.Perrier* in Bull. Soc. Bot. France 83: 32 (1936). TYPE: Madagascar, Mt. Tsaratanana, *H.Perrier* 16497 (holo. P).

DISTRIBUTION: Madagascar: Antsiranana.
HABITAT: shaded, humid rocks.
ALTITUDE: 1500 – 2000 m.
FLOWERING TIME: April – May.
LIFE FORM: lithophyte.
PHYTOGEOGRAPHICAL REGION: III, IV.
DESCRIPTION: {896}.
ILLUSTRATION: {896}.
NOTES: leaves dark green or purplish green; flowers large, purple-red; lip entire, covered by golden hairs.

Polystachya tsinjoarivensis *H.Perrier* in Bull. Soc. Bot. France 83: 34 (1936). TYPE: Madagascar, Tsinjoarivo, *H.Perrier* 16921 (holo. P).

DISTRIBUTION: Madagascar: Antananarivo, Fianarantsoa.
HABITAT: humid, evergreen forest on plateau.
ALTITUDE: 1300 – 2000 m.
FLOWERING TIME: February.
LIFE FORM: terrestrial.
PHYTOGEOGRAPHICAL REGION: III, IV.
DESCRIPTION: {896}.
ILLUSTRATION: {896}, {281}, {484}, {505}
NOTES: similar to *P. rosellata*. Plant small; flowers c. 8 mm long; lip 3-lobed, with a flattened callus at the base.

Polystachya virescens *Ridl.* in J. Linn. Soc., Bot. 21: 474 (1885). TYPE: Madagascar, Ankafana, *Deans Cowan* s.n. (lecto. the watercolour illustrations in Deans Cowan's sketchbook at BM, chosen here).
Dendrorkis virescens (Ridl.) Kuntze, Rev. Gen. Pl. 2: 658 (1891).

DISTRIBUTION: Madagascar: Antananarivo, Antsiranana, Fianarantsoa, Mahajanga, Toamasina.
HABITAT: humid, evergreen forest on plateau.
ALTITUDE: 700 – 1800 m.
FLOWERING TIME: October – April.
LIFE FORM: epiphyte.

PHYTOGEOGRAPHICAL REGION: I, III.

DESCRIPTION: {896}.

ILLUSTRATION: {246}, {896}.

NOTES: very similar to *P. concreta*; plant small; flowers c. 7 mm long, greenish; lip 3-lobed, with a small callus at the base.

Polystachya waterlotii *Guillaumin* in Bull. Mus. Hist. Nat. (Paris) 34: 359 (1928). TYPE: Comoros, without locality, *Waterlot* 17 (holo. P).

DISTRIBUTION: Madagascar: Antananarivo. Comoros.

HABITAT: montane, ericaceous scrub and humid forest.

FLOWERING TIME: June.

PHYTOGEOGRAPHICAL REGION: III.

DESCRIPTION: {896}.

NOTES: a poorly known species close to *P. cornigera* but distinct by the absence of the horn on the lip.

Satyrium amoenum *(Thouars) A.Rich.* in Mém. Soc. Hist. Nat. Paris 4: 31 (1828). TYPE: Réunion, *Thouars* s.n. (holo. P).
Diplectrum amoenum Thouars, Hist. Orchid.: t.21 & 22 (1822).
Satyrium gracile Lindl., Gen. Sp. Orch. Pl.: 338 (1838). TYPE: Madagascar, *Lyall* s.n. (holo. K).

DISTRIBUTION: Madagascar: Antananarivo, Antsiranana, Fianarantsoa. Comoros: Grande Comore. Mascarenes: Réunion, Mauritius.

HABITAT: dry meadows; grassland; rocky outcrops.

ALTITUDE: 1100 – 2000 m.

FLOWERING TIME: October – June.

LIFE FORM: terrestrial.

PHYTOGEOGRAPHICAL REGION: III, IV.

DESCRIPTION: {896}.

ILLUSTRATION: {166}, {246}, {1289}, {1398}.

NOTES: flower whitish-pink c. 12 mm long; spurs c. 30 mm long.

Satyrium amoenum var. **tsaratananae** *H.Perrier* ex *Hermans* **var. nov.** TYPE: Madagascar, Mt. Tsaratanana, *H.Perrier* 16507 (holo. P). See Appendix .
Satyrium amoenum var. *tsaratananae* H.Perrier in Humbert, Fl. Mad. Orch. 1: 168 (1939), *nom. nud.*

DISTRIBUTION: Madagascar: Antsiranana.

HABITAT: montane, ericaceous scrub.

ALTITUDE: c. 2400 m.

FLOWERING TIME: April.

LIFE FORM: terrestrial.

PHYTOGEOGRAPHICAL REGION: IV.

DESCRIPTION: {896}.

NOTES: differs in having a perianth which is smaller than the typical variety with 4 mm long sepals and petals 4 mm and 12 mm long spurs; sepals not apiculate, dorsal sepal is 3-veined, the petals single-veined.

Satyrium baronii *Schltr.* in Bot. Jahrb. Syst. 24: 423 (1897). TYPE: Madagascar, *Baron* s.n. (holo. B†).

DISTRIBUTION: Madagascar: Antananarivo.

HABITAT: grassland, amongst rocky outcrops.

ALTITUDE: 1100 – 2200 m.

FLOWERING TIME: March – May.

LIFE FORM: terrestrial.

PHYTOGEOGRAPHICAL REGION: III, IV.

DESCRIPTION: {896}.

NOTES: flowers purple-pink, c. 6 mm long; sepals and petals spathulate, fused for more than half their length; spurs c. 7 mm long.

Satyrium perrieri *Schltr.* in Beih. Bot. Centralbl. 34 (2): 317 (1916). TYPE: Madagascar, betw. Betafo & Ambohimasina, *H.Perrier* 8122 (XXXV) (holo. B†; iso. P).

DISTRIBUTION: Madagascar: Antananarivo, Fianarantsoa.
HABITAT: grassland; montane ericaceous scrub; rocky outcrops.
ALTITUDE: 1000 – 1600 m.
FLOWERING TIME: March – April.
LIFE FORM: terrestrial.
PHYTOGEOGRAPHICAL REGION: III.
DESCRIPTION: {896}.
ILLUSTRATION: {896}.
NOTES: raceme longly cylindrical, 13 – 25 × 2 cm, narrow and densely flowered; spurs up to 12 mm long.

Satyrium rostratum *Lindl.*, Gen. Sp. Orch. Pl.: 338 (1838). TYPE: Madagascar, *Lyall* (holo. K).
Satyrium gigas Ridl. in J. Linn. Soc., Bot. 22: 126 (1886). TYPE: Madagascar, Ambatovary, Imerina, *Fox* s.n. (holo. BM).

DISTRIBUTION: Madagascar: Antananarivo, Fianarantsoa.
HABITAT: grassland; rocky outcrops; forest and woodland margins.
ALTITUDE: 1200 – 2000 m.
FLOWERING TIME: December – April.
LIFE FORM: terrestrial.
PHYTOGEOGRAPHICAL REGION: III.
DESCRIPTION: {896}.
ILLUSTRATION: {246}, {281}, {504}, {896}.
VERNACULAR NAME: Rasamoala.
NOTES: plant up 90 cm tall; raceme 5 cm in diameter; flowers pink; spurs 35 – 45 mm long.

Satyrium trinerve *Lindl.*, Gen. Sp. Orchid. Pl.: 34 (1838). TYPE: Madagascar, without locality, *Lyall* s.n., (holo. K).
Satyrium atherstonei Rchb.f. in Flora 64: 328 (1881). TYPE from South Africa.
Satyrium zombense Rolfe in Fl. Trop. Afr. 7: 273 (1898). TYPE from Malawi.
Satyrium occultum Rolfe, *loc. cit.* (1898). TYPE from Malawi.
Satyrium nutii Rolfe, *loc. cit.* (1898). TYPE from Zambia.
Satyrium monopetalum Kraenzl., Orchid. Gen. Sp. 1: 662 (1899). TYPE from South Africa.
Satyrium triphyllum Kraenzl., Orchid. Gen. Sp. 1: 660 (1899). TYPE from South Africa.

DISTRIBUTION: Madagascar: Antananarivo, Fianarantsoa, Toamasina. Comoros: Grande Comore. Widespread in tropical; Africa Zambia, Zimbabwe, Malawi, Mozambique.
HABITAT: grassland or marshland, usually in moist localities.
ALTITUDE: 500 – 2000 m.
FLOWERING TIME: December – June.
LIFE FORM: terrestrial.
PHYTOGEOGRAPHICAL REGION: I, III, IV.
DESCRIPTION: {768}, {896}.
ILLUSTRATION: {246}, {281}, {504}, {768}, {1329}.
NOTES: a locally common species; ovary very short, pubescent-papillose; flowers c. 5 mm long, white, the segments fused over almost 2 mm; spurs obtuse, less than 2.2 mm long.

Sobennikoffia fournieriana *(Kraenzl.) Schltr.* in Repert. Sp. Nov. Regni Veg. Beih. 33: 362 (1925). TYPE: Madagascar, cult. Sander & Co. (holo. not located).
Angraecum fournierianum Kraenzl. in Gard. Chron. sér. 3, 15: 808 (1894), *non* André (1896).

DISTRIBUTION: Madagascar.
DESCRIPTION: {646}, {896}.
ILLUSTRATION: {11}?, {1189}.
NOTES: The plant described as *Angraecum fournierae* and illustrated by André {12} is *Angraecum stylosum* Rolfe. The flowers illustrated in {11} refers to this species. Leaves 65 – 70 cm long; flowers large, lateral lobes of the lip rounded; spur the same length as the lip or a little longer.

Sobennikoffia humbertiana *H.Perrier* in Notul. Syst. (Paris) 7: 134 (1938). TYPE: Madagascar, upper basin Onilahy, Andranomiforitra, *Humbert* 7056 (holo. P).

DISTRIBUTION: Madagascar: Fianarantsoa, Toliara.
HABITAT: humid, evergreen forest on plateau; dry, deciduous scrubland.
ALTITUDE: 400 – 1200 m.
FLOWERING TIME: October – November.
LIFE FORM: lithophyte.
PHYTOGEOGRAPHICAL REGION: III, VI.
DESCRIPTION: {896}.
ILLUSTRATION: {50}, {84}, {281}, {470}, {529}, {896}, {1021}, {1291}.
CULTIVATION: {698}.
NOTES: leaves less than 20 cm long; lip mid-lobe elongate, 10 – 12 mm long, the lateral lobules obtuse; spur a third shorter than the lip.

Sobennikoffia poissoniana *H.Perrier* in Notul. Syst. (Paris) 14: 164 (1951). TYPES: Madagascar, Ankorika, *Decary* s.n. & Camp d'Ambre, Sakaramy, *H.Poisson* 43 (both syn. P).

DISTRIBUTION: Madagascar: Antsiranana, Mahajanga.
HABITAT: rocks; coastal vegetation.
ALTITUDE: sea level.
FLOWERING TIME: November – December.
LIFE FORM: epiphyte or lithophyte.
PHYTOGEOGRAPHICAL REGION: I.
DESCRIPTION: {901}.
NOTES: plant small; leaves up to 10 cm long; lip 3-lobed, 2 – 2.3 mm long; spur cylindrical, tapering, 15 – 17 mm long.

Sobennikoffia robusta *(Schltr.) Schltr.* in Repert. Sp. Nov. Regni Veg. Beih. 33: 362 (1925). TYPE: Madagascar, Manongarivo, *H.Perrier* 1653 (holo. P).
Oeonia robusta Schltr. in Ann. Mus. Col. Marseille, sér. 3, 1: 184 (41), t.18 (1913).
Angraecum robustum (Schltr.) Schltr. in Beih. Bot. Centralbl. 33 (2): 437 (1915), *non* Kraenzl. (1930).

DISTRIBUTION: Madagascar: Mahajanga.
HABITAT: seasonally dry deciduous woods and scrub.
ALTITUDE: 1500 – 2000 m.
FLOWERING TIME: November – January.
LIFE FORM: epiphyte or lithophyte.
PHYTOGEOGRAPHICAL REGION: V.
DESCRIPTION: {896}.
ILLUSTRATION: {3}, {29}, {50}, {281}, {457}, {520}, {529}, {896}, {944}, {1041} wrongly identified, {1205}, {1223}, {1259}, {1271}, {1281}.
CULTIVATION: {529}.
NOTES: leaves up to 35 cm long; mid-lobe of the lip shorter and the lateral lobules subacute; spur the same length as the lip or a little longer.

Tylostigma filiforme *H.Perrier* in Notul. Syst. (Paris) 14: 138 (1951). TYPES: Madagascar, E summit of Marojejy massif, Manantenina, *Humbert* 23861 & *Humbert* 23862 (both syn. P).

DISTRIBUTION: Madagascar: Antsiranana.
HABITAT: marshes and in marshy depressions.
ALTITUDE: 1800 – 2000 m.
LIFE FORM: terrestrial.
PHYTOGEOGRAPHICAL REGION: IV.
DESCRIPTION: {901}.
NOTES: similar to *T. perrieri*, but with the sheaths without a blade, the leaves developed above the middle, and a different lip structure.

Tylostigma filiforme subsp. **bursiferum** *H.Perrier* in Notul. Syst. (Paris) 14: 138 (1951). TYPE: Madagascar, Mt. Beandroko, nr. Maroambihy, *Humbert* 23595 (holo. P).

DISTRIBUTION: Madagascar.
HABITAT: marshes.
ALTITUDE: c. 1400 m.
FLOWERING TIME: March – April.
LIFE FORM: terrestrial.
PHYTOGEOGRAPHICAL REGION: III.
DESCRIPTION: {901}.
NOTES: differs from the typical form in its smaller habit and the base of the lip which is transversely purse-shaped.

Tylostigma foliosum *Schltr.* in Repert. Sp. Nov. Regni Veg. Beih. 33: 21 (1924). TYPE: Madagascar, Andringitra massif, *H.Perrier* 14381 (holo. P; iso. K).

DISTRIBUTION: Madagascar: Fianarantsoa.
HABITAT: marshes in peaty and marshy soil.
ALTITUDE: 2000 – 2500 m.
FLOWERING TIME: January.
LIFE FORM: terrestrial.
PHYTOGEOGRAPHICAL REGION: III.
DESCRIPTION: {896}.
ILLUSTRATION: {896}.
NOTES: known only from the type. Leaves lanceolate-linear, 6 – 12 mm wide; raceme densely several-flowered; floral bracts as long as the flowers.

Tylostigma herminioides *Schltr.* in Repert. Sp. Nov. Regni Veg. Beih. 33: 21 (1924). TYPE: Madagascar, Mt. Tsiafajavona, *H.Perrier* 13561 (holo. P).

DISTRIBUTION: Madagascar: Antananarivo.
HABITAT: marshes.
ALTITUDE: c. 2200 m.
FLOWERING TIME: March.
LIFE FORM: terrestrial.
PHYTOGEOGRAPHICAL REGION: IV.
DESCRIPTION: {896}.
ILLUSTRATION: {1223}.
NOTES: known only from the type. Intermediate between *T. hildebrandtii* and *T. tenellum*; leaves towards the middle of the stem; flowers closer together and smaller; lip narrowed in the middle.

Tylostigma hildebrandtii *(Ridl.) Schltr.* in Repert. Sp. Nov. Regni Veg. Beih. 33: 22 (1924). TYPES: Madagascar, Ankaratra, *Hildebrandt* 3860a (syn. BM; isosyn. K); Imerina, *Hilsenberg & Bojer*, Antsirabe, *H.Perrier* 1310 (syn. BM; isosyn. P); Ankafana, Imerina, *Deans Cowan* s.n. (syn. BM).
Habenaria hildebrandtii Ridl. in J. Linn. Soc., Bot. 21: 503 (1885).

DISTRIBUTION: Madagascar: Antananarivo.
HABITAT: marshes.
ALTITUDE: 1500 – 2000 m.
FLOWERING TIME: January – March.
LIFE FORM: terrestrial.
PHYTOGEOGRAPHICAL REGION: III.
DESCRIPTION: {896}.
ILLUSTRATION: {246}, {945}.
NOTES: plant slender; lip glabrous, without ridges; lateral sepals keeled.

Tylostigma madagascariense *Schltr.* in Beih. Bot. Centralbl. 34 (2): 298 (1916). TYPE: Madagascar, Antsirabe, *H.Perrier* 8092 (holo. B†; iso. P).

DISTRIBUTION: Madagascar: Antananarivo.
HABITAT: marshes in peaty and marshy soil.
ALTITUDE: c. 1500 m.
FLOWERING TIME: January.
LIFE FORM: terrestrial.
PHYTOGEOGRAPHICAL REGION: III.
DESCRIPTION: {896}.
ILLUSTRATION: {1223}.
NOTES: leaves narrowly linear or subfiliform; raceme laxly flowered; petals and sepals suborbicular; lip triangular.

Tylostigma nigrescens *Schltr* in Beih. Bot. Centralbl. 34 (2): 298 (1916). TYPE: Madagascar, Mt. Ibity, S of Antsirabe, *H.Perrier* 8106 (*Schlechter* XLVIII) (holo. B†; iso. P).

DISTRIBUTION: Madagascar: Antananarivo.
HABITAT: marshesin peaty and marshy soil; in cypress plantations.
ALTITUDE: 1500 – 2000 m.
FLOWERING TIME: December – February.
LIFE FORM: terrestrial.
PHYTOGEOGRAPHICAL REGION: III.
DESCRIPTION: {896}.
ILLUSTRATION: {896}, {1223}.
NOTES: plant slender up to 60 cm tall with 2 leaves; petals oblong; lip broadly oval, with an obsolete speculum at the base.

Tylostigma perrieri *Schltr.* in Beih. Bot. Centralbl. 34 (2): 298 (1916). TYPE: Madagascar, Central, Ankaratra, *H.Perrier* 8091 (IV) (holo. B†; iso. P).

DISTRIBUTION: Madagascar: Antananarivo, Fianarantsoa.
HABITAT: marshes in peaty and marshy soil.
ALTITUDE: 1500 – 2100 m.
FLOWERING TIME: April – July.
LIFE FORM: terrestrial.
PHYTOGEOGRAPHICAL REGION: III.
DESCRIPTION: {896}.
ILLUSTRATION: {896}, {1223}.
NOTES: plant slender, up to 35 cm tall; petals broadly oval; lip broadly triangular-obtuse, with a kidney-shaped speculum at the base.

Tylostigma tenellum *Schltr.* in Repert. Sp. Nov. Regni Veg. Beih. 33: 23 (1924). TYPE: Madagascar, Manandona, S of Antsirabe, *H.Perrier* 13121 (holo. P).

DISTRIBUTION: Madagascar: Antananarivo.
HABITAT: shaded and humid rocks.
ALTITUDE: c. 1600 m.
FLOWERING TIME: May, December.
LIFE FORM: lithophyte.
PHYTOGEOGRAPHICAL REGION: III.
DESCRIPTION: {896}.
ILLUSTRATION: {896}, {1223}.
NOTES: plant small, 6–11 cm tall, with 3 – 4 leaves, close together at the base; lip broadly oval-obtuse with a semi-rectangular speculum at the base.

Vanilla coursii *H.Perrier* in Rev. Int. Bot. Appl. Agr. Trop. 30: 435 (1950). TYPE: Madagascar, N of Antalaha, *Cours* 6 (holo. P).

DISTRIBUTION: Madagascar: Toliara.
HABITAT: humid, low-elevation forest, on sandy soil.
ALTITUDE: 200 – 300 m.
FLOWERING TIME: January – February.
LIFE FORM: epiphyte.
PHYTOGEOGRAPHICAL REGION: I.
DESCRIPTION: {931}.
NOTES: a poorly known species. Plant carrying leaves; flowers c. 2.5 cm long; lip with an obsolete callus at the base, carrying hairs.

Vanilla decaryana *H.Perrier* in Bull. Mus. Hist. Nat. (Paris), ed. 2, 6: 194 (1934). TYPES: Madagascar, Morondava, *Grevé* 39; Ambovombe, *Decary* 3531; Behara, E Ambovombe, *Decary* 9313; Mahatomotsy, N Ambovombe, *Decary* 9503 (all syn. P).

DISTRIBUTION: Madagascar: Mahajanga, Toliara.
HABITAT: dry, deciduous scrub, among *Didieraceae*.
ALTITUDE: sea level – 700 m.
FLOWERING TIME: October – January.
LIFE FORM: epiphyte.
PHYTOGEOGRAPHICAL REGION: VI.
DESCRIPTION: {149}, {896}.
ILLUSTRATION: {480}, {896}, {957}, {959}.
CULTIVATION: {457}.
VERNACULAR NAME: Vahy amalona.
NOTES: leafless liana; flower white; sepals 2.5 – 3 cm long; central crests of the lip carrying thick hairs from the base to the upper quarter. Said to have aphrodisiac properties.

Vanilla françoisii *H.Perrier* in Notul. Syst. (Paris) 8: 37 (1939). TYPE: Madagascar, Ambilo (S Toamasina), cult. Antananarivo, *E.François* s.n. (holo. P).

DISTRIBUTION: Madagascar: Toamasina.
HABITAT: coastal forest.
FLOWERING TIME: February.
LIFE FORM: epiphyte or terrestrial.
PHYTOGEOGRAPHICAL REGION: I.
DESCRIPTION: {896}, {931}.
NOTES: plant with a slender stem and carrying small leaves; flowers small; sepals c. 2 cm long; lip 18 – 20 mm long, rolled and crenulate at the front edge.

Vanilla humblotii *Rchb.f.* in Gard. Chron. n.s., 23: 726 (1885). TYPE: Comoros, Grande Comore, *Humblot* 1413 (holo. W; iso. P).

DISTRIBUTION: Comoros: Anjouan, Grande Comore, Mayotte. Madagascar ?
HABITAT: rocks; low-elevation vegetation.
ALTITUDE: sea level – 200 m.
FLOWERING TIME: November – December.
LIFE FORM: epiphyte, lithophyte.
DESCRIPTION: {896}, {931}.
ILLUSTRATION: {199}, {1128}, {1298}, {1315}, {1319}.
HISTORY: {8}, {324}, {571}, {1122}, {1128}, {1315}.
CULTIVATION: {1122}, {1128}.
NOTES: plant leafless; flower yellow with a red lip; lip covered with large papillae and hairs, front edges fringed; staminodes with the margins crenulate.

Vanilla madagascariensis *Rolfe* in J. Linn. Soc., Bot. 32: 476 (1896). TYPE: Madagascar, Bomatoe Bay, *Bojer* s.n. (holo. K).

DISTRIBUTION: Madagascar: Antsiranana, Mahajanga, Toliara.
HABITAT: coastal forest; humid evergreen forest; in dry forests in the east.
ALTITUDE: sea level – 800 m.
FLOWERING TIME: June – October.
LIFE FORM: scramblling liana.
PHYTOGEOGRAPHICAL REGION: I, II, V.
DESCRIPTION: {609}, {896}, {931}.
ILLUSTRATION: {281}, {426}, {470}, {672}, {693}, {717}, {719}, {720}, {722}, {741}, {896}, {946}, {1179}.
CULTIVATION: {458}.
VERNACULAR NAME: Amalo.
NOTES: plant leafless; flowers white, the lip tinted red-pink; lip disc carrying 2 crests which bear long fleshy hairs. Said to be an aphrodisiac.

Vanilla montagnacii *Portères* in Bouriquet, Vanillier & Vanille, Encycl. Biol. XIVI: 282 (1954), *nom. nud.*
This species was brought to our attention recently by Anton Sieder in Vienna. It was described by Portères in 1954 without a Latin description and therefore is a *nomen nudum*. It is based on a plant collected by P.R. Montagnac in the Marofandidia forest near Morandava, Tulear province. This specimen has not been traced yet.

Vanilla perrieri *Schltr.* in Repert. Sp. Nov. Regni Veg. Beih. 33: 114 (1924). TYPE: Madagascar, Ankarafantsika, *H.Perrier* 1851 (holo. P).

DISTRIBUTION: Madagascar: Mahajanga, Toliara.
HABITAT: seasonally dry, deciduous woods; in sandy, very dry woods.
ALTITUDE: c. 200 m.
FLOWERING TIME: October, January-February.
LIFE FORM: scrambling liana.
PHYTOGEOGRAPHICAL REGION: V.
DESCRIPTION: {896}, {931}.
ILLUSTRATION: {281}, {896}, {931}, {946}, {959}.
NOTES: plant leafless; flowers bright yellow with an orange-red throat; lip not papillose, just a scattering of hairs in the lower half.

Vanilla planifolia *Andrews*, Bot. Repos: t.538 (1808).
A native of Mexico but widely cultivated for its seed-pods, especially in the North-east of the island. Also a frequent escape from cultivation. Plant a scrambling liana; leaves elliptic-oblong, acute to apiculate; flowers green with a yellowish lip.

Zeuxine madagascariensis *Schltr.* in Ann. Mus. Col. Marseille, sér. 3, 1: 19, t.6 (1913). TYPE: Madagascar, Manongarivo massif, *H.Perrier* 1942 (holo. B†; iso. P).

DISTRIBUTION: Madagascar: Antsiranana.
HABITAT: humid, high-elevation forest.
ALTITUDE: c. 1700 m.
FLOWERING TIME: April.
LIFE FORM: epiphyte or terrestrial.
PHYTOGEOGRAPHICAL REGION: III.
DESCRIPTION: {896}.
ILLUSTRATION: {1205}, {1223}.
NOTES: known only from the type. Plant up to 60 cm tall; flowers downy on the outside, small; lip 10 mm long, inflated-gibbose at the base, with 2 rounded calli at the base.

Dubious Records

Diaphananthe rutila *(Rchb.f) Summerh.*: (as *Aeranthus rutilus* Rchb.f.) mentioned by Voeltzkow {1436} from the Comoros.

Polystachya affinis *Lindl.*, Gen. Sp. Orchid. Pl. 73 (1830). TYPE: Sierra Leone, *Don* 29.36 (holo. K).
Polystachya villosula Schltr. in Beih. Bot. Centralbl. 33 (2): 414 (1915).
Polystachya villosa Cogn. in J. Orch. 7: 139 (1896), *non* Rolfe (1894). TYPE: Madagascar, Cult. *Adde* s.n. (holo. not located).

> NOTES: *Polystachya villosa* described by Cogniaux as originating from Madagascar is *P. affinis* from mainland Africa. The origin of the cultivated plant is without a doubt erroneous and so does not belong in the Madagascan Flora.

Polystachya pubescens *(Lindl) Rchb.f.* (as *Epiphora pubescens* Lindl.) mentioned by Lindley {794} and Sibree {1286} from Madagascar.

Vanilla phalaenopsis *Rchb.f. ex Van Houtte*. listed by Stein {1291} from Madagascar. The type is from the Seychelles. This may be an earlier name for *V. madagascariensis*.

INDEX OF SYNONYMS

Acampe madagascariensis = **Acampe pachyglossa**

Acampe pachyglossa subsp. *renschiana* = **Acampe pachyglossa**

Acampe renschiana = **Acampe pachyglossa**

Acampe rigida = **Acampe pachyglossa**

Aerangis alata = **Aerangis ellisii**

Aerangis anjoanensis = **Aerangis mooreana**

Aerangis avicularia = **?Aerangis rostellaris**

Aerangis buchlohii = **Aerangis rostellaris**

Aerangis buyssonii = **Aerangis ellisii**

Aerangis calligera = **Aerangis articulata**

Aerangis caulescens = **Aerangis ellisii**

Aerangis clavigera = **Aerangis macrocentra**

Aerangis crassipes = **?Aerangis modesta**

Aerangis crassipes = **?Aerangis stylosa**

Aerangis cryptodon sensu H. Perrier = **Aerangis cryptodon**

Aerangis curnowiana sensu H. Perrier = **Aerangis monantha**

Aerangis ellisii var *grandiflora* = **Aerangis ellisii**

Aerangis fastuosa var. *angustifolia* = **Aerangis modesta**

Aerangis fastuosa var. *francoisii* = **Aerangis fastuosa**

Aerangis fastuosa var. *grandidieri* = **Aerangis fastuosa**

Aerangis fastuosa var. *maculata* = **Aerangis fastuosa**

Aerangis fastuosa var. *rotundifolia* = **Aerangis fastuosa**

Aerangis fastuosa var. *vondrozensis* = **Aerangis fastuosa**

Aerangis fuscata sensu H. Perrier = **Aerangis stylosa**

Aerangis ikopana = **Aerangis mooreana**

Aerangis karthalensis = **Aerangis mooreana**

Aerangis malmquistiana = **Aerangis cryptodon**

Aerangis mooreana also see Aerangis pulchella

Aerangis ophioplectron also see Jumellea ophioplectron

Aerangis platyphylla = **Aerangis ellisii**

Aerangis potamophila = **Angraecum potamophila**

Aerangis primulina = **Aerangis ?primulina**

Aerangis pumilio = **Aerangis hyaloides**

Aerangis stylosa sensu H. Perrier = **Aerangis articulata**

Aerangis umbonata = **Aerangis fuscata**

Aerangis venusta = **Aerangis articulata**

Aerangis venusta = **?Aerangis stylosa**

Aeranthes arachnanthus = **Jumellea arachnantha**

Aeranthes biauriculata = **Aeranthes aemula**

Aeranthes brachycentron = **Aeranthes grandiflora**

Aeranthes brachystachyus = **Beclardia macrostachya**

Aeranthes brevivaginans = **Aeranthes ramosa**

Aeranthes calceolus = **Angraecum calceolus**

Aeranthes carpophorus see Angraecum calceolus

Aeranthes comorensis = **Jumellea comorensis**

Aeranthes curnowianus = **Angraecum curnowianum**

Aeranthes englerianus = **Angraecum kraenzlinianum**

Aeranthes gladiator = **Jumellea gladiator**

Aeranthes gladiifolium = **Angraecum mauritianum**

Aeranthes gladiifolius = **Angraecum mauritianum**

Aeranthes gracilis = **Aeranthes subramosa**

Aeranthes grandidierianus = **Neobathiea grandidieriana**

Aeranthes imerinensis = **Aeranthes caudata**

Aeranthes leonis = **Angraecum leonis**

Aeranthes macrostachyus = **Beclardia macrostachya**

Aeranthes meirax = **Angraecum meirax**

Aeranthes ophioplectron = **Jumellea ophioplectron**

Aeranthes pectinatus = **Angraecum pectinatum**

Aeranthes perrieri = **Neobathiea perrieri**

Aeranthes phalaenophorus = **Jumellea phalaenophora**

Aeranthes pseudonidus = **Aeranthes nidus**

Aeranthes pusilla = **Aeranthes setiformis**

Aeranthes ramosus = **Aeranthes ramosa**

Aeranthes ramosus var. *orthopoda* see Aeranthes orthopoda

Aeranthes rigidula = **Aeranthes longipes**

Aeranthes sesquipedalis = **Angraecum sesquipedale**

Aeranthes thouarsii = **Angraecum filicornu**

Aeranthes trichoplectron = **Angraecum trichoplectron**

Aeranthes trifurcus = **Angraecopsis trifurca**

Aeranthes vespertilio = **Aeranthes ramosa**

Aeranthes volucris = **Oeonia volucris**

Aeranthus see Aeranthes

Aerides coriaceum = **Angraecum coriaceum**

Aerides macrostachyum = **Beclardia macrostachya**

Aerobion calceolus = **Angraecum calceolus**

Aerobion caulescens = **Angraecum caulescens**

Aerobion citratum = **Aerangis citrata**

Aerobion crassum = **Angraecum crassum**

Aerobion filicornu = **Angraecum filicornu**

Aerobion gladiifolium = **Angraecum mauritianum**

Aerobion implicatum = **Angraecum implicatum**

Aerobion inapertum = **Angraecum inapertum**

Aerobion maculatum = **Oeceoclades maculata**

Aerobion multiflorum = **Angraecum multiflorum**

Aerobion parviflorum = **Angraecopsis parviflora**

Aerobion pectinatum = **Angraecum pectinatum**

Aerobion superbum = **Angraecum eburneum** subsp. **superbum**

Agrostophyllum seychellarum = **Agrostophyllum occidentale**

Alismorkis centrosis = **Calanthe sylvatica**

Amphorchis lilacina = **Cynorkis ridleyi**

Angorchis articulata = **Aerangis articulata**

Angorchis boutoni = **Microterangis boutonii**

Angorchis brongniartiana (var. *superbum* ?) = **Angraecum eburneum** subsp. **superbum**

Angorchis citrata = **Aerangis citrata**

Angorchis clavigera = **Angraecum clavigerum**

Angorchis cowanii = **Jumellea cowanii**

Angorchis crassa = **Angraecum crassum**

Angorchis cryptodon = **Aerangis cryptodon**

Angorchis curnowiana = **Angraecum curnowianum**

Angorchis eburnea = **Angraecum eburneum**

Angorchis ellisii = **Aerangis ellisii**

Angorchis fastuosa = **Aerangis fastuosa**

Angorchis gladiifolia = **Angraecum mauritianum**

Angorchis hyaloides = **Aerangis hyaloides**

Angorchis implicata = **Angraecum implicatum**

Angorchis maxillarioides = **Jumellea maxillarioides**

Angorchis modesta = **Aerangis modesta**

Angorchis pectangis = **Angraecum pectinatum**

Angorchis pectinata = **Angraecum pectinatum**

Angorchis polystachya = **Oeoniella polystachys**

Angorchis rostrata = **Angraecum rostratum**

Angorchis scottiana = **Angraecum scottianum**

Angorchis sesquipedalis = **Angraecum sesquipedale**

Angorchis spathulata = **Jumellea spathulata**

Angorchis superba = **Angraecum eburneum** subsp. **superbum**

Angorchis teretifolia = **Angraecum teretifolium**

Angorkis physophora = **Microcoelia physophora**

Angraecopsis boutoni = **Microterangis boutonii**
Angraecopsis comorensis = **Angraecopsis trifurca**
Angraecopsis thouarsii = **Angraecopsis trifurca**
Angraecum abietianum = **?Angraecum humblotianum**
Angraecum ambongense = **Lemurella culicifera**
Angraecum anjouanense = **Jumellea anjouanensis**
Angraecum anocentrum = **Angraecum calceolus**
Angraecum aphyllum = **Microcoelia aphylla**
Angraecum articulatum = **Aerangis articulata**
Angraecum avicularium = **Aerangis rostellaris**
Angraecum bathiei = **Angraecum conchoglossum**
Angraecum bathiei subsp. *peracuminatum* = **Angraecum conchoglossum**
Angraecum bosseri = **Angraecum sesquipedale** var. **angustifolium**
Angraecum boutoni = **Microterangis boutonii**
Angraecum brongniartianum = **Angraecum eburneum** subsp. **superbum**
Angraecum buyssonii see Aerangis ellisii
Angraecum calligerum = **Aerangis articulata**
Angraecum carpophorum = **Angraecum calceolus**
Angraecum catati = **Angraecum rutenbergianum**
Angraecum caulescens also see Angraecum multiflorum
Angraecum caulescens var. *multiflorum* = **Angraecum multiflorum**
Angraecum chiloschistae = **Microcoelia exilis**
Angraecum chloranthum = **Angraecum huntleyoides**
Angraecum citratum = **Aerangis citrata**
Angraecum comorense = **Jumellea comorensis**
Angraecum comorense = **Angraecum eburneum** subsp. **superbum**
Angraecum confusum = **Jumellea confusa**
Angraecum cornutum = **Microcoelia cornuta**
Angraecum cowanii = **Jumellea cowanii**
Angraecum crassiflorum = **Angraecum crassum**
Angraecum cryptodon = **Aerangis cryptodon**
Angraecum culiciferum = **Lemurella culicifera**
Angraecum cyclochilum = **Microcoelia cornuta**
Angraecum defoliatum = **Microcoelia aphylla**
Angraecum descendens = **Aerangis articulata**
Angraecum dichaeoides = **Angraecum baronii**
Angraecum divitiflorum = **Microterangis divitiflora**
Angraecum dolichorhizum = **Microcoelia dolichorhiza**
Angraecum dubuyssonii = **Aerangis ellisii**
Angraecum eburneum subsp. *superbum* var. *longicalcar* = **Angraecum longicalcar**
Angraecum eburneum subsp. *typicum* = **Angraecum eburneum**
Angraecum eburneum var. *brongniartianum* = **Angraecum eburneum** subsp. **superbum**
Angraecum eburneum var. *virens* = **Angraecum eburneum**
Angraecum ellisii = **Aerangis ellisii**
Angraecum englerianum = **Angraecum kraenzlinianum**
Angraecum fastuosum = **Aerangis fastuosa**
Angraecum filicornu = **Angraecum sterrophyllum**
Angraecum finetianum = **Angraecum humblotianum**
Angraecum fournierae = **Aerangis stylosa**
Angraecum fournierianum = **Sobennikoffia fournieriana**
Angraecum foxii = **Angraecum rhynchoglossum**
Angraecum fuscatum = **Aerangis spiculata**
Angraecum fuscatum = **Aerangis fuscata**
Angraecum gilpinae = **Microcoelia gilpinae**
Angraecum gladiifolium = **Angraecum mauritianum**
Angraecum gracile = **Chamaeangis gracilis**
Angraecum gracilipes = **Jumellea sagittata**
Angraecum graminifolium = **Angraecum pauciramosum**
Angraecum grandidierianum = **Neobathiea grandidieriana**
Angraecum guillauminii = **Angraecum calceolus**
Angraecum hildebrandtii = **Microterangis hildebrandtii**
Angraecum humblotii = **Angraecum humblotianum**

Angraecum humblotii = **Angraecum leonis**
Angraecum hyaloides = **Aerangis hyaloides**
Angraecum ischnopus = **Angraecum tenuipes**
Angraecum jumelleanum = **Jumellea jumelleana**
Angraecum kimballianum = **Oeoniella polystachys**
Angraecum laggiarae = **Angraecum calceolus**
Angraecum leonii = **Angraecum leonis**
Angraecum lignosum = **Jumellea lignosa**
Angraecum macrocentrum = **Aerangis macrocentra**
Angraecum maculatum = **Oeceoclades maculata**
Angraecum majale = **Jumellea majalis**
Angraecum maxillarioides = **Jumellea maxillarioides**
Angraecum metallicum = **Aerangis stylosa**
Angraecum microphyton = **Angraecum tenellum**
Angraecum modestum = **Aerangis modesta**
Angraecum mooreanum = **Aerangis mooreana**
Angraecum nasutum = **Angraecum pingue**
Angraecum oliganthum = **Microterangis oligantha**
Angraecum pachyrum = **Jumellea pachyra**
Angraecum palmiforme = **Angraecum linearifolium**
Angraecum paniculatum = **Angraecum calceolus**
Angraecum parviflorum = **Angraecopsis parviflora**
Angraecum patens = **Angraecum calceolus**
Angraecum perrieri = **Microcoelia perrieri**
Angraecum physophorum = **Microcoelia physophora**
Angraecum polystachyum = **Oeoniella polystachys**
Angraecum poophyllum = **Angraecum pauciramosum**
Angraecum primulinum = **Aerangis ?primulina**
Angraecum pulchellum = **Aerangis pulchella**
Angraecum ramosum = **Angraecum conchoglossum**
Angraecum ramosum subsp. *typicum* = **Angraecum conchoglossum**
Angraecum ramosum subsp. *bidentatum* = **Angraecum germinyanum**
Angraecum ramosum subsp. *typicum* var. *arachnites* = **Angraecum arachnites**
Angraecum ramosum subsp. *typicum* var. *bathiei* = **Angraecum conchoglossum**
Angraecum ramosum subsp. *typicum* var. *conchoglossum* = **Angraecum conchoglossum**
Angraecum ramosum subsp. *typicum* var. *peracuminatum* = **Angraecum conchoglossum**
Angraecum ramulicolum = **Aerangis pallidiflora**
Angraecum reichenbachianum = **Angraecum scottianum**
Angraecum rhopaloceras = **Angraecum calceolus**
Angraecum robustum = **Sobennikoffia robusta**
Angraecum robustum = **Angraecum kraenzlinianum**
Angraecum rostellare = **Aerangis rostellaris**
Angraecum saccolabioides = **Microterangis humblotii**
Angraecum sanderianum = **Aerangis modesta**
Angraecum sarcodanthum = **Angraecum crassum**
Angraecum spathulatum = **Jumellea spathulata**
Angraecum stylosum also see Sobennikoffia fournieriana
Angraecum stylosum = **Aerangis stylosa**
Angraecum suarezense = **Angraecum curnowianum**
Angraecum subcordatum = **Angraecum curnowianum**
Angraecum superbum = **Angraecum eburneum** subsp. **superbum**
Angraecum tsaratananae = **Angraecum zaratananae**
Angraecum venustum = **?Aerangis articulata**
Angraecum verruculosum = **Angraecum implicatum**
Angraecum virens = **Angraecum eburneum**
Angraecum viride = **Angraecum rhynchoglossum**
Angraecum voeltzkowianum = **Angraecum eburneum** subsp. **superbum**
Angraecum waterlotii = **Angraecum tenellum**
Aplostellis ambigua = **Nervilia simplex**
Arethusa petraea = **Nervilia petraea**
Arethusa simplex = **Nervilia simplex**
Bathiea perrieri = **Neobathiea perrieri**

Beclardia brachystachya = **Beclardia macrostachya**
Beclardia elata = **Cryptopus elatus**
Beclardia erostris = **Beclardia macrostachya**
Beclardia humberti = **Lemurella culicifera**
Beclardia polystachya = **Oeoniella polystachys**
Begonia monophylla = **Nervilia bicarinata**
Benthamia flavida = **Benthamia cinnabarina**
Benthamia graminea = **Cynorkis graminea**
Benthamia herminioides forma *sulfurea* = **Benthamia herminioides**
Benthamia herminioides subsp. *typica* = **Benthamia herminioides**
Benthamia latifolia = **Benthamia bathieana**
Benthamia leandriana = **Cynorkis tryphioides** var. **leandriana**
Benthamia minutiflora = **Benthamia spiralis**
Benthamia nigrescens subsp. *typica* = **Benthamia nigrescens**
Benthamia perrieri = **Benthamia cinnabarina**
Benthamia spiralis var. *dissimulata* = **Benthamia spiralis**
Bicornella gracilis = **Cynorkis lindleyana**
Bicornella longifolia = **Cynorkis graminea**
Bicornella parviflora = **Cynorkis graminea**
Bicornella pulchra = **Cynorkis lilacina** var. **pulchra**
Bicornella schmidtii = **Cynorkis parvula**
Bicornella similis = **Cynorkis graminea**
Bicornella stolonifera = **Cynorkis stolonifera**
Bletia silvatica = **Calanthe sylvatica**
Bletia tuberculosa = **Gastrorchis tuberculosa**
Bolbophyllum see Bulbophyllum
Bolborchis crociformis = **Nervilia simplex**
Bonatea cirrhata = **Habenaria cirrhata**
Bonatea incarnata = **Habenaria incarnata**
Brachycorythis perrieri = **Brachycorythis pleistophylla**
Brownleea madagascarica = **Brownleea coerulea**
Brownleea nelsonii = **Brownleea parviflora**
Brownleea perrieri = **Brownleea parviflora**
Brownleea woodii = **Brownleea parviflora**
Bulbophyllaria pentasticha = **Bulbophyllum pentastichum**
Bulbophyllum abbreviatum = **Bulbophyllum ormerodianum**
Bulbophyllum album = **Bulbophyllum humblotii**
Bulbophyllum ambohitrense = **Bulbophyllum nutans** var. **variifolium**
Bulbophyllum ambongense = **Bulbophyllum rubrum**
Bulbophyllum ambreae = **Bulbophyllum septatum**
Bulbophyllum andrangense = **Bulbophyllum francoisii** var. **andrangense**
Bulbophyllum andringitranum = **Bulbophyllum nutans**
Bulbophyllum befaonense = **Bulbophyllum sarcorhachis** var. **befaonense**
Bulbophyllum bicolor = **Bulbophyllum bicoloratum**
Bulbophyllum bosseri = **Bulbophyllum reflexiflorum**
Bulbophyllum brevipetalum subsp. *majus* = **Bulbophyllum sulfureum**
Bulbophyllum brevipetalum subsp. *speculiferum* = **Bulbophyllum brevipetalum**
Bulbophyllum caeruleolineatum = **Bulbophyllum oxycalyx** var. **rubescens**
Bulbophyllum calamarioides = **Bulbophyllum erectum**
Bulbophyllum chrysobulbum = **Bulbophyllum nutans**
Bulbophyllum clavatum see dubious sp. under Bulbophyllum
Bulbophyllum clavigerum = **Bulbophyllum lemurensis**
Bulbophyllum clavigerum = **Bulbophyllum longiflorum**
Bulbophyllum coeruleum = **Bulbophyllum bicoloratum**
Bulbophyllum comosum = **Bulbophyllum reflexiflorum** subsp. **pogonochilum**
Bulbophyllum compactum = **Bulbophyllum coriophorum**
Bulbophyllum coursianum = **Bulbophyllum rutenbergianum**
Bulbophyllum crenulatum = **Bulbophyllum coriophorum**
Bulbophyllum cylindrocarpum = **Bulbophyllum cylindrocarpum**
Bulbophyllum cylindrocarpum var. *aurantiacum* = **Bulbophyllum cylindrocarpum**
Bulbophyllum cylindrocarpum var. *olivaceum* = **Bulbophyllum cylindrocarpum**
Bulbophyllum fimbriatum = **Bulbophyllum peyrotii**

Bulbophyllum flickingerianum = **Bulbophyllum peyrotii**

Bulbophyllum forsythianum = **Bulbophyllum cataractarum**

Bulbophyllum graciliscapum = **Bulbophyllum perseverans**

Bulbophyllum humblotianum = **Bulbophyllum leoni**

Bulbophyllum implexum = **Bulbophyllum minutum**

Bulbophyllum inauditum = **Bulbophyllum reflexiflorum**

Bulbophyllum johannum = **Bulbophyllum hildebrandtii**

Bulbophyllum laggiarae = **Bulbophyllum humblotii**

Bulbophyllum lichenophyllax var. *microdoron* = **Bulbophyllum afzelii** var. **microdoron**

Bulbophyllum linguiforme = **Bulbophyllum humblotii**

Bulbophyllum lobulatum = **Bulbophyllum erectum**

Bulbophyllum loxodiphyllum = **Bulbophyllum oxycalyx** var. **rubescens**

Bulbophyllum luteolabium = **Bulbophyllum humblotii**

Bulbophyllum maculatum = **Bulbophyllum hildebrandtii**

Bulbophyllum madagascariense = **Bulbophyllum hildebrandtii**

Bulbophyllum malawiense = **Bulbophyllum elliotii**

Bulbophyllum mandrakanum = **Bulbophyllum coriophorum**

Bulbophyllum mangoroanum = **Bulbophyllum ophiuchus**

Bulbophyllum matitanense = **Bulbophyllum pentastichum**

Bulbophyllum matitanense subsp. *rostratum* = **Bulbophyllum pentastichum** subsp. **rostratum**

Bulbophyllum mayae = **Bulbophyllum peyrotii**

Bulbophyllum melanopogon = **Bulbophyllum hildebrandtii**

Bulbophyllum microdoron = **Bulbophyllum afzelii** var. **microdoron**

Bulbophyllum microglossum = **Bulbophyllum moldekeanum**

Bulbophyllum moromanganum see Bulbophyllum moramanganum

Bulbophyllum muscicolum = **Bulbophyllum bryophilum**

Bulbophyllum nigrilabium = **Bulbophyllum maudeae**

Bulbophyllum nitens var. *minus* = **Bulbophyllum nitens**

Bulbophyllum nitens var. *typicum* = **Bulbophyllum nitens**

Bulbophyllum nutans = **Bulbophyllum rutenbergianum**

Bulbophyllum nutans var. *flavum* = **Bulbophyllum nutans**

Bulbophyllum nutans var. *genuium* = **Bulbophyllum nutans**

Bulbophyllum nutans var. *nanum* = **Bulbophyllum nutans**

Bulbophyllum nutans var. *pictum* = **Bulbophyllum nutans**

Bulbophyllum nutans var. *rubellum* = **Bulbophyllum nutans**

Bulbophyllum obtusum = **Bulbophyllum obtusatum**

Bulbophyllum ochrochlamys = **Bulbophyllum vakonae**

Bulbophyllum ophiuchus var. *ankaizinensis* = **Bulbophyllum ankaizinense**

Bulbophyllum ophiuchus var. *pallens* = **Bulbophyllum pallens**

Bulbophyllum peniculus = **Bulbophyllum rutenbergianum**

Bulbophyllum pentastichum = **Bulbophyllum pentastichum**

Bulbophyllum pleurothalloides = **Bulbophyllum conchidioides**

Bulbophyllum pogonochilum = **Bulbophyllum reflexiflorum** subsp. **pogonochilum**

Bulbophyllum pseudonutans = **Bulbophyllum brachystachyum**

Bulbophyllum quinquecornutum = **Bulbophyllum lichenophylax**

Bulbophyllum rhizomatosum = **Bulbophyllum obtusilabium**

Bulbophyllum rictorium = **Bulbophyllum lucidum**

Bulbophyllum ridleyi = **Bulbophyllum multiflorum**

Bulbophyllum robustum = **Bulbophyllum coriophorum**

Bulbophyllum rostriferum = **Bulbophyllum oxycalyx** var. **rubescens**

Bulbophyllum rubescens = **Bulbophyllum oxycalyx** var. **rubescens**

Bulbophyllum rubescens var. *meizobulbon* = **Bulbophyllum oxycalyx**

Bulbophyllum sambiranense var. *ankeranense* = **Bulbophyllum sambiranense**

Bulbophyllum sambiranense var. *typicum* = **Bulbophyllum sambiranense**

Bulbophyllum sarcorhachis var. *beforonense* = **Bulbophyllum sarcorhachis** var. **befaonense**

Bulbophyllum sarcorhachis var. *typicum* = **Bulbophyllum sarcorhachis**

Bulbophyllum schlechteri = **Bulbophyllum metonymon**

Bulbophyllum serratum = **Bulbophyllum septatum**

Bulbophyllum sigilliforme = **Bulbophyllum complanatum**

Bulbophyllum simulacrum = **Bulbophyllum rubrigemmum**

Bulbophyllum spathulifolium = **Bulbophyllum rutenbergianum**

Bulbophyllum theiochlamys = **Bulbophyllum bicoloratum**

Bulbophyllum tsaratananae = **Bulbophyllum zaratananae**
Bulbophyllum tsinjoarivense = **Bulbophyllum nutans**
Bulbophyllum variifolium = **Bulbophyllum nutans** var. **variifolium**
Bulbophyllum zaratananae = **Bulbophyllum metonymon**
Calanthe durani = **Calanthe sylvatica**
Calanthe perrieri = **Calanthe sylvatica**
Calanthe repens var. *pauliani* = **Calanthe repens**
Calanthe sylvatica = **Calanthe madagascariensis**
Calanthe sylvatica var. *alba* = **Calanthe sylvatica**
Calanthe sylvatica forma *humberti* = **Calanthe sylvatica**
Calanthe sylvatica forma *imerina* = **Calanthe sylvatica**
Calanthe sylvatica var. *iodes* = **Calanthe sylvatica**
Calanthe sylvatica var. *lilacina* = **Calanthe sylvatica**
Calanthe sylvatica var. *purpurea* = **Calanthe sylvatica**
Calanthe sylvatica var. *pallidipetala* = **Calanthe sylvatica**
Calanthe violacea = **Calanthe sylvatica**
Calanthe warpuri = **Calanthe madagascariensis**
Callista flavescens = **Polystachya concreta**
Caloglossum flabellatum = **Cymbidiella flabellata**
Caloglossum humblotii = **Cymbidiella falcigera**
Caloglossum magnificum = **Cymbidiella falcigera**
Caloglossum rhodochilum = **Cymbidiella pardalina**
Camilleugenia coccinelloides = **Cynorkis coccinelloides**
Centrosia auberti = **Calanthe sylvatica**
Centrosis sylvatica = **Calanthe sylvatica**
Cestichis caespitosa = **Liparis cespitosa**
Cestichis disticha = **Liparis disticha**
Chamaeangis boutoni = **Microterangis boutonii**
Chamaeangis coursiana = **Microterangis coursiana**
Chamaeangis divitiflora = **Microterangis divitiflora**
Chamaeangis hariotiana = **Microterangis hariotiana**
Chamaeangis hildebrandtii = **Microterangis hildebrandtii**
Chamaeangis humblotii = **Microterangis humblotii**
Chamaeangis oligantha = **Microterangis oligantha**
Chamaeangis pobeguini = **Angraecopsis pobeguini**
Cheirostylis gymnochiloides = **Cheirostylis nuda**
Cheirostylis humblotii = **Cheirostylis nuda**
Cheirostylis micrantha = **Cheirostylis nuda**
Cheirostylis sarcopus = **Cheirostylis nuda**
Cirrhopetalum clavigerum = **Bulbophyllum longiflorum**
Cirrhopetalum longiflorum = **Bulbophyllum longiflorum**
Cirrhopetalum thomasii = **Bulbophyllum longiflorum**
Cirrhopetalum thouarsii = **Bulbophyllum longiflorum**
Cirrhopetalum umbellatum = **Bulbophyllum longiflorum**
Corymbis bolusiana = **Corymborkis corymbis**
Corymbis corymbosa = **Corymborkis corymbis**
Corymbis leptantha = **Corymborkis corymbis**
Corymbis thouarsii = **Corymborkis corymbis**
Corymbis welwitschii = **Corymborkis corymbis**
Corymborkis corymbosa = **Corymborkis corymbis**
Corymborkis disticha = **Corymborkis corymbis**
Corymborkis thouarsii = **Corymborkis corymbis**
Corymborkis welwitschii = **Corymborkis corymbis**
Cranichis luteola = **Polystachya concreta**
Cryptopus elatus subsp. *dissectus* = **Cryptopus dissectus**
Ctenorchis pectinata = **Angraecum pectinatum**
Cymbidiella humblotii = **Cymbidiella falcigera**
Cymbidiella perrieri = **Cymbidiella flabellata**
Cymbidiella rhodochila = **Cymbidiella pardalina**
Cymbidium calcaratum = **Oeceoclades calcarata**
Cymbidium equitans = **Oberonia disticha**
Cymbidium flabellatum = **Cymbidiella flabellata**

Cymbidium humblotii = **Cymbidiella falcigera**

Cymbidium loise-chauvierii = **Cymbidiella falcigera**

Cymbidium rhodochilum = **Cymbidiella pardalina**

Cymbidium umbellatum = **Bulbophyllum longiflorum**

Cynorchis see Cynorkis

Cynorkis amabilis = **Cynorkis angustipetala** var. **amabilis**

Cynorkis andringitrana subsp. *curvicalcar* = **Cynorkis lilacina** subsp. **curvicalcar**

Cynorkis angustipetala var. *typica* = **Cynorkis angustipetala**

Cynorkis bella = **Cynorkis angustipetala** var. **bella**

Cynorkis boiviniana = **Cynorkis lilacina** var. **boiviniana**

Cynorkis calanthoides = **Cynorkis purpurascens**, also see Cynorkis gibbosa

Cynorkis calcarata see Cynorkis squamosa

Cynorkis cuneilabia subsp. *ampullacea* = **Cynorkis ampullacea**

Cynorkis decolorata = **Cynorkis fastigiata**

Cynorkis diplorhyncha = **Cynorkis fastigiata**

Cynorkis exilis = **Cynorkis schlechterii**

Cynorkis fallax = **Cynorkis flexuosa**

Cynorkis fastigiata var. *decolorata* = **Cynorkis fastigiata**

Cynorkis fastigiata var. *diplorhyncha* = **Cynorkis fastigiata**

Cynorkis fastigiata var. *hygrophila* = **Cynorkis fastigiata**

Cynorkis fastigiata var. *laggiarae* = **Cynorkis fastigiata**

Cynorkis fastigiata var. *typica* = **Cynorkis fastigiata**

Cynorkis filiformis = **Cynorkis papillosa**

Cynorkis flexuosa subsp. *fallax* = **Cynorkis fallax**

Cynorkis flexuosa var. *ambongensis* = **Cynorkis flexuosa** var. **bifoliata**

Cynorkis gracilis = **Cynorkis lindleyana**

Cynorkis grandiflora = **Cynorkis uniflora**

Cynorkis grandiflora var. *albata* = **Cynorkis uniflora**

Cynorkis grandiflora var. *purpurea* = **Cynorkis uniflora**

Cynorkis heterochroma = **Cynorkis ridleyi**

Cynorkis hirtula = **Cynorkis filiformis**

Cynorkis hygrophila = **Cynorkis fastigiata**

Cynorkis imerinensis = **Cynorkis rosellata**

Cynorkis inversa = **Cynorkis bathiei**

Cynorkis isocynis = **Cynorkis fastigiata**

Cynorkis laggiarae = **Cynorkis fastigiata**

Cynorkis laggiarae var. *ecalcarata* = **Cynorkis fastigiata**

Cynorkis lilacina var. *typica* = **Cynorkis lilacina**

Cynorkis lilacina × *ridleyi* = **Cynorkis × madagascarica**

Cynorkis longifolia = **Cynorkis graminea**

Cynorkis mesophylla = **Physoceras mesophyllum**

Cynorkis moramangana = **Cynorkis ridleyi**

Cynorkis nigrescens = **Cynorkis baronii**

Cynorkis nigrescens var. *jumelleana* = **Cynorkis baronii**

Cynorkis nutans var. *campenoni* = **Cynorkis nutans**

Cynorkis obcordata = **Cynorkis fastigiata**

Cynorkis oligadenia = **Cynorkis angustipetala** var. **oligadenia**

Cynorkis oxypetala = **Cynorkis angustipetala** var. **oxypetala**

Cynorkis pauciflora = **Cynorkis baronii**

Cynorkis praecox = **Cynorkis purpurascens**

Cynorkis pseudorolfer sphalm = **Cynorkis pseudorolfei**

Cynorkis pulchra = **Cynorkis lilacina** var. **pulchra**

Cynorkis pulchra var. *laxiflora* = **Cynorkis lilacina** var. **laxiflora**

Cynorkis purpurascens = **Cynorkis lowiana**

Cynorkis purpurascens var. *praecox* = **Cynorkis purpurascens**

Cynorkis rhomboglossa = **Cynorkis ridleyi**

Cynorkis similis = **Cynorkis graminea**

Cynorkis speciosa = **Cynorkis angustipetala** var. **speciosa**

Cynorkis triphylla = **Cynorkis fastigiata** var. **triphylla**

Cynorkis tsaratananae = **Cynorkis zaratananae**

Cynosorchis also see Cynorkis

Cynosorchis angustipetala = **Cynorkis angustipetala**

Cynosorchis baronii = **Cynorkis baronii**
Cynosorchis brevicornu = **Cynorkis brevicornu**
Cynosorchis calanthoides = **Cynorkis purpurascens?**
Cynosorchis grandiflora = **Cynorkis uniflora**
Cynosorchis hispidula = **Cynorkis hispidula**
Cynosorchis lilacina = **Cynorkis lilacina**
Cyrtopera biterbaerculata = **Eulophia plantaginea**
Cyrtopera plantaginea = **Eulophia plantaginea**
Cyrtopodium plantaginea = **Eulophia plantaginea**
Cyrtopodium plantagineum = **Eulophia plantaginea**
Dendrobium cultriforme = **Polystachya cultriformis**
Dendrobium flavescens = **Polystachya concreta**
Dendrobium fusiforme = **Polystachya fusiformis**
Dendrobium fusiforme var. *purpurascens* = **Polystachya fusiformis**
Dendrobium fusiforme var. *virescens* = **Polystachya fusiformis**
Dendrobium polystachyum = **Polystachya concreta**
Dendrochilum occultum = **Bulbophyllum occultum**
Dendrorkis anceps = **Polystachya anceps**
Dendrorchis appendiculata = **Polystachya cultriformis**
Dendrorchis estrellensis = **Polystachya concreta**
Dendrorchis extinctoria = **Polystachya concreta**
Dendrorkis jussieuana = **Polystachya concreta**
Dendrorchis minuta = **Polystachya concreta**
Dendrorkis minutiflora = **Polystachya fusiformis**
Dendrorkis rosea = **Polystachya rosea**
Dendrorkis rosellata = **Polystachya rosellata**
Dendrorchis rufinula = **Polystachya concreta**
Dendrorchis shirensis = **Polystachya concreta**
Dendrorchis similis = **Polystachya concreta**
Dendrorchis tesselata = **Polystachya concreta**
Dendrorkis virescens = **Polystachya virescens**
Dendrorchis whightii = **Polystachya concreta**
Dendrorchis zollingeri = **Polystachya concreta**
Diphyes occulta = **Bulbophyllum occultum**
Diplacorchis disoides = **Brachycorythis disoides**
Diplectrum amoenum = **Satyrium amoenum**
Disa compta = **Disa caffra**
Disa fallax = **Disa incarnata**
Disa parviflora = **Brownleea parviflora**
Disa perrieri = **Disa caffra**
Disa rutenbergiana = **Disa buchenaviana**
Disperis afzelii = Doubtful name
Disperis borbonica = **Disperis discifera** var. **borbonica**
Disperis comorensis = **Disperis humblotii**
Disperis hamadryas = **Disperis anthoceras**
Disperis humbertii = **Disperis anthoceras** var. **humbertii**
Dryopeia oppositifolia = **Disperis oppositifolia**
Dryopeia tripetaloides = **Disperis tripetaloides**
Epidendrum brachystachyum = **Beclardia macrostachya**
Epidendrum cespitosum = **Liparis cespitosa**
Epidendrum concretum = **Polystachya concreta**
Epidendrum coriaceum = **Angraecum coriaceum**
Epidendrum distichum = **Oberonia disticha**
Epidendrum equitans = **Oberonia disticha**
Epidendrum macrostachyum = **Beclardia macrostachya**
Epidendrum miniatum = **Polystachya concreta**
Epidendrum polystachys = **Oeoniella polystachys**
Epidendrum resupinatum = **Polystachya concreta**
Epidendrum scriptum = **Graphorkis concolor** var. **alphabetica**
Epidendrum simplex = **Nervilia simplex**
Epidendrum spathulatum   see Oeoniella polystachys
Epidendrum umbellatum = **Bulbophyllum longiflorum**

Epidendrum volucre = **Oeonia volucris**
Epidorkis aphylla = **Microcoelia aphylla**
Epidorkis calceolus = **Angraecum calceolus**
Epidorkis carpophora = **Angraecum calceolus**
Epidorkis caulescens = **Angraecum caulescens**
Epidorkis exilis = **Microcoelia exilis**
Epidorkis gilpinae = **Microcoelia gilpinae**
Epidorkis graminifolia = **Angraecum pauciramosum**
Epidorkis inaperta = **Angraecum inapertum**
Epidorkis multiflora = **Angraecum multiflorum**
Epidorkis parviflora = **Angraecopsis parviflora**
Epidorkis physophora = **Microcoelia physophora**
Epidorkis tenella = **Angraecum tenellum**
Epidorkis viridis = **Angraecum rhynchoglossum**
Epidorkis volucris = **Oeonia volucris**
Epipactis salassia = **Liparis salassia**
Erporkis bracteata = **Platylepis occulta**
Erporkis cryptorpis = **Platylepis densiflora**
Erporkis gymnerpis = **Cheirostylis nuda**
Eulophia aemula = **Eulophia hians** var. **nutans**
Eulophia affines = **Eulophia livingstoniana**
Eulophia alismatophylla = **Oeceoclades alismatophylla**
Eulophia amajubae = **Eulophia hians** var. **nutans**
Eulophia ambaxiana = **Eulophia pulchra**
Eulophia ambongensis = **Oeceoclades perrieri**
Eulophia ambositrana = **Eulophia macra**
Eulophia ambrensis = **Oeceoclades ambrensis**
Eulophia bathiei = **Eulophia hians** var. **nutans**
Eulophia calcarata = **Oeceoclades calcarata**
Eulophia camporum = **Eulophia reticulata**
Eulophia carunculifera = **Eulophia hians** var. **nutans**
Eulophia clavicornis var. *nutans* = **Eulophia hians** var. **nutans**
Eulophia coccifera = **Eulophia pulchra**
Eulophia concolor = **Graphorkis concolor**
Eulophia cordylinophylla = **Oeceoclades cordylinophylla**
Eulophia crinita = **Eulophia hians** var. **nutans**
Eulophia dahliana = **Eulophia pulchra**
Eulophia decaryana = **Oeceoclades decaryana**
Eulophia decurva = **Eulophia hians** var. **nutans**
Eulophia dissimilis = **Oeceoclades lonchophylla**
Eulophia elliotii = **Oeceoclades sclerophylla**
Eulophia emarginata = **Eulophia pulchra**
Eulophia epiphytica = **Paralophia epiphytica**
Eulophia ernestii = **Eulophia hians** var. **nutans**
Eulophia flanaganii = **Eulophia hians** var. **nutans**
Eulophia fractiflexa = **Eulophia clitellifera**
Eulophia galbana = **Eulophiella galbana**
Eulophia galpinii = **Eulophia hians** var. **nutans**
Eulophia gladioloides = **Eulophia hians** var. **nutans**
Eulophia gracilior = **Eulophia livingstoniana**
Eulophia gracillima = **Oeceoclades gracillima**
Eulophia grandibractea = **Eulophia plantaginea**
Eulophia guamensis = **Eulophia pulchra**
Eulophia humbertii = **Oeceoclades humbertii**
Eulophia jumelleana = **Eulophia livingstoniana**
Eulophia lanceata = **Oeceoclades lanceata**
Eulophia laxiflora = **Eulophia hians** var. **nutans**
Eulophia ledienii = **Oeceoclades maculata**
Eulophia leucorhiza = **Eulophia ramosa**
Eulophia lokobensis = **Oeceoclades cordylinophylla**
Eulophia lonchophylla = **Oeceoclades lonchophylla**
Eulophia mackeni = **Oeceoclades maculata**

Eulophia macrostachya = **Eulophia pulchra**
Eulophia maculata = **Oeceoclades maculata**
Eulophia madagascariensis = **Eulophia hians** var. **nutans**
Eulophia medemiae = **Graphorkis medemiae**
Eulophia nelsonii = **Eulophia hians** var. **nutans**
Eulophia novoebudae = **Eulophia pulchra**
Eulophia nutans = **Eulophia hians** var. **nutans**
Eulophia palmicola = **Paralophia palmicola**
Eulophia pandurata = **Oeceoclades pandurata**
Eulophia paniculata = **Oeceoclades calcarata**
Eulophia papuana = **Eulophia pulchra**
Eulophia petiolata = **Oeceoclades petiolata**
Eulophia pseudoramosa = **Eulophia ramosa**
Eulophia pulchra var. *divergens* = **Eulophia pulchra (megistophylla)**
Eulophia purpurascens = **Eulophia hians** var. **nutans**
Eulophia quadriloba = **Oeceoclades quadriloba**
Eulophia robusta = **Eulophia livingstoniana**
Eulophia rouxii = **Eulophia pulchra**
Eulophia schlechteri = **Oeceoclades ambongensis**
Eulophia sclerophylla = **Oeceoclades sclerophylla**
Eulophia scripta = **Graphorkis concolor** var. **alphabetica**
Eulophia scripta var. *concolor* = **Graphorkis concolor**
Eulophia silvatica = **Eulophia pulchra**
Eulophia spathulifera = **Oeceoclades spathulifera**
Eulophia striata = **Eulophia pulchra**
Eulophia tainoides = **Oeceoclades lonchophylla**
Eulophia triloba = **Eulophia hians** var. **nutans**
Eulophia ukingensis = **Eulophia hians** var. **nutans**
Eulophia vaginata = **Eulophia hians** var. **nutans**
Eulophia versicolor = **Eulophia pulchra**
Eulophia vleminckxiana = **Eulophia hians** var. **nutans**
Eulophidium alismatophyllum = **Oeceoclades alismatophylla**
Eulophidium ambongense = **Oeceoclades ambongensis**
Eulophidium analamerense = **Oeceoclades analamerensis**
Eulophidium analavelense = **Oeceoclades analavelensis**
Eulophidium angustifolium = **Oeceoclades angustifolia**
Eulophidium angustifolium subsp. *diphyllum* = **Oeceoclades angustifolia**
Eulophidium boinense = **Oeceoclades boinensis**
Eulophidium cordylinophyllum = **Oeceoclades cordylinophylla**
Eulophidium decaryanum = **Oeceoclades decaryana**
Eulophidium dissimile = **Oeceoclades lonchophylla**
Eulophidium gracillimum = **Oeceoclades gracillima**
Eulophidium ledienii = **Oeceoclades maculata**
Eulophidium lokobense = **Oeceoclades cordylinophylla**
Eulophidium lonchophyllum = **Oeceoclades lonchophylla**
Eulophidium maculatum = **Oeceoclades maculata**
Eulophidium megistophyllum = **Eulophia megistophylla**
Eulophidium nyassanum = **Oeceoclades maculata**
Eulophidium panduratum = **Oeceoclades pandurata**
Eulophidium paniculatum = **Oeceoclades calcarata**
Eulophidium perrieri = **Oeceoclades perrieri**
Eulophidium petiolatum = **Oeceoclades petiolata**
Eulophidium pulchrum = **Eulophia pulchra**
Eulophidium quadrilobum = **Oeceoclades quadriloba**
Eulophidium rauhii = **Oeceoclades rauhii**
Eulophidium roseovariegatum = **Oeceoclades gracillima**
Eulophidium sclerophyllum = **Oeceoclades sclerophylla**
Eulophidium silvaticum = **Eulophia pulchra**
Eulophidium spathuliferum = **Oeceoclades spathulifera**
Eulophidium tainiodes. = **Oeceoclades lonchophylla**
Eulophidium warneckianum = **Oeceoclades maculata**
Eulophiella hamelini = **Eulophiella roempleriana**

Eulophiella peetersiana = **Eulophiella roempleriana**

Eulophiella perrieri = **Eulophiella elisabethae**

Eulophiopsis ecalcarata = **Graphorkis ecalcarata**

Eulophiopsis medemiae = **Graphorkis medemiae**

Forsythmajoria pulchra = **Cynorkis lilacina** var. **pulchra**

Gabertia ellisii = **Grammangis ellisii**

Galeandra angornensis = **Eulophia angornensis**

Galeandra anjoanensis = **Eulophia angornensis**

Gastrochilus aphyllus = **Microcoelia aphylla**

Gastrochilus coriaceus = **Angraecum coriaceum**

Gastrodia madagascariensis = **Didymoplexis madagascariensis**

Gastrorchis françoisii var. *orientalis* see Gastrorchis françoisii

Gastrorchis françoisii var. *pauliani* = **Gastrorchis françoisii**

Gastrorchis humbertii = **Gastrorchis tuberculosa**

Gastrorchis luteus = **Gastrorchis lutea**

Gastrorchis schlechteri = **Gastrorchis humblotii** var. **schlechteri**

Gastrorchis schlechteri var. *milotii* = **Gastrorchis humblotii**

Gastrorchis tuberculosa var. *perrieri* = **Gastrorchis pulchra** var. **perrieri**

Geodorum pictum = **Oeceoclades maculata**

Goodyera nuda = **Cheirostylis nuda**

Goodyera occulta = **Platylepis occulta**

Grammangis ellisii var. *dayanum* = **Grammangis ellisii**

Grammangis falcigera = **Cymbidiella falcigera**

Grammangis fallax = **Grammangis ellisii**

Grammangis pardalina = **Cymbidiella pardalina**

Grammatophyllum ellisii = **Grammangis ellisii**

Grammatophyllum roemplerianum = **Eulophiella roempleriana**

Graphorchis see Graphorkis

Graphorchis bisdahliana = **Eulophia pulchra**

Graphorchis calographis = **Eulophia pulchra**

Graphorchis monographis = **Graphorkis concolor**

Graphorkis aiolographis = **Graphorkis concolor** var. **alphabetica**

Graphorkis beravensis = **Eulophia beravensis**

Graphorkis blumeneana = **Eulophia pulchra**

Graphorkis galbana = **Eulophia galbana**

Graphorkis maculata = **Oeceoclades maculata**

Graphorkis madagascariensis = **Eulophia hians** var. **nutans**

Graphorkis nutans = **Eulophia hians** var. **nutans**

Graphorkis pileata = **Eulophia pileata**

Graphorkis plantaginea = **Eulophia plantaginea**

Graphorkis pulchra = **Eulophia pulchra**

Graphorkis ramosa = **Eulophia ramosa**

Graphorkis reticulata = **Eulophia reticulata**

Graphorkis rutenbergiana = **Eulophia rutenbergiana**

Graphorkis scripta var. *scripta sensu* Senghas = **Graphorkis concolor** var. **alphabetica**

Graphorkis vaginata = **Eulophia hians** var. **nutans**

Gussonea aphylla = **Microcoelia aphylla**

Gussonea aphylla var. *defoliata* = **Microcoelia aphylla**

Gussonea aphylla var. *orientalis* = **Microcoelia aphylla**

Gussonea chilochistae sphalm = **Microcoelia exilis**

Gussonea chiloschistae = **Microcoelia exilis**

Gussonea cornuta = **Microcoelia cornuta**

Gussonea cyclochila = **Microcoelia cornuta**

Gussonea defoliata = **Microcoelia aphylla**

Gussonea dolichorhiza = **Microcoelia dolichorhiza**

Gussonea elliotii = **Microcoelia elliotii**

Gussonea exilis = **Microcoelia exilis**

Gussonea gilpinae = **Microcoelia gilpinae**

Gussonea gilpinae var. *minor* = **Microcoelia gilpinae**

Gussonea macrantha = **Microcoelia macrantha**

Gussonea melinantha = **Microcoelia gilpinae**

Gussonea perrieri = **Microcoelia perrieri**

PLATE 1

A. *Acampe pachyglossa.* JH
C. *Aerangis citrata.* JH

B. *Aerangis articulata.* JH
D. *Aerangis decaryana.* JH

PLATE 2

A. *Aerangis ellisii*. JH
C. *Aerangis fuscata*. JH

B. *Aerangis fastuosa*. JH
D. *Aerangis hyaloides*. JH

PLATE 3

A. *Aerangis macrocentra*. JH   **B.** *Aerangis modesta*. JH
C. *Aerangis monantha*. JH   **D.** *Aerangis mooreana*. JH

PLATE 4

A. *Aerangis pallidiflora*. JH   B. *Aerangis punctata*. JH
C. *Aerangis seegeri*. JH   D. *Aerangis spiculata*. JH

PLATE 5

A. *Aeranthes ecalcarata*. JH
C. *Aeranthes grandiflora*. JH

B. *Aeranthes filipes*. JH
D. *Aeranthes henrici*. JH

PLATE 6

A. *Aeranthes moratii*. JH

C. *Aeranthes peyrotii*. JH

B. *Aeranthes orophila*. JH

D. *Aeranthes ramosa*. JH

PLATE 7

A. *Aeranthes schlechteri.* DD
C. *Angraecopsis parviflora.* JH

B. *Aeranthes virginalis.* JH
D. *Angraecum arachnites.* JH

PLATE 8

A. *Angraecum aloifolium.* JH
B. *Angraecum baronii.* JH

C. *Angraecum calceolus.* JH

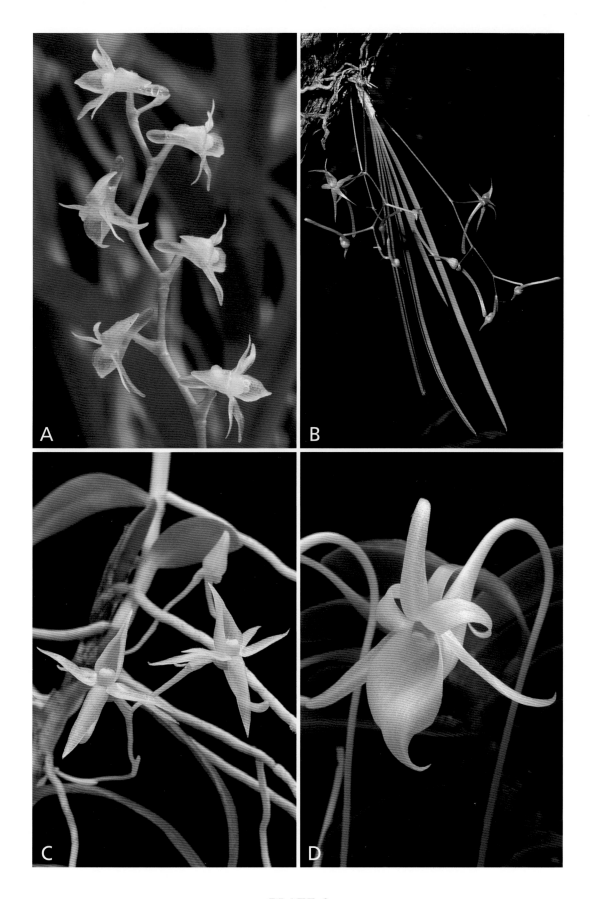

PLATE 9

A. *Angraecum caricifolium*. JH   **B.** *Angraecum chaetopodum*. DD
C. *Angraecum chermezoni*. JH   **D.** *Angraecum clareae*. JH

PLATE 10

Angraecum clavigerum. JH

PLATE 11

A. *Angraecum conchoglossum.* JH   **B.** *Angraecum didieri.* JH
C. *Angraecum danguyanum.* JH

PLATE 12

A. *Angraecum dollii.* JH
C. *Angraecum florulentum.* JH

B. *Angraecum eburneum.* JH
D. *Angraecum germinyanum.* JH

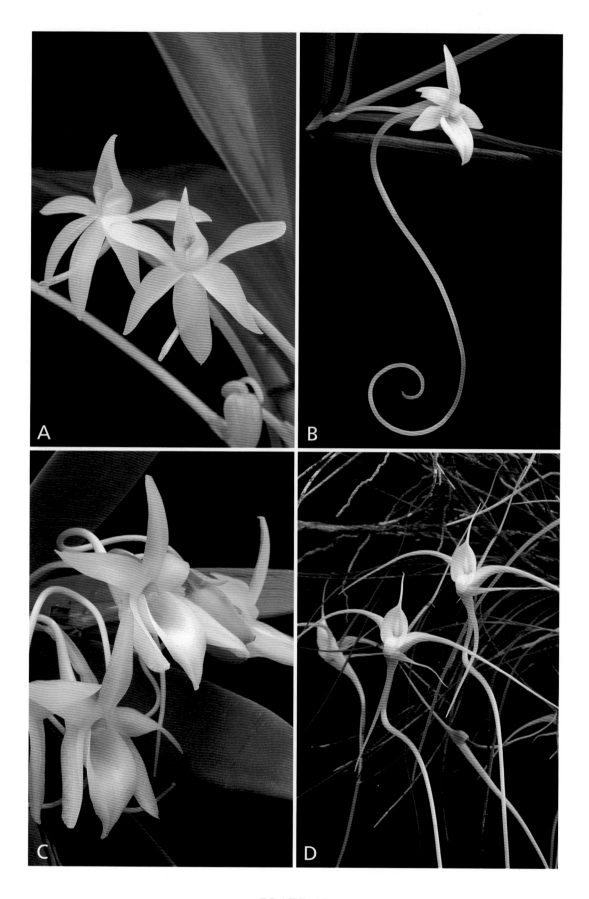

PLATE 13

A. *Angraecum huntleyoides.* JH   **B.** *Angraecum lecomtei.* JH
C. *Angraecum leonis.* JH   **D.** *Angraecum linearifolium.* JH

PLATE 14

A. *Angraecum longicalcar*. DD
C. *Angraecum melanostictum*. JH

B. *Angraecum magdalenae*. JH
D. *Angraecum muscicolum*. DD

PLATE 15

A. *Angraecum obesum.* JH   **B.** *Angraecum peyrotii.* JH
C. *Angraecum pingue.* JH   **D.** *Angraecum protensum.* JH

PLATE 16

Angraecum praestans. DD

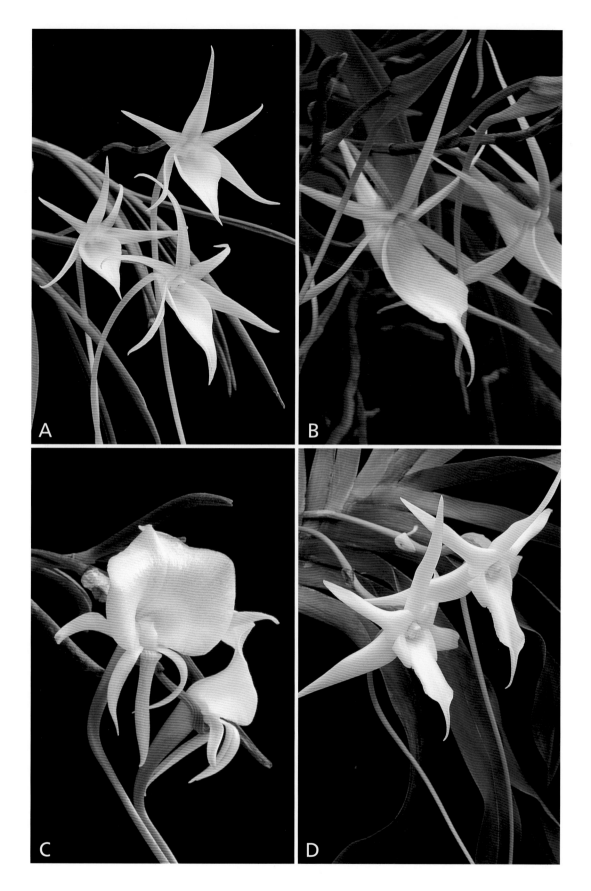

PLATE 17

A. *Angraecum popowii.* JH   **B.** *Angraecum praestans.* JH
C. *Angraecum pseudofilicornu.* JH   **D.** *Angraecum sesquipedale.* JH

PLATE 18

A. *Angraecum rubellum*. JH
B. *Angraecum sororium*. JH

C. *Angraecum urschianum*. JH

PLATE 19

A. *Beclardia macrostachya*. JH   **B.** *Benthamia glaberrima*. DD
C. *Benthamia nivea*. JH   **D.** *Brownleea coerulea*. JH

PLATE 20

A. *Bulbophyllum alexandrae*. JH   B. *Bulbophyllum analamazaotrae*. JH
C. *Bulbophyllum anjozorobeense*. JH   D. *Bulbophyllum auriflorum*. JH

PLATE 21

A. ***Bulbophyllum baronii***. JH
C. ***Bulbophyllum cardiobulbum***. JH
B. ***Bulbophyllum bicoloratum***. JH
D. ***Bulbophyllum conchidiodes***. JH

PLATE 22

A. *Bulbophyllum coriophorum.* JH
C. *Bulbophyllum hamelinii.* JH
B. *Bulbophyllum elliotii.* JH
D. *Bulbophyllum hildebrandtii.* JH

PLATE 23

A. *Bulbophyllum imerinense.* JH
C. *Bulbophyllum molossus.* JH
B. *Bulbophyllum luteobracteatum.* DD
D. *Bulbophyllum nitens.* JH

PLATE 24

A. *Bulbophyllum occlusum.* JH   B. *Bulbophyllum occultum.* DD
C. *Bulbophyllum onivense.* JH   D. *Bulbophyllum pachypus.* JH

PLATE 25

A. *Bulbophyllum pentastichum.* JH
C. *Bulbophyllum vestitum.* JH
B. *Bulbophyllum sulfureum.* JH
D. *Calanthe madagascariensis.* JH

PLATE 26

A. *Cryptopus paniculatus*. JH
B. *Cymbidiella falcigera*. DD

PLATE 27

A. *Cymbidiella falcigera.* DD
C. *Cymbidiella flabellata.* DD

B. *Cymbidiella flabellata.* DD
D. *Cymbidiella pardalina.* JH

PLATE 28

A. *Cynorkis angustipetala.* JH   B. *Cynorkis fastigiata.* JH
C. *Cynorkis elata.* JH

PLATE 29

Cynorkis gibbosa. JH

PLATE 30

A. *Cynorkis flexuosa.* JH
C. *Cynorkis gigas.* JH

B. *Cynorkis gibbosa.* JH
D. *Cynorkis graminea.* JH

PLATE 31

A. *Cynorkis lilacina.* JH   B. *Cynorkis lowiana.* JH

C. *Cynorkis meliantha.* MG   D. *Cynorkis nutans.* JH

PLATE 32

A. *Cynorkis peyrotii.* JH
C. *Cynorkis purpurea.* JH

B. *Cynorkis purpurascens.* JH
D. *Cynorkis ridleyi.* JH

PLATE 33

A. *Cynorkis uncinata.* JH   B. *Cynorkis uniflora.* JH
C. *Cynorkis villosa.* JH   D. *Disa buchenaviana.* JH

PLATE 34

A. *Disa incarnata*. JH

B. *Disperis erucifera*. DD

C. *Disperis lanceolata*. JH

D. *Disperis similis*. JH

PLATE 35

A. *Eulophia beravensis.* JH
C. *Eulophia plantaginea.* JH
B. *Eulophia macra.* JH
D. *Eulophia plantaginea.* JH

PLATE 36

A. *Eulophia pulchra.* JH B. *Eulophia rutenbergiana.* JH

C. *Eulophia galbana.* JH

PLATE 37

Eulophiella elisabethae. JH

PLATE 38

A. *Eulophiella roempleriana.* JH
C. *Gastrorchis françoisii.* JH

B. *Eulophiella roempleriana* on *Pandanus.* PC
D. *Gastrorchis humblotii.* JH

PLATE 39

A. *Gastrorchis lutea*. JH
C. *Gastrorchis pulchra*. JH

B. *Gastrorchis peyrotii*. JH
D. *Gastrorchis pulchra* var. *perrieri*. JH

PLATE 40

A. *Gastrorchis tuberculosa*. JH   B. *Goodyera humicola*. JH
C. *Goodyera afzelii*. JH   D. *Grammangis ellisii*. JH

PLATE 41

A. *Grammangis spectabilis.* JH
B. *Graphorkis concolor* **var.** *alphabetica.* JH

PLATE 42

A. *Habenaria alta*. JH
C. *Imerinaea madagascarica*. JH

B. *Habenaria simplex*. DD
D. *Jumellea comorensis*. JH

PLATE 43

A. *Jumellea densefoliata.* JH
C. *Jumellea lignosa.* JH

B. *Jumellea gracilipes.* DD
D. *Jumellea major.* JH

PLATE 44

A. *Jumellea rigida*. DD   B. *Jumellea spathulata*. JH
C. *Lemurella culicifera*. JH   D. *Lemurella papillosa*. JH

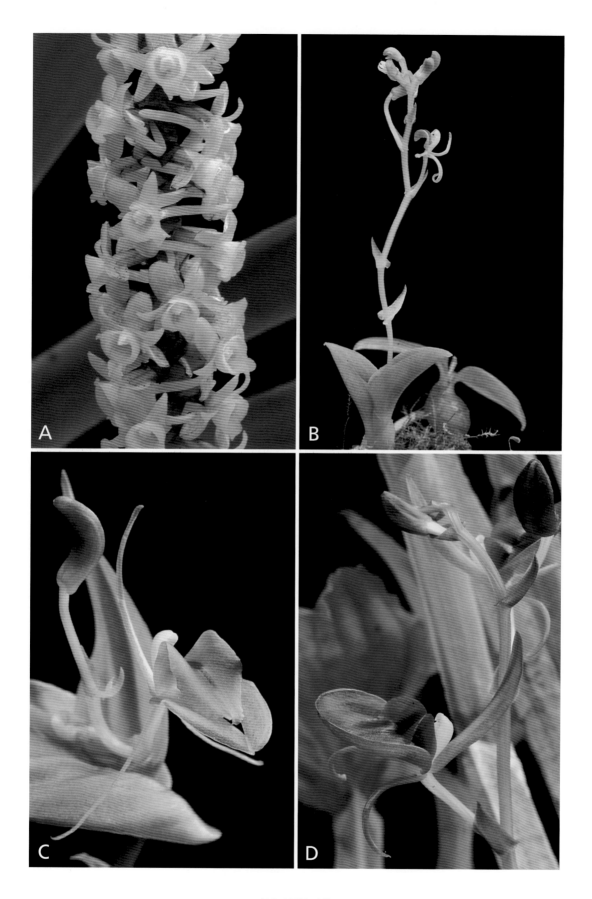

PLATE 45

A. *Lemurorchis madagascariensis.* JH

B. *Liparis bulbophylloides.* JH

C. *Liparis clareae.* JH

D. *Liparis longicaulis.* JH

PLATE 46

A. *Liparis puncticulata.* JH

C. *Microcoelia gilpinae.* JH

B. *Liparis salassia.* JH

D. *Microcoelia macrantha.* JH

PLATE 47

A. *Microcoelia perrieri.* JH

B. *Microterangis divitiflora.* JH

C. *Microterangis hariotiana.* JH

D. *Neobathiea grandidieriana.* JH

PLATE 48

A. *Neobathia perrieri.* DD

C. *Oberonia disticha.* DD

B. *Nervilia simplex.* JH

D. *Oeceoclades angustifolia.* JH

PLATE 49

A. *Oeceoclades antsingyensis.* DD

B. *Oeceoclades boinensis.* JH

C. *Oeceoclades calcarata.* DD

D. *Oeceoclades decaryana.* JH

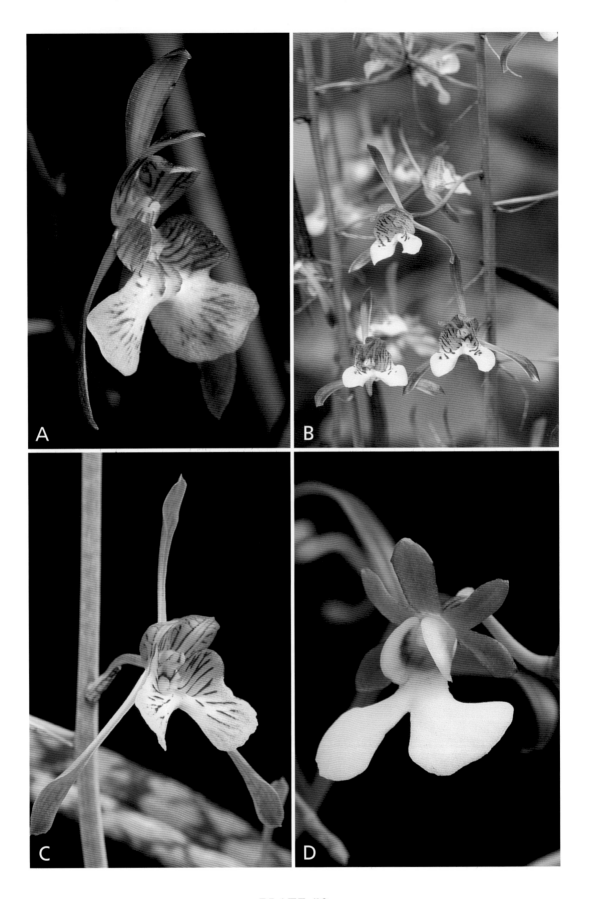

PLATE 50

A. *Oeceoclades hebdingiana*. JH   B. *Oeceoclades petiolata*. JH
C. *Oeceoclades spathulifera*. JH   D. *Oeonia rosea*. JH

PLATE 51

A. *Oeonia brauniana* var. *sarcanthoides*. JH
B. *Oeonia volucris*. JH

C. *Oeoniella polystachys*. JH

PLATE 52

Paralophia epiphytica. JH

PLATE 53

A. *Paralophia epiphytica*. JH
C. *Phaius pulchellus* **var.** *sandrangatensis*. JH
B. *Phaius pulchellus* **var** *pulchellus*. JH
D. *Physoceras violaceum*. JH

PLATE 54

A. *Polystachya clareae.* JH
C. *Polystachya humbertii.* JH
B. *Polystachya cornigera.* JH
D. *Polystachya fusiformis.* JH

PLATE 55

A. *Polystachya tsaratananae.* JH
C. *Satyrium rostratum.* DD

B. *Satyrium amoenum.* JH
D. *Satyrium trinerve.* JH

PLATE 56

A. *Sobennikoffia humbertiana*. JH   B. *Vanilla perrieri*. JH
C. *Sobennikoffia robusta*. JH

PLATE 57

A. *Microcoelia aphylla*. JH
B. *Vanilla madagascarensis*. JH

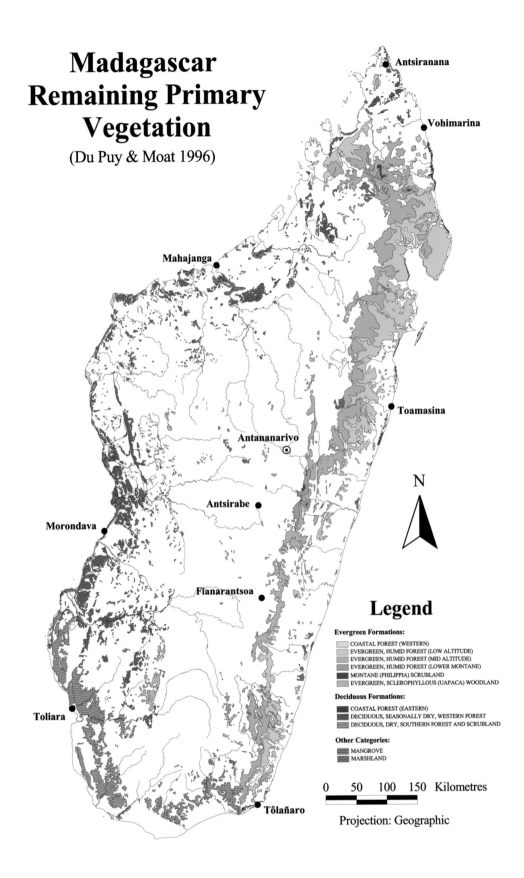

Madagascar
Remaining Primary
Vegetation
(Du Puy & Moat 1996)

Antsiranana

Vohimarina

Mahajanga

Toamasina

Antananarivo

N

Antsirabe

Morondava

Fianarantsoa

Legend

Evergreen Formations:
- COASTAL FOREST (WESTERN)
- EVERGREEN, HUMID FOREST (LOW ALTITUDE)
- EVERGREEN, HUMID FOREST (MID ALTITUDE)
- EVERGREEN, HUMID FOREST (LOWER MONTANE)
- MONTANE (PHILIPPIA) SCRUBLAND
- EVERGREEN, SCLEROPHYLLOUS (UAPACA) WOODLAND

Deciduous Formations:
- COASTAL FOREST (EASTERN)
- DECIDUOUS, SEASONALLY DRY, WESTERN FOREST
- DECIDUOUS, DRY, SOUTHERN FOREST AND SCRUBLAND

Other Categories:
- MANGROVE
- MARSHLAND

Toliara

0   50   100   150   Kilometres

Tôlañaro

Projection: Geographic

PLATE 58

Remaining primary vegetation in Madagascar.

Madagascar Phytogeographical Regions

(Du Puy & Moat after Faramalala 1995 and Humbert 1955)

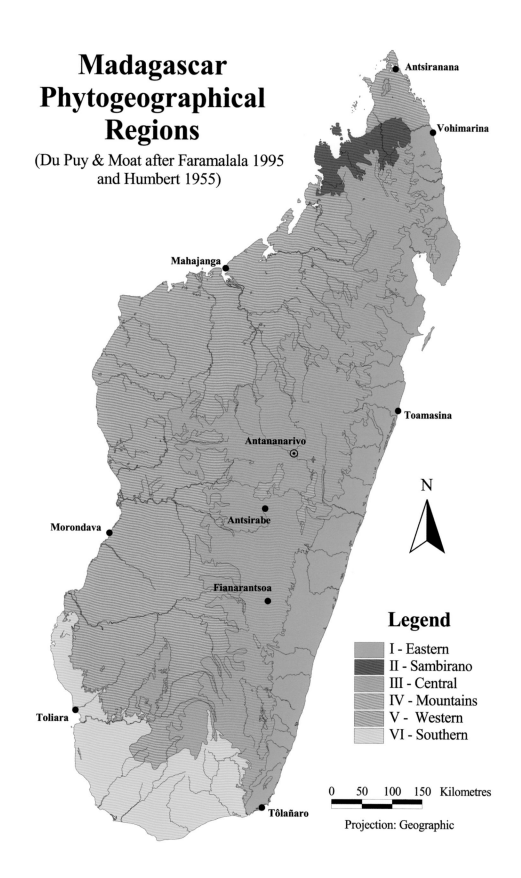

Antsiranana

Vohimarina

Mahajanga

Toamasina

Antananarivo

N

Morondava

Antsirabe

Fianarantsoa

Legend

I - Eastern
II - Sambirano
III - Central
IV - Mountains
V - Western
VI - Southern

Toliara

0   50   100   150   Kilometres

Projection: Geographic

Tôlañaro

PLATE 59

Phytogeographical regions of Madagascar.

PLATE 60

A. NC Madagascar, humid evergreen forest at mid altitude near the Tampoketsa of Ankazobe. JH

B. SC Madagascar, highland savanna dominated by isolated monoliths of basement rock, near Ambalavao. JH

PLATE 61

A. C Madagascar, evergreen sclerophyllous (*Uapaca*) woodland near Antsirabe. JH
B. W Madagascar, *Adansonia grandidieri* near Morondava. JH

PLATE 62

A. **EC Madagascar, montane habitat near Antsirabe**. JH
B. **S Madagascar, littoral forest near Taolanaro**. JH

PLATE 63

A. EC Madagascar, lichen-rich forest near Lakato. JH
B. SW Madagascar, xerophytic scrub near Taolanaro. JH
C. C Madagascar, transition forest, Andohahela. JH

PLATE 64

EC Madagascar, moist evergreen forest near Andasibe. JH

Gussonea physophora = **Microcoelia physophora**
Gymnadenia brachycentra = in schedul? **Cynorkis brachycentra**
Gymnadenia fastigiata = **Cynorkis fastigiata**
Gymnadenia lyallii = **Cynorkis flexuosa**
Gymnadenia muricata = **Cynorkis elegans**
Gymnadenia purpurascens = **Cynorkis purpurascens**
Gymnochilus recurvum = **Cheirostylis nuda**
Gymnadenia rosellata = **Cynorkis rosellata**
Gymnadenia triphylla = **Cynorkis fastigiata** var. **triphylla**
Gymnadenia uniflora = **Cynorkis uniflora**
Gymnochilus nudum = **Cheirostylis nuda**
Gymnochilus recurvum = **Cheirostylis nuda**
Gymnochilus roseum = **Bathiorchis rosea**
Habenaria ankaratrana = **Habenaria hilsenbergii**
Habenaria atra = **Habenaria hilsenbergii**
Habenaria bimaculata = **Cynorkis bimaculata**
Habenaria buchwaldiana = **Platycoryne pervillei**
Habenaria chlorantha = **Benthamia verecunda**
Habenaria cinnabarina = **Benthamia cinnabarina**
Habenaria dauphinensis = **Benthamia dauphinensis**
Habenaria depauperata = **Platycoryne pervillei**
Habenaria diptera = **Habenaria incarnata**
Habenaria disoides = **Brachycorythis disoides**
Habenaria dissimulata = **Benthamia spiralis**
Habenaria elliotii = **Habenaria clareae**
Habenaria filiformis = **Cynorkis papillosa**
Habenaria flammea = **Platycoryne pervillei**
Habenaria glaberrima = **Benthamia glaberrima**
Habenaria graminea = **Cynorkis graminea**
Habenaria gymnochiloides = **Cynorkis gymnochiloides**
Habenaria hildebrandtii = **Tylostigma hildebrandtii**
Habenaria humblotii = **Habenaria incarnata**
Habenaria ichneumoniformis = **Habenaria simplex**
Habenaria imerinensis = **Cynorkis imerinensis**
Habenaria madagascariensis = **Benthamia madagascariensis**
Habenaria mascarensis = **Cynorkis rosellata**
Habenaria minutiflora = **Benthamia spiralis**
Habenaria misera = **Benthamia misera**
Habenaria nigricans = **Habenaria boiviniana**
Habenaria nutans = **Cynorkis nutans**
Habenaria papillosa = **Cynorkis papillosa**
Habenaria perrieri = **Habenaria boiviniana**
Habenaria pervillei = **Platycoryne pervillei**
Habenaria purpurea = **Cynorkis purpurea**
Habenaria regalis = **Megalorchis regalis**
Habenaria rosellata = **Cynorkis rosellata**
Habenaria rutenbergiana = **Habenaria incarnata**
Habenaria simplex = **?Habenaria stricta**
Habenaria spiralis = **Benthamia spiralis**
Habenaria stricta = **Habenaria deanscowaniana**
Habenaria tenerrima = **Cynorkis tenerrima**
Habenaria tenuicaulis = **Platycoryne pervillei**
Haplostellis truncata = **Nervilia simplex**
Helorchis filiformis = **Cynorkis papillosa**
Hemiperis brachcentra = **Cynorkis brachycentra**
Hemiperis tenella = **?Cynorkis frappieri** (cfr. Schltr. 1925)
Hemiperis tenella = **?Cynorkis tenella**
Herminium sigmoideum = **Cynorkis sigmoidea**
Herminium spirale = **Benthamia spiralis**
Hetaeria occulta = **Platylepis occulta**
Holothrix glaberrima = **Benthamia glaberrima**
Holothrix madagascariensis = **Benthamia madagascariensis**

Holothrix schmidtii = **Cynorkis schmidtii**

Iridorchis equitans = **Oberonia disticha**

Jumellea ambongensis = **Jumellea gracilipes**

Jumellea ankaratrana = **Jumellea confusa**

Jumellea curnowiana = **Angraecum curnowianum**

Jumellea curnowiana = **Aerangis monantha**

Jumellea exilipes = **Jumellea gracilipes**

Jumellea ferkoana = **Jumellea lignosa** subsp. **ferkoana**

Jumellea floribunda = **Jumellea brachycentra**

Jumellea henryi = **Jumellea jumelleana**

Jumellea humbertii = **Angraecum ampullaceum**

Jumellea imerinensis = **Jumellea gracilipes**

Jumellea lignosa subsp. *ferkoana* = **Jumellea lignosa**

Jumellea lignosa subsp. *typica* = **Jumellea lignosa**

Jumellea longivaginans var. grandis = **Jumellea longivaginans**

Jumellea meirax = **Angraecum meirax**

Jumellea rutenbergiana = **Angraecum rutenbergianum**

Jumellea serpens = **Angraecum serpens**

Jumellea subcordata = **Angraecum curnowianum**

Jumellea tsaratananae = **Jumellea zaratananae**

Jumellea unguicularis = **Jumellea gracilipes**

Lemuranthe gymnochiloides = **Cynorkis gymnochiloides**

Lemurella ambongensis = **Lemurella culicifera**

Lemurella sarcanthoides = **Oeonia brauniana** var. **sarcanthoides**

Lemurella tricalcariformis = **Oeonia brauniana** var. **sarcanthoides**

Leptocentrum spiculatum = **Aerangis spiculata**

Leptorchis see Leptorkis

Leptorkis bicornis = **Liparis bicornis**

Leptorkis caespitosa = **Liparis cespitosa**

Leptorkis connata = **Liparis ochracea**

Leptorkis disticha = **Liparis disticha**

Leptorkis flavescens = **Liparis flavescens**

Leptorkis longicaulis = **Liparis longicaulis**

Leptorkis longipetala = **Liparis longipetala**

Leptorkis lutea = **Liparis lutea**

Leptorkis ochracea = **Liparis ochracea**

Leptorkis ornithorrhynchos = **Liparis ornithorrhynchos**

Leptorkis parva = **Liparis parva**

Leptorkis puncticulata = **Liparis puncticulata**

Leptorkis purpurascens = **Liparis salassia**

Leptorkis xanthina = **Liparis xanthina**

Limodorum concolor = **Graphorkis concolor**

Limodorum coriaceum = **Angraecum coriaceum**

Limodorum cucullatum = **Eulophia cucullata**

Limodorum eburneum = **Angraecum eburneum**

Limodorum flabellatum = **Cymbidiella flabellata**

Limodorum maculatum = **Oeceoclades maculata**

Limodorum plantagineum = **Eulophia plantaginea**

Limodorum pulchrum = **Eulophia pulchra**

Limodorum scriptum = **Graphorkis concolor** var. **alphabetica**

Limodorum spathulatum see Oeoniella polystachys

Limodorum tuberculosum = **Gastrorchis tuberculosa**

Liparis angustifolia = **Liparis caespitosa**

Liparis auriculate = **Liparis caespitosa.**

Liparis cardiophylla = **Liparis clareae**

Liparis cardiophylla var. *angustifolia* = **Liparis clareae** var **angustifolia**

Liparis comosa = **Liparis caespitosa**

Liparis connata = **Liparis ochracea**

Liparis gregaria = **Liparis disticha**

Liparis hildebrandtiana = **Liparis ochracea**

Liparis latilabris = **Liparis henricii**

Liparis monophylla = **Liparis ambohimangana**

Liparis purpurascens = **Liparis salassia**
Liparis pusilla = **Liparis caespitosa**
Liparis tsaratananae sphalm. = **Liparis zaratananae**
Liparis verecunda = **Liparis henricii**
Lisschilus petiolatus = **Oeceoclades petiolata**
Lissochilus ambongensis = **Oeceoclades perrieri**
Lissochilus ambrensis = **Eulophia ambrensis**
Lissochilus analamerensis = **Oeceoclades analamerensis**
Lissochilus analavelensis = **Oeceoclades analavelensis**
Lissochilus anjoanensis = **Eulophia angornensis**
Lissochilus beravensis = **Eulophia beravensis**
Lissochilus boinensis = **Oeceoclades boinensis**
Lissochilus clitellifera = **Eulophia clitellifera**
Lissochilus cordylinophyllus = **Oeceoclades cordylinophylla**
Lissochilus cornigerus = **Eulophia livingstoniana**
Lissochilus cornigerus var. *minor* = **Eulophia livingstoniana**
Lissochilus decaryanus = **Oeceoclades decaryana**
Lissochilus ecalcaratus = **Graphorkis ecalcarata**
Lissochilus elliotii = **Oeceoclades sclerophylla**
Lissochilus ephipium = **Eulophia ephippium**
Lissochilus fallax = **Eulophia livingstoniana**
Lissochilus faradjensis = **Eulophia livingstoniana**
Lissochilus flexuosus = **Eulophia clitellifera**
Lissochilus galbanus = **Eulophia galbana**
Lissochilus gracilior = **Eulophia livingstoniana**
Lissochilus gracillimus = **Oeceoclades gracillima**
Lissochilus grandidieri = **Eulophia grandidieri**
Lissochilus hebdingianus = **Oeceoclades hebdingiana**
Lissochilus hologlossus = **Eulophia hologlossa**
Lissochilus humbertii = **Eulophia humbertii**
Lissochilus ibityensis = **Eulophia ibityensis**
Lissochilus jumelleanus = **Eulophia livingstoniana**
Lissochilus kranzlini = **Eulophia rutenbergiana**
Lissochilus laggiarae = **Eulophia livingstoniana**
Lissochilus livingstonianus = **Eulophia livingstoniana**
Lissochilus lokobensis = **Oeceoclades cordylinophylla**
Lissochilus lonchophyllus = **Oeceoclades lonchophylla**
Lissochilus macer = **Eulophia macra**
Lissochilus madagascariensis = **Eulophia reticulata**
Lissochilus medemiae = **Graphorkis medemiae**
Lissochilus mediocris = **Eulophia livingstoniana**
Lissochilus megistophyllus = **Eulophia megistophylla**
Lissochilus nervosus = **Eulophia nervosa**
Lissochilus palmicolus = **Paralophia palmicola**
Lissochilus panduratus = **Oeceoclades pandurata**
Lissochilus paniculatus = **Oeceoclades calcarata**
Lissochilus perrieri = **Eulophia perrieri**
Lissochilus petiolatus = **Oececlades petiolata**
Lissochilus pileatus = **Eulophia pileata**
Lissochilus plantagineus = **Eulophia plantaginea**
Lissochilus pulchellus = **Eulophia clitellifera**
Lissochilus pulcher = **Eulophia pulchra**
Lissochilus quadrilobus = **Oeceoclades quadriloba**
Lissochilus ramosus = **Eulophia ramosa**
Lissochilus rehmannii = **Eulophia clitellifera**
Lissochilus rutenbergianus = **Eulophia livingstoniana**
Lissochilus schlechteri = **Oeceoclades ambongensis**
Lissochilus scriptus sensu H. Perrier = **Graphorkis concolor** var. **alphabetica**
Lissochilus spathulifer = **Oeceoclades spathulifera**
Lissochilus vaginatus = **Eulophia hians** var. **nutans**
Lissochilus wendlandianus = **Eulophia wendlandiana**
Listrostachys elliotii = **Microcoelia elliotii**

Listrostachys parviflora = **Angraecopsis parviflora**
Listrostachys polystachys = **Oeoniella polystachys**
Listrostachys trifurca = **Angraecopsis trifurca**
Macroplectrum baronii = **Angraecum baronii**
Macroplectrum calceolus = **Angraecum calceolus**
Macroplectrum didieri = **Angraecum didieri**
Macroplectrum gladiifolium = **Angraecum mauritianum**
Macroplectrum humblotii = **Angraecum humblotianum**
Macroplectrum implicatum = **Angraecum implicatum**
Macroplectrum leonis = **Angraecum leonis**
Macroplectrum madagascariensis = **Angraecum madagascariensis**
Macroplectrum meirax = **Angraecum meirax**
Macroplectrum ochraceum = **Angraecum ochraceum**
Macroplectrum pectinatum = **Angraecum pectinatum**
Macroplectrum ramosum = **Angraecum conchoglossum**
Macroplectrum sesquipedale = **Angraecum sesquipedale**
Macroplectrum xylopus = **Angraecum xylopus**
Malaxis angustifolia = **Liparis cespitosa**.
Malaxis brevifolia = **Oberonia disticha**
Malaxis cespitosa = **Liparis cespitosa**
Malaxis disticha = **Liparis disticha**
Malaxis flavescens = **Liparis flavescens**
Malaxis purpurascens = **Liparis salassia**
Microcoelia melinantha = **Microcoelia gilpinae**
Microstylis atro-ruber = **Malaxis atro-rubra**
Microstylis cardiophylla = **Malaxis cardiophylla**
Microstylis françoisii = **Malaxis françoisii**
Microstylis madagascariensis = **Malaxis madagascariensis**
Microstylis physuroides = **Malaxis physuroides**
Microtheca madagascarica = **Cynorkis lilacina** x **ridleyi**
Monixus clavigera = **Angraecum clavigerum**
Monixus coriaceum see Angraecum coriaceum
Monixus graminifolius = **Angraecum pauciramosum**
Monixus multiflorus = **Angraecum multiflorum**
Monixus polystachys = **Oeoniella polystachys**
Monixus teretifolius = **Angraecum teretifolium**
Monochilus boryi = **Cheirostylis nuda**
Monochilus gymnochiloides = **Cheirostylis nuda**
Mystacidium aphyllum = **Microcoelia aphylla**
Mystacidium calceolus = **Angraecum calceolus**
Mystacidium carpophorum = **Angraecum calceolus**
Mystacidium caulescens = **Angraecum caulescens**
Mystacidium caulescens var. *multiflorum* = **Angraecum multiflorum**
Mystacidium comorense = **Jumellea comorensis**
Mystacidium curnowianus = **Angraecum curnowianum**
Mystacidium dauphinense = **Angraecum dauphinense**
Mystacidium dentiens = **Aeranthes dentiens**
Mystacidium exile = **Microcoelia exilis**
Mystacidium germinyanum = **Angraecum germinyanum**
Mystacidium gilpinae = **Microcoelia gilpinae**
Mystacidium gladiator = **Jumellea gladiator**
Mystacidium gladiifolium = **Angraecum mauritianum**
Mystacidium gracile = **Chamaeangis gracilis**
Mystacidium graminifolium = **Angraecum pauciramosum**
Mystacidium grandidierianum = **Neobathiea grandidieriana**
Mystacidium hariotianum = **Microterangis hariotiana**
Mystacidium inapertum = **Angraecum inapertum**
Mystacidium leonis = **Angraecum leonis**
Mystacidium mauritianum = **Angraecum mauritianum**
Mystacidium multiflorum = **Angraecum multiflorum**
Mystacidium ochraceum = **Angraecum ochraceum**
Mystacidium ophioplectron = **Jumellea ophioplectron**

Mystacidium pectinatum = **Angraecum pectinatum**
Mystacidium pedunculatum = **Angraecopsis parviflora**
Mystacidium phalaenophorum = **Jumellea phalaenophora**
Mystacidium physophorum = **Microcoelia physophora**
Mystacidium sesquipedale = **Angraecum sesquipedale**
Mystacidium tenellum = **Angraecum tenellum**
Mystacidium thouarsii = **Angraecopsis trifurca**
Mystacidium trichoplectron = **Angraecum trichoplectron**
Mystacidium trifurcum = **Angraecopsis trifurca**
Mystacidium viride = **Angraecum rhynchoglossum**
Neobathiea filicornu = **Neobathiea grandidieriana**
Neobathiea gracilis = **Aeranthes schlechteri**
Neobathiea sambiranoensis = **Aeranthes schlechteri**
Neottia salassia = **Liparis salassia**
Nervilia afzelii = **Nervilia petraea**
Nervilia afzelii var. grandiflora = **Nervilia simplex**
Nervilia barklyana = **Nervilia bicarinata**
Nervilia bathiei = **Nervilia simplex**
Nervilia bollei = **Nervilia simplex**
Nervilia commersonii = **Nervilia bicarinata**
Nervilia crispata = **Nervilia simplex**
Nervilia crociformis = **Nervilia simplex**
Nervilia dalbergiae = **Nervilia kotschyi** var. **purpurata**
Nervilia erosa = **Nervilia simplex**
Nervilia françoisii = **Nervilia simplex**
Nervilia ghindiana = **Nervilia bicarinata**
Nervilia humilis = **Nervilia simplex**
Nervilia insolata = **Nervilia renschiana**
Nervilia natalensis = **Nervilia renschiana**
Nervilia perrieri = **Nervilia affinis**
Nervilia pilosa = **Nervilia affinis**
Nervilia prainiana = **Nervilia simplex**
Nervilia reniformis = **Nervilia simplex**
Nervilia renschiana = **Nervilia bicarinata**
Nervilia sakoae = **Nervilia kotschyi**
Nervilia similis = **Nervilia kotschyi**
Nervilia viridiflava = **Nervilia bicarinata**
Notiophrys occulta = **Platylepis occulta**
Oberonia brevifolia = **Oberonia disticha**
Oberonia equitans = **Oberonia disticha**
Oeceoclades mackenii = **Oeceoclades maculata**
Oeceoclades pandurata also see Oeceoclades lanceata
Oeceoclades parviflora = **Angraecopsis parviflora**
Oeceoclades pulchra = **Eulophia pulchra**
Oeceoclades roseovariegata = **Oeceoclades gracillima**
Oeonia auberti = **Oeonia volucris**
Oeonia brachystachya = **Beclardia macrostachya**
Oeonia culicifera = **Lemurella culicifera**
Oeonia elliotii = **Oeonia volucris**
Oeonia erostris = **Beclardia macrostachya**
Oeonia erostris var. *egena* = **Beclardia macrostachya**
Oeonia erostris var. *robusta* = **Beclardia macrostachya**
Oeonia forsythiana = **Oeonia rosea**
Oeonia humblotii = **Oeonia volucris**
Oeonia macrostachya = **Beclardia macrostachya**
Oeonia oncidiiflora = **Oeonia rosea**
Oeonia polystachya = **Oeoniella polystachys**
Oeonia robusta = **Sobennikoffia robusta**
Oeonia subacaulis = **Oeonia brauniana** var. **sarcanthoides**
Oeoniella sarcanthoides = **Oeonia brauniana** var. **sarcanthoides**
Onychium flavescens = **Polystachya concreta**
Ophrys salassia = **Liparis salassia**

Orchiodes nudum = **Cheirostylis nuda**
Orchiodes occultum = **Platylepis occulta**
Orchis fastigiata = **Cynorkis fastigiata**
Orchis mauritiana = **Angraecum mauritianum**
Orchis mauritiana = **Cynorkis fastigiata**
Orchis obcordata = **Cynorkis fastigiata**
Orchis purpurascens = **Cynorkis purpurascens**
Orchis triphylla = **Cynorkis fastigiata** var. **triphylla**
Pectinaria thouarsii = **Angraecum pectinatum**
Peristylus filiformis = **Cynorkis papillosa**
Peristylus glaberrima = **Benthamia glaberrima**
Peristylus gramineus = **Cynorkis graminea**
Peristylus macropetalus = **Benthamia madagascariensis**
Peristylus madagascariensis = **Benthamia madagascariensis**
Peristylus purpureus = **Cynorkis purpurea**
Peristylus spiralis = **Benthamia spiralis**
Perrieriella madagascariensis = **Oeonia madagascariensis**
Phaius fragrans = **Gastrorchis simulans**
Phaius françoisii = **Gastrorchis françoisii**
Phaius françoisii var. *orientalis* see Gastrorchis françoisii
Phaius françoisii var. *pauliana* = **Gastrorchis françoisii**
Phaius geffrayi = **Gastrorchis geffrayi**
Phaius gibbosulus = **Imerinaea madagascarica**
Phaius humblotii = **Gastrorchis humblotii**
Phaius humblotii var. *albiflora* = **Gastrorchis humblotii**
Phaius humblotii var. *rubra* = **Gastrorchis humblotii** var. **rubra**
Phaius humblotii var. *schlechteri* = **Gastrorchis humblotii** var. **schlechteri**
Phaius luteus = **Gastrorchis lutea**
Phaius peyrotii = **Gastrorchis peyrotii**
Phaius pulcher = **Gastrorchis pulchra**
Phaius pulcher var. *perrieri* = **Gastrorchis pulchra** var. **perrieri**
Phaius schlechteri = **Gastrorchis humblotii** var. **schlechteri**
Phaius simulans = **Gastrorchis simulans**
Phaius tuberculatus = **Gastrorchis tuberculosa**
Phaius tuberculatus var. *superba* = **Gastrorchis tuberculosa**
Phaius tuberculosus = **Gastrorchis tuberculosa**
Phaius warpuri = **Gastrorchis tuberculosa**
Phajus see Phaius
Phyllorchis baronii = **Bulbophyllum baronii**
Phyllorchis erecta = **Bulbophyllum erectum**
Phyllorchis hildebrandtii = **Bulbophyllum hildebrandtii**
Phyllorchis minuta = **Bulbophyllum minutum**
Phyllorchis multiflora = **Bulbophyllum multiflorum**
Phyllorchis occlusa = **Bulbophyllum occlusum**
Phyllorchis thompsonii = **Bulbophyllum thompsonii**
Phyllorchis variegata = **Bulbophyllum variegatum**
Phyllorkis longiflora = **Bulbophyllum longiflorum**
Phyllorkis nuphyllis = **Bulbophyllum nutans**
Phyllorkis nutans = **Bulbophyllum nutans**
Platanthera glaberrima = **Benthamia glaberrima**
Platanthera graminea = **Cynorkis graminea**
Platanthera madagascariensis = **Benthamia madagascariensis**
Platanthera madagascariensis = **Benthamia rostrata**
Platycoryne buchananiana = **Platycoryne pervillei**
Platycoryne tenuicaulis = **Platycoryne pervillei**
Platylepis goodyeroides = **Platylepis occulta**
Platylepis humicola = **Goodyera humicola**
Platylepis perrieri = **Goodyera perrieri**
Plectrelminthus spiculatus = **Aerangis spiculata**
Pleurothallis disticha = **Oberonia disticha**
Pogonia barklayana = **Nervilia bicarinata**
Pogonia bicarinata = **Nervilia bicarinata**

Pogonia bollei = **Nervilia simplex**
Pogonia commersonii = **Nervilia bicarinata**
Pogonia crispata = **Nervilia simplex**
Pogonia finetii = **Nervilia simplex**
Pogonia hirsuta = **Nervilia hirsuta**
Pogonia kotschyi = **Nervilia kotschyi**
Pogonia lanceolata = **Nervilia kotschyi** var. **purpurata**
Pogonia leguminosarum = **Nervilia leguminosarum**
Pogonia prainiana = **Nervilia simplex**
Pogonia purpurata = **Nervilia kotschyi** var. **purpurata**
Pogonia renschiana = **Nervilia bicarinata**
Pogonia renschiana = **Nervilia renschiana**
Pogonia sakoae = **Nervilia kotschyi**
Pogonia simplex = **Nervilia simplex**
Pogonia thouarsii = **Nervilia simplex**
Polystachya appendiculata = **Polystachya cultriformis**
Polystachya aurantiaca = **Polystachya concreta**
Polystachya bicolor = **Polystachya rosea**
Polystachy abuchanani = **Polystachya concreta**
Polystachya caquetana = **Polystachya concreta**
Polystachya colombiana = **Polystachya concreta**
Polystachya composita = **Polystachya fusiformis**
Polystachya concreta also see Polystachya aurantiaca
Polystachya concreta var. *exasanguis* = **Polystachya concreta**
Polystachya concreta var. *luteola* = **Polystachya concreta**
Polystachya concreta var. *paniculata* = **Polystachya concreta**
Polystachya concreta var. *tincta* = **Polystachya concreta**
Polystachya cultrata = **Polystachya cultriformis**
Polystachya cultriformis var. *africana* = **Polystachya cultriformis**
Polystachya cultriformis var. *alba* = **Polystachya cultriformis**
Polystachya cultriformis var. *autogama* = **Polystachya cultriformis**
Polystachya cultriformis var. *erubescens* = **Polystachya cultriformis**
Polystachya cultriformis var. *humblotii* = **Polystachya cultriformis**
Polystachya cultriformis var. *nana* = **Polystachya cultriformis**
Polystachya dorotheae = **Polystachya concreta**
Polystachya estrellensis = **Polystachya concreta**
Polysatchya extinctoria = **Polystachya concreta**
Polystachya flavescens = **Polystachya concreta**
Polystachya gabonensis = **Polystachya concreta**
Polystachya gracilis = **Polystachya concreta**
Polystachya hildebrandtii = **Polystachya rosea**
Polystachya hildebrandtii = **Polystachya anceps**
Polystachya huyghei = **Polystachya concreta**
Polystachya hypocrita = **Polystachya concreta**
Polystachya jussieuana = **Polystachya concreta**
Polystachya kraenzliniana = **Polystachya concreta**
Polystachya latifolia = **Polystachya concreta**
Polystachya lepidantha = **Polystachya concreta**
Polystachya lettowiana = **Polystachya concreta**
Polystachya luteola = **Polystachya concreta**
Polystachya mauritiana = **Polystachya concreta**
Polystachya mauritiana var. *anceps* = **Polystachya anceps**
Polystachya minuta = **Polystachya concreta**
Polystachya minutiflora = **Polystachya fusiformis**
Polystachya mukandaensis = **Polystachya concreta**
Polystachya nitidula also see Polystachya aurantiaca
Polystachya nitidula = **Polystachya concreta**
Polystachya plehniana = **Polystachya concreta**
Polystachya praealta = **Polystachya concreta**
Polystachya reichenbachiana = **Polystachya concreta**
Polystachya rigidula = **Polystachya concreta**
Polystachya rufinula = **Polystachya concreta**

Polystachya siamensis = **Polystachya concreta**

Polystachya shirensis = **Polystachya concreta**

Polystachya similis = **Polystachya concreta**

Polystachya tessellata = **Polystachya concreta**

Polystachya tricruris = **Polystachya concreta**

Polystachya villosa = **Polystachya affinis**

Polystachya villosula = **Polystachya affinis**

Polystachya affinis = **Not a Madagascan species**

Polystachya zanguebarica = **Polystachya concreta**

Polystachya zeylanica = **Polystachya concreta**

Polystachya zollingeri = **Polystachya concreta**

Rhaphidorhynchus aphyllus = **Microcoelia aphylla**

Rhaphidorhynchus articulatus = **Aerangis articulata**

Rhaphidorhynchus chiloschistae = **Microcoelia exilis**

Rhaphidorhynchus citratus = **Aerangis citrata**

Rhaphidorhynchus curnowianus = **Aerangis monantha**

Rhaphidorhynchus fastuosus = **Aerangis fastuosa**

Rhaphidorhynchus gilpinae = **Microcoelia gilpinae**

Rhaphidorhynchus macrostachyus = **Beclardia macrostachya**

Rhaphidorhynchus macrostachyus var. *brachystachyus* = **Beclardia macrostachya**

Rhaphidorhynchus modestus = **Aerangis modesta**

Rhaphidorhynchus modestus var. *sanderianus* = **Aerangis modesta**

Rhaphidorhynchus ophioplectron = **Jumellea ophioplectron**

Rhaphidorhynchus perrieri = **Microcoelia perrieri**

Rhaphidorhynchus pobeguini = **Angraecopsis pobeguini**

Rhaphidorhynchus spiculatus = **Aerangis spiculata**

Rhaphidorhynchus stylosus = **Aerangis stylosa**

Rhaphidorhynchus stylosus sensu Finet = **Aerangis cryptodon**

Rhaphidorhynchus umbonatus = **Aerangis fuscata**

Rolfeella glaberrima = **Benthamia glaberrima**

Saccolabium aphyllum = **Microcoelia aphylla**

Saccolabium coriaceum = **Angraecum coriaceum**

Saccolabium hariotianum = **Microterangis hariotiana**

Saccolabium humblotii = **Microterangis humblotii**

Saccolabium microphyton = **?Angraecum tenellum**

Saccolabium parviflorum = **Angraecopsis parviflora**

Satyrium atherstonei = **Satyrium trinerve**

Satyrium calceatum = **Disa buchenaviana**

Satyrium gigas = **Satyrium rostratum**

Satyrium gracile = **Satyrium amoenum**

Satyrium gramineum = **Cynorkis graminea**

Satyrium monopetalum = **Satyrium trinerve**

Satyrium nutii = **Satyrium trinerve**

Satyrium occultum = **Satyrium trinerve**

Satyrium praealtum = **Habenaria praealta**

Satyrium rosellatum = **Cynorkis rosellata**

Satyrium spirale = **Benthamia spiralis**

Satyrium triphyllum = **Satyrium trinerve**

Satyrium zombense = **Satyrium trinerve**

Serapias salassia = **Liparis salassia**

Solenangis aphylla = **Microcoelia aphylla**

Solenangis cornuta = **Microcoelia cornuta**

Spiranthes africana = **Benthamia spiralis**

Stelis micrantha see Liparis disticha

Stellorkis aplostellis = **Nervilia simplex**

Stichorkis cestichis = **Liparis cespitosum**

Stichorkis disticha = **Liparis disticha**

Zeuxine boryi = **Cheirostylis nuda**

Zeuxine gymnochiloides = **Cheirostylis nuda**

Zeuxine sambiranoensis = **Cheirostylis nuda**

Zygoglossum umbellatum = **Bulbophyllum longiflorum**

APPENDIX

Notes on the nomenclature of Orchids from Madagascar, the Mascarenes and the Comoros

Introduction

In preparing the *The Orchids of Madagascar* {281}, a number of nomenclatural problems related to the orchids of the region emerged. In that work information on the orchids of the island was summarised and cross-referenced but no attempt was made to resolve the nomenclatural inconsistencies that came to light. This appendix aims to rationalise some of the anomalies that were exposed.

Much of the taxonomic work on the orchids of Madagascar by Henri Perrier de la Bâthie and Rudolf Schlechter was undertaken between the two World Wars (1918 – 1939), especially towards the end of that period. This had a considerable effect on their work; communication was not always very effective, inter-herbarium loans were difficult and the publication of scientific papers was often erratic. In addition, conventions regarding the publication of Latin diagnoses after 1935 had been accepted and published but had not always been fully understood or implemented by botanists (International Code of Botanical Nomenclature 1935).

A number of taxa were described by Perrier de la Bâthie after 1st January 1935 but, lacking a Latin diagnosis, are considered invalid under Article 36 of the International Code of Botanical Nomenclature, 2000. The materials upon which Perrier's names were based have been re-examined and most are validated here. A few of his named varieties, in our opinion, do not warrant taxonomic recognition. Where this occurs such names are included in the synonymy of accepted taxa.

New Synonyms and Combinations

Aeranthes henricii Schltr. var. ***isaloensis*** H.Perrier ex Hermans **var. nov.** a varietate typica bracteis triangularibus acutis, floribus minoribus, lamina tranverse elliptica et columa distincta differt. Typus: Madagascar, Isalo, Oct. 1924, *H.Perrier* 16895 (holo. P!). (Fig. 1).
Aeranthes henricii Schltr. var. *isaloensis* H.Perrier in Humbert, Fl. Mad. Orch. 2: 138 (1941), *nom. nud.*

Other collections: Madagascar, Analavelona forest, N of Fiherenana, 950 – 1250 m, March 1934, *Humbert* 14217 (P!, K!, BM!, TANA!); Analavelona Massif, Tulear, March 1972, *Morat* 3397 (P!).

Angraecum acutipetalum Schltr. var. ***analabeensis*** H.Perrier ex Hermans **var. nov.** a var. typica periantho breviore, calcari breviore et anthera ad apicem late indentata et marginibus acute angulatis distinguenda. Typus: Analabe forest, N Imerina, *H.Perrier* 18516 (holo. P). (Fig. 2).
Angraecum acutipetalum var. *analabeensis* H.Perrier, Fl. Mad. Orch. 2: 229 (1941), *nom. nud.*

Angraecum acutipetalum Schltr. var. ***ankeranae*** H.Perrier ex Hermans **var. nov.** a var. typica flore minore, calcari breviore, disco labelli leviter pubescenti et anthera ad apicem indentata differt. Typus: Madagascar, Ankeramadinika, *E.Francois* in *H.Perrier* 18517 (holo. P!). (Fig. 3).
Angraecum acutipetalum var. *ankeranae* H.Perrier, Fl. Mad. Orch. 2: 230 (1941), *nom. nud.*

Bulbophyllum henrici Schltr. var. ***rectangulare*** H.Perrier ex Hermans **var. nov.** a varietate typica sepalis connatis biapiculatis, labello minore subrectangulari vix papillato distinguenda. Typus: Madagascar, Farafangana, Ikongo Massif, Oct. 1926, *Decary* 5785 (holo. P!). (Fig. 4).
Bulbophyllum henrici var. *rectangulare* H.Perrier in Notul. Syst. (Paris) 6 (2): 113 (1937), *nom. nud.*

Bulbophyllum ophiuchus Ridl. var. ***baronianum*** H.Perrier ex Hermans **var. nov.** a varietate typica floribus minoribus, sepalo dorsali lato obtuso, sepalis lateralibus connatis, labello suborbiculari et columni pede non dilatato bene distinguenda. Typus: Madagascar, without locality, *Baron* 4468 (holo. K!; iso. BM!, P!). (Fig. 5).
Bulbophyllum ophiuchus var. *baronianum* H.Perrier in Humbert, Fl. Mad. Orch. 1: 374 (1939), *nom. nud.*

Bulbophyllum pantoblepharon Schltr. var. ***vestitum*** H.Perrier ex Hermans **var. nov.** a varietate typica bracteolis parvis triangularibus crassis, pedunculo bracteis 5 – 6 sterilibus ad apicem dilatatis pedunculum obtegentibus, sepalis ellipticis, petalis minoribus labello dissimilis, stelidiis acutis et apiculo antherae carente bene distinguenda. Typus: Madagascar, Ankeramadinika, *François* 4 in Herb. *H.Perrier* 18646 (holo. P!). (Fig. 6).

Bulbophyllum pantoblepharon var. *vestitum* H.Perrier in Humbert, Fl. Mad. Orch. 1: 324 (1939), *nom. nud.*

Bulbophyllum pentastichum (Pfitzer ex Kraenzl.) Schltr. subsp. ***rostratum*** H.Perrier ex Hermans **subsp. nov.** a subspecie typica racemo magis curvato et floribus rubro-verrucosis differt. Typus: Madagascar, Fort Dauphin district, Vinanibe, Oct. 1932 *Decary* 10.865 (holo. P!)

Bulbophyllum matitanense subsp. *rostratum* H.Perrier in Humbert, Fl. Mad. Orch. 1: 418 (1939), *nom. nud.*

Other collections: Madagascar, Soanierana, littoral forest S of Fort Dauphin, Oct. 1932 *Decary* 10924 (P!); Toliara prov., between Fort Dauphin and Evatra, Jan. 2000, *Hermans* 5031 (K!).

Bulbophyllum sambiranense Jum. & H.Perrier var. ***latibracteatum*** H.Perrier ex Hermans **var. nov.** a varietate typica pseudobulbis approximatis orbicularibus, foliis latioribus et bracteis flores occultantibus differt. Typus: Madagascar, Centre, Tampoketsa between the Ikopa and the Betsiboka, forest of the eastern slopes, Aug. 1924, c. 1500 m, *H.Perrier* 16724 (holo. P!). (Fig. 7).

Bulbophyllum sambiranense var. *latibracteatum* H.Perrier in Notul. Syst. (Paris) 6 (2): 86 (1937), *nom. nud.*

Other collections: Madagascar, Sambirano, lower mountains of the Sambirano valley, c. 600 m, Apr. 1924, *H.Perrier* 16521 (P!).

Bulbophyllum sarcorhachis Schltr. var. ***flavomarginatum*** H.Perrier ex Hermans **var. nov.** a varietate *befoana* (Schltr.) H.Perrier pseudobulbis latioribus ovato-oblongis, foliis latioribus, racemo griseo-viridi, sepalo dorsali valde obovato, et sepalis latioribus connatis brunneis ad marginem pallide luteis differt. Typus: Madagascar, Mt. Tsaratanana, between 2200 and 2600 m, lichen forest, April 1924, *H.Perrier* 16511 (holo. P!). (Fig. 8).

Bulbophyllum sarcorhachis var. *flavomarginatum* H.Perrier in Notul. Syst. (Paris) 6 (2): 110 (1937), *nom. nud.*

Other collections: Madagascar, Centre, Sambirano Mountains, Manongarivo Massif, c. 1200 m, May 1909, *H.Perrier* 8006 (P!); Manongarivo reserve, Bekolosy, valley towards the falls of the Bekolosy, 1100 m, edge of forest, epiphyte, July 1994, *Derleth* 15 (G, K!, P!).

Bulbophyllum zaratananae Schltr. subsp. ***disjunctum*** H.Perrier ex Hermans **subsp. nov.** a subspecie typica floribus leviter majoribus usque 9 cm longis, marginibus columnae infra rostellum rectis, pede columnae plano et stigmatibus stelidiisque acutis distinguenda. Typus: Madagascar, Farafangana, Ikongo Massif, Oct. 1926, *Decary* 5611 (holo. P!; iso. L). (Fig. 9).

Bulbophyllum tsaratananae subsp. *disjunctum* H.Perrier in Notul. Syst. (Paris) 6 (2): 59 (1937), *nom. nud.*

Cynorkis laeta Schltr. var. ***angavoensis*** H.Perrier ex Hermans **var. nov.** a varietate typica foliis numerosis (7 – 8) angustioribus (4 – 8 mm latis) differt. TYPE: Madagascar, Angavo massif, *R.Decary* 7344 (lecto. P; isolecto. TAN both selected here). (Fig. 10).

Cynorkis laeta var. *angavoensis* H.Perrier in Humbert, Fl. Mad. Orch. 1: 131 (1939), *nom. nud.*

Other collections: Madagascar, Tampoketse d'Ankazobe, March 1930, *Decary* 7363 (P!, K!, BM); Angavo Massif, near Ankazobe, March 1930, *Decary* 7344 (P!); Ambohimalaza, on the Tampoketse d'Ankazoba, May 1930, *Decary* 7730 (P!); Ambohitantely forest, Tampoketse d'Ankazobe, *Ursch* s.n. (P!, TANA!); Antananarivo Prov., near Ankazobe, May 2001, *Hermans* 413 (K!).

Cynorkis lilacina Ridl. var. ***comorensis*** H.Perrier ex Hermans **var. nov.** a varietate typico labello crasso lobis angustis ornato, lobis lateralibus uninerviis ferenti et calcari recto ad apicem obtusum angustato differt. Typus: Comoros, Grande Comore, May 1850, *Boivin* s.n. (holo. P!). A single plant mixed in with the type of *C. lilacina* var. *boiviniana* (Kraenzl.) H.Perrier – P80968. (Fig. 11).

Cynorkis lilacina var. *comorensis* H.Perrier in Humbert, Fl. Mad. Orch. 1: 92 (1939), *nom. nud.*

Cynorkis lilacina Ridl. var. **tereticalcar** H.Perrier ex Hermans **var. nov.** a varietate typica calcari subcylindraceo plus minusve elongato saepe ovarium aequanti differt. Typus: Madagascar, Andringitra Massif, c. 1800 m, moss forest and undergrowth, Feb. 1922, *H.Perrier* 14571 (holo. P!). (Fig. 12).
Cynorkis lilacina var. *tereticalcar* H.Perrier in Humbert, Fl. Mad. Orch. 1: 91 (1939), *nom. nud.*

Other collections: Madagascar, Mt. Tsaratanana, *Philippia* Savaka, 2000 m, March 1924, *H.Perrier* 16519 (P!); without locality, *Le Myre de Vilers* s.n. (P80978) (P!).

Cynorkis fastigiata Thouars. var. **ambatensis** Hermans **var. nov.** a varietate typica floribus parvis, ovario glabro et lobo medio rostelli integro differt. Typus: Madagascar, Sambirano nr. Ambato peninsula, dry slopes, *H.Perrier* 1857 (*Schlechter* 90) (holo. P!). (Fig. 13).
Cynorchis fastigiata var. *ambatensis* H.Perrier Fl. Mad. Orch. 1: 141 (1939), *nom. nud.*

Jumellea lignosa (Schltr.) Schltr. subsp. **acutissima** H.Perrier ex Hermans **subsp. nov.** a subspecie typica angulis supernis alarum columnae longe acutis, calcari valde breviore 6 cm longo, anthera ad apicem late indentata, retinaculo angustiore elongato et lobo medio rostelli 0.6 mm longo valde acuto quam alis breviore differt. Typus: Central, Ankeramadinika, moss forest, epiphyte, c. 1400 m, Mar. 1928, *E.François* in Herb. *H.Perrier* 18518 (holo. P!). (Fig. 14).
Jumellea lignosa var. *acutissima* H.Perrier in Notul. Syst. (Paris) 7: 59 (1938), *nom. nud.*

Jumellea lignosa (Schltr.) Schltr. subsp. **latilabia** H.Perrier ex Hermans **subsp. nov.** a subspecie typica planta usque 80 cm alta, foliis 12 – 14 cm longis 2.2 – 2.6 cm latis, labello late ovato acuto 17-nervo, callo labellum aequanti, dente medio rostelli alas fere aequanti, et alis columnae ad basin obtusis differt. Typus: Madagascar, Central, Andasibe, on the Onive, forest, c. 1000 m, epiphyte, by streams, Feb. 1925, *H.Perrier* 17129 (holo. P). (Fig. 15).
Jumellea lignosa var. *latilabia* H.Perrier in Notul. Syst. (Paris) 7: 60 (1938), *nom. nud.*

Jumellea lignosa (Schltr.) Schltr. subsp. **tenuibracteata** H.Perrier ex Hermans **subsp. nov.** a subspecie typica foliis longioribus usque 20 cm longis, bracteis sterilis ecarinatis minoribus, et bracteis fertilibus brevioribus, neque carinatis neque ancipitibus differt. Typus: Madagascar, Central, Mt. Tsaratanana, lichen forest, epiphyte, pendant, c. 2000 m, Jan. 1923, *H.Perrier* 15331 (holo. P!). (Fig. 16).
Jumellea lignosa var. *tenuibracteata* H.Perrier in Notul. Syst. (Paris) 7: 59 (1938), *nom. nud.*

Liparis clareae Hermans var. **angustifolia** H.Perrier ex Hermans **var. nov.** a varietate typica foliis angustioribus, floribus 3 – 5, sepalis longioribus, petalis binervosis, labello maiore lamina rectangulari, disco ad basin distincte calloso, columna excelsiori 5.5 mm et anthera minore ad apicem latiore obtusioreque differt. Typus: Madagascar, area of the confluence of the Mangoro & Onive, eastern forest, c. 700 m, Feb. 1925 *H.Perrier* 17018 (holo. P!). (Fig. 17).
Liparis cardiophylla var. *angustifolia* H.Perrier in Notul. Syst. (Paris) 5: 244 (1936), *nom. nud.*

Liparis perrieri Schltr. var. **trinervia** H.Perrier ex Hermans **var. nov.** a varietate typica sepalis lateralibus 5-nervis, labello ad apicem breve apiculato trinervio nervo medio integro callum distinctum prominentissimum conicum ferenti et lobo apici antherae subrectangulari distinguenda. Typus: Madagascar, area of the confluence of the Mangoro & Onive, 700 m, Feb. 1925, *H.Perrier* 17136 (holo. P!). (Fig. 18).
Liparis perrieri var. *trinervia* H.Perrier in Notul. Syst. (Paris) 5: 251 (1936), *nom. nud.*

Neobathiea hirtula H.Perrier var. **floribunda** H.Perrier ex Hermans **var. nov.** a varietate typica habitu maiore, caule longiore, foliis usque 11 × 2 mm, caule omni inflorescentias 6 – 10 et 30 – 50 cm longas 6 – 10-floribus ornatas ferenti, sepalo dorsali 18 mm longo, sepalis lateralibus 21 – 22 mm longis et lobo medio rostelli quam alis breviore differt. Typus: Madagascar, Boina, Ankarafantsika reserve, April 1933, *Ursch* 20 (holo. P!). (Fig. 19).
Neobathiea hirtula var. *floribunda* H.Perrier in Notul. Syst. (Paris) 7: 50 (1938), *nom. nud.*

Satyrium amoenum (Thouars) A.Rich. var. **tsaratananae** H.Perrier ex Hermans **var. nov.** a varietate typica sepalis petalisque minoribus (4 mm longis), sepalis non apiculatis, sepalo dorsale trinervio, petalis uninerviis, calcari 12 mm longo differt. Type: Madagascar, Mt. Tsaratanana, ericaceous scrub of *Philippia*, 2400 m, April 1924, *H.Perrier* 16507 (holo. P!). (Fig. 20).
Satyrium amoenum var. *tsaratananae* H.Perrier in Humbert, Fl. Mad. Orch. 1: 168 (1939), *nom. nud.*

Perrier's Unpublished Taxa

In the orchid account for the *Flore de Madagascar*, Perrier de la Bâthie {896} referred to his revision of Madagascan *Cynorkis*, that included a number of new species and sections, in volume 5 of the *Archives de Botanique (Bulletin Mensuel) de Caen* (1931). Four volumes of this journal, edited by Perrier, R. Viguier and H. Chermezon, appeared but, unfortunately, no copies of the fifth volume have been found despite extensive searches in botanical libraries in England, Belgium and France. It is apparent that this important volume was never published. Several years after its supposed publication date, Perrier de la Bâthie {885} himself referred to a number of the *Cynorkis* taxa that were in that unpublished treatment as '*Perrier in Viguier Arch. Bot. inédit*'. Other papers referred to as being published in the same volume have also not been located (J.-N. Labat, pers. comm.). Thus, it seems certain that Perrier's revision of the genus *Cynorkis* in Madagascar was never published. It would have contained a total of four new sections, nine new species, four subspecies, six new varieties, five new combinations and one new name. After examination of Perrier's various type specimens and those of related entities, the following sections, species, subspecies and varieties are formally described as follows:

Cynorkis section *Gibbosorchis* H.Perrier ex Hermans **sect. nov.** e sectionibus ceteris rostello in dimidio basale connato sed ad apicem trilobato, lobo medio integro antice ascendenti differt. Typus: *C. gibbosa* Ridley.
Cynorkis sect. *Gibbosorchis* H.Perrier in Humbert, Fl. Mad. Orch. 1: 124 (1939), *nom. nud.*

The species of this section have the base of their petals and lateral sepals free or more or less fused to the base, and a rostellum that is entire for some distance in front of the anther, then divided into three lobes, the side lobes at least 1 mm long, the middle lobe obliquely ascending at the front.

Cynorkis section *Imerinorchis* H.Perrier ex Hermans **sect. nov.** e sectionibus aliis rostello in medio profunde emarginato lobo medio dentiformi vel laminato inter loculis antherae ad filum adnato, sepalis lateralibus atque petalis ad basim labelli plus minusve connatis differt. Typus: *Cynorkis imerinensis* (Ridley) Kraenzl.
Cynorkis sect. *Imerinorchis* H.Perrier in Humbert , Fl. Mad. Orch. 1: 145 (1939), *nom. nud.*

This section has flowers with the lateral sepals and petals more or less fused at the base of the lip, and a three lobed rostellum, the two side lobes divided up to the base of the anther filament, the mid-lobe folded towards the base, reduced to a small protrusion, or more or less developed in a blade narrowly placed against the filament between the two anther chambers.

Cynorkis section *Lowiorchis* H.Perrier ex Hermans **sect. nov.** e sectionibus aliis lobo medio rostelli bilobato, lobo superiore reducto vel lobulato antice ascendenti, lobo inferiore membranaceo canaliculato vel inflexo distinguenda. Typus: *Cynorkis lowii* Rchb.f.
Cynorkis sect. *Lowiorchis* H.Perrier in Humbert, Fl. Mad. Orch. 1: 117 (1939), *nom. nud.*

Species of Sect. *Lowiorchis* have flowers in which the 3-lobed rostellum is place in front of the anther and has long or short arms, always entire for some distance. The rostellum mid-lobe is more or less deeply divided, the upper lobe reduced to 1 or 2 small obtuse protrusions or, otherwise developed into an obliquely ascending blade, the lower lobe membranous and forming a gutter or inflexed lobe.

Cynorkis section *Monadeniorchis* H.Perrier ex Hermans **sect. nov.** e sectionibus ceteris labello simplici carinato rostello elongatissimo antice carinato atque decrescenti atque carinam ad centrum parvam excavatam ad apicem truncatam rostratam ornato, caudiculis et canalibus antherae ad viscidium affixis differt. Typus: *Cynorkis monadenia* H.Perrier.
Cynorkis sect. *Monadeniorchis* H.Perrier in Humbert, Fl. Mad. Orch. 1: 161 (1939), *nom. nud.*

This section has flowers in which the rostellum is very elongate, narrowed partly into a keel and entire at the front, with a small central hollowed keel on top, truncate at the front and extended by a small rostrum. The anther caudicles are united by a common viscidium. The lip is entire, folded into a narrow keel and very concave below.

Cynorkis ampullacea H.Perrier ex Hermans **sp. nov.** affinis *C. bimaculata* (Ridley) H.Perrier sed folio unico et calcari ampulliformi obovoideo ad apicem quam ad basin 3plo latiore satis differt. Typus: Madagascar, Centre, slopes at the side of Lake Andraikibo near Antsirabe, 1400 m, *H.Perrier* 17895 (holo. P!). This species is illustrated in Veyret {1433}.
Cynorkis ampullacea (H.Perrier) H.Perrier in Fl. Mad. Orch. 1: 112 (1939), *comb. inval.*

Cynosorchis cuneilabia subsp. *ampullacea* H.Perrier in Arch. Bot. Bull. Mens. 5: 51 (1931), *nom. inval.*

Labellum obtriangulare vel late obovatum ad apicem trisinuatum vel breviter trilobulatum; calcar inflatum obovatum ampulliforme; caudiculae polliniarum ad apicem distincte glandulosae; rostellum subintegrum ad apicem bidentatum vel tridentatum.

Other collections: Madagascar, Ranomafana area, Mar. 1970, *Veyret* 1264 (P!); a few km from Ambohimasoa, Mar. 1970, *Veyret* 1269 (P!); PK365-7 on Route Nationale 7, Mar. 1970, *Veyret* 1261 (P); cultivated, 1971, *Veyret* s.n. (P!).

Cynorkis angustipetala Ridl. var. **moramangensis** H.Perrier ex Hermans **var. nov.** a varietate typica floribus extus sparse glandulosis et rostello non apiculato quam brachiae longiore differt. Typus: Madagascar, Moramanga, grassland, Feb. 1925, *H.Perrier* 16890 (holo. P!). (Fig. 21).
Cynosorchis angustipetala var. *moramangensis* H.Perrier in Humbert, Fl. Mad. Orch. 1: 144 (1939), *nom. nud.*

Other collections: Madagascar, Moramanga, *Decary* 15324 (P!); Moramanga area, July 1942, *Decary* 17925 (P!).

Cynorkis angustipetala Ridl. var. **tananarivensis** H.Perrier ex Hermans **var. nov.** a varietate typica floribus pauceglandulosis et anthera obtuse apiculata non sulcata differt. Typus: Madagascar, Antananarivo area, *H.Perrier* 16896 (holo. P!). (Fig. 22).
Cynorchis angustipetala var. *tatanarivensis* H.Perrier in Humbert, Fl. Mad. Orch. 1: 143 (1939), *nom. nud.*

Other collections: Madagascar, Tamatave area, on wet rock, Mar. 1925, *H.Perrier* s.n. (P!).

Cynorkis decaryana H.Perrier ex Hermans **sp. nov.** affinis *C. tenella* Ridley sed planta plus quam 25 cm alta, foliis latioribus 12 – 18 mm latis, bracteis quam ovario brevioribus, sepalo dorsali obtuso et lobo medio labelli 3plo latiore quam longiore satis differt. Typus: NE Madagascar, Ranolania, Mananara area, Sept. 1920, *Decary* 74 (holo. P!).
Cynosorkis decaryana H.Perrier in Fl. Mad. Orch. 1: 105, fig. VIII, 26 – 31 (1939), *nom. nud.*

Herba terrestris. Folia 5 ovato-lanceolata. Racemus elongatus laxe circa 15-florus; ovarium quam bracteae fertiles duplo longior; sepalum dorsale obtusum; labellum trilobatum lobo medio fere 3-plo latiore quam longiore ad apicem dente triangulari acuto ornato.

Cynorkis ericophila H.Perrier ex Hermans **sp. nov.** affinis *C. stenoglossa* Kraenzl. sed planta gracilis caule quam folio 5 – 6plo longiore, racemo brevi 3 – 6-floro, petalis 3 mm latis 4 – 5-nerviis, labello 5-nervio, calcari 6 mm longo ad apicem clavato reflexo differt. Type: Madagascar, Mt. Tsaratanana, ericaceous scrub, 2400 m, April 1924, *H.Perrier* 16517 (holo. P!). (Fig. 23).
Cynosorchis ericophila H.Perrier in Fl. Mad. Orch. 1: 157 (1939), *nom. nud.*

Planta gracilis; folium unicum lanceolatum longe petiolatum; flores rosei papillosi; sepalum dorsale cucullatum, sepala lateralia maiores asymmetrica late ovata obtusa; labellum oblongum in tertio inferno utrinque dente obtuso ornatum; calcar 6 mm longum ad apicem dilatatum; anthera obtusa, caudiculae breves sub polliniis dilatatae, brachiae rostelli crassae.

Cynorkis fimbriata H.Perrier ex Hermans **sp. nov.** affinis *C. lowiana* Rchb.f. sed calcare ad apicem abrupte dilatato, labello 5-lobato lobis basalibus parvis, lobis aliis grandis fimbriato-dentatis et rostello medio reflexo brachiis lateralibus subcylindraceis truncatis satis distinguenda. Typus: Madagascar, Mt. Tsaratanana, 2000 m, April 1924, *H.Perrier* 16492 (holo. P!).
Cynosorchis fimbriata H.Perrier in Fl. Mad. Orch. 1: 122, fig. X, 12 – 16 (1939), *nom. nud.*

Caulis atque pedunculus ad basin bractea sterili quam folio unico longiore inclusi; racemus brevis; bracteae fertiles amplae quam ovarium 2 – 3plo breviores; flores 4 – 8 grandes clare purpurei, sepala obtusa; sepalum dorsale ovatum, concavum; sepala laterales fere semiorbiculares; petala angusta curvata obtusa; labellum 5-lobatum lobo medio bifido ad apicem fimbriato-lacerato; calcar anguste cylindraceum in tertio apicali dilatatum; anthera retusa; rostellum bifidum.

Other collections: Madagascar, Marivorahona Massif, SW of Manambato, 2100 – 2240 m, *Humbert & Capuron*, 25773 (P!, BR!, K!); without locality, *Randriansolo* 109 (MO, TANA!).

Cynorkis hologlossa Schltr. var. **angustilabia** H.Perrier ex Hermans **var. nov.** a varietate typica habitu elatiore, bracteis quam ovario 3-plo brevioribus, sepalo dorsali breviore, labello breviore ad basin aliquantum dilatato, calcare breviore, stigmatibus in partibus apicalibus discretis, brachiis rostelli brevioribus. Typus: Madagascar, Mt. Tsaratanana, ericaceous scrub, c. 2600 m, April 1924, *H.Perrier* 16516 (holo. P!). (Fig. 24).

Cynosorchis hologlossa var. *angustilabia* H.Perrier in Fl. Mad. Orch. 1: 155 (1939), *nom. nud.*

Cynorkis pinguicularioides H.Perrier ex Hermans **sp. nov.** affinis *C. saxicola* Schltr. sed calcari 1.2 ? 1 mm scrotiformi obtussisimo et foliis parvis lanceolatisque distinguenda. Typus: Madagascar, Pic St.-Louis, nr. Fort-Dauphin, July 1932, *Decary* 10008 (holo. P!).

Cynosorchis pinguicularioides H.Perrier in Fl. Mad. Orch. 1: 96, fig. VII, 16 – 18 (1939), *nom. nud.*

Planta parva; folia 3 – 4 brevia lanceolata radicalia; racemus laxe 4 – 10-florus; bracteae acutae quam ovario breviores; flores parvi purpurei, papillosi; sepalum dorsale concavum obtusum quam sepalis lateralibus aliquantum brevius; labellum 5-lobum lobis obtusis, 3 apicalibus aequalibus, quam 2 basalibus maioribus; calcar scrotiforme; anthera retusa; rostellum 0.7 – 1 mm longum ad apicem subintegrum.

Other collection: Madagascar, Toliara Prov., NW of Fort Dauphin, Andohahela reserve, 200 – 500 m, May 1992, *S.Malcomber et al.* (K!, MO).

Cynorkis quinqueloba H.Perrier ex Hermans **sp. nov.** affinis *C. coccinelloideo* Schltr. sed planta robustior caule in medio trifoliata, foliis oblongis, racemo 7 – 10-floro, perianthio papilloso et sepalo dorsali 5-nervo distinguenda. Typus: Madagascar, SE, Beampingaratra massif, Mt. Papanga, 1400 – 1576 m, Nov. 1928, *Humbert* 6373 (holo. P!). (Fig. 25).

Cynosorchis quinqueloba H.Perrier in Fl. Mad. Orch. 1: 77 (1939), *nom. nud.*

Planta gracilis terrestris; folia 3 – 4 basalia elliptica vel oblonga cuspidato-acuta ad basin angustata; racemus brevis 7 – 10-florus ad apicem pubescens 5 – 6 bracteis sterilibus parvis intectus; flores extus subtiliter papillosi; sepalum dorsale late ovatum obtusum; sepala lateralia maiora; petala late ovata valde obtusa; labellum obtuse 5-lobatum lobis lateralibus quam lobo medio duplo brevioribus; staminodia spathulata obtusa quam anthera duplo brevioribus; rostellum tridentatum, dente medio leviter longiore.

Other collection: Madagascar, Andringitra RN, 50 km S of Ambalavao, 1975 m, Jan. 1987, *Nicoll* 244 (K!, MO)).

Cynorkis quinquepartita H.Perrier ex Hermans **sp. nov.** affinis *C. quinqueloba* H.Perrier ex Hermans sed folio unico basali petiolato et calcari clavato satis differt. Typus: Madagascar, Mt. Ambre, 1200 m, Sept. 1926, *H.Perrier* 17738 (holo. P!).

Cynosorchis quinquepartita H.Perrier in Fl. Mad. Orch. 1: 78, fig. VI, 8 – 9 (1939), *nom. nud.*

Planta terrestris tubera nulla radicibus fasciculatis villosis leviter crassis; folium unicum vel raro dua, anguste lanceolatum acutum anguste elongato-petiolatum; flores violacei extus leviter glandulosi; sepalum dorsale orbiculare concavum; sepala lateralia maiora obovato-oblonga; petala ovata obtusa per tertium partem longitudinis ad labellum adnata; labellum 5-lobatum lobis lateralibus quam lobis aliis longioribus; calcar cylindraceum latum; rostellum tridentatum; brachiae stigmaticae in dimidio superiore liberae quam rostello breviores.

Cynorkis raymondiana H.Perrier ex Hermans **sp. nov.** affinis *C. sigmoidea* Kraenzl. sed calcari clavato 1.5 – 1.8 mm longo distinguenda. Typus: Madagascar, nr. Fort-Dauphin, littoral forest, Oct. 1932, *Decary* 10142 (holo. P!). (Fig. 26).

Cynosorchis raymondiana H.Perrier in Fl. Mad. Orch. 1: 97 (1939), *nom. nud.*

Planta glabra; tubera 2 ovata; radices villosae; folia 2(– 3) radicalia oblongo-lanceolata cuspidato-acuta; racemus gracilis 3 – 4 vaginis remotis obtectus; flores roseo-albi minores; sepalum dorsale ovatum obtusum concavum; sepala lateralia petalaque semiovata obtusa; labellum ad basin angustatum inaequaliter 5-lobatum lobo medio integro, lobis lateralibus infimis parvis triangulatis acutis divergentibus aliis oblongis acutis angustioribus; calcar clavatum; rostellum ad apicem tridentatum.

Other collections: Madagascar, nr. Fort-Dauphin, littoral forest, July 1932, *Decary* 10.019 (P!); Ambomdrombe area, moss forest, c. 1200 m, April 1941, Herb. Jard. Bot. Tananarivo s.n. (P!)

Cynorkis spatulata H.Perrier ex Hermans **sp. nov.** affinis *C. hispidula* Ridley et *C. elata* Rolfe sed foliis 2 late ovatis vel ovato-lanceolatis 20 mm quidem latis, inflorescentia 2 – 6-flora, floribus grandibus, rostello 9 mm longo calcari quam ovarium aequanti distinguenda. Typus: Madagascar, nr. Fort Dauphin, on rock amongst the dunes, June 1926, *Decary* 4070 (holo. P!).

Cynosorchis spatulata H.Perrier in Fl. Mad. Orch. 1: 125, fig. XI, 1 – 6 (1939), *nom. nud.*

Tubera 2 oblonga; folia 2 radicalia prostrata ovata vel ovato-lanceolata breviter acuta leviter ad basin angustata atque rotundata; pedunculus 2-vaginatus; racemus 2 – 6-florus; rhachis brevis ovarium aequans; bracteae angustae quam ovarium quartum partem breviores; flores grandes rosei purpureo-maculati; sepalum dorsale cucullatum ovatum acutum; sepala lateralia curvata ad basin angustissima; petala angustiora falcata; labellum integrum angustum ad basin rectangulare in parte inferiore angulatum in medio indentatum ad apicem dilatato-spathulatum rotundatum; calcar ad basin cylindraceum ad apicem leviter dilatatum; anthera retusa.

Other collections: Madagascar, Ambovombe district, Elakelaka, June 1924, *Decary* 2851 (P!); nr. Fort Dauphin, wet grassland, June 1926, *Decary* 4011 (P!); nr. Fort-Dauphin, wet grassland, June 1926, *Decary* 14137 (P!).

Cynorkis stenoglossa H.Perrier var. **pallens** H.Perrier ex Hermans **var. nov.** a varietate typica floribus albis vel lilaceo-tinctis, calcare quam ovario breviori, viscidiis angustis et brachiis stigmaticis connatis differt. Typus: Madagascar, Ambre forest, 1200 m, Sept. 1926, *H.Perrier* 17733 (holo., P!). (Fig. 27).
Cynosorchis stenoglossa var. *pallens* H.Perrier in Fl. Mad. Orch. 1: 159 (1939), *nom. nud.*

Cynorkis tenuicalar Schltr. subsp. **andasibeensis** H.Perrier ex Hermans **subsp. nov.** a subspecie typico habitu maiore 40 – 50 cm elato, foliis maioribus, 11 – 15 × 1 – 2 cm, et racemo 4 – 6 bracteis sterilibus ornatis distinguendo. Typus: Madagascar, Andasibe, by the Onive, moss forest, 1000 m, Feb. 1925, *H.Perrier* 17147 (holo. P!). (Fig. 28).
Cynosorchis tenuicalcar subsp. andasibeensis H.Perrier in Fl. Mad. Orch. 1: 153 (1939), *nom. nud.*

Cynorkis tenuicalar Schltr. subsp. **onivensis** H.Perrier ex Hermans **subsp. nov.** a subspecie typica inflorescentia floribusque sparse pubescentibus, labello ovato-lanceolato vel oblanceolato plus minusve elongato ad medium vel ad apicem dilatato, callo destituto, viscidio breviter fusiformi acuto, lobo medio rostelli quam anthera breviore et aque quam brachiae rostelli dimidio longiore. Typus: Madagascar, Andasibe forest (Onive), lichen forest, c. 1300 m, Feb. 1925, *H.Perrier* 17146 (holo. P!). (Fig. 29).
Cynosorchis tenuicalar subsp. onivensis H.Perrier in Fl. Mad. Orch. 1: 153 (1939), *nom. nud.*

Cynorkis uncinata H.Perrier ex Hermans **sp. nov.** affinis *C. purpurascenti* Thouars sed planta epiphytica etuberosa radicibus 1 – 3 crassis, folio unico scapus aequanti, bracteis grandis longe acuminatis recurvatis quam floribus longioribus, loculis antherae connectivo 1 mm lato separatis distinguenda. Typus: Moramanga area, on *Pandanus*, Feb. 1930, *Decary* 7269 (holo. P!).
Cynosorchis uncinata H.Perrier in Fl. Mad. Orch. 1: 137, fig. XII, 3 – 4 (1939), *nom. nud.*

Planta grandis epiphytica; tubera 1 – 3 inflata asymmetrica; folium unicum late ovato-lanceolatum acutum inflorescentia aequans; pedunculus glaber 1 – 2-vaginatus; racemus dense 7 – 15-florus, subglobosus; bracteae flores aequans; flores grandes rosei; ovarium leviter glandulosum; sepalum dorsale ovatum obtusum concavum; sepala lateralia arcuata apiculata vel acuta; petala angusta lineari-lanceolata obtusa; labellum 4-lobatum, lobis lateralibus subrectangularibus ad apicem subcrenatis, lobis apicalibus cuneiformibus vel obovatis ad apicem crenulatis; calcar subcylindraceum; anthera grandis breviter apiculato-acuta filo lato caudiculis rigidis; viscidium parvum ovatum; staminodia spathulata.

This fine species has been confused with *C. purpurascens* Thouars, and it was illustrated and described as such by J. D. Hooker (1902) in *Curtis' Botanical Magazine* (t. 7852) The illustration identified as *Cynorkis uncinata* (fig 23b) in Du Puy *et al.* {281} is of another species whose true identity is still under discussion. It is possible that this species has been described earlier, under a different name. However, in the absence of detailed analysis, it is prudent to accept Perrier's name for this taxon and validate it. Also see note under *C. uncinata*, p. 167.

Other collections: Madagascar, Forest S of Moramanga, on *Pandanus*, Feb. 1930, *Decary* 7173 (P!); Forest S of Moramanga, Feb. 1930, *Decary* 7062 (P!); Moramanga area, *Decary* 7269 (P!); Manjahe, Tamatave district, March 1941, *Decary* 16995 (P!); Zahamena reserve, March 1941, *Decary* 17002 (P!); Ambatovy, North of the Mokarana Massif, on *Pandanus*, 1997, *Rakatomalaza et al.* 1237 (P!, MO); East, Fiananarantsoa Prov., Ranomafana National Park, vicinity of Ampasina, 1150 m, Mar. 1995, *D.Turk et al.* 710 (P!); Toliara prov., Cap St Luce, edge of the Andohahela reserve, 610 m, Feb. 1995, *D.J.DuPuy et al.* M866 (K!); Toamasina Prov., Andasibe area, *Hermans* 4327 (K!).

Illustrations

All illustrations, except fig. 17, drawn by Linda Gurr after pencil sketches by Johan Hermans. These drawings are intended as an broad aid to identify major features. Fig. 17 drawn by Oliver Q. Whalley after sketches by Johan Hermans.

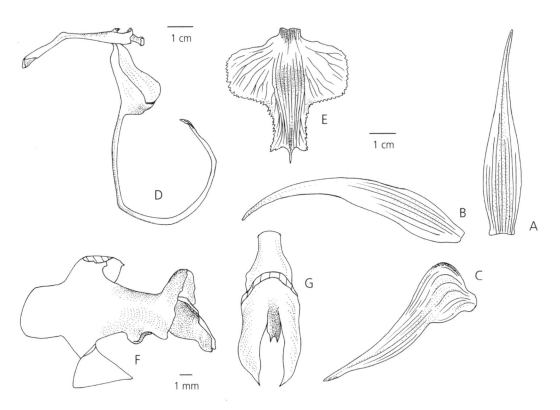

Fig. 1. *Aeranthes henricii* Schltr. var. *isaloensis* H.Perrier ex Hermans *var. nov.* A dorsal sepal, B lateral sepal, C petal, D column, spur and mentum, E lip, F column, side, G column top. Based on *H.Perrier* 16895.

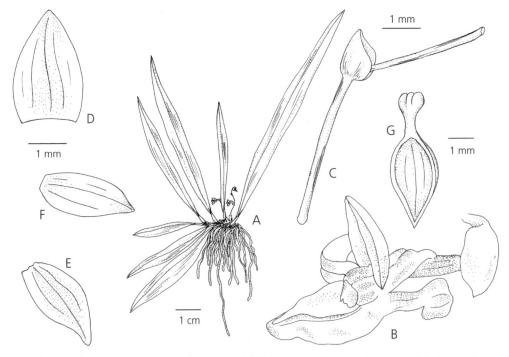

Fig. 2. *Angraecum acutipetalum* Schltr. var. *analabeensis* H.Perrier ex Hermans *var. nov.* A habit, B flower, C peduncle and floral bract, D dorsal sepal, E lateral sepal, F petal, G lip and spur. Based on *H.Perrier* 18516.

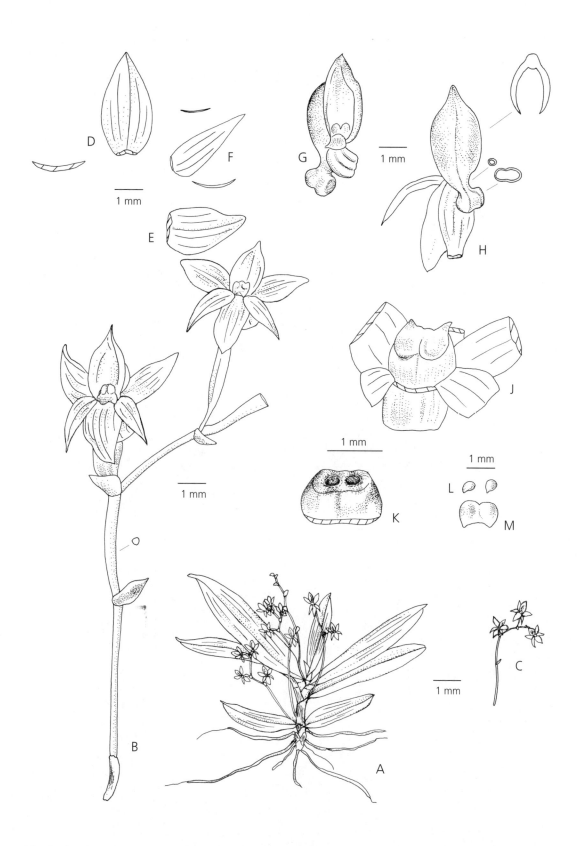

Fig. 3. *Angraecum acutipetalum* Schltr. var. *ankeranae* H.Perrier ex Hermans *var. nov.* A habit, B inflorescence, C inflorescence, D dorsal sepal, E lateral sepal, F petal, G lip and spur, H lip and spur, back, J column with anther cap, K column without anther cap, L pollinia, M anther cap. Based on *H.Perrier* 18517.

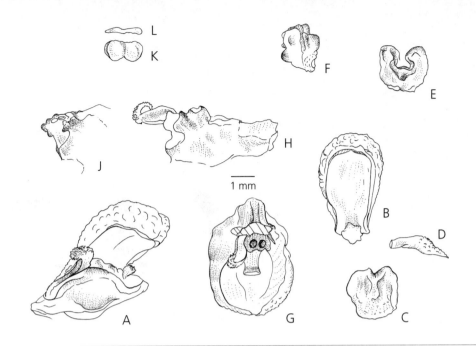

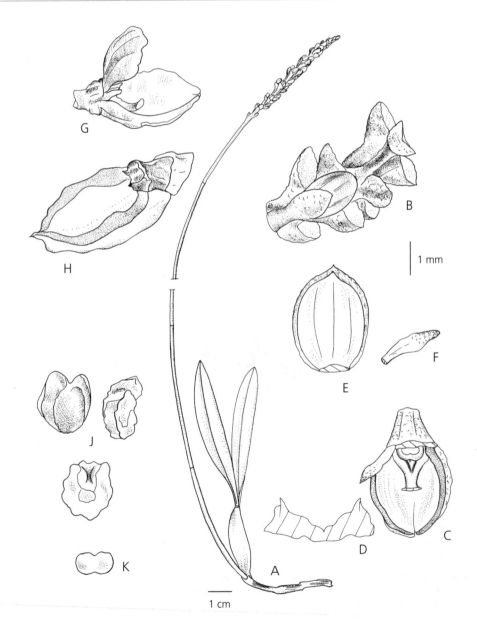

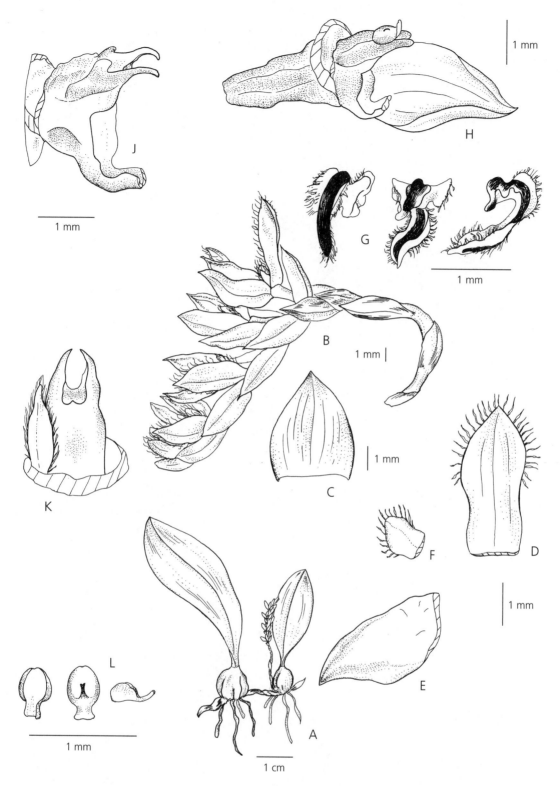

Fig. 4 (opposite, top). *Bulbophyllum henrici* Schltr. var. *rectangulare* H.Perrier ex Hermans *var. nov.* A flower, B dorsal sepal, C lip, top, D petal, E lip, under, F lip, side, G fused lateral sepals, petal and column, H column, side and petal, J column, side, K anther cap, L anther cap, side. Based on *Decary* 5785.

Fig. 5 (opposite, bottom). *Bulbophyllum ophiuchus* Ridl. var. *baronianum* H.Perrier ex Hermans *var. nov.* A plant and inflorescence, B part of peduncle and floral bracts, C lateral sepals and petal, D cross-section of lateral sepals, E dorsal sepal, F petal, G cross-section of flower with dorsal sepal lifted, lip removed, H lateral sepals and column, J lip: front, side, back, K anther cap. Based on *Baron* 4468.

Fig. 6 (above). *Bulbophyllum pantoblepharon* Schltr. var. *vestitum* H.Perrier ex Hermans *var. nov.* A plant and inflorescence, B inflorescence, C floral bract, D dorsal sepal, E lateral sepal, F petal, G lip: fron, back, side, H column, and lateral sepal, side, J column, K column and petal, top, L anther cap. Based on *H.Perrier* 18646.

Fig. 7. *Bulbophyllum sambiranense* Jum. & H.Perrier. var. *latibracteatum* H.Perrier ex Hermans *var. nov.* A habit, B part of inflorescence, C dorsal sepal, D lateral sepal, interior, E lateral sepal, exterior, F petal, G column, side, sepals and petals, H column, side, J floral bract, K lip, L anther cap, M pollen, N leaves. Based on *H.Perrier* 16724.

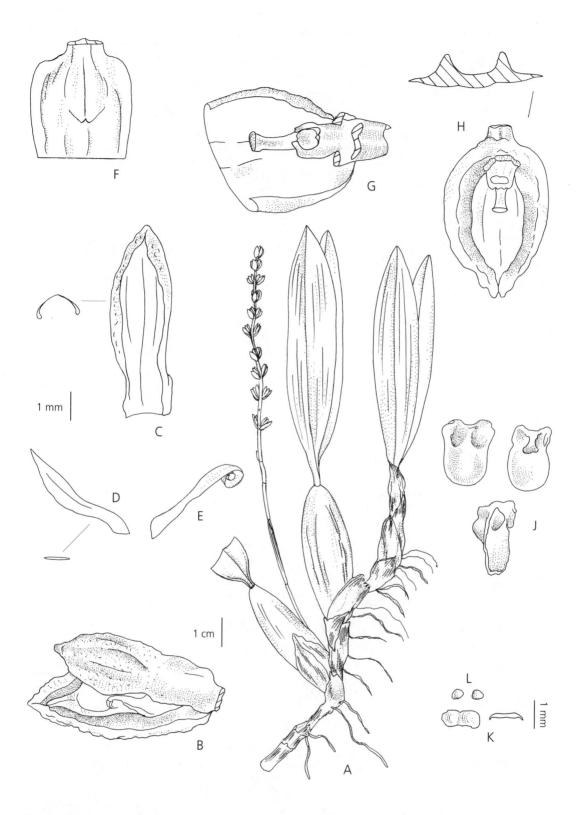

Fig. 8. *Bulbophyllum sarcorhachis* Schltr. var. *flavomarginatum* H.Perrier ex Hermans *var. nov.* A habit, B flower, C dorsal sepal, D petal uncurled, E petal, natural position, F fused lateral sepal, under, G base of fused lateral sepals with column, H fused lateral sepal, top, J lip, K anther cap, L pollen. Based on *H.Perrier* 16511.

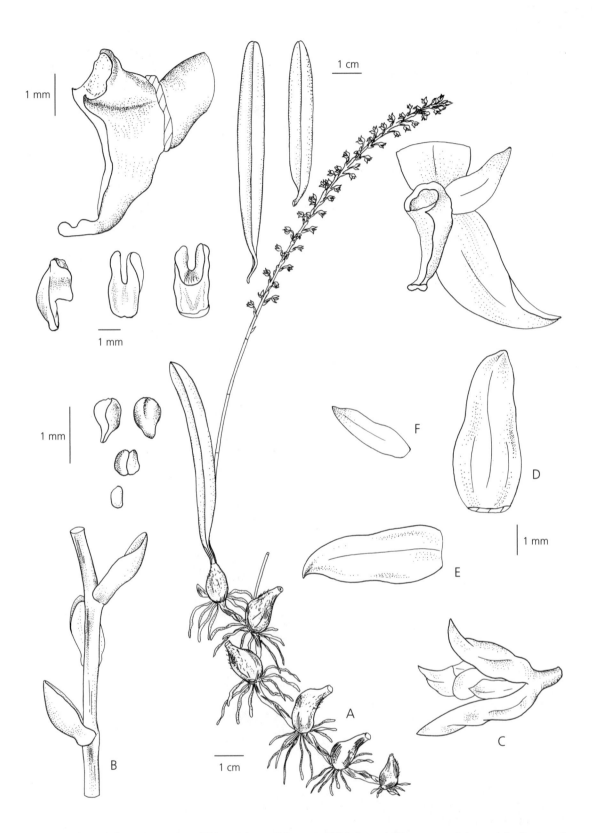

Fig. 9. *Bulbophyllum zaratananae* Schltr. subsp. *disjunctum* H.Perrier ex Hermans *subsp. nov.* A plant and inflorescence, B peduncle and floral bracts, C flower, D dorsal sepal, E lateral sepal, F petal, G column and part of perianth, H column, side, J lip: side, top, under, K anther cap, L pollen, M leaves. Based on *Decary* 5611.

Fig. 10. *Cynorkis laeta* Schltr. var. *angavoensis* H.Perrier ex Hermans *var. nov.* A plant and inflorescence, B cross-section of flower stem, C flower, D floral bract, E dorsal sepal, F petal, G lateral sepals, H column, side, J column, top, K spur, L lip, side, M lip, N pollinia. Based on *Decary* 7344.

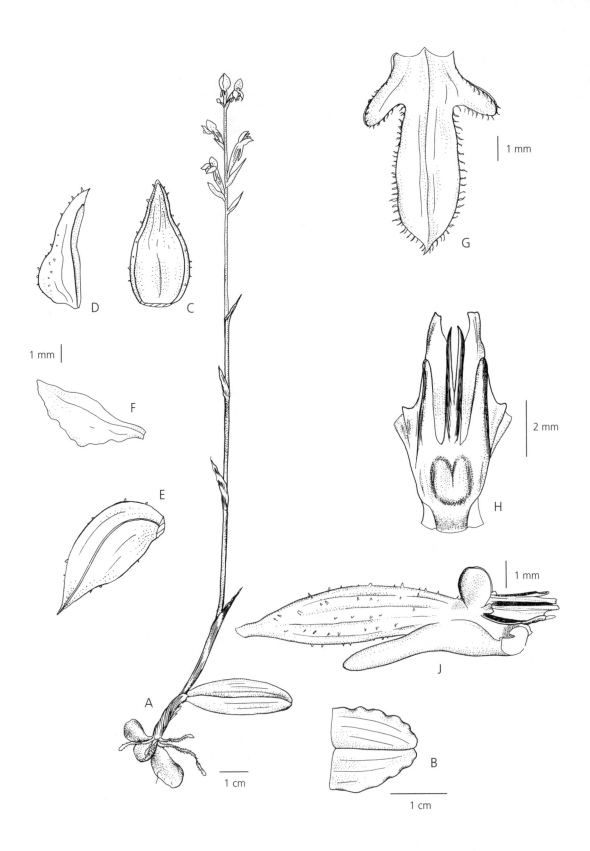

Fig. 11. *Cynorkis lilacina* Ridl. var. *comorensis* H.Perrier ex Hermans *var. nov.* A plant habit, B leaf tip, C dorsal sepal, D dorsal sepal, side, E lateral sepal, F petal, G lip, H column, top, J ovary, column and spur. Based on *Boivin* s.n.

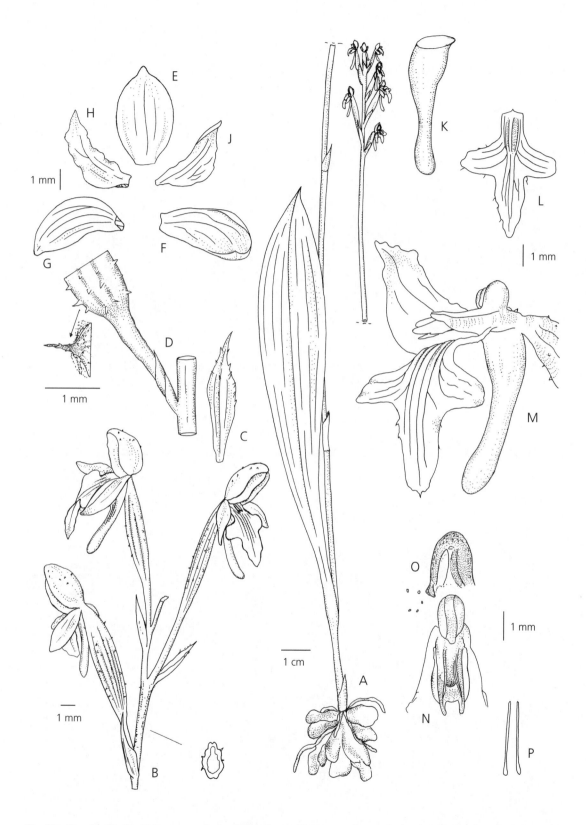

Fig. 12. *Cynorkis lilacina* Ridl. var. *tereticalcar* H.Perrier ex Hermans *var. nov.* A plant and inflorescence, B part of inflorescence, C floral bract, D base of pedicellate ovary and detail of spine, E dorsal sepal, F lateral sepal, interior, G lateral sepal, exterior, H petal, interior, J petal exterior, K spur, L lip, M flower without dorsal sepal and petal, N column, top, O anther and pollen, P pollinarium. Based on *H.Perrier* 14571.

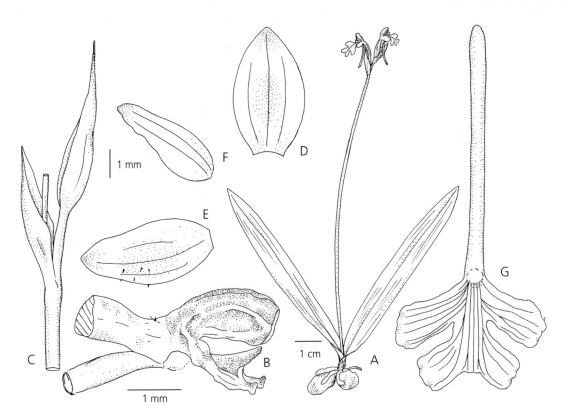

Fig. 13. *Cynorkis fastigiata* Thouars var. *ambatensis* Hermans *var. nov.* A habit, B column and spur, side, C floral bracts, D dorsal sepal, E lateral sepals, F petals, G lip. Based on *H.Perrier* 1857.

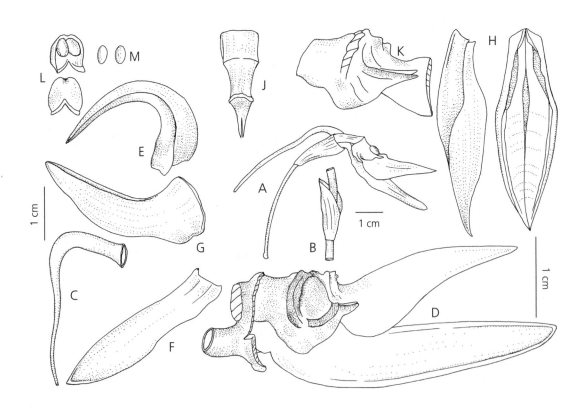

Fig. 14. *Jumellea lignosa* (Schltr.) Schltr. subsp. *acutissima* H.Perrier ex Hermans *subsp. nov.* A flower with dorsal sepal and lip removed, side, B floral bract, C spur, D column, side, E dorsal sepal, F petal, G lateral sepals, H lip, J column, front, K column, side, L anther cap, M caudicles. Based on *E.François* in Herb. *H.Perrier* 18518.

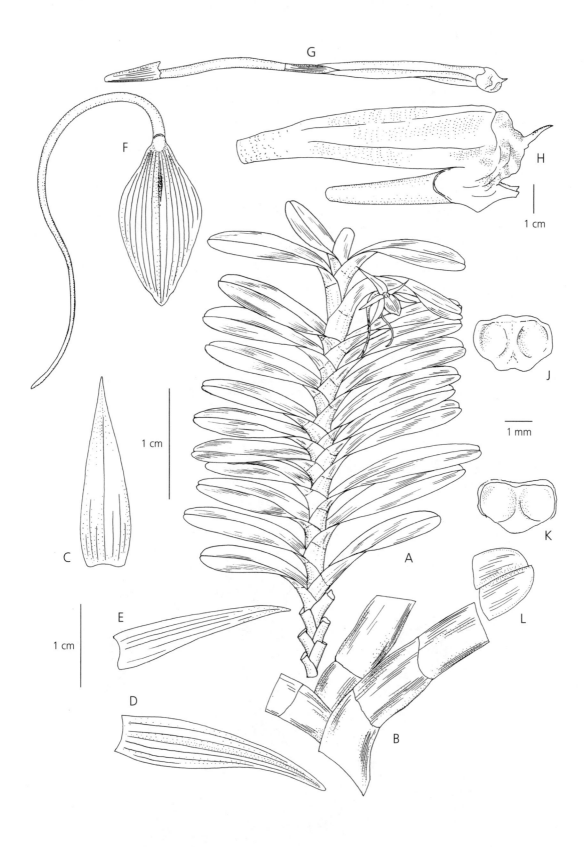

Fig. 15. *Jumellea lignosa* (Schltr.) Schltr. subsp. *latilabia* H.Perrier ex Hermans *subsp. nov.* A plant and flower, B leaf base and stem, C dorsal sepal, D lateral sepal, E petal, F lip and spur, G column, ovary and pedicel, H column, side, J anther cap, interior, K anther cap, exterior, L leaf tip. Based on *H.Perrier* 17129.

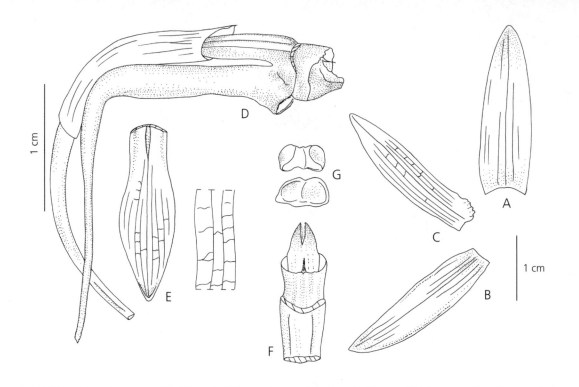

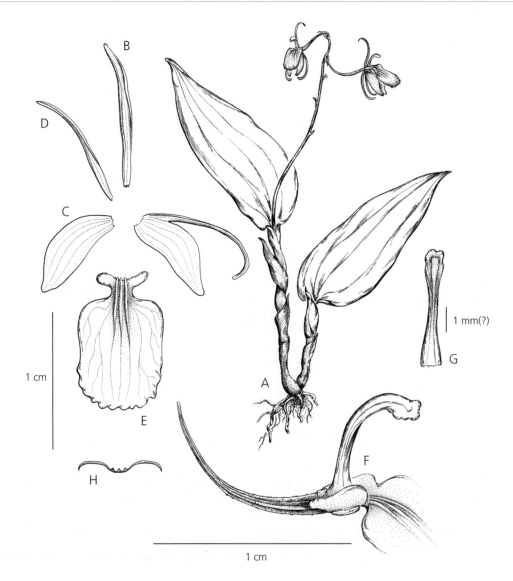

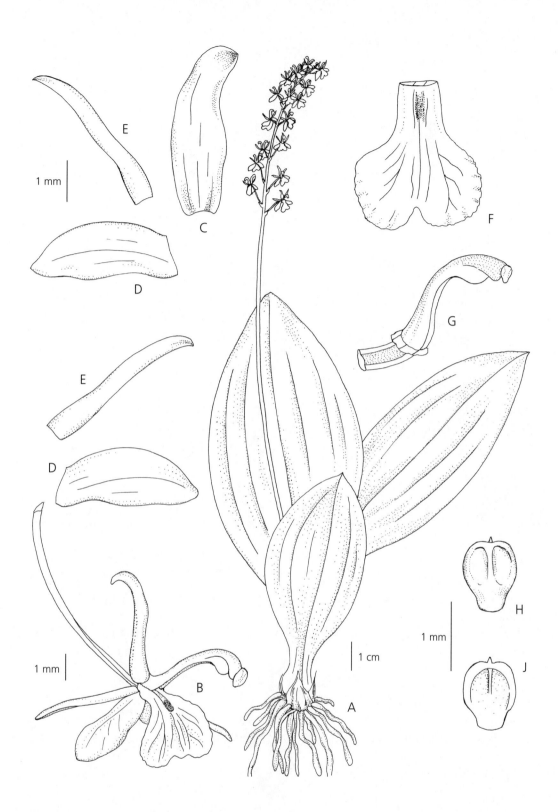

Fig. 16 (opposite, top). *Jumellea lignosa* (Schltr.) Schltr. subsp. *tenuibracteata* H.Perrier ex Hermans *subsp. nov.* A dorsal sepal, B lateral sepal, C petal, D column and spur, E lip, veination, F column, top, G anther cap. Based on *H.Perrier* 15331.

Fig. 17 (opposite, bottom). *Liparis clareae* Hermans *nom. nov.* var. *angustifolia* H.Perrier ex Hermans *var. nov.* A habit. B dorsal sepal, C lateral sepals, D petals, E lip, F column and ovary, G column, H lip profile. Based on *H.Perrier* 17018.

Fig. 18 (above). *Liparis perrieri* Schltr. var. *trinervia* H.Perrier ex Hermans *var. nov.* A habit, B flower, C dorsal sepal, D lateral sepal, E petal, F lip, G column, H anther cap, exterior, J anther cap, interior. Based on H.Perrier 17136.

Fig. 19. *Neobathiea hirtula* H.Perrier var. *floribunda* H.Perrier ex Hermans *var. nov.* A habit, B flower, C dorsal sepal, D lateral sepal, E petal, F column, spur and ovary, G lip, H column, top. Based on *H.Perrier* 1438.

1 cm

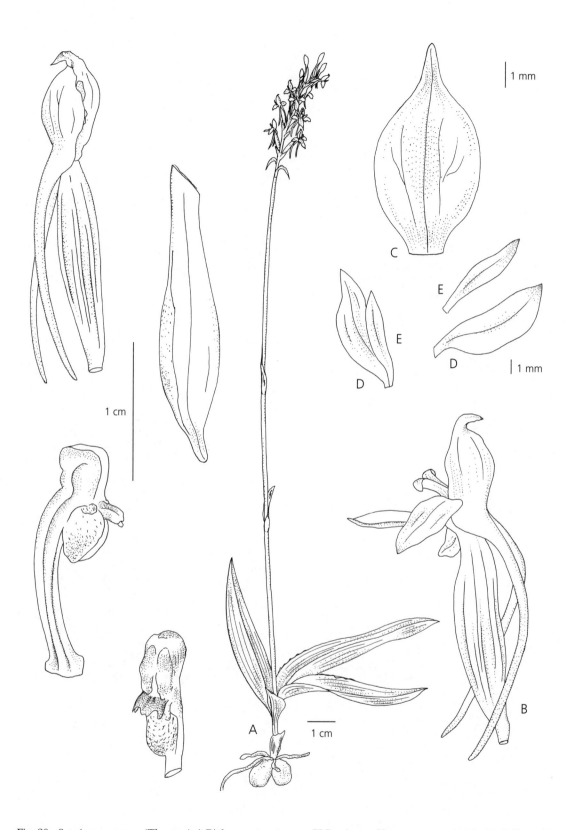

Fig. 20. *Satyrium amoenum* (Thouars) A.Rich. var. *tsaratananae* H.Perrier ex Hermans *var. nov.* A habit, B flower, C dorsal sepal, D lateral sepal and petal, E petal, F lip, G spurs and ovary, side, H column and ovary, J column. Based on *H.Perrier* 16507.

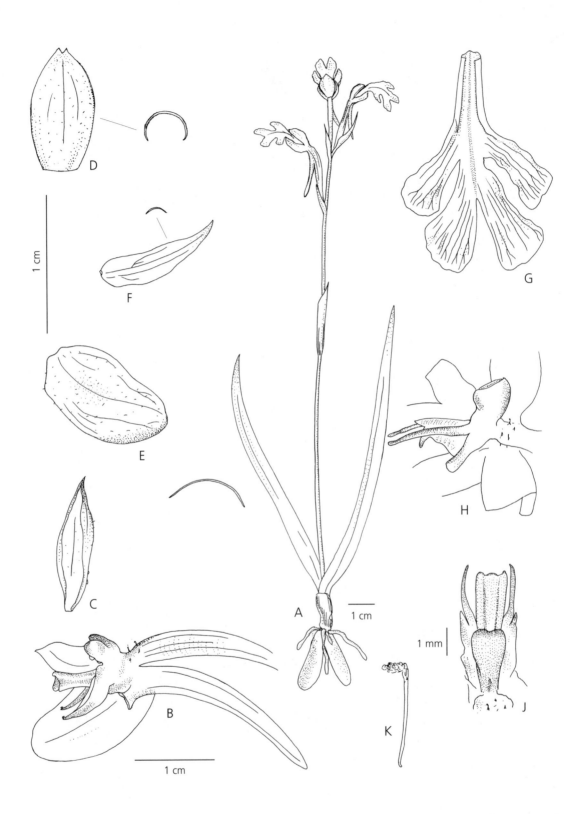

Fig. 21. *Cynorkis angustipetala* Ridl. var. *moramangensis* H.Perrier ex Hermans *var. nov.* A habit, B column and spur, side, C floral bract, D dorsal sepal, E lateral sepal, F petal, G lip, H column side, based on *Decary* 17925, J column top, based on *Decary* 17925, K pollinia, based on *Decary* 17925. Based on *H.Perrier* 16890.

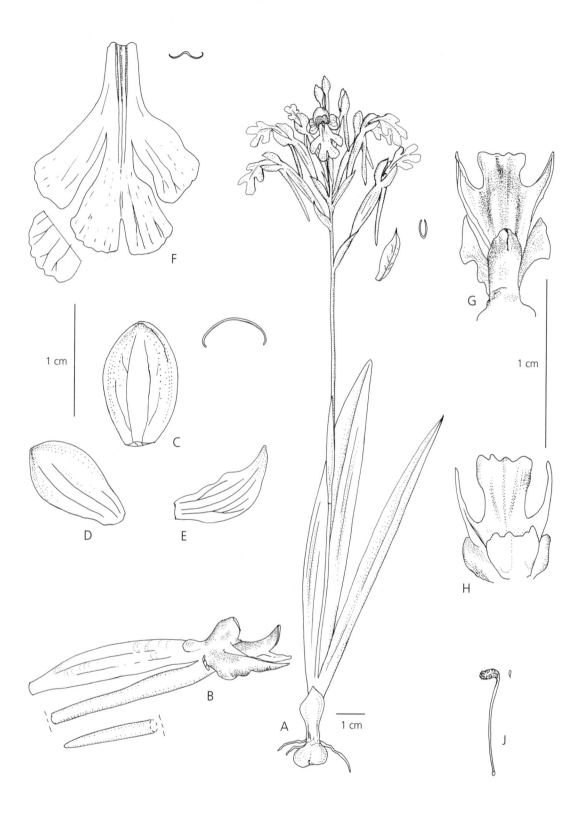

Fig. 22. *Cynorkis angustipetala* Ridl. var. *tananarivensis* H.Perrier ex Hermans *var. nov.* A habit, B column, spur and ovary, C dorsal sepal, D lateral sepal, E petal, F lip, G column top, H column, top, based on *H.Perrier* s.n. Herb. Alleizette (P), J pollinia. Based on *H.Perrier* 16896.

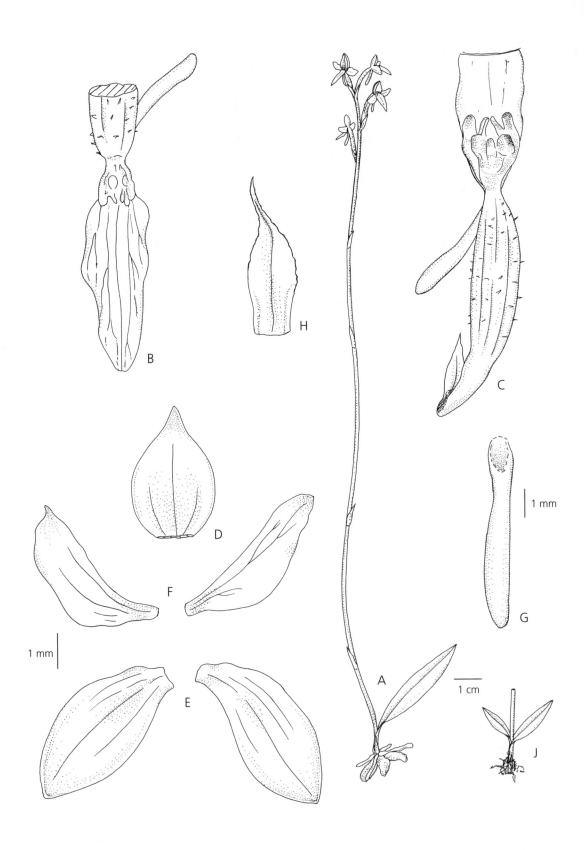

Fig. 23. *Cynorkis ericophila* H.Perrier ex Hermans *sp. nov.* A habit. B lip and column, C ovary and column, D dorsal sepal, E lateral sepals, F petals, G spur, H floral bract, J base of plant. Based on *H.Perrier* 16517.

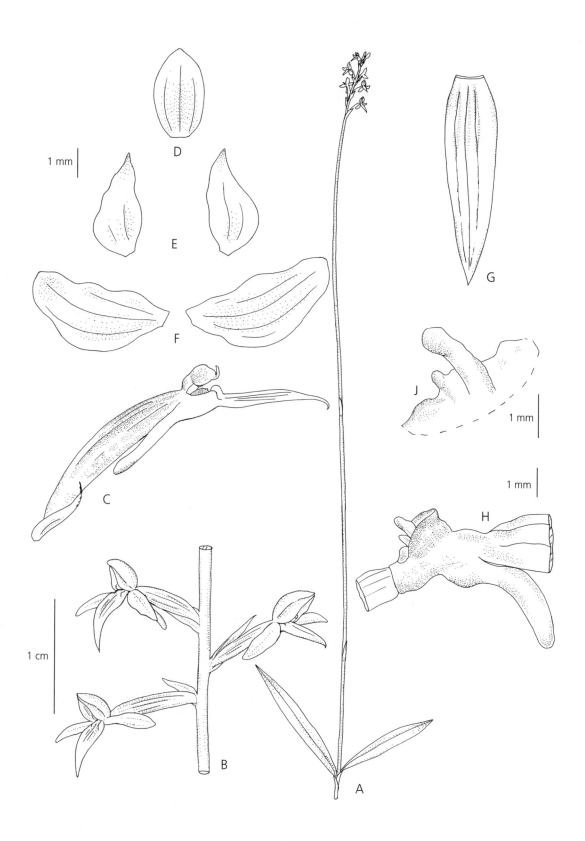

Fig. 24. *Cynorkis hologlossa* Schltr. var. *angustilabia* H.Perrier ex Hermans *var. nov.* A habit, B part of inflorescence, C pedicellate ovary, column, lip and spur, D dorsal sepal, E petals, F lateral sepals, G lip, H column and spur, J rostellar arms. Based on *H.Perrier* 16516.

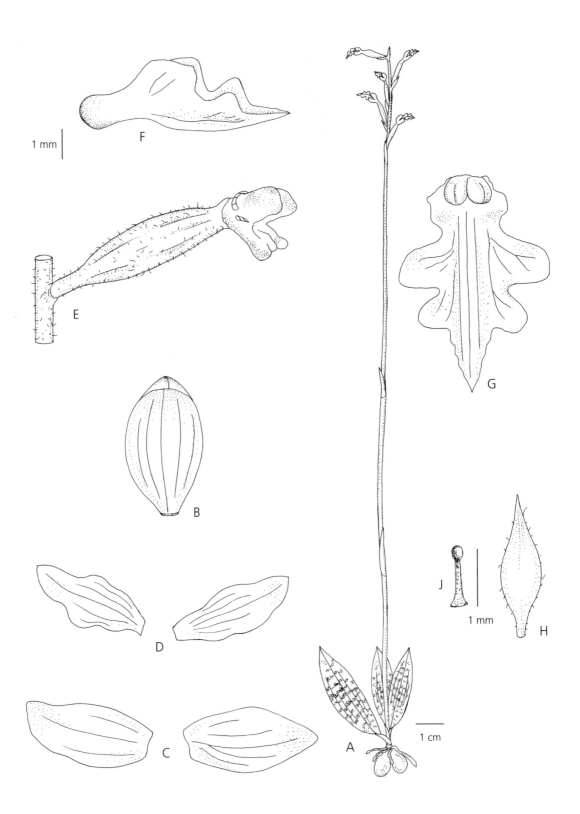

Fig. 25. *Cynorkis quinqueloba* H.Perrier ex Hermans *sp. nov.* A habit, B dorsal sepal, C lateral sepals, D petals, E pedicellate ovary and column, F lip and spur, side, G lip, H floral bract, J detail of hair on ovary. Based on *Humbert* 6373.

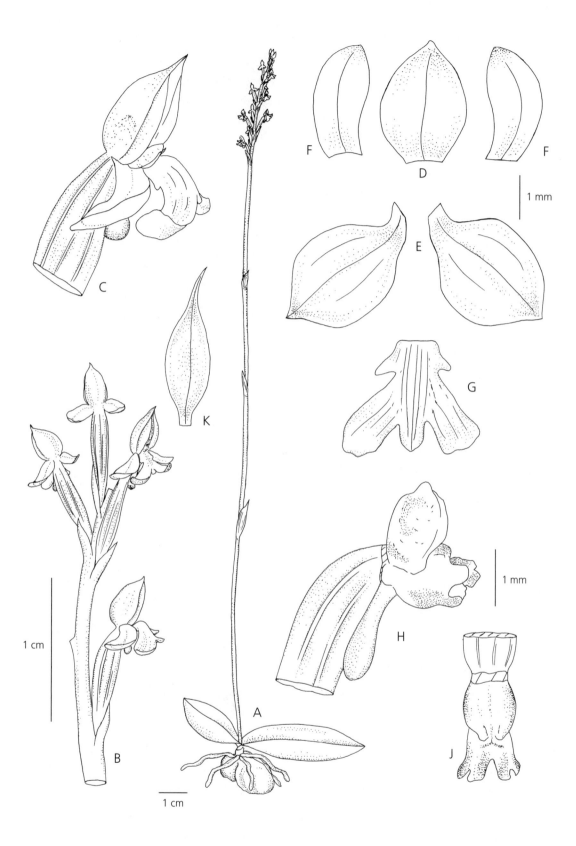

Fig. 26. *Cynorkis raymondiana* H.Perrier ex Hermans *sp. nov.* A habit, B tip of inflorescence, C flower side, D dorsal sepal, E lateral sepals, F petals, G lip, H column and spur, side, J column, top. Based on *Decary* 10142.

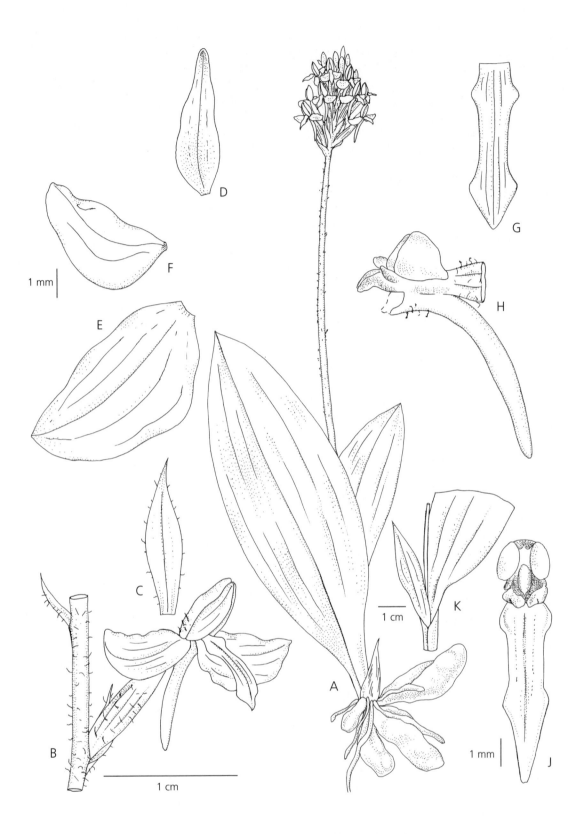

Fig. 27. *Cynorkis stenoglossa* H.Perrier var. *pallens* H.Perrier ex Hermans *var. nov.* A habit, B flower, C floral bract, D dorsal sepal, E lateral sepal, F petal, G lip, H column and spur, J column and lip, top, K base of inflorescence. Based on *H.Perrier* 17733.

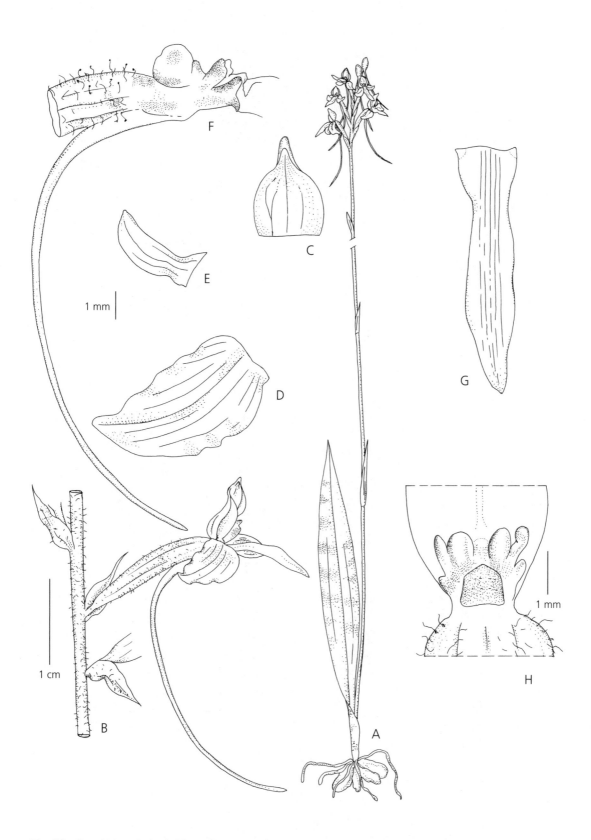

Fig. 28. *Cynorkis tenuicalar* Schltr. subsp. *andasibeensis* H.Perrier ex Hermans *subsp. nov.* A habit, B flower, C dorsal sepal, D lateral sepals, E petals, F column and spur, side, G lip, H column, top. Based on *H.Perrier* 17147.

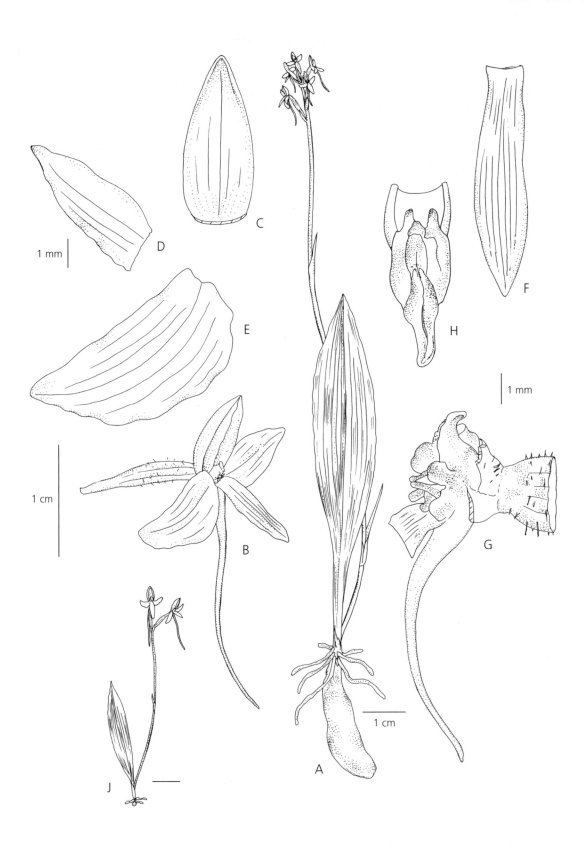

Fig. 29. *Cynorkis tenuicalar* Schltr. subsp. *onivensis* H.Perrier ex Hermans *subsp. nov.* A habit. B flower, C dorsal sepal, D lateral sepal, E petal, F lip, G column and spur, H column, top, J plant. Based on *H.Perrier* 17146.

BIBLIOGRAPHY

This Bibliography attempts to cover all the literature references relating to the orchids of Madagascar and most of those to the surrounding islands. It links directly to the checklist of orchid species, the bibliographic reference numbers being quoted in the checklist in curly brackets.

References are drawn from the broadest spectrum of literature, giving on the one hand summaries of travelogues, horticultural awards, history and cultural information and on the other references to the original descriptions, taxonomic revisions and studies in pollination and morphology.

All journal and personal names except H.G. Reichenbach (Rchb.f.) are entered in full, names following the *Dictionary of Authors of Plant Names* (Brummitt & Powell, 1992).

More than 200 new references have been added to the Bibliography since the first edition, those included before have been reduced to a literature citation only, new references are accompanied by a short annotation.

1. **Alexander**, H. G. (1905). *Phaius tuberculosus. The Garden* 67: 287 & t.1274.

2. **Allibert**, Claude. (1995). *Histoire de la Grande Isle Madagascar.* Gervais Clouzier. Paris. **Flacourt**, Étienne. 1658, annotated version, Éditions Karthalia, Paris: 656 pp. (Also see Flacourt).

3. **American Orchid Society**. (1970 – 2004). **Awards Quarterly.** *Awards Quarterly* 1 – 34.

4. **American Orchid Society**. (2001). *Awards Quarterly* 32, 1: 25.
Cynorkis fastigiata var. *decolorata* is illustrated.

5. **Ames**, Oakes. (1908). **Illustrations and studies of the family *Orchidaceae.*** *Orchidaceae*, fasc. 2, 228 pp.

6. **Anderson**, Edgar. (1956). **The Star of Bethlehem orchid.** *American Orchid Society Bulletin* 25, 12: 818 – 820.

7. **Andersen**, Geert. (1999). **Vanille — Die Außergewöhnliche Orchidee.** *Journal für den Orchideenfreund* 6, 2: 94 – 99.
The history, cultivation and preparation of culinary vanilla is described in detail. Illustrations show the chemical composition of vanillin, *Vanilla planifolia* flowers and plants and preparation of the pods on the Comoro Islands.

8. **André**, Eduard. (1885). **Nouvelles Orchidées de Madagascar.** *Revue Horticole* 57: 294 – 295.

9. **André**, Eduard. (1888). *Angraecum sanderianum. Revue Horticole* 60: 516.

10. **André**, Eduard. (1891). *Phaius humblotii. Revue Horticole* 63: 204.

11. **André**, Eduard. (1894). *Angraecum fournierianum. Revue Horticole* 66: 581 – 582.

12. **André**, Eduard. (1896). *Angraecum fournierae. Revue Horticole* 68: 256 – 257.

13. **André**, Eduard. (1903). *Eulophiella peetersiana. Revue Horticole* 75: 403 – 405.

14. **Andriantiana**, Jacky Lucien. (2003). **Plants at risk/Les plantes menacées.** *Angraecum longicalcar. Ravintsara*: 16.
Conservation issues of the species are discussed.

15. **Aoyama**, Mikio. (1994). **Developments in orchid cytology.** in **Pridgeon**, Alec M., ed., *Proceedings of the 14th World Orchid Conference.* HMSO, Edinburgh: 74 – 77.

16. **Arditti**, Joseph. (1971). *Vanilla:* **an historical vignette.** *American Orchid Society Bulletin* 40: 610 – 613.

17. **Arditti**, Joseph. (1976). **Post-pollination phenomena in orchid flowers.** *Orchid Review* 84, 999: 261 – 268.

18. **Arditti**, Joseph. (1976). **Post-pollination phenomena in orchid flowers.** In K. Senghas, ed., *Proceedings of the 8th World Orchid Conference.* DOG, Frankfurt: 528 – 533.

19. **Arditti**, Joseph. (1981). *Cymbidiella rhodochilum. Orchid Review* 89, 1053: 223.

20. **Arditti**, Joseph. (1992). *Fundamentals of orchid biology.* Wiley, New York.

21. **Arends**, J. C. & **Van Der Laan**, F. M. (1983). **Cytotaxonomy of the monopodial orchids of the African and Malagasy regions.** *Genetica* 62: 81 – 94.

22. **Argus**. (1894). *Dies Orchidianae, Eulophiella elisabethae. Orchid Review* 2, 16: 101.

23. **Argus**. (1894). *Dies Orchidianae. Orchid Review* 2, 20: 31.

24. **Argus**. (1900). **Notes.** *Orchid Review* 8: 34 – 35.

25. **Argus**. (1901). *Dies Orchidianae. Orchid Review* 9, 99: 65 – 68.

26. **Arnold**, Ralph E. (1931). *Phaius* **species and hybrids.** *Orchid Review* 39, 452: 45 – 48.

27. **Arnold**, Ralph E. (1935). *Angraecum citratum. Orchid Review* 43, 507: 273.

28. **Arnold**, Ralph E. (1950). *Angraecum ellisii. Orchid Review* 58, 687: 123 – 124.

29. **Asahi Shimbum Sha**. (1975 – 1978). **Asahi hyakka sekai no shokbutsu. Botany of the World,** Tokyo, 8.

30. **Attenborough**, David. (1995). *The private life of plants.* BBC Books, London: 320 pp.

31. **Aublet**, M. Fusée. (1775). *Histoire des plantes de La Guiane Française* 2. P.F. Didot, London & Paris: 160 pp.

32. **Bailes**, Christopher, ed. (1986). *Angraecum sesquipedale* **'Bromesberrow Place' CCC/RHS.** *Orchid Review* 94, 1118: 406 & 410.

Baillon, H. E., see Rolfe 1098.

Baker, John Gilbert, see Moore Spencer Le Marchant.

33. **Baker**, Margaret L. & **Baker**, Charles O. (1991). *Orchid species culture. Pescatorea, Phaius, Phalaenopsis, Pholidota, Phragmipedium, Pleione.* Timber Press, Portland, Oregon: 250 pp.

34. **Ball**, John S. (1978). *Southern African epiphytic orchids.* Conservation Press (PTY) Ltd., Johannesburg: 248 pp.

35. **Ballif**, Otto. (1886). *Phaius tuberculosus. L'Orchidophile.* 93 – 94.

36. **Ballif**, Otto. (1891). *Angraecum du Buyssoni. Journal des Orchidées* 2: 237.

37. **Ballif**, Otto. (1895). **Petites Nouvelles et Petite Correspondance.** *Journal des Orchidées,* 6: 180. A remarkable *Grammangis ellisii* is mentioned, with 5 spikes and 147 flowers.

38. **Barker**, J. T. (1912). **Calender of operations for November.** *Orchid Review* 20 239: 325 – 328.

39. **Baron**, Richard. (1881). **Genera of Madagascar plants.** *Antananarivo Annals* 5: 94 – 97.

40. **Baron**, Richard. (1881). **A list of Madagascar plants.** *Antananarivo Annals* 5: 109 – 114.

41. **Baron**, Richard. (1883). **Genera of Madagascar plants.** *Antananarivo Annals* 7: 96 – 102.

42. **Baron**, Richard. (1889 – 1890). **The Flora of Madagascar.** *Journal of the Linnean Society* 25: 246 – 294.

43. **Baron**, Richard. (1893). *Eulophiella elisabethae*, **the new Madagascan orchid.** *Antananarivo Annals* 17, 1: 117 – 118. (Also see 429 & 430).

44. **Baron**, Richard. (1901). **Compendium des plantes Malagaches.** *Revue de Madagascar* 3,10: 714 – 725. (Dorr 271 places the Orchideae in Fasc. XXI. 8 (8): 714 – 725, 1906).

45. **Bateman**, James, ed. (1867). *A second century of orchidaceous plants.* L. Reeve & Co., London.

46. **Bauer**, F. & **Lindley**, J. (1830 – 1838). *Illustrations of orchidaceous plants.* James Ridgway & Sons, London.

47. **Baumann**, W. (1991). *Aerangis citrata* **(Thouars) Schlechter 1916.** *Die Orchidee.* 43, 5: Orchideenkartei 659 – 660.

48. **Baumann**, W. (1992). *Aerangis articulata* **(Rchb.f.) Schlechter 1914.** *Die Orchidee* 43, 1: Orchideenkartei 671 – 672.

49. **Bayley Balfour**, Isaac. (1878). **An account of the petrological, botanical and zoological collections made in Kerguelen's Land and Rodriguez.** *Orchidaceae.* *Philosophical Transactions of the Royal Society of London* 168 (extra volume): 371 pp.

50. **Bechtel**, Helmut, **Cribb**, Phillip & **Launert**, Edmund. (1992). *The manual of cultivated orchid species,* ed. 3: 585 pp., 864 colour photos. Blandford, London.

Beckman,Steven D, see Perrier de la Bâthie 896.

51. **Beckton**, Bertram J. (1943). **Aeranthus.** *Orchid Review* 51, 595: 23 – 24.

52. **Beentje**, Henk J. (1998). **J. M. Hildebrandt (1847 – 1881): notes on his travels and plant collections.** *Kew Bulletin* 53, 4: 835 – 856.
A concise biography of Hildebrandt. His itineraries are outlined, a map shows some of the localities that he visited.

53. **Beer**, J. G. (1854). *Praktische Studien ander Familie der Orchideen.* Carl Gerold & Sohn. Vienna.

54. **Bell**, Sandra. (1996). **The orchid paintings of Marianne North.** *Kew Magazine* 13, 3: 157 – 163.

55. **Bentham**, George & **Hooker** Joseph Dalton. (1883). *Genera plantarum.* 3. L. Reeve & Co., London.

56. **Bentham**, George. (1881). **Notes on** *Orchidae. Journal of Linnean Society Botany* 18: 281 – 360.

57. **Berliocchi**, Luigi. (2000). *The orchid in lore and legend.* Timber Press. Portland, Oregon: 184 pp.
Jumellea fragrans and *Angraecum sesquipedale* are mentioned.

58. **Bernhardt**, Peter. (1989). *Wily violets & underground orchids.* William Morrow & Co., New York: 255 pp., 12 photos.

59. **Blowers**, John W., ed. (1958). *Cynorchis incinata. Orchid Review* 66, 785: 257.

60. **Blowers**, John W., ed. (1959). *Eulophiella* **Rolfei.** *Orchid Review* 67, 789: 72.

61. **Blowers**, John. (1960). **Royal Horticultural Society awards.** *Orchid Review* 68, 810: 399 – 340.
A short note on *Angraecum comorense* on receiving an award of merit.

62. **Blowers**, John W., ed. (1961). *Aerangis citrata. Orchid Review* 69, 818: 266.

63. **Blowers**, John W., ed. (1962). *Grammangis ellisii. Orchid Review* 70, 830: 264.

64. **Blowers**, John W., ed. (1964). *Jumellea sagittata. Orchid Review* 72, 851 p. 183.

65. **Blowers**, John W., ed. (1964). *Aeranthes ramosa. Orchid Review* 73, 870: 377 – 378.

66. **Blowers**, John W., ed. (1966). *Gastrorchis (Phaius) tuberculosa. Orchid Review* 74, 874: 119 – 120.

67. **Blowers**, John W., ed. (1966). *Angraecum compactum* **'Master Hugh' AM/RHS.** *Orchid Review* 74, 881: 372 – 373.

68. **Blowers**, John W., ed. (1967). *Angraecum teretifolium* **Ridley.** *Orchid Review* 75, 884: 49.

69. **Blowers**, John W., ed. (1967). *Cynorkis gibbosa* **Ridley.** *Orchid Review* 75, 886: 138.

70. **Blowers**, John W., ed. (1967). *Aerangis citrata* **'Lydart' AM/RHS.** *Orchid Review* 75, 890: 292.

71. **Blowers**, John W., ed. (1967). *Jumellea papangensis* **'Fincham' AM/RHS.** *Orchid Review* 75, 891: 323.

72. **Blowers**, John W., ed. (1967). *Phaius schlechteri* **'Fincham' AM/RHS.** *Orchid Review* 75, 891: 325 – 326.

73. **Blowers**, John W. (1968). *Angraecum viguieri* **'Sherwood' AM/RHS.** *Orchid Review* 76, 900: 182.

74. **Blowers**, John W., ed. (1968). *Aeranthes grandiflora* **'Fincham' AM/RHS.** *Orchid Review* 76, 903: 281.

75. **Blowers**, John W., ed. (1968). *Angraecum compactum* **'Superbe' AM/RHS.** *Orchid Review* 76, 904: 310 – 311.

76. **Blowers**, John W., ed. (1968). *Angraecum superbum* **'Pine Close' AM/RHS.** *Orchid Review* 76, 906: 373.

77. **Blowers**, John W., ed. (1969). *Aeranthes henricii* **'Peyrot' AM/RHS.** *Orchid Review* 77, 910: 119 – 120.

78. **Blowers**, John W. (1990). *Oeonia oncidiflora* **'Clare' AM/RHS.** *Orchid Review* 98, 1159: 161 & 165.

79. **Blowers** John W. (1992). **Royal Horticultural Society awards.** *Orchid Review* 100, 1185: 254 & 261.

80. **Blowers**, John W. (1993). **Report on Royal Horticultural Society awards.** *Orchid Review* 101 1194: 276.

81. **Blowers**, John W. (1995). **Royal Horticultural Society awards.** *Orchid Review* 103, 1203: 117 & 111.

82. **Blowers**, John W. (1996). **Royal Horticultural Society awards.** *Orchid Review* 104, 1211: 251 – 253.

83. **Blowers**, John W. (1996). **Royal Horticultural Society awards.** *Orchid Review* 104, 1212: 309 & 312.

84. **Blowers**, John W. (1997). **Royal Horticultural Society awards.** *Orchid Review* 105, 1217: 265 – 269.

85. **Blowers**, John W. (1997). **Royal Horticultural Society awards.** *Orchid Review* 105, 1218: 329 – 331.

86. **Blowers**, John. (1998). **Royal Horticultural Society awards.** *Orchid Review* 106, 1224: 329.
A specimen plant of *Eulophia pulchra* 'Clare' is pictured on the back cover.

87. **Blowers**, John. (1999). **Royal Horticultural Society awards.** *Orchid Review* 107, 1227: 137 – 138.
Gastrorchis tuberculosa 'Clare', with a short discussion.

88. **Blowers**, John & **Hermans**, Clare. (2000). **Royal Horticultural Society awards.** *Orchid Review* 108, 1232: 76.
Neobathiea grandidieriana is described and illustrated.

89. **Blowers**, John. (2000). **Royal Horticultural Society awards.** *Orchid Review* 108, 1233: 137 – 139.
A short description of *Angraecum sesquipedale* with a photograph on the front cover.

90. **Blume**, Carl Ludwig. **Reinwardt**, in (1823). *Catalogus van Eenige der Merkwaardigste zoo in - als uit - heemsche gewasssen te vinden in 's lands plantentuin te Buitenzorg.* Holland.

91. **Blume**, Carl Ludwig. (1825). *Bijdragen tot de Flora van Nederlandsch Indië.* Batavia.

92. **Blume**, Carl Ludwig. (1852). **Museum botanicum lugduno-batavium** 2, 12.

93. **Blume**, Carl Ludwig. (1857). *re. Phaius stuppeus. Orch. Arch. Ind.:* 14.

94. **Blume**, Carl Ludwig. (1858). *Collection des orchidées les plus remarquables de l'Archipel Indien et du Japon.* 1. Sulpke, Amsterdam.

95. **Bock**, I. (1996). *Jumellea confusa* (Schltr.) Schltr. 1915. *Die Orchidee* Orchideenkartei 47, 5 centre.

96. **Bockemühl**, L. & **Senghas**, Karlheinz. (1977). *Angraecum sesquipedale* **Thouars 1822.** *Die Orchidee.* Orchideenkartei 28.

97. **Bockemühl**, L. (1994). *Angraecum sesquipedale* **Thou. 1822.** *Caesiana* 3 centre pages.

98. **Bois**, D. (1893). *Les orchidées manuel de l'amateur.* J. B. Baillière & Fils. Paris: 323 pp.

99. **Boiteau**, Pierre. (1937). **Richesses floristiques et fauniques de Madagascar.** *La Revue de Madagascar* 20: 53 – 74.

100. **Boiteau**, Pierre. (1942). **Plantes nouvelles de Madagascar.** *Bulletin de L'Academie Malgache* 24: 81 – 90.
Three new species are described: *Physoceras australe, Bulbophyllum ceriodorum* and *Cynorchis ambondrombensis.*

101. **Boiteau**, Pierre. (1979). *Precis de Matiere Medicale Malgache.* La Librairie de Madagasacar: 97 pp.

102. **Boiteau**, Pierre & **Allorge-Boiteau**, Lucile. (1993). *Plantes médicinales de Madagascar.* ACCT & Éditions Karthala, Paris: 135 pp.

103. **Bojer**, Wenceslas. (1837). *Hortus mauritianus, ou énumération des plantes exotiques indigènes, qui croissent a L'Ile Maurice.* 456 pp. (orchids p. 310 – 323). Mauritius.

104. **Bolus**, Harry. (1889). **Contributions to South-African botany part IV.** *Journal of the Linnean Society* 25: 171 – 172.

105. **Boni**, U. (1993). **Due** *Bulbophyllum* **dal Madagascar.** *Orchidee Spontanee & Coltivate* 6, 2: 31 – 33.

105b. **Bordes**, Robert (2006). *Cynorkis purpurascens* **Thouars.** *L'Orchidophile*, 37, 168: 97 – 103. Notes on the genus *Cydorkis* in Madagascar, linked with field observations.

106. **Bory** de St. Vincent. (1804). *Voyages dans les quatres principales Iles des mers d'Afrique, fait par ordre du goverment (1801 – 02) avec l'histoire de la traversie du Captaine Boudin jusqu'au Port Louis de l'Ille Maurice.* F. Buisson, Paris, 3 vols.

107. **Bosquet**, G. (1883). **Les** *Angraecum. Revue de l'Horticulture Belge* 9, 8: 186 – 188.

108. **Bosser**, Jean. (1965). **Contribution à l'étude des** *Orchidaceae* **de Madagascar V. Revision de quelques sections du genre** *Bulbophyllum* **à Madagascar.** *Adansonia* sér.2, 5, 3: 375 – 410.

109. Bosser, Jean. (1966). **Contribution a l'étude des** *Orchidaceae* **de Madagascar VI. Etablissement d'une nouvelle espèce de** *Calanthe. Adansonia* sér. 2, 6, 3: 399 – 404.

110. Bosser, Jean. (1969). **Contribution a l'étude des** *Orchidaceae* **de Madagascar VII.** *Adansonia* sér. 2, 9, 1: 135 – 137.

111. Bosser, Jean. (1969). **Contribution a l'étude des** *Orchidaceae* **de Madagascar VIII. Sur trois nouvelles espèces du genre** *Habenaria. Adansonia* sér. 2, 9, 2: 293 – 298.

112. Bosser, Jean & **Morat**, Philippe. (1969). **Contribution a l'étude des** *Orchidaceae* **de Madagascar IX. Les genres** *Grammangis* **et** *Eulophiella. Adansonia* sér. 2, 9, 2: 299 – 309.

113. Bosser, Jean. (1969). **Contribution a l'étude des** *Orchidaceae* **de Madagascar X.** *Adansonia* sér. 2, 9, 3: 343 – 359.

114. Bosser, Jean. (1969). **Contribution a l'étude des** *Orchidaceae* **de Madagascar XI. Sur les affinites des genres** *Cryptopus* **et** *Neobathiea. Adansonia* sér. 2, 9, 4: 539 – 547.

115. Bosser, Jean. (1970). **Contribution a l'étude des** *Orchidaceae* **de Madagascar XII.** *Jumellea* **et** *Angraecum* **nouveaux.** *Adansonia* sér. 2, 10, 1: 95 – 110.

116. Bosser, Jean & **Veyret**, Yvonne. (1970). **Contribution a l'étude des** *Orchidaceae* **de Madagascar XIII. Sur un** *Cynorkis* **et un** *Eulophia* **nouveau.** *Adansonia* sér. 2, 10, 2: 213 – 217.

117. Bosser, Jean. (1970). **Contribution a l'étude des** *Orchidaceae* **de Madagascar XIV. Le genre** *Lemurella* **Schlechter.** *Adansonia* sér. 2, 10, 3: 367 – 373.

118. Bosser, Jean. (1971). **Contribution a l'étude des** *Orchidaceae* **de Madagascar XV. Nouvelles especes du genre** *Aeranthes. Adansonia* sér.2, 11, 1: 81 – 93.

119. Bosser, Jean. (1971). **Contribution a l'étude des** *Orchidaceae* **de Madagascar XVI. Espèces nouvelles du genre** *Bulbophyllum. Adansonia* sér. 2, 11, 2: 325 – 335.

120. Bosser, Jean. (1971). **Contribution a l'étude des** *Orchidaceae* **de Madagascar XVII. Revision de genre** *Phaius. Adansonia* sér.2, 11, 3: 519 – 543.

121. Bosser, Jean. (1972). **Contribution à l'étude des** *Orchidaceae* **de Madagascar et des Mascareignes XVIII.** *Adansonia* sér. 2, 12, 1: 73 – 78.

122. Bosser, Jean. (1974). **Contribution a l'étude des** *Orchidaceae* **de Madagascar XIX. Une nouvelle espece du genre** *Eulophiella* **Rolfe.** *Adansonia* sér. 2, 14, 2: 214 – 218.

123. Bosser, Jean. (1976). **Le genre** *Hederorkis* **Thou. (***Orchidaceae***) aux Mascareignes et aux Seychelles.** *Adansonia* sér. 2, 16, 2: 225 – 228.

124. Bosser, Jean. (1980). **Contribution a l'étude des** *Orchidaceae* **de Madagascar XX. B. Un** *Physoceras* **et un** *Cynorkis* **nouveaux de Madagascar.** *Adansonia* sér. 2, 20, 3: 258 – 261.

125. Bosser, Jean. (1984). **Contribution à l'étude des** *Orchidaceae* **de Madagascar et des Mascareignes XXI. Sur l'identité du genre** *Perieriella* **Schlechter.** *Adansonia* sér. 4, 6, 3 p. 369 – 372.

126. Bosser, Jean. (1987). **Contributions a l'étude des** *Orchidaceae* **de Madagascar et des Mascareignes XXII.** *Adansonia* sér.4, 9, 3: 249 – 254.

127. Bosser, Jean. (1988). **Contribution a l'étude des** *Orchidaceae* **de Madagascar et des Mascareignes XXIII.** *Adansonia* sér. 4, 10, 1: 19 – 24.

128. Bosser, Jean. (1989). **Contribution a l'étude des** *Orchidaceae* **de Madagascar et des Mascareignes XXIV.** *Adansonia* sér. 4, 11, 1: 29 – 38.

129. Bosser, Jean. (1989). **Contribution à l'étude des *Orchidaceae* de Madagascar et des Mascareignes XXV. *Oeonia* Lindley.** *Adansonia* sér. 4, 11, 2: 157 – 165.

130. Bosser, Jean. (1989). **Contribution a l'étude des *Orchidaceae* de Madagascar et des Mascareignes XXVI.** *Adansonia* sér. 4, 11, 4: 369 – 382.

131. Bosser, Jean. (1992). ***Cymbidiella flabellata.*** *Flowering Plants of Africa* 52, 1: pl. 2042.

132. Bosser, Jean. (1992). ***Cymbidiella pardalina.*** *Flowering Plants of Africa* 52, 1: pl. 2043.

133. Bosser, Jean. (1992). ***Grammangis ellisii.*** *Flowering Plants of Africa* 52, 1: pl. 2044.

134. Bosser, Jean. (1992). ***Phaius humblotii* var. *ruber.*** *Flowering Plants of Africa* 52, 1: pl. 2045.

135. Bosser, Jean. (1994). **Orchidées de Madagascar.** *Orchidées. Culture et Protection* 18: 9 – 13.

136. Bosser, Jean & **Cribb** Phillip. (1996). **An extraordinary saprophyte in the genus *Habenaria* (*Orchidaceae*) from Madagascar.** *Adansonia* sér.3, 18, 3 – 4: 335 – 337.

137. Bosser, Jean (1997). **Contribution à l'étude des *Orchidaceae* de Madagascar et des Mascareignes. XXVII.** *Adansonia* sér.3, 19, 2: 181 – 188.

138. Bosser, Jean. (1998). **Contribution à l' étude des *Orchidaceae* de Madagascar et des Mascareignes. XXVIII.** *Adansonia*, sér.3, 20, 2: 281 – 283.
The description of *Cynorkis bardotiana*.

139. Bosser, Jean. (2000). **Contribution à l'étude des *Orchidaceae* de Madagascar et des Mascareignes. XXIX. Révision de la section *Kainochilus* du genre *Bulbophyllum*.** *Andansonia*, sér.3, 22, 2: 167 – 182.
A revision of *Bulbophyllum* section *Kainochilus*.

140. Bosser, Jean. (2000). **Contribution à l'étude des *Orchidaceae* de Madagascar et des Mascareignes. XXX. Description d'une nouvelle espèce d'Oeonia de Madagascar.** *Andansonia*, sér. 3, 22, 2: 231 – 233.
The new species *Oeonia curvata* is described.

141. Bosser, Jean & **Morat**, Philippe. (2001). **Contribution à l'étude des *Orchidaceae* de Madagascar et des Mascareignes XXXI.** *Adansonia*, sér.3, 23, 1: 7 – 22.
The characteristics used by Garay & Taylor and by Summerhayes to distinguish the genera *Oeceoclades* and *Eulophia* are discussed. Species of *Oeceoclades* are described and revised: *O. peyrotii, O. furcata, O. flavescens, O. longibracteata, O. filifolia, O. humbertii, O. ambrensis. Eulophiella galbana* is re-classified.

142. Bosser, Jean & **Cribb**, Phillip. (2001). **Trois nouvelles espèces de *Bulbophyllum* (*Orchidaceae*) de Madagascar.** *Adansonia*, sér.3, 23, 1: 129 – 135.
Three new *Bulbophyllum* are described in the section *Ploiarium: Bulbophyllum perreflexum, B. ranomafanae, B. turkii.*

143. Bosser, Jean. (2002). **Contribution à l'étude des *Orchidaceae* de Madagascar, des Comores et des Mascareignes XXXII. Un *Cynorkis* nouveau des Comores et un *Eulophia* nouveau de La Réunion.** *Adansonia*, sér.3, 24, 1: 21 – 25.
The new epiphytic *Cynorkis comorensis* is described from the Comoro Islands. *Eulophia borbonica* from La Réunion is also described.

144. Bosser, Jean & **Cribb**, Phillip (2003). **Contribution a l'étude des *Orchidaceae* de Madagascar, des Comores et des Mascareignes XXXIV. *Bathiorchis*, nouveau genre monotypique de Madagascar.** *Adansonia*, sér.3, 25, 2: 229 – 231.
A new monotypic genus of Physureae from Madagascar is described. Originally placed by Perrier de la Bâthie in the genus *Gymnochilus, Bathiorchis rosea* is close to *Platylepis* but has a column that is morphologically different.

145. Bosser, Jean. (2004). **Contribution à l'étude des *Orchidaceae* de Madagascar et des Mascareignes. XXXIII.** *Andansonia*, sér. 3, 26, 1: 53 – 61.
Bulbophyllum labatii, B. ambatoavense, Cynorkis subtilis are described and illustrated. *Nervilia simplex* and *N. crociformis* are shown to represent the same taxon from the Mascarenes. New synonyms are established for taxa of *Bulbophyllum, Aerangis, Angraecum* and *Oeceoclades* from Madagascar and the Mascarenes.

146. Botanical Register, (likely by Lindley). (1840). ***Angraecum gladiifolium.*** *Botanical Register* t. 68.

147. Bourgeat, Emile. (1993). ***Aerangis* Mon Amour.** *L' Orchidée* 11, 44: 136 – 137.

148. Bouriquet, Gilbert. (1946). **La vanillier et la vanille à Madagascar.** *Revue Internationale de Botanique Appliquée et d'Agriculture Tropicale* 26: 283 – 284, 398 – 404.

149. Bouriquet, Gilbert, ed. (1954). **Le vanillier et la vanille dans le Monde.** *Encyclopédie Biologique.* Paul le Chevalier, Paris: 748 pp., 4 colour pl.

149b. Bournérias, Janine & Marcel. (1997). **Voyage d'études de la S.F.O. à L'ile de la Réunion et à l'ile Maurice du 29 Septembre au 13 Octobre 1994,** *L'Orchidophile*, 28, 129: 197 – 224.
A diary of a trip by members of the French orchid society to Mauritius and Réunion.

150. Boyle, Frederick. (1893). ***About orchids a chat.*** Chapman & Hall, London: 250 pp.

151. Boyle, Frederick. (1901). ***The Woodlands orchids.*** Macmillan, London: 271 pp.

152. Bradt, Hilary, **Schuurman**, Derek, **Garbutt**, Nick. (1996). ***Madagascar wildlife. A visitor's guide.*** Bradt Publications, Chalfont St. Peter, 138 pp.

153. Braem, Guido J. (1991). ***Angraecum popowii* Braem**. *Schlechteriana* 2, 4: 163 – 166. a similar illustration in Orchideen der Welt, Verlag Ritschel, Gladenbach (1997): 9.

154. Braid, K. W. (1926) **XL. Angraecoid orchids.** *Kew Bulletin,* 8 : 323 – 357.
A taxonomic history of the angraecoids. A key to the genera is given together with an alphabetical list of synonyms, based on the work of Schlechter.

155. Britt, Adam. (2003). **Vital work in Madagascar reaches a new phase.** *Kew*, Autumn issue: 41.
An update on the Madagascar orchid-conservation appeal. Negotiations with various partner organisations are outlined.

155a. Britt, Adam. (2004). **Madagascar. Growing Orchids for Conservation.** *Kew Scientist*, 25: 6.
A note on fieldwork for the Madagascar Threatened Plant Project. Target species include *Eulophiella roempleriana.*

155b. Britt, Adam, **Clubbe**, Colin, **Ranarivelo**, Tianjanahary. (2004). **Conserving Madagascar's plant diversity: Kew's Madagascar Threatened Plant Project.** *Curtis's Botanical Magazine*, 21, 4: 258 – 264.
An outline of a conservation project managed by the Royal Botanic Gardens, Kew. Target orchids are listed.

156. Brown, Robert. (1810). ***Prodromus Florae Novae Hollandiae et Insulae Van-Diemen.*** Richard Taylor, London: 331 pp.

157. Buchenau, Franz. (1896). **Christian Rutenberg's Ende.** *Abhandlungen des naturwissenschaftlichen Verein Bremen.*
A short biographical note, including a report by Dr. Voeltzkow about Rutenberg's murder. Most of the information was supplied by Hildebrandt.

158. Bulletin De L'Académie Malagache, the secretary. (1928). **Au sujet de deux orchidées nouvelles des genres *Nervilia* et *Gastrorchis*.** *Bulletin de l'Académie Malagache* n. sér. 11, xi.

159. Bultel. (1913). **Les *Eulophiella*.** *Le Jardin* 27: 71 – 72.

160. **Burberry**, H. A. (1900). **Calendar of operations for March.** *Orchid Review* 8, 87: 83 – 87.

161. **Burrage**, Albert C. (1930). *The Orchidvale Collection.* Burrage, Mass.: 188 pp.

162. **Burvenick**, Fr., ed. **Collecteurs d'orchidées.** *Revue Horticole Belge et Étrangère* 18, 1: 18.
A report from The Gardeners' Chronicle regarding the search for the scarlet Cymbidium (*Cymbidiella pardalina*). It is said that three collectors went in search for it; two died of a fever, the other one is on his way home but is not very well.

163. **Butzin**, Friedhelm. (1975). **Neue Namen und neue Kombinationen in der Orchideengattung** *Eulophia.* *Willdenowia* 7, 3: 587 – 590.

164. **Butzin**, Friedhelm (1981). **Typenstudien im Berliner Orchideen-Herbar: Diverse markierte Typen.** *De Herbario Berolinensi Notulae,* 15.
The type specimens that survive in the Berlin Herbarium are listed, a few from Madagascar, all described by Perrier: *Benthamia humbertii, Cynorkis andohahelensis, Jumellea brevifolia, Lissochilus analavensis, L. humbertii, Polystachya humbertii.*
Previous listings by the same author: (1978). **In Berlin vorhandene Typen von Schlechters Orchideenarten,** *De Herbario Berolinensi Notulae* 2 and (1978). **Typenstudien im Berliner Orchideen-Herbar: Arten von Kraenzlin und Mansfeld,** *De Herbario Berolinensi Notulae,* 14 did not include any Madagascan material and so it must be concluded that all were destroyed in the 1943 fire.

165. **Cabanis** Yvon, **Chabouis** Lucette, **Chabouis** Francis. (1969 – 1970). *Végétaux et groupement végétaux de Madagascar et des Mascareignes.* Bureau pour le Développement de la Production Agricole. Tananarive, 4 vols. 1128 pp.
After an introduction to the islands, short descriptions and notes are given plus a line drawing of the most commonly seen plants of Madagascar and the Mascarenes.

166. **Cadet**, Th. (1981). *Fleurs et plantes de la Réunion et de l'île Maurice.* Editions du Pacifique, Times Editions, Singapore. 1981.
A discussion and description of the various plants and their habitat on the islands.

167. **Cadet**, Janine. (1989). *Les orchidées de la Réunion.* Nouvelle Impr. Dionysienne, Réunion.

168. **Cadet**, Th. (1979). **Notes sur les orchidées de l'Ile de la Réunion.** *L'Orchidophile* 10, 37: 1295 – 1308.

169. **Cadet**, Th. (1989). *Fleurs et plantes de la Réunion et de l'île Maurice.* Delachaux et Niestlé, Neufchâtel, Paris: 98 – 108.

170. **Cambornac**, Michel. (1994). **La vanille de Madagascar, Améliorations génétiques.** *Orchidees, Culture et Protection* 18: 17 – 18.

171. **Canals**, Hervé. (1985). *Angraecum sesquipedale* Thou. *L'Orchidophile* 16, 68: 897.

172. **Candolle**, Augustin De. (1901). *Plantae madagascariensis ab Alberto Mocquerysio lectae.* Bulletin *de L'Herbier Boissier* 2e. sér. 1, 6: 549 – 587.
Before listing the plants collected by Albert Mocquerys in 1898, a short history of work on the flora of Madagascar is given. The works by Baron, Ridley and Drake del Castillo are mentioned. Affinities between the African and Asiatic floras are discussed.

173. **Carlsward**, Barbara S. (2001). **Eccentric epiphytes.** *Orchid Review,* 109, 1241: 302 – 304.
The anatomy and function of leafless epiphytes is discussed, including *Microcoelia exilis.*

173b. **Carlsward**, Barbara, **Whitten**, Mark, **Williams**, Norris & **Bytebier**, Benny (2006). **Molecular Phylogenetics of** *Vandeae* (*Orchidaceae*) **and the evolution of leaflessness.** *American Journal of Botany,* 93, 5: 770 – 786.
A detailed analysis of the phylogenetics of angraecoids. *Solenangis aphylla* and *S. cornuta* are transferred to *Microcoelia.*

173c. Carlsward, Barbara, **Stern,** William & **Bytebier**, Benny (2006). **Comparative vegetative anatomy and systematics of the angraecoiods with an emphasis on the leafless habit.** *Botanical Journal of the Linnean Society,* 151: 165 – 218.
The anatomy and morphology of the angraecoid orchids is discussed.

174. Carrière, Élie Abel & **André**, Eduard. (1885). *Angraecum leonis. Revue Horticole* 57: 436.

175. Carrière, Élie Abel. (1887). *Angraecum grandidierianum. Revue Horticole* 59: 42.

176. Carrière, Élie Abel. (1887). *Eulophia megistophylla. Revue Horticole* 59: 87

177. Carrière, Élie Abel. (1887). *Angraecum fuscatum. Revue Horticole* 59: 235 – 6.

178. Carrière, Élie Abel & **André** Eduard. (1894). **Chronique horticole.** *Angraecum Fournierianum. Revue Horticole* 66: 418.

179. Carter, Brad. (1996). **Growing eulophias.** *Orchids (AOS)*: 1057. Also in *UCI Arboretum Quarterly,* 4, 2 (1996). and on *World Wide Web.*

180. Carvill, J. (1890). *Angraecum sesquipedale. Gardeners' Chronicle* ser. 3, 7: 136.

181. Castillon, Jean-Bernard. (1983). **Un** *Aeranthes* **spectaculaire de Madagascar.** *L'Orchidophile* 59: 490 – 491.

182. Castillon, Jean-Bernard. (1984). **Orchidées rares ou nouvelles de l'Ile de Réunion.** *L'Orchidophile* 15, 61: 589 – 591.

183. Castle, Lewis. (1887). *Orchids. Their structure, history and culture.* Journal of Horticulture Office, London, 123 pp., 21 ill.

184. Cavestro, William. (1980). **The culture of African** *Aerangis. American Orchid Society Bulletin* 49, 7: 747 – 750.

185. Cavestro, William. (2004). **The genus** *Cryptopus* **Lindl.** *Orchid Review* 112, 1259: 287 – 291.
Notes on the members of the genus *Cryptopus*: *C. elatus* and *C. paniculatus* are illustrated.

185a. Cavestro, William. (2005) **Oeonia, Cryptopus and Neobathiea: three angraecoid genera.** *Orchid Review,* 113, 1261: 34 – 38.
Notes on the members of the genus *Neobathiea.*

186. Chapman, H. J. (1901). **Notes on orchids. A new** *Phaius. The Gardeners' Magazine* 44: 116.

187. Charles, Emerson. (1963). *Grammangis ellisii. Orchid Digest* 27, 1: 4.

188. Charlesworth. (1905). *Descriptive catalogue.* Charlesworth 320 pp.
A nurseryman's catalogue, including a few species from Madagascar, prices are given.

189. Charlesworth & Co. (1922). *Orchids.* Charlesworth & Co.: 74 pp.

190. Chazal, Malcy de. (1989). *The medicinal plants of Mauritius.* The Schoolhouse Gallery, Berwickshire: 125 pp.

191. Chiron, Guy. (1994). *Aerangis* **notes de culture.** *Orchidees, Culture et Protection* 18: 19 – 24.

192. Chittenden, F. J., ed. (1914). *Cymbidiella humblotii.* Journal of the Royal Horticultural Society London 39: fig. 121.

193. Christensen, Carl. (1932). **The Pteridophyta of Madagascar.** *Dansk Botanik Arkiv* 7, 2: V – XV.

194. Christenson, Eric A. (2003). *Oeceoclades monophylla.* **A rare species enters Horticulture.** *Orchids (AOS)* 72, 11: 860 – 861.

It is reported that *Oeceoclades monophylla* has been brought into cultivation. It is rejected that the Mascarene plants could be a subspecies or synonym of *O. maculata*.

195. Clegg, G. S. (1946). ***Angraecum eburneum.*** *Orchid Review* 54, 639: 134.

196. Clements, Mark A. (1984). **Catalogue of Australian *Orchidaceae.*** *Australian Orchid Research* 1: 99.

197. Cogniaux, Alfred. (1893). **Études de botanique élémentaire sur les orchidées.** *Le Journal des Orchidées* 4 & 5: 81 – 85.

198. Cogniaux, Alfred Céléstin. (1896). ***Polystachya villosa.*** *Le Journal des Orchidées* 7: 139.

199. Cogniaux, Alfred Céléstin & **Goossens** A. (1896). ***Dictionaire iconographique des orchidées.*** Brussels.

200. Cogniaux, Alfred, ed. (1897). **Orchidées nouvelles ou récemment introduites.** *Chronique Orchidéenne* 1, 2: 12.

201. Cogniaux, Alfred, ed. (1897). **Orchidées nouvelles ou récemment introduites.** *Chronique Orchidéenne* 1, 3: 21.

202. Cogniaux, Alfred. (1897). **L'*Angraecum eburneum* et les Espèces affines.** *Supplement au Dictionaire Iconographique des Orchidées* 1, 4: 29 – 30.

203. Cogniaux, Alfred, ed. (1898). **Petites notes.** *Chronique Orchidéenne* 1, 17: 129.

204. Cooper, Dave. (1995). **Displaying fragrant orchids.** *American Orchid Society Bulletin* 64, 1: 24 – 29.

205. Cooper, E. (1938). ***Cymbidiella rhodochila.*** *Orchid Review* 46, 545: 333 – 334.

206. Cordemoy, Jacob de. (1895). ***Flore de l'Ile de la Réunion.*** Klincksieck, Paris, France. Orchidées: 165 – 262. Also reprint Histor. Nat. Class. 94 (1972).

207. Cordemoy, Jacob de. (1899). **Orchidées de la Réunion.** *Revue Génerale Botanique* 11: 409 – 429, pl. 6 – 11.

208. Costantin, Julien. (1913). ***Atlas en couleurs des orchidées Cultivées.*** Librarie Générale de l'Enseignement, Paris. 91 pp., 30 pl.

209. Cours, Gilbert. (1948). **La vanille dans la region d'Antalaha.** *Bulletin Agric. Madagascar* 6: 3 – 5.

210. Cowan & Co. (c. 1930). **Special offer of orchids.** *Trade Catalogue* 8.

211. Cribb, Phillip. (1975). ***Aeranthes caudata.*** *Curtis's Botanical Magazine* 180, 3, t. 685.

212. Cribb, Phillip. (1975). **A note on the identity of the Madagascan orchid *Aeranthes caudata*** Rolfe. *Adansonia* ser.2, 15, 2: 195 – 197.

213. Cribb, Phillip. (1977). **Kew und seine Orchideen.** *Die Orchidee* 28, 2: 76 – 79.

214. Cribb, Phillip. (1977). ***Phaius françoisii.*** *Curtis's's Botanical Magazine* 181, no.4: t. 738.

215. Cribb, Phillip. (1978). **Studies in the genus *Polystachya* in Africa.** *Kew Bulletin* 32: 743 – 766.

216. Cribb, Phillip. (1981). **Calanthe madagascariensis.** *Flowering Plants of Africa* 46: pl. 1830.

217. Cribb, Phillip. (1980). ***Aeranthes henricii* Schlechter.** *Orchid Review* 88, 1043: 167.

218. Cribb, Phillip & **Stewart**, Joyce. (1983). ***Aerangis mooreana (Orchidaceae).* An overlooked orchid.** *Orchid Review* 91, 1077: 218 – 219.

219. Cribb, Phillip. (1984). *Orchidaceae*: **Part 2.** in R. M. Polhill, ed., **Flora of Tropical East Africa,** *Orchidaceae*. A.A. Balkema, Rotterdam: 411 pp., b/w. ill.

220. Cribb, Phillip. (1985). **Orchids and conservation in tropical Africa and Madagascar.** *Bothalia* 14, 3 – 4: 1013 – 1014.

221. Cribb, Phillip. *Orchidaceae*. **Part 3.** (1989). in R. M. Polhill, ed., **Flora of Tropical East Africa.** A.A. Balkema, Rotterdam: 651 pp., b/w. illustations/colour photos.

222. Cribb, Phillip & **Bailes**, Chrisopher. (1992). *Orchids. A romantic history.* Running Press, Philadelphia. 63 pp.

223. Cribb, Phillip & **Pridgeon**, Alec, eds. (1996). *Angraecum sesquipedale* **revisited.** *Orchid Research Newsletter* 27: 4 – 5.

224. Cribb, P.J. & **Ormerod**, P. (2001). *Goodyera* **R. Br.** (*Orchidaceae*): **a new generic record for tropical Africa.** *Kew Bulletin*, 56: 501 – 502.
Goodyera afzelii is newly recorded from Mozambique, it also occurs in Madagascar. A key to the Spiranthoideae and a description of the species are included.

225. Cribb, Phillip J., **Du Puy**, David & **Bosser**, Jean. (2002). **An unusual new epiphytic species of** *Eulophia* (*Orchidaceae*) **from the southeastern Madagascar.** *Adansonia*, sér. 3, 24, 2: 169 – 172.
Eulophia epiphytica from southeastern Madagascar is described.

226. Cribb, Phillip & **Tibbs**, Michael. (2004). *A very Victorian passion. The orchid paintings of John Day 1863 to 1888.* Blacker Publishing & Royal Botanic Gardens, Kew, 2004: 464 pp.
An overview of the life and times of John Day. Many of his watercolours are reproduced together with extensive notes on the plants' history. A large number of biographies and historical notes on growers and nurseries are contained in two appendices.

226b. Cribb, Phillip, **Roberts**, David & **Hermans**, Johan. (2005). **Distribution, Ecology and Threat to Selected Madagascan Orchids.** *Selbyana*, 26, 1,2, 125 – 133.
Research on the distribution patterns of Madagascan orchids is described, implications on conservation status is outlined.

229. Crippa, Eugenio. (1995). *Madagascarum Equilibrio Precario. Orchis* 92: 23 – 27.

230. Curtis, Charles H., ed. (1933). **Notes.** *Orchid Review* 41, 486: 357.

231. Curtis, Charles H., ed. (1934). **Notes.** *Orchid Review* 42, 489: 65 – 66.

232. Curtis, Charles H., ed. (1942). *Angraecum ellisii. Orchid Review* 50, 583: 2.

233. Curtis, Charles H. (1948). **The House of Veitch.** *Journal of the Royal Horticultural Society*, 73, 9: 284 – 290.
A general history of the firm of Veitch. Amongst the collectors employed is Charles Curtis who collected *Angraecum sesquipedale* in Madagascar. A short biographical note.

235. Curtis, Charles H. (1950). *Orchids — their description and cultivation.* Putnam, London: 274 pp.

236. Curtis, Charles H., ed. (1955). *Grammangis ellisii. Orchid Review* 63, 748: 158.

237. Curtis, Charles H., ed. (1956). *Phaius humblotii. Orchid Review* 64, 762: 235 – 236.

238. Curtis, Charles H., ed. (1957). *Cymbidium rhodochilum. Orchid Review* 65, 771: 178.

239. Dacco Hensemberger, Giulia. (1992). *Aerangis fastuosa* (**Rchb.f.**) **Schltr.** *Orchidee Spontanee & Coltivate* 5, 4: 88.

240. Dandouau, A. (1909). *Catalogue alphabétique des noms Malgaches de végétaux.* Imprimerie Officielle, Antananarivo.

241. Darwin, Charles. (1904). *The various contrivances by which orchids are fertilised by insects.* Ed. 2. John Murray, London: 300 pp., 38 drawings.

242. Darwin, Charles. (1959). **Orchid classics — 2. Observations on** *Angraecum sesquipedale.* *American Orchid Society Bulletin* 28, 4254: 261 – 262.

243. Davies, Kevin L. (1996). **The orchid collection of J. J. Neale, 1854 – 1919.** *Orchid Review* 104, 1210: 240 – 243.

244. Day, John. (1863 – 1887). **Scrapbooks.** Archives, Library of The Royal Botanic Gardens, Kew.

245. Dean, Richard Ed. (1880). *Angraecum scottianum. The Floral Magazine* new series: t. 422.

246. Deans Cowan, William. (1880). *Drawings of Madagascar orchids.* Archives, The Natural History Museum, London. Photographic copy at Kew.

247. De Bosschere, Ch. (1893). **Deux orchidées nouvelles.** *Revue de l'Horticulture Belge* 19, 3: 53 – 55.

248. De Candolle, A. (1864). *Prodromus Systematis Universalis. Regno Vegetabilis:* 15, Part 1.

249. Decary, Raymond. (1926). **Une mission scientifique dans le Sud-Est de Madagascar.** *Bulletin de l'Académie Malgache* n. sér., 9: 80 – 86.

250. Decary, Raymond. (1930) (Tananarive 1931). **Contribution à l'étude de la végétation de Madagascar. La Flore de la ville de Tananarive.** *Bulletin de l'Académie Malgache* n. sér., 13: 127 – 149.

251. De Kerkhove, O. de Denterghem. (1889). **Les** *Angraecum. Revue de l'Horticulture Belge* 15, 10: 217 – 220.

252. De Kerckhove, O. de Denterghem. (1890). **Les** *Phaius. Revue de l'Horticulture Belge* 16: 264 – 269.

253. De Kerckhove, O. de Denterghem. (1894). *Le livre des orchidées.* Masson, Hoste, Gent - Paris.

254. De Kerckhove, O. de Denterghem. (1895). *Eulophiella elisabethae. Revue de l'Horticulture Belge* 21, 8: 181 – 183.

255. De Lanessan J. L. (1886). *Les plantes utiles des Colonies Françaises.* Imprimerie Nationale, Paris: p. 62 – 74.

256. Delchevalerie, G. (1889). *Les orchidées, culture, propagation, nomenclature.* Librarie Agricole, Paris: 132 pp., 3 engravings.

256b. Demissew, Sebsebe, **Cribb**, Phillip & **Rasmussen**, Finn (2004). *Field guide to Ethiopian orchids.* Royal Botanic Gardens. Kew: 300 pp.
A well-illustrated field guide, a few species also occur in the Madagascar area.

257. Demoly, Jean Pierre. (1990?). **Les Cèdres, an exceptional botanical garden**: 268 – 272.
Includes a chapter on Madagascan orchids species, *Oeceoclades spathulifera* and *Eulophia macra* are described and illustrated.

258. De Priori, Patrizio. (1996): **La cultivazione del genere** *Aerangis. Orchis* 107: 3 – 5.

259. De Puydt, Emile. (1880). *Les orchidées, histoire iconographique.* Editions J. Rothschild, Paris: 348 pp.

260. Descheemaeker, A. (1986). *Ravi-Maitso.* Imprimerie Saint-Paul, Ambositra: numerous b/w. photos. French translation (1990).

261. De Vallia, Mas. (1895). **Petite notes sur les orchidées d'amateur.** *Le Journal des Orchidées* 6: 167 & 261.

262. De Vallia, Mas. (1896) **Petites notes sur les orchidees d'amateur.** *Journal des Orchidées*, 6: 357 – 358.
Notes on *Angraecum sesquipedale* and *A. scottianum.*

263. Dillon, Gordon W., ed. (1943). ***Angraecum sesquipedale.*** *American Orchid Society Bulletin* 11, 10: 355. cover photo.

264. Dillon, Gordon W., ed. (1944). ***Angraecum eburneum*** *American Orchid Society Bulletin* 13, 7: 216 & 219, frontispiece.

265. Dillon, Gordon W., ed. (1953). ***Angraecum sesquipedale.*** *American Orchid Society Bulletin* 22, 12: 872, cover photo.

266. Dillon, Gordon W., ed. (1963). **American Orchid Society awards.** *American Orchid Society Bulletin* 32, 2: 84 & 132.

267. Dillon, Gordon W., ed. (1963). ***Cymbidiella rhodochila.*** *American Orchid Society Bulletin* 32, 11: 884 & 926.

268. Dillon, Gordon W., ed. (1966). **American Orchid Society awards.** *American Orchid Society Bulletin* 35, 8, suppl.: 8.

269. Dillon, Gordon W., ed. (1967). **American Orchid Society awards.** *American Orchid Society Bulletin* 36, 3: 240.

270. Dockrill, Alec W. (1964). **A new combination for a North Queensland orchid.** *The Orchidian* 1, 9: 106 – 107.

271. Dorr, Laurence. (1987). **Rev. Richard Baron's compendium des plantes Malgaches.** *Taxon* 36, 1: 39 – 46.

272. Dorr, Laurence. (1997). ***Plant collectors in Madagascar and the Comoro Islands.*** Royal Botanic Gardens, Kew. 524 pp. (Also on CD-ROM).

273. Douglas, J. (1885). ***Aeranthus leonis.*** *Gardeners' Chronicle* n. ser. 23, p. 142.

274. Du Buysson, François. (1878). ***L'orchidophile.*** Auguste Goin, Paris: 536 pp.

275. Ducharte, Pierre. (1862). **Note sur l'*Angraecum sesquipedale* et l'*Oncidium splendidum.*** *Bulletin de la Société Botanique de France* 9: 30 – 36.

Du Petit-Thouars, see Thouars.

276. Dupont, Mrs. Pierre S. (1936). ***Eulophiella* Rolfei.** *American Orchid Society Bulletin* 4, 4p. 73.

277. Du Puy,, David. (1995). **Madagascar. Growing orchids for conservation.** *Kew Scientist* 8: 8.

278. Du Puy, David J. (1996). ***Cynorkis gibbosa.*** *Curtis's Botanical Magazine* 13, 3: tab. 296, 124 – 129.

279. Du Puy, David J. & **Bell**, Sandra. (1996). **The Madagascar orchid project.** *Curtis's Botanical Magazine* 13, 3: 163 – 164.

280. Du Puy, David. (1997). **Orchid conservation project — mapping Madagascar's vegetation.** *Orchids (American Orchid Society)* 66: 1053 & 1055.

281. Du Puy, David, **Cribb**, Phillip, **Bosser**, Jean, **Hermans**, Johan & Clare. (1999). ***The orchids of Madagascar, annotated checklist and annotated bibliography.*** Royal Botanic Gardens, Kew: 424 pp., 136 photographs.
The previous edition of this book.

282. Durand, Théophile Alexis & **Schinz**, Hans. (1895). ***Conspectus Florae Africae.*** 5.

283. Dyer, R. A. (1949). *Eulophia dissimilis. Flowering Plants of Africa*, 27, t.1066.
The description of *Eulophia dissimilis* together with a full page watercolour illustration. It is compared with *E. mackenii*.

284. Ecott, Tim. (2004). *Vanilla. Travels in search of the luscious substance.* Michael Joseph, London: 278 pp.
A very thoroughly researched history of culinary vanilla. The author visits vanilla producers in Madagascar.

284b. Ecott, Tim. (2004). **The mysterious history of vanilla.** Kew, Autumn (2004): 17 – 19.
A well-illustrated history of culinary vanilla.

285. Eggli, Urs. (1999). **Nachtblühende Sukkulenten und ihre Blütenbesucher.** Die Sukkulentenwelt, 3: 11 – 33.
The pollination of various night flowering succulents is discussed, including that of *Angraecum sesquipedale* and *A. sororium. Angraecum compactum* is illustrated with its pollinating moth. The pollination process is described.

286. Ellis, John E. (1873). *Life of William Ellis missionary to the South Seas and to Madagascar.* John Murray, London, 1873: 310 pp.

287. Ellis, William. (1858). *Three visits to Madagascar during the years 1853 – 1854 – 1856.* John Murray, London: 476 pp. Also, Harper & Brothers, New York (1859): 514 pp.

288. Ellis, William. (1859). **Madagascar orchids.** *Gardeners' Chronicle*: 100.

Farmer, Sir John, editor of *Gardeners' Chronicle*, see 366.

289. Ferko, P. (1912). *Eulophiella hamelinii. Orchid World* 2, 11: 254 – 257.

290. Ferko, P. (1913). *Eulophiella hamelinii. Orchid World* 4, 3: 59 – 60.

291. Fibeck, Werner & **Wendhut**, Barbara. (2003). **Orchideenflora der Seychellen — Eine Migrationsanalyse.** *Die Orchidee*, 54, 4: 50 – 58.
A summary of the biogeography of the orchid flora of the Seychelles.

292. Finet, Achille. (1907). **Classification et énumération des orchidées africanes de la tribu des Sarcanthées, d'après les collections du Muséum de Paris.** *Bulletin de la Société Botanique de France* 54, 2, Mem. 9: 1 – 65, pl. 1 – 12.

293. Finet, Achille. (1909). **Orchidée nouvelle de Madagascar.** *Notulae Systematicae* 1: 89 – 90.

294. Finet, Achille. (1911). **Orchidées nouvelles ou peu connues.** *Notulae Systematicae* 2: 23 – 27.

295. Fitch, Charles Marden (2004). **Adaptable Angraecums.** *Orchids (AOS.)* 73, 9: 666 – 671.
A well-illustrated feature on Angraecoid orchids, with special emphasis on the hybrids.

296. Fitzgerald, R. D. (1883). **New Australian orchids.** *Journal of Botany* 21: 204.

297. Flacourt, Étienne. 1658 (2nd ed. 1661). *Histoire de la Grande Isle Madagascar.* Gervais Clouzier. Paris. Reproduced in an annotated version by **Allibert**, Claude (1995). Éditions Karthalia, Paris: 656 pp.
An early history of Madagascar by Flacourt who was based in a French settlement at Fort Dauphin. He describes the island, the inhabitants and also the fauna and flora. Chapter 36 describes the plants, some descriptions may refer to orchids.

298. Flourens. (1849). **Éloge.** *Mém. Acad. Sci. Paris*, xx: xii.

298a. Fournel, Jacques, **Micheneau**, Clare, **Pailler**, Thierry (2005). *Aeranthes adenopoda* **H.Perrier, une orchidée nouvelle pour l'île de la Réunion.** *L'Orchidophile*, 36, 164: 7 – 10.
Aeranthes adenopoda is recorded for the first time from Réunion where it is rare. With photographs and a map.

299. **Forster**, Johann Georg. (1786). *Florulae Insularum Australium Prodromus.* Christian Dieterich, Gottingen: 103 pp.

300. **Fowlie**, Jack A. (1968). **An account of** *Eulophiella elisabethae. Orchid Digest* 32: 218 – 219.

301. **Fowlie**, Jack A. (1968). **Some notes on** *Eulophiella elisabethae. Orchid Digest* 32: 219.

302. **Fowlie**, Jack A. (1969). **The curious tryst of the star orchid of Madagascar (***Angraecum sesquipedale***).** *Orchid Digest* 33: 208 – 210.

303. **François**, Edmond. (1924). **Une belle orchidée de Madagascar. Le** *Gastorchis humblotii. Revue Horticole* 96: 106 – 107.

304. **François**, Edmond. (1927). **La Flore de Madagascar, les orchidées.** *Revue Horticole,* 99: 515 – 518.

305. **François**, Edmond. (1928). **Deux orchidées nouvelles de Madagascar.** *Revue Horticole* 100: 304 – 306.

306. **François**, Edmond. (1932). **Deux orchidées de Madagascar.** *Revue Horticole* 104: 72 – 74.

307. **François**, Edmond. (1934). **Les produits malagaches: la vanille.** *La Revue de Madagascar* 5 p. 37 – 56.

308. **François**, Edmond. (1937). **Plantes de Madagascar.** *Mémoires de l'Académie Malgache* 24: 71 pp., 23 pl.

309. **François**, Edmond. (1941). **Les mille aspects de la flore de Madagascar.** *La Revue de Madagascar* 30: 7 – 45.

310. **Frappier**, Charles de Monbenoist. (1880). *Orchidées de l'Ile de la Réunion, Catalogue des espèces indigènes découvertes jusqu'a ce jour.* St. Denis, Réunion: 16 pp.

311. **Fraser**, Hedvika. (1988). **Dreaming of orchids.** *Hortus* 2, 1: 86 – 95.

312. **Friis**, I. & **Rasmussen**, Finn N. (1974). **The two alternative systems of nomenclature proposed and used for orchids by Du Petit-Thouars, with special regard to** *Corymborkis* Thouars. *Taxon* 24: 307 – 318.

313. **Garay**, Leslie A. (1972) **On the systematics of monopodial orchids I.** *Botanical Museum Leaflets Harvard University,* 23, 4: 149 – 212.

314. **Garay**, Leslie A. (1973). **Systematics of the genus** *Angraecum* **(***Orchidaceae***).** *Kew Bulletin* 28, 3: 495 – 516.

315. **Garay**, Leslie A. (1974). **On the systematics of the monopodial Orchids II.** *Botanical Museum Leaflets of Harvard University* 24, 10: 369 – 370.

316. **Garay**, Leslie A. &, **Sweet** Herman R. (1974). *Terminología de orchidearum Jacquinii. Orquideologia* 9, 3: 200 – 210.

317. **Garay**, Leslie A. (1976). **The cultivated species of** *Cymbidiella. Orchid Digest* 40, 5: 192 – 193.

318. **Garay**, Leslie A. & **Taylor** Peter. (1976). **The genus** *Oeceoclades* Lindley. *Botanical Museum Leaflets of Harvard University* 24, 9: 249 – 274.

319. **Garay**, Leslie A., **Hamer** Fritz, **Siegerist** Emly S. (1994). **The genus** *Cirrhopetalum* **and the genera of the** *Bulbophyllum* **alliance.** *Nordic Journal of Botany* 14, 6; 609 – 646.

320. **Garay**, Leslie A. (1999). **Orchid species currently in cultivation.** *Harvard. Papers in Botany,* 4, 1: 295 – 299

Includes the description of *Bulbophyllum hapalanthos*. The new species is similar in habit and columnar structure to *B. lakatoense*.

321. Garay, Leslie A. & **Romero-Gonzalez**, Gustavo A. (1999). *Schedulae Orchidum* **II**. *Harvard. Papers in Botany*, 4, 2: 475 – 488
The sections of the genus *Liparis* are analysed, a key is given.

322. Garden. W.G., ed. (1884). *Phaius tuberculosus.* *The Garden* 26: 46, t. 449.

323. Garden. W.G., ed. (1889). *Angraecum citratum.* *The Garden* 35, 1: 284.

324. Garden. W.G., ed. (1889). *Vanilla humbloti.* *The Garden* 35, 1: 338.

325. Garden. (E. J.). (1898). *Angraecum sesquipedale.* *The Garden* 53 6.

326. Garden. ed. (1898). *Angraecum sesquipedale.* *The Garden* 53: 40.

327. Garden. ed. (1898). *Angraecum eburneum.* *The Garden* 53: 147.

328. Garden. (S.). (1898). **Societies and exhibitions.** *The Garden* 53 223 & 369.

329. Garden. ed. (1898). *Eulophiella Elizabethae.* *The Garden* 53, 369.

330. Garden. (H. J. C.). (1898). *Phaius tuberculosus* **Hybrids.** *The Garden* 53: 298.

331. Garden. signed C. (1898). **Orchids.** *Eulophiella peetersiana.* *The Garden* 53: 379.

332. Garden. ed. (1902). *Cymbidium rhodochilum.* *The Garden* 56: 383.

333. Garden The, Editor. (1908). *Angraecum germinyanum.* *Garden*, 72: 250.
A short note regarding the species on receiving an Award of Merit from the RHS for Sir Trevor Lawrence.

334. Gardeners' Chronicle. (by L.W.J.) (1862). **Angraecums.** *Gardeners' Chronicle.* 1148 – 1149.

335. Gardeners' Chronicle, M. T. Masters, ed. (1866). *Angraecum sesquipedale.* *Gardeners' Chronicle.* 52.

336. Gardeners' Chronicle, M. T. Masters, ed.. (1867). **Plant portraits.** *Angraecum citratum.* *Gardeners' Chronicle* 6: 126.

337. Gardeners' Chronicle, M. T. Masters, ed. (1872). **Obituary.** *Gardeners' Chronicle.* 806.

338. Gardeners' Chronicle, M. T. Masters, ed. (1873). **Angraecum eburneum.** *Gardeners' Chronicle.* 217, fig. 46.

339. Gardeners' Chronicle, M. T. Masters, ed. (1873). *Angraecum sesquipedale.* *Gardeners' Chronicle.* 255, fig.53.

340. Gardeners' Chronicle, M. T. Masters, ed. (1873). *Angraecum sesquipedale.* *Gardeners' Chronicle.* 1078.

341. Gardeners' Chronicle, M. T. Masters, ed. (1874). **Floral Committee notes.** *Gardeners' Chronicle* n. ser. 1: 254.

342. Gardeners' Chronicle, M. T. Masters, ed. (1874). *Angraecum sesquipedale.* *Gardeners' Chronicle* n. ser. 1: 346.

343. Gardeners' Chronicle, M. T. Masters, ed. (1874). *Angraecum ellisii.* *Gardeners' Chronicle* n. ser. 1: 446.

344. Gardeners' Chronicle, M. T. Masters, ed. (1875). **Respecting** *Angraecum ellisii*. *Gardeners' Chronicle* n. ser. 3: 277 – 278.

345. Gardeners' Chronicle, M. T. Masters, ed. (1880). *Angraecum citratum*. *Gardeners' Chronicle* n. ser. 13: 338.

346. Gardeners' Chronicle, M. T. Masters, ed. (1881). **Orchid notes and gleanings.** *Angraecum humbloti*. *Gardeners' Chronicle* n. ser. 15 p. 202.

347. Gardeners' Chronicle. W. B. H. (1882). **List of garden orchids.** *Gardeners' Chronicle* n. ser.18: 565 – 566, fig.101.

348. Gardeners' Chronicle. M. T. Masters, ed. (1883). *Angraecum sesquipedale*. *Gardeners' Chronicle* n. ser.19: 377.

349. Gardeners' Chronicle, M. T. Masters, ed. (1884). **Orchids at Laurie Park.** *Gardeners' Chronicle* n. ser. 20: 520, fig.104.

350. Gardeners' Chronicle, W.F.M. (1885). *Angraecum eburneum* **at Birdhill.** *Gardeners' Chronicle* n. ser. 23: 147.

351. Gardeners' Chronicle, M. T. Masters, ed. (1885). *Angraecum fastuosum*. *Gardeners' Chronicle* n. ser. 23: 533.

352. Gardeners' Chronicle, M. T. Masters, ed. (1885). *Angraecum fastuosum*. *Gardeners' Chronicle* n. ser. 23: 533, fig. 96.

353. Gardeners' Chronicle, M. T. Masters, ed. (1885). *Aeranthus leonis*. *Gardeners' Chronicle* n. ser. 23: 80 – 81.

354. Gardeners' Chronicle, M. T. Masters, ed. (1886). **Two new orchids.** *Gardeners' Chronicle* n. ser. 26: 173.

355. Gardeners' Chronicle, ed. (1890). *Angraecum sesquipedale*. *Gardeners' Chronicle* ser. 3, 7: 11.

356. Gardeners' Chronicle. F. R. (1890). *Angraecum sesquipedale*. *Gardeners' Chronicle* ser. 3, 7: 547.

357. Gardeners' Chronicle. (by Rolfe *fide* Hooker,1892). (1892). **Orchid notes and gleanings.** *Gardeners' Chronicle* ser. 3, 11: 619, fig. 88.

358. Gardeners' Chronicle, . M. T. Masters, ed. (1892). **Awards.** *Gardeners' Chronicle*, (3rd series). 12: 535.
Report from the RHS Scientific Committee on a monstrous form of *Angraecum sesquipedale*, with one of the lateral petals spurred, making a supernumerary lip.

359. Gardeners' Chronicle, M. T. Masters, ed. (1892). *Angraecum sesquipedale*. *Gardeners' Chronicle* ser. 3, 12: 123.

360. Gardeners' Chronicle. M. T. Masters, ed. (1893). *Cynorchis*. *Gardeners' Chronicle* ser. 3, 13: 197.

361. Gardeners' Chronicle, M. T. Masters, ed. (1896). **Obituary.** *Gardeners' Chronicle*, (3rd series). 20: 345.
A short obituary of the collector Richard Curnow who worked for the firm of H. Low & Co.

362. Gardeners' Chronicle, M. T. Masters, ed. (1897). **New or noteworthy plants.** *Gardeners' Chronicle* ser. 3, 21: 245.

363. Gardeners' Chronicle, M. T. Masters, ed. (1898). *Eulophiella peetersiana*. *Gardeners' Chronicle* ser. 3, 23: 201 – 202, fig. 200.

364. Gardeners' Chronicle, M. T. Masters, ed. (1901). *Phaius warpuri*. *Gardeners' Chronicle* ser. 3, 38: 82.

365. Gardeners' Chronicle, M. T. Masters, ed. (1904). *Bulbophyllum hamelinii*. *Gardeners' Chronicle* ser. 3, 36: 118 & 124, fig. 51.

366. Gardeners' Chronicle, John Farmer, ed. (1908). **Our supplementary illustration.** *Cynorchis lowiana*. *Gardeners' Chronicle* ser. 3, 43: 184 – 185.

367. Gardeners' Chronicle, Frederick Keeble, ed. (1913). **Awards.** *Angraecum recurvum*. *Gardeners' Chronicle* ser. 3, 54: 367, p. 374, fig. 132.

368. Gardeners' Magazine, ed. (1913). *Angraecum recurvum*. *The Gardeners' Magazine* 16: 899.

369. Gardeners' Magazine, ed. (1892). *Disa incarnata*. *The Gardeners' Magazine* 35: 157.

370. Gardeners' Magazine, signed C. C. (1895). *Aeranthus grandiflorus*. *The Gardeners' Magazine* 38: 764.

371. Gardeners' Magazine, signed C. (1898). **Notes on orchids.** *Eulophiella peetersiana*. *The Gardeners' Magazine* 41: 371.

372. Gardeners' Magazine. (1900). **Notes on orchids.** *The Gardeners' Magazine* 43: 491.

373. Gardeners' Magazine. (1901). **New plants, & c.** *The Gardeners' Magazine* 44: 104.

374. Gardening World, ed. (1892). **Orchid notes and gleanings.** *The Gardening World* 8: 232.

375. Gardening World, ed. (1893). **Orchid notes and gleanings.** *The Gardening World* 9: 360.

376. Gardening World, ed. (1893). **Orchid notes and gleanings.** *The Gardening World* 9: 488.

377. Gardening World, ed. (1893). **Orchid notes and gleanings.** *The Gardening World* 9: 663.

378. Gardening World, ed. (1904). *Phaius humblotii albus*. *The Gardening World*, 21 (new ser.): 359 & 361.

379. Gardening World, ed. (1904). **Hybrid** *Phaius* **with parents.** *The Gardening World*, 21 (new series): 670 – 671, pl.4927.

380. Garnier, Max. (1892). **Le Grand Exposition de L'Horticulture Internationale.** *Le Journal des Orchidées*, 3: 95, fig. 11.

381. Garnier, Max. (1897). **Nouveautés.** *La Semaine Horticole*, 1, 15: 158 – 159.

382. Garreau De Loubresse, Xavier. (1992). **Les** *Oeceoclades* **(Madagascar, Comores, La Réunion).** *L'Orchidophile*, 23, 101: 69-p. 72.

383. Garreau De Loubresse, Xavier. (1994). *Oeceoclades*. *L'Orchidée*, 12, 104: 105 – 107.

384. Geerinck, Daniel. (1984, 1992): *Orchidaceae* part 1. 296 pp.; part 2: 483 pp. **Flore d'Afrique Central (Zaire, Rwanda, Burundi). Spermatophytes.** Jardin Botanique National de Belgique, Brussels, Belgium. Additional authors *Bulbophyllum*: J. J. Vermeulen, *Nervilia*: B. Pettersson.

385. Gellert, Magdalene. (1923). **Anatomische Studien über den Bau der Orchideenblüte.** *Repertorium Specierum novarum regni vegetabilis* 25: 1 – 66.

386. Gerlach, Günter. (1995). *Oeceoclades antsingyensis*, **eine neue Orchidee aus dem Westen Madagaskars.** *Journal für den Orchideenfreund*, 2, 2: 62 – 65.

387. Giraud, Michel. (2003). **Fiche de culture,** *Angraecum leonis* *L'Orchidophile*, 34, 157: 175 – 176. Distribution, habitat and cultural conditions are given for *Angraecum leonis*.

388. Giraud, Michel. (2003). **Fiche de culture.** *Aeranthes grandiflora*. *L'Orchidophile*, 34, 158: 297 – 298. A few basic facts are given about the species mainly concentrated on cultivation.

389. Glaubitz, W. (1955). *Angraecum sesquipedale* Thou. *Die Orchidee*, 6, 3: 74 – 77.

390. Glaw, Frank & **Vences**, Miguel.(1994). **A fieldguide to the amphibians and reptiles of Madagascar.** M. Vences & F. Glaw Verlag GbR, Köln: 480 pp.

391. Godefroy-Lebeuf, Alexandre, ed. (1881). **Plantes directement de Madagascar par M. Humblot arrivées en parfait état.** *L'Orchidophile*, 1, 6: facing p. 120.
An advertisement for recently imported plants by Léon Humblot and their numbers. It is also stated that Mr. Humblot on his return from Madagascar has brought back plants that people had ordered beforehand.

392. Godefroy-Lebeuf, Alexandre, ed. (1881). **Nouveautés. Les orchidées de Madagascar.** *L'Orchidophile*, 1, 7: 145 – 147.
A large consignment of orchids from Madagascar is reported to have arrived in Marseille, they were probably sent by M. Humblot. It is reported that the last importation of plants from Humblot sold at Stevens for 50.000 fr.. It is hoped that amongst the plants will be *Angraecum sanderianum*, which, according to Humblot, is like *A. sesquipedale* but has pink flowers. There also is likely to be *Grammatophyllum roemplerianum.*

393. Godefroy-Lebeuf, Alexandre, ed. (1885). *Angraecum superbum.* *L'Orchidophile*, 5, 49: 168.

394. Godefroy-Lebeuf, Alexandre, ed. (1885). **Nouveautés.** *L'Orchidophile*, 5, 50: 196 – 197.

395. Godefroy-Lebeuf, Alexandre. (1886). *Angraecum elatum.* *L'Orchidophile*, 6: 80 – 81, & plate.

396. Godefroy-Lebeuf, Alexandre. (1886). *Phaius tuberculosus.* *L'Orchidophile*, 6, 64: 276.

397. Godefroy-Lebeuf, Alexandre. (1886). *Grammatophyllum ellisi.* *L'Orchidophile*, 6, 66: 352.

398. Godefroy-Lebeuf, Alexandre. (1887). *Angraecum du Buyssonii.* *L'Orchidophile*, 7, 7: 280 – 281.

399. Godefroy-Lebeuf, Alexandre. (1888). **Nouveautés.** *L'Orchidophile*, 8, 86: 200.

400. Godefroy-Lebeuf, Alexandre. (1891). *Angraecum Buyssoni.* *L'Orchidophile*, 11, 124: 282 – 286.

401. Godefroy-Lebeuf, Alexandre, ed. (1892). *Angraecum superbum.* *L'Orchidophile*, 12, 138: 367 – 368.

Goldblatt, see Thouars 1398.

402. Gomes, Carol K. (1992). **Maui Orchid Society's second FCC award.** *Hawaii Orchid Journal*, 21, 2: 2 – 3.

403. Goss, Gary J. (1977). **The reproductive biology of the epiphyte orchids of Florida. 6.** *Polystachya flavescens* **(Lindley) J. J. Smith.** *American Orchid Society Bulletin*, 46, 11: 990 – 994.

404. Gower, W. H. (1889). *Eulophia megistophylla.* *The Garden*, 35 part 1: 62.

405. Graf, Alfred Byrd. (1978). *Tropica. Color encyclopedia of exotic plants.* Roehrs Co., East Rutherford, N. J.: p. 1120.

406. Greenaway, Paul. (1997). *Madagascar & Comoros.* 3rd edition. Lonely Planet Publications. Australia: 441 pp.
A travellers' guide to Madagascar. Orchids are mentioned on page 30 – 31.

407. Gregg, Alan (1999). **Orchids throughout the seasons at Swansea Botanical Gardens.** *Orchid Review*, 107, 1227: 155 – 158.
Angraecum scottianum is mentioned and pictured.

408. Greuter, W. *et al.*, eds. (2000) *International Code of Botanical Nomenclature* **(St Louis Code).** Koeltz Scientific Books. Königstein: 474 pp.

409. Griessen, A. E. P. (1899). **Unrecorded discovery of** *Eulophiella peetersiana.* *Orchid Review*, 7, 81: 258.

410. Griggs, Pat. (1997). **More orchids for Madagascar.** *Kew*, pp. 4 – 5.

411. Grignan, G. T. (1900). **Orchidées, Les Eulophiella.** *Le Jardin*, 14: 23.

412. Grignan, G. T. (1901). **Plantes nouvelles ou peu connues.** *Le Jardin*, 15: 76 – 277.

413. Grignan, G. T. (1901). *Phaius fragrans.* *Le Jardin*, 15: 359.

414. Grimm, Markus & **Ruckstuhl**, Anita. (1998). **Die Orchideenflora der Seychellen.** *Die Orchidee*, 49, 3: 105 – 112.
A report on the orchid habitats on the Seychelles together with a broad outline of the islands' climate and a listing of endemic orchids and plants in cultivation.

415. Grindon, Leo.(1872) **The Fairfield orchids.** Bradbury, Evans & Co. London.

416. Grubb, Roy & Ann. (1966). *Selected orchidaceous plants.* Roy & Ann Grubb, Caterham, Surrey: Four parts. 147 pp.

417. Guého, Joseph. (1988). *La Végétation de l'Île Maurice.* Editions de l'Ocean Indien: 57 pp.
A general introduction to the floral of Mauritius. The different habitats and vegetation zones are described.

418. Guillaumin André. (1920). **Les espèces cultivèes du genre** *Listrostachys.* *Bulletin du Muséum National d'Histoire Naturelle*, 20: 574 – 577.

419. Guillaumin, André. (1928) **Plantes nouvelles ou critiques des serres du Museum.** *Bulletin du Muséum National d'Histoire Naturelle*, 34: 359 – 362.

420. Guillaumin, André & **Manguin**, Émile. (1934). **Floraison observèes dans les serres du Museum en 1934.** *Bulletin du Muséum National d'Histoire Naturelle*, ser. 2, 6: 520 – 524.

421. Guillaumin, André & **Manguin**, Émile. (1940). **Floraison observées dans les serres du Museum en 1939 & 1940.** *Bulletin du Muséum National d'Histoire Naturelle*, ser. 2, 12: 456 – 460.

422. Guillaumin, André. (1963). **Une orchidée nouvelle de Madagascar.** *Bulletin du Muséum National d'Histoire Naturelle*, ser. 2, 5 tome 35: 521.

423. Guillaumet, Jean-Louis.(1984). **The vegetation: an extraordinary diversity.** In Alison **Jolly**, Philippe **Oberlé**, Roland **Albignac**, eds, *Key environments.* **Madagascar**. Pergamon Press, Oxford etc.: 27 – 54, 2 maps, 15 photos.

424. Gunn, Spence (2002). **Trouble on the Ark.** *Kew.* 22 – 25.
Conservation problems with the orchids of Madagascar are discussed, a few examples are quoted and a fund-raising scheme is introduced. *Angraecum longicalcar* is mentioned.

425. Hackett, W. (1904). *Cymbidium rhodochilum.* *Flora & Sylva*, 2: 40.

426. Hagaster, Eric & **Dumont**, Vinciane, eds. (1996). *Status survey and conservation action plan.* **Orchids.** IUCN, Gland & Cambridge: 153 pp. **Bosser** J. M., **Du Puy** David & **Phillipson** Pete. pp. 103 – 107.

427. Hajasoa, Randriamahazo. (2001). **The malagasy orchid farm: des orchidées en héritage.** *Revue de l'Ocean Indien:* 43 – 44.
A history of the Malagasy Orchid Farm, started by Remi Andriamaharo, includes a photograph of the family and Fred Hillerman.

428. Hall, A. V. (1965). **Studies of the South African species of** *Eulophia.* *Journal of South African Botany, Supplementary volume*, 5: 248.

429. Hamelin, L. & **Linden**, Lucien. (1893). **M. Hamelin's experiences in Madagascar. Discovery of the famed** *Eulophiella. The Gardening World*, 9: 708 – 710.

430. Hamelin, L. (1893). *Eulophiella elisabethae. The Gardening World*, 9, p. 742.

431. Hamilton, Robert M. (1990). **Flowering months of orchid species under cultivaton.** in **Arditti** Joseph, ed., Orchid Biology, Reviews and Perspectives, V: 265 – 405. Timber Press. Portland Oregon.

432. Hammond, William. (1842). *A catalogue of orchidaceous plants.* Simms and Dinham, Manchester: 46 pp.

433. Hariot, P. (1902). **Plantes nouvelles ou peu connues**, *Cymbidium rhodochilum. Le Jardin*, 16: 351.

434. Hattori, Yoshitoshi & **Goda**, Hiroyuki. (1990). *Yoran; hinshu to saibai kanri. Orchids, variety and culture.* Hoikusha, Osaka: 232 pp.

435. Hawkes, Alex D. (1950). **The major genera of cultivated orchids — I.** *American Orchid Society Bulletin*, 19, 3: 142 – 144.

436. Hawkes, Alex D. (1952). **The genus** *Grammangis. Orchid Journal*, 1, 3: 136 – 138.

437. Hawkes, Alex D. (1952). *Gastrorchis,* **a fascinating** *Phaius* **ally.** *Orchid Journal*, 1, 5 – 6: 245.

438. Hawkes, Alex D. (1952). **Eulophiellas.** *Orchid Journal*, 1, 12: 504 – 505.

439. Hawkes, Alex D. (1952). **The Madagascar species of** *Cymbidium. Orchid Journal*, 1, 12: 508 – 509.

440. Hawkes, Alex D. (1953). **The genera and species of cultivated orchids. Part II.** *Orchid Journal* 2, 4: 176 – 178.

441. Hawkes, Alex D. (1961). **Some Little-known orchids. The genus** *Eulophiella. American Orchid Society Bulletin*, 10, 30: 809 – 811.

442. Hawkes, Alex D. (1963). *Angraecum Veitchii. Orchid Review*, 71, 843: 311 – 312.

443. Hawkes, Alex D. (1963). **The genus** *Grammangis. Orchid Digest*, 27, 9: 401 – 403.

444. Hawkes, Alex D. (1965). *Encyclopedia of cultivated orchids.* Faber & Faber, London: 602 pp., 140 b/w., 24 colour pl.

445. Hawkes, Alex D. (1966). **Notes on** *Bulbophyllum* **1.** *Phytologia*, 13, 5: 308 – 309.

446. Hemsley, W. B. (1882). **How to distinguish orchids out of flower.** *The Gardeners' Chronicle*, (new series). 17: 341 – 342, fig.52.

447. Hemsley, William Botting. (1904). *Cymbidium rhodochilum. Curtis's Botanical Magazine*, 130 t.7932 – 7933.

448. Hennessy, Esmée Franklin. (1985). *Corymborkis corymbis. Flowering Plants of Africa*, 48 pl. 1908.

449. Hennesy, Esmé & **Zank**, Hugh. (1985) *Jumellea arachnantha* (Reichb.f.) Schltr. *South African Orchid Journal*, 16, 2: 54 – 55 & Front cover.

450. Heriz-Smith, Shirley. (2002). *The House of Veitch.* Shirley Heriz-Smith, Diss, 182 pp.
A history of the House of Veitch, including that of their collectors. *Angraecum citratum* (p. 96) is mentioned. A short biography of Charles Curtis's (p. 111) is given including his journey to Madagascar and Mauritius.

451. Hermans, Johan & Clare. (1994). **Meandering through Madagascar, introduction &** *Eulophiella roempleriana. Orchid Review,* 102, 1195: 11 – 17.

452. Hermans, Johan & Clare. (1994). **Cohabitation in Madagascar.** *Orchid Review,* 102, 196: 64 – 69.

453. Hermans, Johan & Clare. (1994). **Above the fires. The orchids of the Ibity Massif, Madagascar.** *Orchid Review,* 102, 1197: 125 – 128.

454. Hermans, Johan & Clare. (1994). *Orchids in Madagascar.* in Guide to Madagascar 4th. Edition, in Bradt Hilary. Bradt Publications. Bucks. U.K: p. 296.

455. Hermans, Johan & Clare. (1994). **Scrambling in Madagascar. The genus** *Oeonia. Orchid Review,* 102, 1199: 268 – 271.

456. Hermans, Johan & Clare. (1994). **A tale of two tombs.** *Orchid Review,* 102, 1200: 295 – 298.

457. Hermans, Johan & Clare. (1995). **Orchids in the spiny forest.** *Orchid Review,* 103, 1201: 44 – 49.

458. Hermans, Clare & Johan. (1995). **The perfumed isle.** *Vanilla* **in Madagascar.** *Orchid Review,* 103, 1204: 174 – 180.

459. Hermans, Johan & Clare. (1995). **Northern exposure in Madagascar. Orchids in the extreme.** *Orchid Review,* 103, 1206: 282 – 287.

460. Hermans, Johan. (1996). **New hybrid** *Aërangis. Orchid Review,* 104, 1207, 3 & 63.

461. Hermans, Johan. (1996). *Microcoelia gilpinae* (Rchb.f.) & S. Moore & Summerh, '**Clare**' BC/RHS. *Orchid Review,* 104, 1208: 77.

462. Hermans, Johan & Clare. (1996). **North-eastern Madagascar. Part 1. Sambava. Orchids in paradise.** *Orchid Review,* 104, 1208: 78 – 84.

463. Hermans, Johan & Clare. (1996). **North-eastern Madagascar. Part 2. Andapa & Marojejy. Orchids in hell.** *Orchid Review,* 104, 1209: 141 – 146.

464. Hermans, Johan & Clare. (1996). **Extremes of adaptation.** *Kew.* 19 – 21.

465. Hermans, Clare & Johan. 1211 (1996). **The orchid men of Tana. A journey ends.** *Orchid Review,* 104: 293 – 298.

466. Hermans, Johan & **Cribb**, Phillip. (1997). **A new species of** *Angraecum* (*Orchidaceae*) **from Madagascar.** *Orchid Review,* 105, 1214: 108 – 111.

467. Hermans, Clare & Johan. (1997). *Eulophiella elisabethae,* **a pretty tall story.** *Orchid Digest,* 61, 2: 64 – 71.

468. Hermans, Johan & Clare. (1997). *Oeonia.* **A unique genus from Madagascar.** *Orchids (AOS).* 66: 1050 – 1054.

469. Hermans, Johan & Clare. (1998). **The glorious novelties of the Reverend William Ellis.** *Orchid Digest,* 62, 3: 112 – 115.

470. Hermans, Johan, **Eggli**, Urs, **Supthut**, Diedrich J, **Theisen**, Inge. (1998). **Mehr als nur Vanille: Sukkulente Orchideen in Madagascar.** *Die Sukkulentenwelt.* 2: 1 – 34.
A general overview of the succulent orchids of Madagascar, including a special feature on culinary vanilla. Succulence and the general features of orchids are explained, followed by a description of the different habitats and the various genera. Conservation problems and cultivation are also described.

471. Hermans, Johan. (1999). *Phaius pulchellus* **var.** *sandrangatensis*. *Curtis's Botanical Magazine*, 16, no. 4 (Nov. 1999). t.375, p. 362 – 366.

The taxonomy, history and cultivation of the species are discussed. Illustrations and a full description of the variety are provided and the variability of the species is considered. Its position in the *Phaius-Gastrorchis* complex is outlined.

472. Hermans, Clare & Johan. (2000). **Royal Horticultural Society awards.** *Orchid Review*, 108, 1233: 139.

A short description of *Disa buchenaviana* 'Madagascan Sky' on the occasion of it gaining a Botanical Certificate.

473. Hermans, Johan. (2000). **Aubert Aubert du Petit Thouars.** *Orchid Review*, 108, 1233: 146 – 155.

An in depth biography of Thouars, a few orchids connected with him are described.

474. Hermans, Clare & Johan. (2000). **Royal Horticultural Society awards.** *Orchid Review*, 108, 1235: 264 – 271.

Benthamia nivea, Angraecum clavigerum, Eulophia pulchra var. *divergens* are described and illustrated.

475. Hermans, Clare & Johan. (2000). **Royal Horticultural Society awards.** *Orchid Review*, 108, 1236: 329 – 337.

Eulophia pulchra var. *divergens* is correctly identified as *Eulophia megistophylla* and *Cryptopus elatus* are described and illustrated.

476. Hermans, Johan & **La Croix** I. (2001) *Angraecum clareae* **Hermans, la Croix & P.J. Cribb sp. nov.** *Orchid Review*, 109, 1237: 43 – 46.

The description of *Angraecum clareae*.

477. Hermans, Johan. (2001). *Oeceoclades decaryana*. *Curtis's Botanical Magazine*, 18, pl. 405: 2 – 5.

A description and discussion of the species to accompany a plate by Christobel King. The same plate also appeared on the cover of the 2000 Kew-Curtis's Botanical Calender.

478. Hermans, Johan. (2001). **De Succulente Orchideeën van Madagascar.** *Venusschoentje*, 22, 1: 10 – 15.

A general description of the various orchid habitats of Madagascar and their inhabitants. The eastern forests and highlands are outlined.

479. Hermans, Johan. (2001). **De Succulente Orchideeën van Madagascar. Deel 2. Het hooggebergte.** *Venusschoentje*, 22, 2: 33 – 37.

The high mountain habitats, western area, Isalo and Zombitse are discussed, typical orchids are mentioned.

480. Hermans, Johan. (2001). **De Succulente Orchideeën van Madagascar. Deel 3.** *Venusschoentje*, 22, 3: 58 – 61.

The orchids of the southern region are discussed.

481. Hermans, Clare & Johan. (2001). **RHS awards. January-February 2001** *Orchid Review*, 109, 1239: 137 – 139.

Gastrorchis pulchra var. *perrieri* and *G. steinhardtiana*.are briefly mentioned.

482. Hermans, Johan. (2001). **De Succulente Orchideeën van Madagascar. Deel 4.** *Venusschoentje*, 22, 4: 84 – 87.

The orchids of northern Madagascar are discussed and illustrated: *Acampe pachyglossa, Angraecum eburneum, A. sororium, Cynorkis uncinata, Oeonia rosea, Grammangis spectabilis, Cymbidiella falcigera, Gastrorchis humblotii, Graphorkis concolor* var. *alphabetica*.

483. Hermans J., **Cribb**, P. J. & **Bosser** J. (2002). **A distinctive new species of *Angraecum* from Madagascar.** *Orchid Review*, 110, 1242: 22 – 23.

The description of *Angraecum platycornu*, with a line drawing.

484. Hermans, Johan. (2002). **The orchids of Madagascar, 350 years of exploration.** In Jean **Clark** *et al.*, eds, Proceedings of the 16th World Orchid Conference, Vancouver: 326 – 331. Vancouver Orchid Society, Vancouver.

A short introduction to the various habitats is followed by a history of orchid exploration on the island. Etienne Flacourt, Thouars, Lyall, Baron, Deans Cowan, Humblot, Hamelin, Warpur, Perrier de la Bâthie, Schlechter, Jumelle, Decary, Bosser and Senghas are mentioned in some detail.

485. Hermans, Johan. (2002). **Photography.** In Jean Clark *et al.*, eds, *Proceedings of the 16th World Orchid Conference*, Vancouver: 332 – 333. Vancouver Orchid Society, Vancouver.
An essay on how the author takes photographs of orchids in various places, *Microterangis hariotiana* is shown on its side on page 486.

486. Hermans, Johan. (2002). *Aerangis ellisii* **var.** *grandiflora. Curtis's Botanical Magazine*, 19, 3, t.449: 178 – 183.
Two varieties of *Aerangis ellisii*, var. *ellisii* and var. *grandiflora* are compared and the latter is illustrated. Their history, taxonomy, habitat and cultivation are discussed and a description of var. *grandiflora* is provided.

487. Hermans, Clare & Johan. (2003). **Royal Horticultural Society awards.** *Orchid Review*, 111, 1250: 75.
An Award of Merit to *Cynorkis gibbosa* is reported.

488. Hermans, Clare & Johan. (2003). **Royal Horticultural Society awards.** *Orchid Review*, 111, 1250: 78 – 81.
An Award of Merit to *Cynorkis lowiana* is reported, also see 1275.

489. Hermans, Clare & Johan. (2003). **How to grow them.** *Orchid Review*, 111, 1250: 82 – 85.
Cultivation and habitat of *Cynorkis gibbosa* are discussed.

490. Hermans, Clare & Johan. (2003). **Royal Horticultural Society awards, 13 March at the European Orchid Conference.** *Orchid Review*, 111, 1252: 201 – 201.
An Award of Merit to *Gastrorchis* Elizabeth Castle (*G. pulchra* × *G. tuberculosa*) is reported, illustrated on the front cover, but showing a wilted, closing flower.

491. Hermans, Johan. (2003). **A striking new species of** *Polystachya* **(***Orchidaceae***) from Madagascar.** *Orchid Review*, 111, 1254: 354 – 357.
The description of *Polystachya clareae*, with a note on cultivation.

492. Hermans, Clare & Johan. (2003). **Royal Horticultural Society awards.** *Orchid Review*, 111, 1254: 328 – 332.
An Award of Merit to *Gastrorchis francoisii* is reported. A Botanical certificate is reported for *Polystachya clareae*.

493. Hermans, Johan & **Bosser** Jean. (2003). **A new species of** *Aeranthes* **(***Orchidaceae***) from the Comoro Islands.** *Adansonia*, sér. 3, 25, 2: 215 – 217.
Aeranthes campbelliae is described from the Comoro Islands.

494. Hermans, Johan & **Cribb** Phillip, eds. (2003). *Proceedings of the European Orchid Conference & Show.* Naturalia Publications, Turriers. 349 pp.
Eulophiella roempleriana is shown on the front cover and page 49. *Angraecum sesquipedale* is shown on p. 51.

495. Hermans, Clare & Johan (2004). **RHS Awards.** *Orchid Review* 112, 1259: 265 – 269
Angraecum germinyanum from the Comoro Islands is illustrated on it receiving a cultural commendation.

495a. Hermans, Johan. (2004). *Angraecum praestans. Curtis's Botanical Magazine*, 21, 4, t.510: 248 – 250.
The history, taxonomy, habitat and cultivation of *Angraecum praestans* from Madagascar are discussed. Accompanying an illustration by Christobel King.

495b. Hermans, Johan. (2004). *Paralophia epiphytica. Curtis's Botanical Magazine*, 22, 1, t.519: 47 – 52.
The the history, taxonomy, habitat and cultivation of the recently described *Eulophia epiphytica* from Madagascar are discussed. A description is given. It is placed in the new genus *Paralophia*.

496. Hermans, Clare & Johan (2005). **RHS Orchid Awards.** *The Orchid Review*, 113, 2264: 238.
Angraecum compactum is illustrated on receiving a Botanical certificate.

497. Hermans, Johan & **Cribb**, Phillip. (2005). **The *Angraecum germinyanum* complex.** *Orchid Review*, 113, 111: 90 – 97.
There is considerable confusion over the identity of *Angraecum germinyanum sens. lat.* which occurs in Madagascar and surrounding islands. A maze of nomenclature has developed over time; this paper aims to rationalise these names, based on the examination of herbarium specimens and living plants. From the very widespread and variable complex three distinct species are recognised: *Angraecum germinyanum* which may be endemic to the Comoro Islands, the very variable *Angraecum conchoglossum* from Madagascar and la Réunion and the morphologically constant *Angraecum arachnites* from Madagascar. This artificial division recognises easily identifiable and constant characteristics of populations that have developed independently. *Angraecum conchiferum* Lindl. from East Africa may be interpreted as a fourth distinct population. The status of the genus *Bonniera*, endemic to the island of la Réunion, is also re-assessed.

497b. Hermans, Johan & **Cribb**, Phillip (2005). **Orchids of Madagascar, A Conservation Challenge.** *Orchids (AOS)*, 74, 11, 836 – 847.
A summary of the orchid conservation challenges in Madagascar, some RBG Kew projects are highlighted.

498. Herndon, Christopher N. (1996). ***Eulophiella*: jewels of Madagascar.** *Orchids* (AOS): 155.

499. Herndon, Christopher N. (1996). ***Cymbidiella*: a study in contrasts.** *Orchids* (AOS): 390 – 397.

500. Herndon, Christopher N. (1997). **Colorful angraecoids.** *Orchids* (AOS): 816 – 823.

501. Herndon, Christopher N. (1998). **Preserving Madagascar's orchid heritage.** *Orchids* (AOS). 67, 2: 151 – 161.

502. Herndon, Christopher N. (1999). ***Grammangis*, Madagascan orchid of distinction.** *Orchids (AOS)*. 68, 9: 926 – 931.
A general overview of *Grammangis* with special emphasis on *G. ellisii*. Cultural requirements and taxonomy are outlined.

503. Hervouet, Chantal & Jean-Michel (1995). **Les orchidées des Seychelles.** *L'Orchidophile*, 26, 118: 176 – 180.
Notes on the orchid flora of the Seychelles.

504. Hervouet, Chantal & Jean Michel. (2001) **Voyage de la SFO à Madagascar.** *L'Orchidophile*, 32, 148: 155 – 170.
An account of a trip by the French Orchid Society to Madagascar. The itinerary and the various orchids encountered are discussed. A number of species are illustrated.

505. Hervouet, Jean Michel. (2003). **Deux semaines troublées à Madagascar.** *L'Orchidophile*, 34, 158: 253 – 265.
A travelogue of a trip to Madagascar in February 2002 by a group of orchid enthusiasts . Political unrest caused a number of logistical problems but the party still managed to visit Perinet, Mantadia, the Pangalines and Isle St. Marie. A number of species are discussed and illustrated.

506. Hesse, Robert H. (1977). ***Aerangis curnowiana* Perrier. A splendid, miniature orchid.** *American Orchid Society Bulletin*, 46, 3: 211 – 213.

507. Highfield, Roger (2002). **Half the world's plant species 'facing extinction'.** *Daily Telegraph*, 1 November: 7.
Plant conservation issues are discussed, including projects at the Royal Botanic Gardens, Kew. *Aeranthes henrici* is illustrated.

508. Hillerman, Fred E. (1974). **The exotic Angraecums-I.** *American Orchid Society Bulletin*, 43,1: 39 – 42.

509. Hillerman, Fred E. (1974). **Adventures in growing African and Madagascan species.** *Orchid Digest*, 38, 2: 48 – 50. Reproduced from the *Newsletter of the San Francisco Orchid Society.*

510. Hillerman, Fred E. (1974). **Some thoughts on the culture of Madagascan angraecoids-II.** *American Orchid Society Bulletin*, 43, 10: 907 – 911.

511. Hillerman, Fred E. (1975). **Modest cousins of the exotic Angraecums.** *American Orchid Society Bulletin*, 44, 1: 32 – 34.

512. Hillerman, Fred E. (1975). **Angraecum sesquipedale: the comet orchid.** *Orchid Digest*, 39, 3: 104 – 105.

513. Hillerman, Fred E. (1976). **The Angraecoids-pearls from Madagascar.** *American Orchid Society Bulletin*, 45, 1: 47 – 52.

514. Hillerman, Fred E. (1976). **Hybridising with Madagascan angraecoids.** *Orchid Review*, 84, 991: 15 – 19.

515. Hillerman, Fred E. (1976). *Angraecum sororium*-**A new** *Angraecum* **star.** *American Orchid Society Bulletin*, 45, 2: 113 – 114.

516. Hillerman, Fred E. (1976). *Angraecum viguieri*-**A brown** *Angraecum?* *American Orchid Society Bulletin*, 45, 11: 1005 – 1006.

517. Hillerman, Fred. (1976). **The tiny** *Angraecum rutenbergianum.* *American Orchid Society Bulletin*, 45, 12: 1097 – 1098.

518. Hillerman, Fred E. (1978). *Aeranthes neoperrieri*: **leprechaun from Madagascar.** *American Orchid Society Bulletin*, 47, 4: 310 – 311.

519. Hillerman, Fred E. (1978). **At last: some small Angraecums.** *American Orchid Society Bulletin*, 47, 5: 404 – 407.

520. Hillerman, Fred E. (1979). **New and pleasant surprises from Madagascar.** *American Orchid Society Bulletin*, 48, 4: 335 – 340.

521. Hillerman, Fred E. (1980). *Angraecum longicalar*-**the spectacular** *Angraecum.* *American Orchid Society Bulletin*, 49, 3: 234 – 236.

522. Hillerman, Fred E. (1980). **Small Madagascan orchids for windows; under lights and other small places.** *American Orchid Society Bulletin*, 49, 4: 379 – 383.

523. Hillerman, Fred E. (1980). *Angraecum sesquipedale* and *Angraecum magadalenae*-**The King and Queen of "Angraecia".** *American Orchid Society Bulletin*, 49, 10: 1107 – 1109.

524. Hillerman, Fred E. (1981). *Angraecum sororium* & **Madagascar revisited-Part 1. Vanquished by Vazimba.** *American Orchid Society Bulletin*, 50, 4: 388 – 392.

525. Hillerman, Fred E. (1981). **Madagascar revisited Part – 2. The disappearing forest.** *American Orchid Society Bulletin*, 50, 6: 669 – 673.

526. Hillerman, Fred E. (1981). **Madagascar revisited Part – 3. A "superb(um)" collecting trip on Saint Marie Island.** *American Orchid Society Bulletin*, 50, 9: 1063 – 1069.

527. Hillerman, Fred E. (1982). **Angraecoid orchids for outdoor growing.** *American Orchid Society Bulletin*, 51, 3: 262 – 266.

528. Hillerman, Fred E. (1985). **The germinyanum gang from Madagascar.** *American Orchid Society Bulletin*, 54, 5: 574 – 579.

529. Hillerman, Fred E. & **Holst**, Arthur W. (1986). **An introduction to the cultivated angraecoid orchids of Madagascar.** Timber Press, Oregon: 302 pp., 95 line drawings, 36 colour pl, maps.

530. Hillerman, Fred E. (1990). **A cultural manual for *Aerangis* orchid growers.** Fred Hillerman, California: 32p. 29 pl.

531. Hillerman, Fred E. (1992). **A cultural manual for angraecoid orchid growers.** Fred Hillerman, Grass Valley, California: 47 pp.

532. Hillerman, Fred E. (1993). **Angraecoid orchids.** *American Orchid Society Bulletin*, 62, 9: 915 – 917.

533. Hillerman, Fred E. (1993). *Angraecum.* in Growing Orchids, AOS. West Palm Beach: 38 – 39.

534. Hillerman, Fred E. (1994). **Hybridizing angraecoids.** *American Orchid Society Bulletin*, 63, 8: 890 – 895.

535. Hillerman, Fred. (1994). **Madagascar. La foresta scomparsa.** *Orchis*: 88

536. Hillerman, Fred E. (No date) **A cultural manual for angraecoid orchid growers.** Fred Hillerman, California: 34 pp.

537. Hochreutiner, B. P. G. (1908). ***Sertum Madagascariense-étude de deux collections des plantes recoltées à Madagascar par M. M. J. Guillot & H. Rusillon.*** Geneva, Switzerland: 101 pp. numerous drawings. (Also *Ann. Cons. Jard. Bot. Geneve* 12e. année; 56).

538. Höck, F. (1904). **Allgemeine Pflanzengeographie.** *Just's Botanischer Jahresbericht*, 32, 2 (Leipzig 1906): 225 – 427.

539. Hogg, Robert, ed. (1894). **Certificates and awards.** *Journal of Horticulture and Cottage Gardener*, 28 (3rd series): 207.

540. Holttum, R. E. (1957). **Henry Nicholas Ridley.** *Taxon*, 6, 1: 1 – 6.

540a. Hommes & Plantes Editor. (2005). **Une dynastie d'Orchidéistes.** Hommes & Plantes, 52.
A history of the Vacherot and Lecoufle families, including photographs of Marcel Lecoufle and *Cryptopus elatus.*

541. Hooker, Joseph Dalton. (1867). ***Angraecum citratum.*** *Curtis's Botanical Magazine*: t.5624.

542. Hooker, Joseph Dalton. (1876). ***Eulophia macrostachya.*** *Curtis's Botanical Magazine*, (3rd series) 32: t.6246.

543. Hooker, Joseph Dalton. (1883). ***Angraecum scottianum.*** *Curtis's Botanical Magazine*,(3rd series) 39: t.6723.

544. Hooker, Joseph Dalton. (1883). ***Angraecum modestum.*** *Curtis's Botanical Magazine*: t.6693.

545. Hooker, Joseph Dalton. (1889). ***Angraecum germinyanum.*** *Curtis's Botanical Magazine*: 7061.

546. Hooker, Joseph Dalton. (1891). ***Angraecum fastuosum.*** *Curtis's Botanical Magazine*, 117: t.7204.

547. Hooker, Joseph Dalton. (1892). ***Cirrhopetalum thouarsii.*** *Curtis's Botanical Magazine*: t.7214.

548. Hooker, Joseph Dalton. (1892). ***Disa incarnata.*** *Curtis's Botanical Magazine*: t.7243.

549. Hooker, Joseph Dalton. (1893). ***Brownleea coerulea.*** *Curtis's Botanical Magazine*, ser. 3, 119: t.7309.

550. Hooker, Joseph Dalton. (1894). ***Eulophiella elisabethae.*** *Curtis's Botanical Magazine*, (3rd series). 120, 95: t.7387.

551. Hooker, Joseph Dalton. (1897). *Cynorchis purpurascens. Curtis's Botanical Magazine.* t.7551.

552. Hooker, Joseph Dalton. (1897). *Cynorchis grandiflora. Curtis's Botanical Magazine,* 123: t.7564.

553. Hooker, Joseph Dalton. (1898). *Eulophiella peetersiana*, Kraenzlin. *Curtis's Botanical Magazine.* t.7612, t.7613.

554. Hooker, Joseph Dalton. (1901). *Calanthe madagascariensis. Curtis's Botanical Magazine,* 57: t.7780.

555. Hooker, Joseph Dalton. (1902). *Cynorchis villosa. Curtis's Botanical Magazine.* t.7845.

556. Hooker, Joseph Dalton. (1902). *Cynorchis purpurascens. Curtis's Botanical Magazine,* 128: t.7852.

557. Hooker, William Jackson. (1825). *Polystachya luteola. Exotic Flora.* II William Blackwood, Edinburgh: t.103.

558. Hooker, William Jackson & **Arnott** Walker. (1832). **The botany of Captain Beechey's voyage.** H. G. Bohn. London: p. 71.

559. Hooker, William Jackson. (1834). **New or rare *Orchideae.*** *London Journal of Botany,* (*Hooker's Journal*): 44 – 49, t.115 – 117.

560. Hooker, William Jackson. & **Smith** John. (1849). *Pesomeria tetragona. Curtis's Botanical Magazine,* series t.4442.

561. Hooker, William Jackson, ed. (1854). *Angraecum eburneum. Curtis's Botanical Magazine.* t.4761.

562. Hooker, William Jackson, ed. (1859). *Angraecum sesquipedale. Curtis's Botanical Magazine,* t.5113.

563. Hooker, William Jackson, ed. (1860). *Angraecum eburneum* var. *virens. Curtis's Botanical Magazine,* 86: t.5170.

564. Hooker, William Jackson, ed. (1860). *Grammatophyllum ellisii. Curtis's Botanical Magazine.* t.5179.

565. Hooker, William Jackson. (1873). *Aeranthes arachnites. Curtis's Botanical Magazine* 99 t.6034.

566. Humbert, Henri. (1928). **Vegetation des hautes montagnes de Madagascar.** In *Contribution à l'Étude du Peuplement des Hautes Montagnes*, Soc. de Biogéographie. L. Paul Lechevalier, Paris II.

Humbert, Henri, ed. (1939, 1941). Flore de Madagascar: 49e. Famille-Orchidees, see Perrier de La Bâthie 896.

567. Humbert, H. (1942). **Notice sur Paul Danguy.** *Bulletin de la Société Botanique de France,* 89, 12: 249 – 252.
A short biography of the botanist Paul Danguy, including a portrait.

568. Humbert, Henri & **Leandri**, Jaques. (1954). **Cinquante ans de recherches botaniques à Madagascar.** *Bulletin de l'Académie Malgache,* Special Cinquantenaire: 33 – 42.

569. Humbert, Henri, ed. (1955). **Une merveille de la nature à Madagascar. Première exploration botanique du Massif du Marojejy et des satellites.** *Mémoires de L' Institut Scientific de Madagascar.* 6.

570. Humbert, H. & **Cours** Darne G., eds. (1965). **Notice de la Carte Madagascar. Description des types de vegetation.** *Extrait des travaux de la section scientifique et technique de l'Institut Français de Pondichéry,* 6: 45 – 84.

571. Humblot, Léon. (1881 – 1886). **Humblot letters to Sander.** Manuscript letters at the Royal Botanic Gardens, Kew (Sander Archive no.260 – 369).

572. Humblot, Léon. (1882 – 1885). **Humblot's notebooks.** Mss. MHN, Paris.
Listings of Humblot's botanical collections on the Comoros and in Madagascar. Notes re. Humblot's numbering systems accompany the lists.

573. Hunt, David, ed. (1981). *Orchids from Curtis's Botanical Magazine.* Curwen books, London: 121 pp., 32 pl.

574. Hunt, P. Francis. (1968). **African orchids XXXII.** *Kew Bulletin*, 22, 3: 490 – 492.

575. Hunt, P. Francis. (1970). **Notes on Asiatic orchids V.** *Kew Bulletin*, 24, 1: 75 – 99.

576. Hunt, P. Francis. (1971). *Angraecum magdalenae.* Curtis's Botanical Magazine, 178, 2: t.591.

577. Hunt, P. Francis. & **Grierson** Mary A. (1973). *Orchidaceae.* Bourton Press. 1973: 144p.

578. Hurst, C. C. (1899). **Curious crosses.** *Orchid Review*, 7, 73: 15.

579. Hutton, G. T., ed. (1968). *Angraecum viguieri.* The Journal of the Orchid Society of Great Britain, 17, 3: 72.

580. Ilgenfritz, Margaret M. (1976). **Remarkable orchid species for the amateur.** In Karlheinz Senghas, ed., *Proceedings of the 8th World Orchid Conference.* D.O.G., Frankfurt: 545 – 549.

581. Ilgenfritz, Margaret M. (1982). **Orchid species for the amateur.** In Joyce Stewart & C. N. Van der Merwe, eds, *Proceedings of the 10th World Orchid Conference.* South African Orchid Council: 115 – 120.

582. Immelman, Kathleen Leonore & **Hardy** David Spencer. (1986). *Cirrhopetalum umbellatum.* Flowering Plants of Africa, 49: pl.1940.

583. Index Kewensis. (1900). *Cirrhopetalum kamesinum. Index Kewensis Plantarum Phanerogmarum* Supplement 2: 202.

584. Ishida, Genjiro. (2000). **Pollination mechanism of *Cynorkis fastigiata. Orchidaceae.*** *Bulletin of the Hiroshima Botanical Garden*, 19: 7 – 10.
The pollination mechanism of *Cynorkis fastigiata* cultivated at the Hiroshima Botanical Garden is outlined.

585. Ives, P., ed. (1989). **Catalogue of living plant collections Part 2.** Royal Botanic Gardens, Kew: 163p.

586. Jacquin, Nicolaus Joseph. (1760). **Enumeratio systematica plantarum.** Haak, Leiden: 30, 13.

587. Jansen, Jo. (1996). **Het Geslacht *Cymbidiella.*** *Orchideeën* 4: 64; reproduced from *Orchika* Nov. 1995.

588. Japan Orchid Society. (1972). *The orchids, culture and breeding.* Seibundo-Shinkosha, Tokyo: 798 pp.

589. Jekyll, Miss & **Cook**, E. T., eds. (1900). **New and rare plants.** *The Garden*, 58, 1497: 60.

590. Jekyll, Miss & **Cook**, E. T., eds. (1900). **An artist's note-book.** *Angraecum humblotii. The Garden*, 58, 1516: 433.

591. Jennings, Samuel. (1875). *Orchids, and how to grow them in India and other tropical climates.* Reeve & Co., London. Plate III.

592. Johnson, Liz. (2003). *Eulophiella roempleriana.* In Johan **Hermans** & Phillip **Cribb**, eds., *Proceedings of the European Orchid Conference & Show.* Naturalia Publications, Turriers: 30 – 31.
A short note on the Madagascan species, concentrating on the cultivar Bromsberrow Place that was chosen as the most interesting orchid at the show.

593. Johnston, Frank D. (1971). **Three interesting orchids flowering in the Lyon Orchid Garden, Foster Gardens, Honolulu.** *Orchid Review*, 79, 932: 56 – 58.

594. Jones, H. G. (1960). **The culture of orchid species in the West Indies (5).** *Orchid Review*, 68, 801: 95 – 96.

595. Jones, Keith. (1967). **The chromosones of orchids: II** *Vandae* Lindl. *Kew Bulletin*, 21, 1: 151 – 156.

596. Jonsson, Lars. (1981). **A monograph of the genus** *Microcoelia*. *Acta Universitatis Upsaliensis, Symbolae Botanicae Upsalensis*, 23, 4.

597. Jonsson, L. (1992). ***Orchidaceae, Microcoelia.*** *Distributiones Plantarum Africanum*, 39: 1285 – 1311.
A series of maps show the distribution of African plants. This issue contains the following from Madagascar: *Microcoelia exilis, M. physophora, M. perrieri, M. decaryana, M. bispiculata, M. macrantha, M. gilpinae, M. dolichorrhiza, M. elliotii.*

598. Journal of Horticulture, ed. (1896). **Orchids.** *Angraecum leonis.* *Journal of Horticulture and Cottage Gardener*, 32 (3rd series): 439, fig.71.

599. Journal of Horticulture (by H. R. R.). (1899). **Orchids.** *Journal of Horticulture and Cottage Gardener* (3rd series). 38, 1: 48 – 49.

600. Journal of Horticulture, ed. (1899). ***Angraecum virens.*** *Journal of Horticulture and Cottage Gardener*, 39: 515.

601. Journal of Horticulture, (by H. R. R.). (1900). ***Angraecum citratum.*** *Journal of Horticulture and Cottage Gardener*, 40: 281.

602. Journal of Horticulture, visitor. (1900). ***Angraecum (Aeranthus) leonis.*** *Journal of Horticulture and Cottage Gardener*, 40: 301 & 303.

603. Journal of Horticulture, ed. (1900). ***Angraecum sesquipedale.*** *Journal of Horticulture and Cottage Gardener*, 41: 393.

604. Journal of Horticulture, ed. (1901). **Orchids.** *Angraecum fastuosum.* *Journal of Horticulture and Cottage Gardener*, 42 (3rd series): 297.

605. Journal of Horticulture, ed. (1901). **Orchids.** *Angraecum sanderianum.* *Journal of Horticulture and Cottage Gardener*, 43 (3rd series): 459.

606. Journal of Horticulture. W. (W. Watson?). (1902). ***Cymbidium rhodocheilum.*** *Journal of Horticulture and Cottage Gardener*, (3rd series). 45: 49.

607. Journal of Horticulture (1906). **Orchids.** *Phaius tuberculosus* var. *superbus.* *Journal of Horticulture and Cottage Gardener*, ed, 52 (3rd series).1: 154 – 155.

608. Journal of Horticulture, (1912). ***Angraecum sesquipedale.*** *The Journal of Horticulture and Cottage Gardener*, 64 (3rd series.). 1: 393 – 394.

609. Jumelle, Henri & **Perrier de la Bâthie**, Henri. (1912). **Une vanille aphylle de Madagascar.** *Revue Géneral Botanique Paris*, 24, 282: 198 – 199.

610. Jumelle, Henri & **Perrier de la Bâthie**, Henri. (1912). **Les** *Nervillia* **et les** *Bulbophyllum* **du Nord-Ouest de Madagascar.** *Annales du Faculté des Sciences Marseille*, 21 2: 187 – 216.

611. Jumelle, Henri. (1916). **Catalogue descriptif des collections botaniques du Musée Colonial de Marseille.** *Annales du Musée Colonial de Marseille*, (3rd. ser.) 24, 1 (1916). 112 pp.
A catalogue of useful plants from the French colonies, including Madagascar and la Réunion. Vanilla and Fahan are mentioned.

612. Junginger, Bernd. (1997). **Der Splitter des Gondwanalandes.** *Journal für den Orchideenfreund* 4, 1: 33 – 36.

613. Kaiser, Roman. (1993). *The scent of orchids olfactory and chemical investigation.* Elseviers Science, Amsterdam, Netherlands: 259 pp., also *Vom Duft der Orchideen.* (1993). Editions Roche, Basel, Switzerland: 265 pp.

614. Karasawa, Kohji. (1992). **Orchid atlas. Vol 8** *Vanda*, *Phalaenopsis.* Orchid Atlas Publishing Society, Tokyo.

615. Katoh, Mitsuharu & **Yoshio**, Futakuchi. (1974). *Orchids in colour.* Heibonsha Ltd., Tokyo, 2 vols.

Katz, H. J. & **Simmons**, J. T., see Schlechter 1214.

617. Kendal Mercury The, editor. (1907). **The Late Rev. R. Baron.** *Kendal Mercury*, Oct. 18.

618. Kennedy, George. (1961). **Orchid collecting in Madagascar.** *Orchid Digest*, 25, 8: 413 – 416.

619. Kennedy, George C. (1962). **Orchid collecting in the Gran Comoro.** *American Orchid Society Bulletin*, 31, 5: 375 – 380.

620. Kennedy, George C. (1972). **Notes on the genera** *Cymbidiella* **and** *Eulophiella* **of Madagascar.** *Orchid Digest*, 36: 120 – 122.

621. Kennedy, George C. (1976). *Angraecum sororium.* *American Orchid Society Bulletin*, 45, 8: 708 – 709.

622. Kennedy, George C. (1977). **Peloric orchids.** *Orchid Digest* 41, 5: 169.

623. Kennedy, Ruth. Mrs. George C. (1971). **Orchid collecting in Madagascar.** *Orchid Digest*, 35: 161 – 165.

Kerckhove, see de Kerckhove.

624. Keruadren-Aymonin, Monique & **Aymonin**, Gérard. (1969). **Les explorations et les collections botaniques du Professeur Henri Humbert.** *Travaux du Laboratoire de 'La Jaysinia' à Samoëns*, 3: 11 – 33.

625. Kew Bulletin ed. (1898). **New garden plants of the year 1897.** *Angraecum mooreanum.* *Kew Bulletin of Miscellaneous Information.* Appendix II p. 39.

626. Kew Bulletin ed. (1903) **New garden plants of the year 1902.** *Listrostachys bracteosa.* *Kew Bulletin of Miscellaneous Information*, App. III: 93.

627. Kew Bulletin. (1908). **New garden plants of the year 1907.** *Kew Bulletin of Miscellaneous Information*, App. III.

628. Kew Bulletin. (1928). **Miscellaneous notes.** *Kew Bulletin (Bulletin of Miscellaneous Information* 9: 383.
A short obituary of Charles Curtis who collected in Mauritius and Madagascar for Messrs. James Veitch.

629. Kijima, Takashi. (1988). **Orchids. Wonders of nature.** Salamander Books. London. New York: 205 pp.

630. Killick, D. J. B. (1992). *Eulophiella elisabethae.* *Flowering Plants of Africa*, 52, 1: pl. 2041.

631. Kittredge, W. (1985). *Botanical Museum Leaflets of Harvard University.*30 (2): 100 (1985).

632. Klaassen, A (1983). **Over** *Angráecum.* *Orchideeën*, 45, 5: 161 – 163.

633. **Klinge**, J. (1899). **Diagnoses orchidearum novarum.** *Acta Horti Petropolitani*, 17: 140, t.2.

634. **Kluge**, Manfred, **Brulfert**, Jeanne & **Vinson**, Bettina. (1995). **Signification biogeographique des processus d'adaptation photosynthetique. II: L'exemple des orchidees malgaches.** in **Biogéographie de Madagascar — Biogeography of Madagascar** in **Lourenço**, Wilson R. Actes du Colloque International Biogéographie de Madagascar. Société de Biogéographie, Muséum, ORSTOM, Paris: 157 – 163.
Crassulacean Acid Metabolism, a water saving mode of photosynthesis was found to be common in about 50% of epiphytic orchids in Madagascar.

635. **Kluge**, Manfred, **Vinson**, Bettina. (1995). **Der Crassulaceen — Säurestoffwechsel bei Orchideen Madagaskars: Analyse einer ökologischen Anpassung der Photosynthese.** Rundgespräche der Kommission für ökologie, 10 — Tropenforschung, Verlag Dr. Dietrich Pfeil, München: 159 – 171.
The incidence and modes of CAM in epiphytic orchids of Madagascar are screened by a survey of *Angraecum*, *Bulbophyllum* and leafless orchids.

636. **Kluge**, Manfred, **Vinson**, Bettina & **Ziegler**, H. (1997). **Ecophysiological studies on orchids of Madagascar: incidence and plasticity of crassulacean acid metabolism in species of the genus Angraecum** Bory. *Plant Ecology*, 135: 43 – 57.
An investigation into the incidence of CAM in the genus *Angraecum* of Madagascar, specifically in *A. sesquipedale* and *A. sororium*.

637. **Knight**, Gary. (1994). *Aerangis punctata* (J.Stewart). *Orchids Australia*, 6, 6: 59.

638. **Koechlin**, Jean; **Guillaumet**, Jean-Louis & **Morat** Philippe. **Flore et végétation de Madagascar.** (1974). Flora & Vegetation Mundi V, J. Cramer, Germany: 687 pp., b/w photos. Also J. Cramer/A. R. G. Gantner,Vaduz, Liechtenstein (1997).

639. **Koopowitz**, Harold. (1966). **Pollination and floral morphology in four African orchids.** In Proceedings of the Fifth World Orchid Conference. Long Beach, California: 238 – 244.

640. **Kotschy**, Theodor. (1864). Reichenbach f. in. **Mittheilungen aus den Vilgegenden.** *Oesterreichische Botanische Zeischrift*, 14, 11: 338.

641. **Kraenzlin**, Fritz. (1882). **Orchidaceen (Reliquiae Rutenbergiana).** *Abhandlungen Herausgegeben vom Naturwissenschaftlichen Vereine zu Bremen* 7: 254 – 263.

642. **Kraenzlin**, Fritz. (1886). **Beiträge zur Flora von Kamerun.** *Orchidaceae.* *Botanische Jahrbücher für Systematik Pflanzengeschichte* 7: 333 – 334.

643. **Kraenzlin**, Fritz. (1891). *Acampe madagascariensis.* *Gardeners' Chronicle* ser. 3, 10: 608.

644. **Kraenzlin**, Fritz. (1892). **Beitrage zu einer Monographie der Gattung Habenaria** Willd. *Botanische Jahrbücher für Systematik Pflanzengeschichte* 16: 52 – 223.

645. **Kraenzlin** Fritz. (1893). *Orchidaceae africanae.* *Botanische Jahrbücher für Systematik Pflanzengeschichte* 17: 49 – 68.

646. **Kraenzlin**, Fritz. (1894). **New or noteworthy plants.** *Angraecum fournierianum* Krzl. *Gardeners' Chronicle* ser. 3, 15: 808.

647. **Kraenzlin**, Fritz. (1894). **New or noteworthy plants.** *Gardeners' Chronicle* ser. 3, 16: 592.

648. **Kraenzlin**, Fritz (1894). **Re Corymbis leptantha.** *Abhandlungen Herausgegeben vom Naturwissenschaftlichen Vereine zu Bremen* 45 & 53.

649. **Kraenzlin**, Fritz. (1896). **New or noteworthy plants.** *Bolbophyllum multiflorum.* *Gardeners' Chronicle* ser. 3, 19: 294.

650. **Kraenzlin**, Fritz. (1897). *Eulophiella peetersiana.* Gardeners' Chronicle ser. 3, 21: 182.

651. **Kraenzlin**, Fritz. (1897). *Bolbophyllum ptiloglossum.* Gardeners' Chronicle ser. 3, 21: 330.

652. **Kraenzlin**, Fritz. (1897). *Eulophia wendlandia* **Kraenzlin** (sect. *Pulchrae*). Gardeners' Chronicle ser. 3 22: 262.

653. **Kraenzlin**, Fritz. (1897). *Orchidacea africanae* **II.** Botanische Jahrbücher für Systematik Pflanzengeschichte 22: 17 – 31.

654. **Kraenzlin**, Fritz. (1897). *Mystacidium hariotianum* **Kraenzlin.** Journal de Botanique 11: 153 – 154.

655. **Kraenzlin**, Fritz. (1900). *Xenia Orchidacea, Beiträge zur Kenntniss der Orchideen,* **3.** Leipzig, 300 pp.

656. **Kraenzlin**, Fritz. (1899 – 1901). *Orchidaceae africanae.* Botanische Jahrbücher für Systematik Pflanzengeschichte 28: 2 – 179.

657. **Kraenzlin**, Fritz. (1901). *Orchidearum Genera et Species.* **1,** *Apostasieae, Cypripedieae* **&** *Ophrydea.* Mayer & Mueller, Berlin.

658. **Kraenzlin**, Fritz. (1902). **Beiträge zur Flora von Afrika. XXIV.** Botanische Jahrbücher für Systematik Pflanzengeschichte 33: 60-.

659. **Kraenzlin**, Fritz. (1904). *Orchidaceae africanae* **VII.** Botanische Jahrbücher für Systematik Pflanzengeschichte 33: 63 – 72.

660. **Kraenzlin**, Fritz. (1905). *Orchidaceae africanae* **IX.** Botanische Jahrbücher für Systematik Pflanzengeschichte 36: 114 – 119.

661. **Kraenzlin**, Fritz. (1908). *Reliquiae Pfitzerianae.* Orchis 2, 12: 134 – 136.

662. **Kraenzlin**, Fritz. (1909). *Orchidaceae africanae* **X.** Botanische Jahrbücher für Systematik Pflanzengeschichte 43: 397.

663. **Kraenzlin**, Fritz. (1922). **Über einige Orchideen.** Mitteilungen aus dem Herbar des Instituts für allgemeine Botanik 5: 237 – 240.

664. **Kraenzlin**, Fritz. (1923). **Über zwei** *Polystachya*-**Arten.** Annalen des Naturhistorischen Museum in Wien 36: 5 – 6.

665. **Kraenzlin**, Fritz. May 1926). **Monographie der Gattung** *Polystachya.* Fedde Repertorium Novarum Specierum 39: 1 – 136.

666. **Kraenzlin**, Fritz. (1928). **Ein neues** *Bulbophyllum* **aus Madagaskar.** Repertorium Specierum Novarum Regni Vegetabilis 25, 4 – 6: 55 – 56.

667. **Kraenzlin**, Fritz. (1930). *Reliquiae quaedam Reichenbachianae.* Annalen des Naturhistorischen Museum in Wien 44: 323 – 327.

668. **Kraenzlin**, Fritz. (no date). Manuscript in Herb. Berol.

669. **Kuhnt**, Otto. (1954). *Angraecum sesquipedale.* Die Orchidee 5, 3: 65 – 66.

670. **Kuntze**, Otto. (1891 – 1893). *Revisio generum plantarum cum enumeratione plantarum exoticarum.* **Monocotyledons, 169.** *Orchidaceae.* Arthur Felix, Leipzig: 3 vols.

671. **Kuntze**, Otto. (1894). **Nomenclatur Studien.** Bulletin de l'Herbier Bossier 2: 459.

672. **Labe**, Philippe. (2002). **Les orchidées de Madagascar.** Orchidées. Culture et Protection, Publ. AFCPO, 52: 8 – 18.

An overview of the orchids of Madagascar, accompanied by drawing by the author. Key characteristics and some cultural advice are given.

673. Lackner, C. (1891). *Phaius tuberculosus.* *Gartenflora* 40: 33 – 34, t.1339.

674. Lacroix, Alfred M. (1822). **Notice historique sur Alfred Grandidier.** *Mémoires de L'Académie des Sciences de L'Institut de France* ser. 2, 58: 1 – 58.

675. Lacroix, Alfred. (1916). **Notice historique sur Bory de Saint-Vincent.** *Mémores de L'Academie des Science de L'Institut de France*, 2e. sér. 24: 1 – 75.
An extensive biography of Bory de Saint-Vincent. With a bibliography and list of works published by the explorer.

676. La Croix, I. F., **La Croix**, E. A. S., **La Croix**, T. M., **Hutson**, J. A. & **Johnston-Stewart**, N. G. B.(1983). **Malawi orchids. Volume 1. Epiphytic orchids.** National Fauna Preservation Society of Malawi: 152 pp., 100 Drawings.

677. La Croix, I. F., **La Croix**, E. A. S. & **La Croix**, T. M. (1991). *Orchids of Malawi.* A. A. Balkema, Rotterdam, Netherlands: 358 pp.

678. La Croix, Isobyl. (1993). **The genus *Angraecum*.** *Orchid Review* 101, 1193: 253 – 256.

679. La Croix, Isobyl. (1993). **The angraecoid orchids. Part III. Mainly jumelleas and *Aeranthes*.** *Orchid Review* 101, 1194: 279 – 281.

680. La Croix, Isobyl. (1994). **The angraecoid orchids. IV.** *Orchid Review* 102, 1196: 91 – 96.

681. La Croix, Isobyl. (1994). **Orchids in Malawi; their affinities and distribution.** in Pridgeon Alec M. ed., *Proceedings of the 14th World Orchid Conference.* HMSO, Edinburgh: 217 – 224.

682. La Croix, Isobyl & **Cribb**, Phillip J. (1995). ***Orchidaceae*. Part 1.** In G. V. Pope, ed., *Flora Zambesiaca*. 320 pp., 108 drawings, 56 photos. Flora Zambesiaca Managing Committee, London.

683. La Croix, Isobyl. (1997). **African terrestrial orchids — *Eulophia*.** *Orchid Review* 105, 1214: 81 – 86.

684. La Croix, Isobyl & **Eric**. (1997). **African orchids in the wild and in cultivation.** Timber Press, Portland, Oregon: 379 pp.

685. La Croix, I & **Cribb**, P. J. (1998). **Flora Zambesiaca. *Orchidaceae*.** In G. V. **Pope**, ed., Flora Zambesiaca 11, 2, Royal Botanic Gardens, Kew: 569 pp.
The flora is continued with the subfamily Epidendroideae, some species overlap with the Madagascan flora:

686. La Croix, Isobyl, **Bosser**, Jean & **Cribb**, Phillip J. (2002). **The genus *Disperis* (*Orchidaceae*) in Madagascar, the Comores, the Mascarenes and the Seychelles.** *Andansonia*, sér. 3, 24, 1: 55 – 87.
The genus is revised and a key is provided for the identification of the 22 species that are currently recognised. Six new Madagascan species and a new variety from Réunion are described. Most species are illustrated.

687. La Croix, Isobyl (2004). *Microcoelia*, **African jewels.** *Orchid Digest*, 68, 2: 79 – 85.
A general introduction to the genus *Microcoelia*; taxonomy, distribution and habitat are discuss. A number of Madagascan species are mentioned and illustrated.

688. Lamarck, De, J.-B. & **Poiret**.(1783 – 1817). *Encylopedie methodique botanique.* Paneckoucke. Paris: 13 vols.

Lanessan, see de Lanessan 98 & 255.

689. Lavergne, Roger. (1980). *Fleurs de Bourbon.* Cazal. St. Denis, Réunion: 218 pp.

690. **Lavergne**, Roger. (1980). **Une orchidée medicinale de La Réunion, le Faham.** *L'Orchidophile* 12, 41: 1535 – 1538.

691. **Lavergne**, Roger (1982). **Orchidées menacées de La Réunion.** *L'Orchidophile*, 14, 50: 17 – 21. A number of orchids from Réunion are presented with short notes and line drawings.

692. **Lavergne**, R. & **Véra**, R. (1989). *Médecine traditionelle et pharmacopée. Étude ethnobotanique des plantes utilisées dans la pharmacopée traditionelle à la Réunion.* Agence de Cooperation Culturelle et Technique: 236 pp.

693. **Lawler**, Leonard (1984). *Ethnobotany of the Orchidaceae.* In Joseph Arditti, ed., *Orchid Biology Reviews and Perspectives*, 3:.52 – 62. Cornell University Press, Ithaca. London, United Kingdom.

694. **Lebl**, M. (1894). *Eulophiella elisabethae. Neubert's Deutsches Garten Magazin* 47, 1: 2.

695. **Le Bon Silvestre**. (1831). **Funérailles de M. Aubert Dupetit-Thouars.** *Institut de France Academie Royale des Sciences*: 1 – 4.

696. **Lecoufle**, Marcel. (1964). **Notes about** *Cymbidiella. Orchid Review* 72, 853: 232 – 236.

697. **Lecoufle**, Marcel and **Perkins**, Brian L. (1964). **Correspondence.** *Orchid Review* 72, 855: 326 – 327.

698. **Lecoufle**, Marcel. (1965). **Orchids outside in France.** *Orchid Review* 73, 862: 126.

699. **Lecoufle**, Marcel. (1965). *Cryptopus elatus. American Orchid Society Bulletin* 34, 4: 327 – 328.

700. **Lecoufle**, Marcel. (1966). **Orchids of Madagascar.** In *Proceedings of the Fifth World Orchid Conference.* Long Beach, California: 233 – 237.

701. **Lecoufle**, Marcel. (1967). *Jumellea fragrans* — **the Faham of Réunion.** *American Orchid Society Bulletin* 36, 7: 587 – 588.

702. **Lecoufle**, Marcel. (1967). *Oeonia volucris. Orchid Review* 75, 893: 383 – 384.

703. **Lecoufle**, Marcel. (1970). *Cymbidiella rhodochila. L'Orchidophile* 1: 4 – 6.

704. **Lecoufle**, Marcel. (1970). *Eulophiella roempleriana. L'Orchidophile* 2: 13 – 16.

705. **Lecoufle**, Marcel. (1970). **Voyage de quelque membres de la S. F. O. aux Illes Australes de L'Ocean Indien.** *L'Orchidophile* 18: 398 – 405.

706. **Lecoufle**, Marcel. (1970). **Voyage de quelque membres de la S. F. O. aux Illes Australes de L'Ocean Indien, 2me. partie.** *L'Orchidophile* 18: 429 – 439.

707. **Lecoufle**, Marcel. (1971). *Eulophiella roempleriana. Orchid Review* 79, 932: 59 – 60.

708. **Lecoufle**, Marcel. (1971). *Eulophiella roempleriana. Orchid Digest* 35 : 166.

709. **Lecoufle**, Marcel. (1971). *Eulophiella roempleriana. American Orchid Society Bulletin* 40, 7: 578 & 597 – 599.

710. **Lecoufle**, Marcel. (1976). **Orchids of Madagascar.** In Karlheinz Senghas, ed., *Proc. of the 8th World Orchid Conference.* D.O.G., Frankfurt, 1976: 249 – 252.

711. **Lecoufle**, Marcel J. (1980). **Orchids of Madagascar.** in Kashemsanta Sukshom M.R., ed., *Proceedings of the 9th World Orchid Conference.* Amarin Press, Bangkok: 169 – 171.

712. **Lecoufle**, Marcel. (1981). **Voyage à Madagascar en Octobre 1980.** *L'Orchidophile* 48: 1892 – 1896.

713. **Lecoufle**, Marcel. (1981). **Pollination de l'***Angraecum sesquipedale. L'Orchidophile* 48: 1897 – 1899.

714. Lecoufle, Marcel. (1982). **The *Angraecum sesquipedale* and pollinization.** *Orchid Digest* 46 3: 102.

715. Lecoufle, Marcel. (1982). **Voyage à Madagascar Sambava à Antalaha.** *L'Orchidophile* 50: 22 – 26.

716. Lecoufle, Marcel. (1982). **Voyage à Madagascar Diego Suarez ou Antsirananae.** *L'Orchidophile* 54: 197 – 201.

717. Lecoufle, Marcel. (1982). **Voyage à Madagascar en 1980, Nosy Bé.** *L'Orchidophile* 55: 296 – 298.

718. Lecoufle, Marcel. (1984). **Notes on *Angraecum eburneum* and its varieties.** *Orchid Review* 92, 1083: 34.

719. Lecoufle, Marcel. (1985). **Voyage au Cap Est a Madagascar.** *l'Orchidophile* 16, 66: 801 – 810.

720. Lecoufle, Marcel. (1985). **Voyage au Cap Est a Madagascar (Suite et Fin).** *l'Orchidophile* 16, 68: 887 – 894.

721. Lecoufle, Marcel. (1985). *Oeceoclades ecalcarata.* *l'Orchidophile* 16, 68: 908.

722. Lecoufle, Marcel. (1994). **Madagascan orchids.** in Pridgeon Alec M. Editor, *Proceedings of the 14th. World Orchid Conference.* HMSO. Edinburgh: 381- 382.

723. Lecoufle, Marcel. (1996). ***Nepenthes* and orchids in the north-east of Madagascar.** *Orchid Review* 104, 1212: 355 & 356. a similar text appears in French in *Nepenthes* et Orchidées au Nord-est de Madgascar. (1997). *L'Orchidophile* 127: 117 – 121

724. Lecoufle, Marcel. (1996 – 7) **Les orchidées de Madagascar et leur culture.** *Hommes & Plantes. Revue trimestrielle du Conservatoire français des collections vegétales specialisées,* 20: 21 – 28.
A history of the author's work with Madagascan orchids, his travels and the Lecoufle nursery on the occasion of it becoming the French national collection of Madagascan orchids.

725. Lecoufle, Marcel.(1997). **Les angraecoides.** Proceedings of Orchidée 97, Genève: 38 – 39.

726. Lecoufle, Marcel.(1997). *Nepenthes* **et orchidées au Nord-est de Madagascar.** *L'Orchidophile,* 28, 127: 117 – 121.
The author travels to Fort Dauphin and Tamatave to find a new *Nepenthes.* He also discovers *Cymbidiella flabellata* and *Bulbophyllum lecouflei.*

· **727. Lecoufle**, Marcel. (1999). **Orchidées et plantes rares de la collection Marcel Lecoufle.** Rieunier & Bailly-Ponnery, Paris: 22 pp.
A catalogue for the sale of divisions of stock plants from the Marcel Lecoufle collection. The aim is to raise funds for building a Marcel Lecoufle Foundation or Museum. A short biographical note of Mr. Lecoufle is followed by the sales listing of plants, many from Madagascar.

728. Lecoufle, Marcel. (2001). *Neobathiea spatulata* **H. Perrier.** *Orchid Review,* 109, 1242: 380.
A short note on *Neobathiea spatulata* collected by the author on Montagne des Français.

729. Lecoufle, Marcel. (2002). **Variegated orchids.** *Orchid Review,* 110, 1242: 19 – 21.
Some variegated orchids are discussed and illustrated, including: *Aerangis citrata* and *Angraecum sesquipedale.*

730. Lecoufle, Marcel. (2004). **Nos amis les orchidéistes.** *L'Orchidophile,* 35, 161.125 – 137.
An informal interview with Marcel Lecoufle describing his long career with orchids.

731. Ledien, F. (1911). **Orchideen aus Madagaskar.** *Orchis* 5, 5: 114 – 116, t.14 – 15.

732. Legée, Georgette. (1972). **Aubert du Petit Thouars et la botanique de son temps, d'après J. P. Flourens.** In 97e.Congres National des Sociétes Savantes, Nantes. Science T1, 1: 47 – 70.
A detailed biography of du Petit Thouars based on the work of J. P. Flourens 298. With an extensive bibliography.

733. Leistner & **Plessis**, eds. (1992). **The flowering plants of Africa.** *National Botanical Institute* 52, part 1.

734. Lemaire, Charles. (1859). **Miscellanées, Plantes recommandées.** *L'Illustration Horticole* 6: 53 – 55.

735. Lemaire, Charles, ed. (1862). **Miscellanées.** *L'Illustration Horticole* 8: 68 – 71.

736. Lemaire, Charles, ed. (1862). **Miscellanées. Mode de transport des voyageurs et des marchandises à travers les forêts de Madagascar.** *L'Illustration Horticole* 9: 1 – 2.

737. Lemaire, Charles, ed. (1866). **Plantes recommandées,** *Angraecum sesquipedale*. *L'Illustration Horticole* 13 pl. 475.

738. Lemcke, Karsten (1999). *Bulbophyllum bosseri* **nom. nov.** — **Berichtgung eines Schechterschen Vershehens.** *Die Orchidee,* 50, 6: 663.
Bulbophyllum inauditum is given the new name *B. bosseri*.

739. Leroux, Armand. (1996). *Aerangis articulata* (Rchb.f.) Schltr. *L'Orchidophile* 27, 121: 79.

740. Lewis, Beverley A., **Phillipson**, Peter B., **Andrianarisata** Michèle, **Rahajasoa** Grace, **Rakotomalaza** Pierre J., **Randriambololona** Michel & **Mcdonagh** John F. (1996). **A study of the botanical structure, composition, and diversity of the eastern slopes of the Réserve Naturelle Intégrale d'Andringitra, Madagascar.** *Fieldiana (Zoology)* new series, 85: 24 – 72.
A listing of orchids found in the Andringitra Massif (Appendix I), many from old records, short locality notes are given.

741. License, Olivier. (2003). **Présentation de la collection du genre Vanilla du Centre for Exonomic Botany, Royal Botanic Gardens, Kew (G.-B).** *L'Orchidophile,* 34, 156: 61 – 68.
The history of vanilla and its properties are explored through the collections at Kew.

742. Limier, Franz. (1994). *Angraecum* **Bory: un genre majeur de la flore indigène à La Réunion.** *Orchidées. Culture et Protection* 18: 5 – 8.

743. Linden, Jean. (1854 – 1860). **Pescatorea. Iconographie des orchidées.** Bruxelles, Belgium. Also reprinted in French and translated into English, German and Portugese by Naturalia Publications, Turriers, France (1994).

744. Linden, Lucien, **Rodigas**, Émile & **Linden**, Jean, eds. (1885 – 1906). *Lindenia. Iconographie des orchidees*. Gent. Also reprint by Naturalia Publications (S. A. Transfaire) Turriers, France (1992), including the English edition. A series of colour lithographs of orchids cultivated at that time, including most of the more flambuoyant Malagasy species. Text describing the horticultural history by a number of authors, see under J. Linden (745 – 750, 751, 752), L. Linden (753, 758) and Rolfe (1056, 1060, 1061, 1066, 1067).

745. Linden, Jean, ed. (1885). *Aeranthus leonis* **Lindley.** *Lindenia* 1 (1): (121 – 122), plate 37.

746. Linden, Jean, ed. (1886). *Angraecum ellisi* **Williams.** *Lindenia* 2 (1): 91 – 92 (286), plate 92.

747. Linden, Jean, ed. (1887). *Aeranthes grandiflora* **Lindl.** *Lindenia* 3 (1): 29 – 30 (286 – 287), plate 109.

748. Linden, Jean, ed. (1888). *Angraecum sesquipedale* **Thouars.** *Lindenia* 4 (2): 65 (66), plate 175.

749. Linden, Jean, ed. (1889). *Angraecum eburneum* **Thouars var.** *superbum*. *Lindenia* 5 (2): 91 (66), pl. 236.

750. Linden, Jean, ed. (1890). *Phaius humbloti* **Rchb.f.** *Lindenia* 6 (2): 31 – 32 (305), pl. 254.

Linden, Jean, ed. (1891). *Eulophiella elisabethae*. *Lindenia,* see Rolfe, 1060.

751. Linden, Jean, ed. (1896). *Phaius* **Marthae.** *Lindenia* 12 (4): (256 – 7), pl. 561.

752. Linden, Jean, ed. (1897). *Phaius* **Normani.** *Lindenia* 13, 4 (426 – 427), pl. 618.

753. Linden, Lucien. (1892). *Eulophiella elisabethae. Lindenia* 8, 3: 90 (217 & 224).

754. Linden, Lucien. (1892). **Culture du** *Grammatophyllum ellisi. Le Journal des Orchidées* 3: 340 – 342.

755. Linden, Lucien. (1892). **Le** *Grammatophyllum ellisii*, **ou mieux** *Grammangis ellisi. Journal des Orchidées*, 3: 187 – 188.
A short description of the species.

756. Linden, Lucien. (1893). **Les orchidees a l'Exposition Internationale Quinqennale de 1893 a Gand.** *Journal des Orchidées*, 4: 54.
Eulophiella elisabethae, displayed at the Show is described.

757. Linden, Lucien. (1893). **Petites nouvelles et petite correspondance.** *Journal des Orchidées*, 4 supplement.
The controversy over the origin and introduction of *Eulophiella elisabethae* is reported, Mr. Linden is adamant that the species was introduced by M. A. Sallerin and not Hamelin. Further acrimonious correspondence in vol.5: 84, 178.

758. Linden, Lucien. (1893). *Eulophiella elisabethae. Lindenia* 9,: 253 & 257.

759. Linden, Lucien & **Gardening World** The, ed. (1893). *Eulophiella elisabethae. Gardening World* 9: 756.

760. Linden, Lucien. (1893). **Plantes nouvelles introduites.** *L'Illustration Horticole* sér. 5, 7: 39.

761. Linden, Lucien.(1894). *Les Orchidées Exotiques.* Linden. Bruxelles & Octave Doin. Paris: 1019 pp., 140 Drawings.

762. Linden, Lucien. (1896). **Petite nouvelles et petite correspondance**, **les orchidées de Madagascar.** *Journal des Orchidées* 7: 212.

763. Linden, Lucien. (1898). *Phaius* × **Normani.** *Semaine Horticole* 2, 64, 175.

764. Linder, H. Peter. (1981). **Taxonomic studies of the** *Disinae*, **I. A revision of the genus** *Brownleea* Lindley. *Journal of South African Botany* 47, 1: – 48.

765. Linder, H. Peter. (1981). **Taxonomic studies on the** *Disinae*; **III. A revision of** *Disa* **excluding section** *Micranthae. Contributions from the Bolus Herbarium* 9, 1: 130 – 131.

766. Linder, H. Peter. (1981). **Taxonomic studies in the** *Disinae (Orchidaceae)* **IV. A revision of** *Disa* Berg. **section** *Micranthae* Lindley. *Bulletin Jardin Botanique National Belgique* 51: 296 – 299.

767. Linder, H. P. (1981). *Orchidaceae, Disa. Distributiones Plantarum Africanum*, 22: 727 – 751.
The distribution of *Disa incarnata* is mapped.

768. Linder, H. P. & **Kurzweil**, H. (1999). *Orchids of Southern Africa.* A. A. Balkema,Rotterdam: 492 pp.
A thorough revision of the orchids of Southern Africa, including chapters on Geography, origin of the flora, biology, classification, history, conservation, economic use and cultivation. The Taxonomic account gives detailed information on every species of the region plus where available, photos, line drawings and maps. A complete list of synonyms is given, most are illustrated, each have descriptions. A number also occur on Madagascar.

769. Lindley, John. (1821). *Collectanea Botanica.* R.& A. Taylor. London: tt.41.

770. Lindley, John. (1822). *Angraecum maculatum. Botanical Register* 8: t.618.

771. Lindley, John. (1824). *Aeranthes grandiflora. Botanical Register* 10: t.817.

772. **Lindley**, John. (1824). *Spiranthes cernua. Botanical Register* 10: t.823.

773. **Lindley**, John. (1824). *Brassia caudata. Botanical Register* 10: t.832.

774. **Lindley**, John. (1824). *Polystachya puberula. Botanical Register* 10: t.851.

775. **Lindley**, John. (1825). *Liparis foliosa. Botanical Register* 11: t.882.

776. **Lindley**, John. (1830 – 1840). *The Genera & Species of Orchidaceous Plants.* Ridgeways, London. Also, reprint A. Asher & Co., Amsterdam (1963).

777. **Lindley**, John. (1832). *Herminium cordatum. Botanical Register* 18: t.1499.

778. **Lindley**, John. (1832). *Angraecum eburneum. Botanical Register* 18: t.1522.

779. **Lindley**, John. (1837). *Cynorchis fastigiata. Botanical Register* 23: t.1998.

780. **Lindley**, John. (1837). **Notes upon some genera and species of** *Orchidaceae* **in the collection formed by Mr. Drége, at the Cape of Good Hope.** *Companion to the Botanical Magazine* 2, 19: 202.

781. **Lindley**, John. (1838). *Pesomeria tetragona. Botanical Register* 24: misc.6, t.4 – 5.

782. **Lindley**, John. (1838). *Cirrhopetalum thouarsii. Botanical Register* 24: t.11.

783. **Lindley**, John. (1838). *Polystachya zeylanica. Botanical Register* 24: t.78.

784. **Lindley**, John. (1838). *Aetheria occulta. Botanical Register* 24: 179.

785. **Lindley**, John. (1840). *Angraecum gladiifolium. Botanical Register,* t.68
A coloured engraving and short description of the species, said to be 'like the greater part of the genus it is a plant of little beauty'.

786. **Lindley**, John. (1842). **Notes upon Cape** *Orchidaceae. Hooker's London Journal of Botany* 1: 14 – 18.

787. **Lindley**, John. (1846). *Cirrhopetalum thouarsii. Botanical Register* 32: t.4237.

788. **Lindley**, John. (1847). **New Garden Plant.** *Botanical Register* 33: t.19.

789. **Lindley**, John. (1849). **Memoranda concerning some new plants recently introduced into gardens otherwise than through the Horticultural Society.** *Journal of the Horticultural Society* 4: 263.

790. **Lindley**, John & **Paxton**, Joseph, eds. (1850 – 1851). **Gleanings and original memoranda.** *Paxton's Flower Garden* 1: 25, f. 9 & 10.

791. **Lindley**, John. (1854). *Folia orchidacea,* **1.** V. J. Matthews, London. Reprint A. Asher & Co., Amsterdam (1964).

792. **Lindley**, John. (1857). **New plants.** *Angraecum sesquipedale. Gardeners' Chronicle.* 253.

793. **Lindley**, John. (1857). **The orchidology of India.** *Journal of the Linnean Society,* 1, 4: 189.

794. **Lindley**, John. (1858). **New plants.** *Epiphora pubescens* Lindl. *Gardeners' Chronicle.* 437.

795. **Lindley**, John. (1862). **West African tropical orchids.** *Journal of the Linnean Society, Botany,* 1: 123 – 140.

796. **Linné**, Carl von.(1753). *Species plantarum.* II. (Facsimile: Ray Society, London,1959).

797. **Louvell**, M. (1931). *Les plantes ornamentales et curieuses de Madagascar.* Gouvernement Général de Madagascar, Publiè à l'Occasion de l'Exposition Coloniale Internationale, Paris.

798. Lückel, Emil, **Fessel**, Hans, **Röth**, Jürgen. (1994). **Neues aus dem Department Systematik/Bestimmungszentrale.** *Die Orchidee* 45, 2: 75 – 76.

799. Mackay, John. (1902). **Calendar of operations for July.** *Orchid Review* 10, 115: 198.

800. Mackay, John. (1903). **Calendar of operations for January.** *Orchid Review* 11, 121: 27.

801. Macoda, P. (2000). *Cymbidiella pardalina. Orchideeën*, 6 : 106 – 180.
A concise history of the species.

Mansfeld, Rudolf, see Schlechter, R. 1223

802. Mantin, Georges. (1889). *Angraecum hyaloides* **(H. Rchb.f.).** *L'Orchidophile* 2, 9: 347 – 349.

803. Maréchal, Roland. (2000). **Sur quelques orchidées au XIXe siècle. Les** *Angraecum* **malgaches de Louis Fournier.** *L'Orchidophile*, 31, 143: 192 – 193.
A short note on *Angraecum fournierianum* and *A. fournieriae* with colour reproductions of the plates in Revue horticole and a reproduction of the text by André 11 and Maron 805.

804. Maron, Ch. (1888). **Curieux mode de multiplication de l'***Angraecum leonis. L'Orchidophile* 8, 80: 29 – 30.

805. Maron, Ch. (1894). *Angraecum fournierianum. Revue Horticole* 66: 475.

806. Martin, Ph. Ch. (1993). **Les** *Aeranthes. L' Orchidée* 11, 43: 119 – 125.

807. Martin, Ph. Ch. (1993). **Les** *Aeranthes* **(2).** *L' Orchidée* 11, 44: 161 – 167.

808. Martin, Ph. Ch. (1996). *Angraecum. L'Orchidée* 14, 56: 143 – 146.

809. Martin, Ph. Ch. (1996). *Oeceoclades cordylinophylla.* **Une redécouverte.** *L'Orchidée* 14, 56: 154 – 156.

810. Martinet, H. (1893). *Eulophiella elisabethae. Le Jardin* 7, 95: 111, fig.43.

811. Mason, Maurice. (1960). **Orchid collecting in Madagascar.** *Journal of the Amateur Orchid Growers' Society* 9, 2: 65 – 66.

Masters, M. T., editor of the Gardeners' Chronicle, see 335 – 365.

813. McClelland, Thomas B. (1955). *Eulophiella* **Rolfei.** *American Orchid Society Bulletin* 24, 10: 673.

814. McDonald, Lester. (1958). **Angraecums for the collector.** *Orchid Digest* 22: 225 – 227.

815. Mcinerny, G. J. (2002). **Prioritising priorities: quantitative inference of threat and extinction from sighting data.** BSc Ecology Dissertation, School of Biological Sciences, University of East Anglia, Norwich: 76 pp.
Records for a number of *Angraecum* species are analysed.

816. Moir, May & Goodale. (1963). **Madagascar-the unknown.** *Orchid Review* 71, 837 – 838: 68 – 73 & 104 – 108.

817. Moir, W. W. G. (1963). *Angraecum eburneum. Orchid Review* 71, 841: 229 & 233 – 234.

818. Moir, W. W. G. (1963). **Cymbidiellas in Madagascar.** *Orchid Review* 71, 843: 297, 300 – 301.

819. Moir, W. W. G. (1963). *Aerangis ellisii. American Orchid Society Bulletin* 32, 10: 803 – 806.

820. Moir, W. W. G. (1963). *Aerangis ellisii* **or** *Aerangis articulata? Orchid Review* 71, 846: 382 – 383.

821. Moir, W. W. G. (1964). *Oeoniella polystachys. Orchid Review* 72, 847: 31.

822. **Moir**, W. W. G. (1979). **Madagascar's angraecoids.** *Orchid Review* 87, 1031: 155 – 159.

823. **Möllers Deutsche Gärtner-Zeitung**, ed. (1937). *Cynorchis purpurascens*, **Eine schöne Orchidee von Madagascar.** *Möllers Deutsche Gärtner-Zeitung*, 52, 1: 3.
A note on a cultivated plant.

824. **Moore**, Spencer Le Marchant. (1876). **Notes on Mascarene orchidology.** *Journal of Botany* 14: 289 – 292.

825. **Moore**, Spencer Le Marchant. (1876) **On the orchids collected at the Island of Bourbon, during the transit of Venus expedition, by Dr. I. B. Balfour.** *Journal of Botany* 14: 292 – 293.

826. **Moore**, Spencer Le Marchant. (1877). **Order XCIV. *Orchideae*.** in Baker, John Gilbert, ed., **Flora of Mauritius and the Seychelles.** London.

Moore, Spencer Le Marchant (also see Bailey Balfour, 49).

827. **Moreau**, Laurent. (1912). **Etude du développement et de l'anatomie des *Pogonia* malgaches.** *Revue Génerale Botanique* 24: 8 – 112, figs 2 – 8.

828. **Moreau**, Laurent. (1913). *Thèses Faculté Sciences Paris*, série A, 733: 1503.

829. **Morel**, Ch. (1885). *Culture des orchidées.* Dusacq, Paris: 195 pp.

830. **Moritzi**, Alexandre. (1845 – 1846). **Systematisches verzeichiss der von H. Zollinger in den Jahren 1842 – 1844, auf Java gesamelten pflanzen.** Solothurn.

831. **Morris**, Brian. (1968). **The epiphytic orchids of the Shire Highlands, Malawi.** *Proceedings of the Linnean Society London* 179, 1: 51 – 66.

832. **Neirynck**, Rik. (2001). *Angraecum sesquipedale.* *Het Venusschoentje*, 22, 3: 62 – 64.
A note on the species, its pollination and cultivation are discussed.

832b. **Neirynck**, Rik. & **Herremans**, Marck (2006). *Aerangis karthalensis*, **a new species from the Comoro Islands.** *Caesiana*, 26, 2: 1 – 26.
Aerangis karthalensis, a new species closely related to *A. mooreana*, is described in detail and illustrated.

833. **Neuling**, H. (1999). **Mitteilungen aus dem Tagebuche von Dr. Christian Rutenberg.** *Veröffentlicht in der Weser-Zeitung*, Jan.
A transcript of Rutenberg's diaries, with route maps.

834. **Nicholson**, G. E. (1967). **Orchids in flower at Kew, August 1967.** *Orchid Review* 75, 892: 349.

835. **Nilsson**, L. Anders, **Jonsson** Lars, **Rason** Lydia & **Randrianjohany** Emile. (1985). **Monophily and pollination mechanisms in *Angraecum arachnites* Schlechter (*Orchidaceae*) in a guild of long-tongued hawk-moths (*Sphingidae*) in Madagascar.** *Biological Journal of the Linnean Society* 26: 1 – 19.

836. **Nilsson**, L. Anders & **Jonsson** Lars. (1985). **The pollination specialization of *Habenaria decaryana* in Madagascar.** *Adansonia* 7, 2: 161 – 166.

837. **Nilsson**, L. Anders, **Jonsson**, Lars, **Rason**, Lydia & **Randrianjohany**, Emile. (1986). **The pollination of *Cymbidiella flabellata* (*Orchidaceae*) in Madagascar: a system operated by sphecid wasps.** *Nordic Journal of Botany* 6, 4: 411 – 422.

838. **Nilsson**, L. Anders, **Jonsson**, Lars, **Ralison**, L. & **Randrianjohany**, E. (1987). **Angraecoid orchids and hawkmoths in central Madagascar: specialised pollination systems and generalist foragers.** *Biotropica* 19, 4: 310 – 318.

839. **Nilsson**, L. Anders & **Rabakonandrianina**, Elizabeth. (1988) **The pollination of *Aerangis*.** *Botanical Journal of the Linnean Society* 97: 9 – 61.

840. **Nilsson**, L. Anders. (1988). **The evolution of flowers with deep corolla tubes.** *Nature*, 334 p. 147 – 148 translated into Dutch: (1994) *Orchideeën*, 3: 9.

841. **Nilsson**, L. Anders. (1992). **Long pollinnia on eyes: hawk-moth pollination of** *Cynorkis uniflora* **Lindley (***Orchidaceae***) in Madagascar.** *Botanical Journal of the Linnean Society* 109: 145 – 160.

842. **Nilsson**, L. Anders, **Rabakonandrianina**, Elisabeth & **Pettersson** Börge (1992). **Exact tracking of pollen transfer and mating in plants.** *Nature*, 360, 6405: 666 – 668.
A study of pollen transfer by hawkmoths in a population of *Aerangis ellisii*. Microtags was used to track pollen.

843. **O'Brien**, James. (1888). **Orchids at Messrs. Hugh Low & Co.** *Gardeners' Chronicle* ser. 3, 3: p. 237.

844. **O'Brien**, James. (1888). *Angraecum modesta*. *Gardeners' Chronicle* ser. 3, 3: 428.

845. **O'Brien**, James. (1899). **New or noteworthy plants,** *Grammatophyllum roemplerianum*, *Eulophiella roempleriana* Rchb.f. *Gardeners' Chronicle* ser. 3, 26: 353.

846. **O'Brien**, J. (1923). **Three Madagascar cymbidiums.** *Gardeners' Chronicle* ser. 3, 73: 7 – 188.

847. **Oliver**, S. P. (1874). **Forest scenery in Madagascar.** *Gardeners' Chronicle*, new ser. 1: 633 – 634.

848. **Orchidee**, ed. (1977). **Orchideenbewertung.** *Die Orchidee* 28, 4: 8.

849. **Orchidee**, ed. (1994). **Orchideenbewertung der Sweizerischen Orchideen-Gesellschaft (S.O.G.) 1993.** *Die Orchidee* 45, 2: 105 – 110.

850. **Orchid Digest, The.** (1998). **Gallery. Magnificent orchids of Madagascar.** *Orchid Digest*, 62, 4: 168 – 169.
Photographs by Fred Hillerman of a variety of Madagascan orchids:

851. **Orchid Review**, ed. (1962 – 1964). **Royal Horticultural Society awards.** *Orchid Review* 70 – 71.

852. **Ormerod**, Paul. (1996). **Notes on the orchids of New Guinea and the Pacific. Part 1.** *Australian Orchid Review*, 61, 4: 38.
Cynorkis fastigiata appears to have been accidentally introduced earlier in the century in Fiji and the Home Islands and is now well established. A new nomenclature is proposed for the species.

853. **Ormerod**, Paul. (2002). **Taxonomic changes in** *Goodyerinae* (*Orchidaceae*: *Orchidoideae*). *Lindleyana*, 17, 4: 189 – 238.
A variety of taxonomic changes and observations are made, affecting *Goodyera nuda*.

853b. **Ormerod**, Paul. (2006). *Notulae Goodyerinae* (**III**). *Taiwania*, 51, 3: 153 – 161.
Notes on *Goodyera*, the new species *Goodyera goudotii* is created.

854. **Otto**, Friedrich & **Dietrich**, Albert, eds. (1838). *Cirrhopetalum thomasii* **Lindley.** *Allgemeine Gartenzeitung und Zeitschrift* 6, 22: 174.

855. **Otto** Friedrich & **Dietrich** Albert, eds. (1846). *Cirrhopetalum thouarsii* **Lindley.** *Allgemeine Gartenzeitung und Zeitschrift* 14, 30: 239.

856. **Otto** Friedrich & **Dietrich** Albert, eds. (1841). *Angraecum gladiifolium* **Thouars.** *Allgemeine Gartenzeitung und Zeitschrift* 9, 13: 104.

857. **Palacky**, J. (1906). *Catalogus plantarum Madagascariensum*. **I: Monocotyledonae**: 4 – 17. Prague.

858. **Palacky**, J. (1907). *Catalogus plantarum Madagascariensum*. **II, Monocotyledonae**: 37. Prague.

859. **Palacky**, J. (1907) *Catalogus plantarum Madagascariensum*. **III: Monocotyledonae**: 76 – 77. Prague.

860. **Pasetti**, Mario. (1996). **La specie:** *Aerangis citrata* (Thouars) *Orchis* 107: 5 – 6.

861. **Pasfield**, **Oliver** Captain S. (1892). **Dr. Robert Lyall.** *Gardeners' Chronicle,* 12, 2: 519 – 221. A biography of the diplomat and plant collector.

862. **Paulian De Felice**, L. (1958). **Les orchidées malagaches.** *Revue de Madagascar* n. ser., 2, 2: 47 – 52.

863. **Paxton**, Joseph, ed. (text attributed to J. Lindley and to J. D. Hooker). (1849) **New and rare plants in flower —** *Angraecum eburneum. Paxton's Magazine of Botany* 16: 90.

864. **Pen**, J. (1994). **Les orchidées angraecoides.** *Les Orchidophiles Réunis de Belgique* 9, 31: 10 – 14.

865. **Perkins**, Brian L. (1961). *Angraecum comorense. Orchid Review* 69, 811: 8 – 9.

866. **Perkins**, Brian L. (1963). *Angraecum compactum. Orchid Review* 71, 840: 190 – 191 & 214.

867. **Perkins**, Brian L. (1964). *Angraecum sesquipedale. Orchid Review* 72, 849: 76 – 77.

868. **Perkins**, Brian L. (1964). *Angraecum magdelenae. Orchid Review* 72, 852: 193 – 195.

869. **Perkins**, Brian L. (1966). *Angraecum viguieri* Schlechter. *Orchid Review* 74, 878: 245 – 246.

870. **Perkins**, Brian L. (1966). *Aerangis modesta-***and others.** *Orchid Review* 74, 881: 364 – 365.

871. **Perkins**, Brian L. (1967). *Angraecum eburneum* **s.sp.** *xerophilum* **Perrier.** *Orchid Review* 75, 889: 240 – 242.

872. **Perkins**, Brian L. (1968). *Jumellea sagittata* **Perrier and other species.** *Orchid Review* 76, 896: 71 – 73.

873. **Perrier de la Bâthie**, Henri. (1921). **La végétation malagache.** *Annales du Musée Colonial de Marseille* sér.3, 9: 1 – 270, 4 maps.

874. **Perrier de la Bâthie**, Henri. (1927). **Le Tsaratanana, l'Ankaratra et l'Andringitra.** *Mémoires de l'Académie Malgache* 3: 68.

875. **Perrier de la Bâthie**, Henri. (1929). **Note au sujet des noms de deux orchidées de Madagascar.** *Archives de Botanique (Bulletin Mensuel) de Caen* 3: 197 – 198.

876. **Perrier de la Bâthie**, Henri. (1929) **Note sur un nouveau** *Gastrorchis.* *Bulletin de l'Académie Malgache* n. sér., 12: 33 – 35.

877. **Perrier de la Bâthie**, Henri. (1930). *Catalogue des plantes de Madagascar publié par L'Academie Malagache. Orchidaceae d'après R. Schlechter.* Colonie de Madagascar et Dépendances, Paris: 1 – 60.

878. **Perrier de la Bâthie**, Henri. (1931, unpublished). *Archives de Botanique (Bulletin Mensuel) de Caen* 5.

879. **Perrier de la Bâthie**, Henri. (1934). **La genre** *Benthamia* (Orchidées). *Bulletin de la Société Botanique de France* 81: 25 – 38.

880. **Perrier de la Bâthie**, Henri. (1934). **Les vanilles de Madagascar.** *Bulletin du Muséum National d'Histoire Naturelle* sér. 2, 6: 192 – 197.

881. **Perrier de la Bâthie**, Henri. (1934). *Ambrella,* **genre nouveau d'Orchidées (Angraecoidés) de Madagascar.** *Bulletin de la Société Botanique de France* 81: 655 – 657.

882. **Perrier de la Bâthie**, Henri. (1935). **Les eulophias de Madagascar.** *Bulletin de la Société Botanique de France* 82, 3 & 4: 146 – 161.

883. Perrier de la Bâthie, Henri. (1936). *Biogéographie des Plantes de Madagascar.* Société d'Editions Geographiques, Maritimes et Coloniales, Paris.

884. Perrier de la Bâthie, Henri. (1936). **Notes sur quelques orchidées de Madagascar.** *Bulletin de la Société Botanique de France* 83, 1 & 2: 22 – 35.

885. Perrier de la Bâthie, Henri. (1936). **Notes sur quelques *Habenaria* en Madagascar.** *Bulletin de la Société Botanique de France* 83, 8 & 9: 579 – 585.

886. Perrier de la Bâthie, Henri. (1936). **Les *Disperis* de Madagascar, des Comores et des Mascareignes.** *Notulae Systematicae* 5, 3: 217 – 229.

887. Perrier de la Bâthie, Henri. (1936). **Les Liparidinées de Madagascar.** *Notulae Systematicae* 5, 4: 231 – 260.

888. Perrier de la Bâthie, Henri. (1937). **Les Bulbophyllums de Madagascar.** *Notulae Systematicae* 6: 41 – 124.

889. Perrier de la Bâthie, Henri. (1938). **Sarcanthae nouvelle ou peu connues de Madagascar.** *Notulae Systematicae* 7, 1: 29 – 48.

890. Perrier de la Bâthie, Henri. (1938). **Sarcanthae nouvelle ou peu connues de Madagascar.** *Notulae Systematicae* 7, 2: 49 – 65.

891. Perrier de la Bâthie, Henri. (1938). **Sarcanthae nouvelle ou peu connues de Madagascar (fin).** *Notulae Systematicae* 7, 3: 105 – 139.

892. Perrier de la Bâthie, Henri. (1938). ***Bulbophyllum* nouveau de Madagascar.** *Notulae Systematicae* 7, 3: 139 – 143.

893. Perrier de la Bâthie Henri. (1939). **Les orchidées de la Region malagache (Variation; Biologie & Distribution).** *Mémoires du Muséum d'Histoire Naturelle* n.s. 10, 5: 237 – 297.

894. Perrier de la Bâthie, Henri. (1939). **Orchidées et Palmiers nouveau de Madagascar.** *Notulae Systematicae* 8: 32 – 48.

895. Perrier de la Bâthie, Henri. (1939). **Trois monocotyledones nouvelles de Madagascar: *Disperis ankarensis.*** *Notulae Systematicae* 8: 128 – 131.

896. Perrier de la Bâthie, Henri. (1939,1941). **49ᵉ· Famille. — Orchidées. I & II.** 864 pp., 121 pls. in Humbert Henri, ed., **Flore de Madagascar.** Tananarive Imprimerie Officielle, Madagascar. Also, Reprint Margaret M. Ilgenfritz, Monroe Michigan: (1980), 2 vols.

Beckman, Steven D. (1981). *Flora of Madagascar. Orchids.* Steven D. Beckman Pub., California, USA: 542 pp., line drawings.

897. Perrier de la Bâthie, Henri. (1940). **Orchidées utiles ou ornamentales de Madagascar.** *Revue International de Botanique Appliquée et d'Agriculture Tropicale* 20: 497 – 502.

898. Perrier de la Bâthie, Henri. (1941). ***Bulbophyllum* nouveau de Madagascar.** *Notulae Systematicae* 9: 145 – 146.

899. Perrier de la Bâthie, Henri. (1950). **Plantes nouvelles; rares ou critiques des Serres du Museum.** *Bulletin du Museum* ser. 2, 22, 1: 114.

900. Perrier de la Bâthie, Henri. (1950). **Une nouvelle *Vanille* de Madagascar.** *Revue Internationale de Botanie Appliquée et Agriculture Tropicale.* 30, 333 – 334: 435 – 436.

901. Perrier de la Bâthie, Henri. (1951). **Orchidées de Madagascar et des Comores. Nouvelles observations.** *Notulae Systematicae* 14, 2: 138 – 165.

902. **Perrier de la Bâthie**, Henri. (1955). **Les orchidées du Massif Marojejy et de ses avant-monts.** *Memoires de l'Institute de Scientifique de Madagascar* ser. B, 6: 3 – 271. in Humbert 569.

903. **Persoon**, C.H. (1807) *Synopsis Plantarum seu Enchididium Botanicum.* J.G. Cottam, Paris: pp. 512 – 513.

904. **Peters**, Hendrelien. (2000) **Treasures in paradise.** *South African Orchid Journal,* 28, 4: 135 – 139. Orchids inhabiting Isle St. Marie and Ile aux Nattes and their habitat are described superficially:

905. **Petitjean**, A. (1995). *Madagascar par sa Flore.* OCPFM. Antananarivo, 48 pp.

906. **Petitjean**, Alain & Michèle. (2003). *Beautiful orchids. Fascinantes orchidées du Sud Ouest de l'Océan Indien.* Alain & Michèle Petitjean. La Reunion, 143 pp.
A general overview of some of the orchids found on the Indian Ocean Islands. A chapter on the orchid family preceeds an account of a limited number of species, all have information on synonyms, common names, distribution, habitat, flowering time. All are illustrated, often whole plants and habitats are also shown.

Petit-Thouars, see Thouars.

Petterson, also see 384 & 682.

907. **Petterson**, Börge.(1989). **Pollination in the African species of** *Nervilia. Lindleyana,* 4, 1: 33 – 41.

908. **Pettersson**, Börge. (1990). **Studies in the genus** *Nervilia* **in Africa.** *Nordic Journal of Botany* 9, 5: 487 – 497.

909. **Pettersson**, Börge. (1990). *Nervilia (Orchidaceae) in Africa. Acta Univsitatis Upsaliensis, Comp. Sum. Upps. Diss. Fac. Sc.* 281 pp.

910. **Pettersson**, Börge. (1991). **The genus** *Nervilia (Orchidaceae)* **in Africa and the Arabian Peninsula.** *Orchid Monographs* 5: 1 – 69.

911. **Petterson**, Börge & **Nilsson**, L. Anders. (1993). **Floral variation and deceit pollination in** *Polystachya rosea (Orchidaceae)* **on an inselberg in Madagascar.** *Opera Botanica.* 121: 237 – 245. Also, a summary of this article in Dutch in: (1995). **Stuif es in** *Orchideeën* 5: 9.

912. **Pfitzer**, Ernst Hugo Heinrich. (1889). *Orchidaceae. Engler et Prantl, Pflanzenfamien* 2, 6.

Pfitzer, Ernst Hugo Heinrich. (1908). re *Bulbophyllum pentasticha;* see 661.

913. **Phillipson**, P. B. & **Condy**, G. (1997). *Oeceoclades decaryana. Flowering Plants of Africa,* 55: 48 – 53.

914. **Piers**, Frank (1915) *A book of East African orchids.* Patwa Publications, Nairobi: 112 pp.
A guide to the orchids of East Africa. A few species with a Madagascar connection are mentioned.

915. **Piers**, Frank. (1955). **Vanillas.** *American Orchid Society Bulletin* 24, 5: 299 – 302.

916. **Piers**, Frank. (1957). **Large-flowered angraecums.** *American Orchid Society Bulletin* 26, 8: 556 – 557.

917. **Piers**, Frank (translated by W. Haber). (1958) **Zwerg und Riese.** *Die Orchidee* 9, 3: 74 – 75.

918. **Piers**, Frank. (1959). *Orchids of East Africa.* Nairobi, 148 pp.
A detailed work on the orchids of East Africa. A number of species with a Madagascan connection are mentioned:

919. **Piers**, Frank. (1965). *Angraecum sororium. American Orchid Society Bulletin* 34, 12: 1091 – 1092.

920. **Piers**, Frank. (1966). *Gastrorchis humblotii. American Orchid Society Bulletin* 35, 12: 987 – 988.

921. **Piers**, Frank. (1967). **Through the letter slot.** *American Orchid Society Bulletin* 36, 8: 684.

922. **Piers**, Frank. (1968). **Orchids of East Africa.** J. Cramer, Lehre: 304 pp., 116 photos.

923. **Pilonchéry**, Anne. (1999). **L'Ile de la Réunion et la culture de la vanille.** *Bullentin Mensuel de la Société Linnéenne de Lyon,* 68, 5: 96 – 99.
Commercial vanilla production on la Réunion is discussed: Description of the plant, hand fertilisation, harvesting, cultivation, propagation, preparation of the pods.

924. **Pinciroli**, Lupo Aimone. (1995). **Gli *Angraecum.*** *Orchis* 92: 6 – 11.

925. **Pléione.** (1890). **Culture du *Phajus tuberculosus.*** *L'Orchidophile* 10, 105: 48 – 49.

926. **Podzorski**, A. C. & **Cribb**, Phillip. (1980). **A revision of *Polystachya* section *Cultriformes.*** *Kew Bulletin* 34,1: 147 – 186.

Poiret see Lamarck 688.

927. **Poisson**, M. H. (1911). **François Geay, voyageur naturaliste.** *Bulletin du Museum d'Histoire Naturelle Paris* 17, 3: 86 – 90.

928. **Poisson**, H. (1912). **Recherches sur la Flore Méridionale de Madagascar.**
Augustin Challamel, ed. Librarie Maritime et Coloniale, Paris, pp. 172 – 195.

929. **Poisson**, M. H. (1917). **Humblot, naturaliste-voyageur.** *Bulletin du Museum d'Histoire Naturelle Paris* 23, 4: 216 – 218.

930. **Pole-Evans**, Illtyd Buller. (1938). ***Brownleea coerulea.*** *Flowering Plants of South Africa* 18: t.702.

931. **Portères**, Roland. (1954). **Le vanillier et la vanille dans le Monde.** In G.Bouriquet ed, Encyclopédie Biologique. Editions Lechevalier, Paris.

932. **Pottinger**, Mollie. (1977). **The mighty miniatures — angraecoids.** *Orchid Review* 85, 1010: 234 – 235.

933. **Pottinger**, Mollie. (1980). **Orchid superstars *Eulophiella roempleriana* 'Bromesberrow Place' AM/RHS.** *Orchid Review* 88, 1043: 135 – 36 & 259.

934. **Pottinger**, Mollie, **Bailes** Christopher & **Menzies** David. (1981). **Orchids flowering at Kew.** *Orchid Review* 89,1057: 364.

935. **Pottinger**, Mollie. (1982). **African orchids — The genus *Angraecum.*** *Orchid Review* 90,1059: 30 – 34.

936. **Pottinger**, Mollie. (1982). **African orchids — *Acampe, Aeranthes, Ancistrorhynchus, Angraecopsis* and *Ansellia.*** *Orchid Review* 90. 1062: 132 – 134.

937. **Pottinger**, Mollie. (1982). **African orchids — *Eulophiella* and *Eurychone.*** *Orchid Review* 90, 1067: 299 – 301.

938. **Pottinger**, Mollie. (1982). **African orchids — genera from 'G' to 'M'.** *Orchid Review* 90, 1068: 327 – 329.

939. **Pottinger**, Mollie, **Bailes**, Christopher & **Menzies**, David. (1983). **Orchids flowering at Kew.** *Orchid Review* 91, 1074: 111 – 113.

940. **Pottinger**, Mollie, **Bailes**, Christopher & **Menzies**, David. (1983). **Orchids flowering at Kew.** *Orchid Review* 91,1080: 319 – 320.

941. **Pottinger**, Mollie. (1983) ***African orchids a personal view.*** HGH Publications, Wokingham: 62 pp.

Pourret, see De Candolle 248.

942. Pranaranara, Jakadal. (cfr. **Neirynck** Rik). (2001). **Nieuw zoogdiertje in Madagascar.** *Het Venusschoentje*, 22, 1, inside back cover.
A photograph shows a Sifaka lemur suspended from the spur of *Angraecum sesquipedale*. The text claims that it is a new species of Lemur (*Aprilisicus apricus*) that seems to feed on the nectar of the orchid. The article was published around the 1st of April.

943. Preston-Mafham, Ken. (1991) *Madagascar: a natural history.* Facts On File, Oxford & New York, 224 pp.

944. Pridgeon, Alec, Editor. (1992). *The illustrated encyclopedia of orchids.* Headline, Weldon, Sydney, Australia: 304 pp.

945. Pridgeon, Alec M., **Cribb** Phillip J., **Chase**, Mark W. & **Rasmussen** Finn, N., eds. (2001). *Genera orchidacearum, Volume 2 Orchidoideae* (Part one). Oxford University Press, Oxford: 416 pp.
A complete revision of the orchid family. Each genus is described, and illustrated. Distribution, Infrageneric treatment, Derivation of name, Cytogenetics, Phylogenetics, Ecology, Pollination and Taxonomic literature are all discussed at length.
The following Madagascan genera are included: *Benthamia* p. 261, *Brachycorythis* p. 265, *Brownleea* p. 8, *Cynorkis* p. 276, *Disa* p. 33, *Disperis* p. 26, *Habenaria* p. 298-, *Megalorchis* p. 315, *Physoceras* p. 345, *Platycoryne* p. 350, *Satyrium* p. 50, *Tylostigma* p. 379.

946. Pridgeon, Alec M., **Cribb**, Phillip J., **Chase**, Mark W. & **Rasmussen**, Finn N., eds. (2003). *Genera orchidacearum, Volume 3 Orchidoideae* (Part two). Oxford University Press, Oxford: 358 pp.
The following Madagascan genera are included: *Cheirostylis* p. 77, *Goodyera* p. 94, *Platylepis* p. 136, *Zeuxine* p. 150, *Galeola* p. 313, *Vanilla* p. 320 (including an essay on its phytochemistry).

947. Pring, George H. (1959). *Eulophiella roempleriana.* *American Orchid Society Bulletin* 28, 6: 400 – 402.

948. Protheroe & Morris. (1893). **The largest and most valuable lot of imported orchids.** Advertisement of Auction.
The sale of a batch of *Eulophiella elisabethae* is announced. The plants were collected by Hamelin who claims that every plant was collected. A short description and several quotes from Hamelin. With an accompanying engraving. With an accompanying letter from F. Sander & Co giving an even more elaborate story of the species.

949. Pynaert, ed. (1892). **Le *Phaius tuberculosus*.** *Revue de l'Horticulture Belge* 18: 145 – 146.

950. Rain, Patricia. (1983). *Vanilla cookbook.* Celestial Arts, Berkley California: 124 pp., 17 colour pl.

951. Rajeriarison, Charlotte. (date?) **Madagascar plant.** *Species*, 41: 15 & 18.
A short note on an orchid documentation project; populations were recorded in the Highlands of Madagascar. White-flowered forms of *Cynorkis gibbosa*, *Cynorkis uniflora* and *Cynorkis flexuosa* are recorded.

952. Ramsey, Carl T. (1965). **Darwin and the star orchid of Madagascar.**
American Orchid Society Bulletin 34, 12: 1056 – 1062.

953. Rand, Edward Sprague Jr. (1876). *Orchids. A description of the species and varieties grown at Glen Ridge, near Boston.* Hurd & Stoughton, New York.

954. Rasmussen, Finn N. (1977). **Revision of *Corymborkis*.** *Botanisk Tidsskrift* 71, 3 – 4: 161 – 170.

955. Rasmussen Finn N. (1978). **Observations on the genus *Angraecopsis* (*Orchidaceae*).** *Botany* 2: 137 – 144.

956. Rasmussen, Finn. N. (1979). **Nomenclatural notes on Thouars work on Orchids.** *Botaniska Notiser* 132: 385 – 392.

957. Rauh, Werner. (1962). **Bemerkungen zur Orchideenflora Madagaskar.** *Die Orchidee* 13, 5: 166 – 173.

958. Rauh, Werner. (1995). *Succulents and xerophytic plants of Madagascar.* Volume One. Strawberry Press, Mill Valley: 343 pp., 1011 ill.

959. Rauh, Werner. (1998). *Succulents and xerophytic plants of Madagascar.* Volume Two. Strawberry Press, Mill Valley: 385 pp.
Volume two covers the succulent plants of south and southwest of the island. A few orchids are illustrated.

960. Redpath, Janet & **Hunt**, Francis P. (1972). **Index to African orchids I – XXX by V. S. Summerhayes.** *Kew Bulletin* 27, 2: 337 – 369.

961. Reeve, H., ed. (1893). **Philibert Commerson, naturalist.** *Edinburgh Review.* 321 – 353.

962. Regel, Eduard. (1872). *Angraecum sesquipedale. Gartenflora* 21: t.744.

963. Regel, Eduard von. (1891). *Aeranthus brachycentron* **Rgl.** *Gartenflora* 1891: 40.

964. Reichenbach, H. G.(1847). **Orchidiographische Beiträge.** *Linnaea* 20: 678 – 681 or *Beitrage zur Pflanzenkunde* 4.

965. Reichenbach, H. G. (1849). **Ueber zwei merkwürdige Orchideen.** *Botanische Zeitung* 7: 868.

966. Reichenbach, H. G. (1855). *Symbolae orchidaceae. Bonplandia* 3: 15 &16.

967. Reichenbach, H. G. (1855). *Symbolae orchidaceae. Bonplandia* 3, 15: 221.

968. Reichenbach, H. G. (1856). *Stipulae orchidacea Reichenbachianae. Bonplandia* 4, 20: 324.

969. Reichenbach, H. G. (1863). *Xenia Orchidacea, Beiträge zur Kenntniss der Orchideen* : vols. 1 & 2 & atlas. Leipzig.

970. Reichenbach, H. G. (1860). **Orchidographic Streitfrage.** *Hamburger Garten und Blumenzeitung* 16: 520 – 521.

971. Reichenbach, H. G. (1861). **Empfehlenswerthe oder neue Pflanzen.** *Bonplandia* 9, 5 – 6: 81.

972. Reichenbach, H. G. (1861 – 1865). *Orchideae.* in *Walpers, Annales Botanices Systematicae.*

973. Reichenbach, H. G. (1864). *Orchideae. Walpers, Annales Botanices Systematicae* 6: 650.

974. Reichenbach, H. G.. (1864). *Orchideae. Walpers Annales Botanices* 6: 899.

975. Reichenbach, H. G. (1865). **Dr. Welwitsch's Orchideen aus Angola.**
Flora 48: 77 – 191.

976. Reichenbach, H. G. (1867). **Dr. Welwitsch's Orchideen aus Angola** (end). *Flora* 50: 113 – 115.

977. Reichenbach, H. G. (1872). **Neue Orchideen, entdecht und gesammelt von Herrn Gustav Mann.** *Flora* 55: 278.

978. Reichenbach, H. G. (1872). **New garden plants.** *Angraecum articulatum. Gardeners' Chronicle.* 73.

979. Reichenbach, H. G. (1872). **New garden plants.** *Eulophia scripta.*
Gardeners' Chronicle. 1003.

980. Reichenbach, H. G. (1877). *Orchideae Roezlianae novae seu criticae. Linnaea.* 41: 62.

981. Reichenbach, H. G. (1877). *Grammatophyllum roemplerianum. Gardeners' Chronicle* n. ser. 7: 240.

982. Reichenbach, H. G. (1878). **New garden plants.** *Angraecum hildebrandtii* **n.sp.** *Gardeners' Chronicle* n. ser. 9: 725.

983. Reichenbach, H. G. (1878). **New garden plants.** *Eulophia scripta* **Lindley.** *Gardeners' Chronicle* n. ser. 10: 332 – 333.

984. Reichenbach, H. G. (1878). **New garden plants.** *Gardeners' Chronicle* n. ser. 10: 333.

985. Reichenbach, H. G. (1878). *Angraecum scottianum* **n.sp.** *Gardeners' Chronicle* n. ser. 10: 556.

986. Reichenbach, H.G. & **Moore**, S. (1877). **On a collection of ferns made by Miss Helen Gilpin in the interior of Madagascar.** *Journal of the Linnean Society* 16: 206.

987. Reichenbach, H. G. (1880). *Angraecum hyaloides.* *Gardeners' Chronicle* n. ser. 13: 264.

988. Reichenbach, H.G. (1880). **New garden plants.** *Gardeners' Chronicle* n. ser. 14: 326.

989. Reichenbach, H.G. (1880). *Phajus humblotii.* *Gardeners Chronicle* n. ser. 14: 812.

990. Reichenbach, H.G. (1880). **New garden plants.** *Calanthe sylvatica.* *Gardeners Chronicle* n. ser. 14: 812.

991. Reichenbach, H.G. (1881). *Orchideae Hildebrandtianae.* *Otia Botanica Hamburgensia* 2, 6. Hamburg.

992. Reichenbach, H.G. (1881). *Novitiae Africanae.* *Otia Botanica Hamburgensia* 2, 8. Hamburg.

993. Reichenbach, H.G. (1881). *Angraecum hyaloides.* *Gardeners Chronicle* n. ser. 15: 136.

994. Reichenbach, H.G. (1881). **New garden plants,** *Phaius tuberculosus* Bl. *Gardeners Chronicle* n. ser. 15: 341.

995. Reichenbach, H.G. (1881). *Angraecum fastuosum.* *Gardeners' Chronicle* n. ser. 16: 748.

996. Reichenbach, H.G. (1881). *Angraecum fastuosum.* *Gardeners' Chronicle* n. ser. 16: 844.

997. Reichenbach, H.G. (1881). *Orchideae Hildebrandtianae.* *Botanische Zeitung* 39: 447 – 450.

998. Reichenbach, H. G. *f.* (1882). *Cyrtopera plantaginea.* *Gardeners' Chronicle* n. ser. 17: 700
A short note on the species published by Thouars in 1822. Now flowered in Bohemia with Baron Hruby under the care of M. Schopetz. Collected in Madagascar by Humblot and introduced by Sander. With a short description.

999. Reichenbach, H. G. *f.* (1882). *Orchideae describuntur.* *Flora,* 65, 34: 531 – 535.
The description of *Lissochilus ephippium* affects the Madagascan flora.

1000. Reichenbach, H.G. (1882). *Angraecum descendens.* *Gardeners' Chronicle* n. ser. 17: 558.

1001. Reichenbach, H.G. (1882). **New garden plants.** *Gardeners' Chronicle* n. ser. 17: 732 – 733.

1002. Reichenbach, H.G. (1882). *Angraecum fuscatum.* *Gardeners' Chronicle* n. ser. 18: 488.

1003. Reichenbach, H.G. (1883). **New garden plants.** *Gardeners' Chronicle* n. ser. 19: 306 – 307.

1004. Reichenbach, H.G. (1884). *Eulophia pulchra* (Lindl.) *divergens,* **n. var.** *Gardeners' Chronicle* n. ser. 22: 102.

1005. Reichenbach, H.G. (1885). **New garden plants.** *Gardeners' Chronicle* n. ser. 23: 726.

1006. Reichenbach, H.G. (1885). **New garden plants,** *Angraecum florulentum, Eulophia megistophylla.* *Gardeners' Chronicle* n. ser. 23: 787.

1007. Reichenbach, H.G. (1885). **Comoren-Orchideen Herrn. Léon Humblot.** *Flora* 68, 20: 377 – 382.

1008. Reichenbach, H.G. (1885). **Comoren-Orchideen Herrn. Léon Humblot.** *Flora* 68, 30 – 31: 535 – 544.

1009. Reichenbach, H.G. (1886). **New garden plants.** *Phaius tuberculosus* **Bl.** *Gardeners' Chronicle* n. ser. 25: 362.

1010. Reichenbach, H.G. (1887). **Plants new or noteworthy.** *Angraecum avicularium. Gardeners' Chronicle*, series 3, 1: 40 – 41.

1011. Reichenbach, H.G. (1887). *Angraecum calligerum. Gardeners' Chronicle* ser. 3, 2 : 552.

1012. Reichenbach, H.G. (1888). *Aeranthus grandidierianus* Rchb.f. *The Gardeners' Chronicle* ser. 3: 72.

1013. Reichenbach, H.G. (1888). **New or noteworthy plants.** *Gardeners' Chronicle* ser. 3, 3: 168.

1014. Reichenbach, H.G. (1888). *Aeranthus trichoplectron. Gardeners' Chronicle* ser. 3, 3: 264.

1015. Reichenbach, H.G. (1888). **New or noteworthy plants.** *Gardeners' Chronicle* ser. 3, 3: 424.

1016. Reichenbach, H.G. (1888). **New or noteworthy plants.** *Gardeners' Chronicle* ser. 3, 4: 91.

1017. Reichenbach, H.G. (1888). *Orchideae describuntur. Flora* 71,10: 149 – 155.

1018. Reichenbach, H.G. (1888 – 1894). *Reichenbachia.* F. Sander, St. Albans. 48 plates per volume.

1019. Reichenbach, H.G. (1890). *Angraecum scottianum* Rchb.f. *Xenia Orchidacea* 3, 4: 75, t. 239.

1020. Reinikka, Merle A., ed. (1968). **American Orchid Society awards.**
American Orchid Society Bulletin 37, 12: 1083.

1021. Reinikka, Merle A., ed. (1969). **American Orchid Society awards.**
American Orchid Society Bulletin 38, 7: 625.

1022. Reinikka, Merle A. (1969). **The FCCs in 1968.** *American Orchid Society Bulletin* 38, 11: 999.

1023. Reinikka, Merle A. (1995). *A history of the orchid.* Timber Press, Portland, Oregon: 324 pp.

Reinwardt, see Blume 90.

1024. Rensch, K. (1894). **Das Denkmal J. M. Hildebrandts.** *Gartenflora* 43, 11: 285 – 289.

1025. Rentoul, J. N. (1980 – 1991).*Growing orchids.* Lothian, Melbourne, Australia, 7 vols.

1026. Rice Rod. (1998). **Orchids of the Forgotten Land. 1.** *Australian Orchid Review,* 63, 4: 4 – 9.
A general introduction to Angraecoid orchids, some cultural advice is included.

1027. Rice Rod. (1998). **Orchids of the forgotten land. 2.** *Australian Orchid Review,* 63, 5: 10 – 11.
A few more angraecoids are mentioned.

1027b. Rice Rod. (2006). **An overview and three new species of the Solenangis aphylla alliance.** *Oasis,* 2.
An overview of *Solenangis*, sone new names are created, including *Solenangis cyclochila* (Schltr.) R.Rice with a line drawing.

1028. Richard, Achile (1828). **Monographie des orchidées des Iles de France et Bourbon.** *Memoires de la Societé d' Histoire Naturelle de Paris*, Extrait 4.

1029. Richter, Walter (1958). *Die schoensten aber sind Orchideen.* Neumann Verlag, Radebeul: 280 pp.

1030. Ridley, Henry N. (1883). **Descriptions and notes on new or rare monocotyledonous plants from Madagascar, with one from Angola.** *Journal of the Linnean Society* 20: 329 – 338.

1031. Ridley, Henry N. (1885). **The orchids of Madagascar.** *The Journal of the Linnean Society* 21: 456 – 523.

1032. Ridley, Henry N. (1886). **On Dr. Fox's collection of orchids from Madagascar, along with some obtained by the Rev. R. Baron from the same Island.** *Journal Linnean Society, Botany,* 22: 116 – 127.

1033. Ridley, Henry N. (1887). **A monograph of the genus *Liparis*.** *Journal of the Linnean Society* 22: 244 – 297.

1034. Ridley, Henry N. (1908). **On a collection of plants from Gunong Tahan, Pahang.** *Journal of the Linnean Society* 38: 325.

1035. Rittershausen, Wilma. (1975). **'Twelve scented jewels'. The fairytale angraecums.** *Orchid Review* 83, 979: 11 – 14.

1036. Rittershausen, Wilma. (1978). **Yesterday's orchids, today -a foot and a half.** *Orchid Review* 86, 1015: 12 – 13.

1037. Rittershausen, Wilma, ed. (1981). *Grammangis ellisii* **'Palewell' CCC/RHS.** *Orchid Review* 89, 1055: 280, 331 & 366.

1038. Rittershausen, Wilma, ed. (1984). *Eulophiella roempleriana* **'Bromesgrove Place' CCC-AM/RHS.** *Orchid Review* 92, 1089: 208 & 235.

1039. Rittershausen, Wilma. (1989). **The latest awards from the Royal Horticultural Society.** *Orchid Review* 97, 1143: 29 – 31.

1040. Rittershausen, Wilma. (1989). *Graphorkis ecalcarata* **AM/RHS.** *Orchid Review* 97, 1150: 265 – 266.

1041. Rittershausen, Wilma. (1990). *Sobennikoffia humbertiana* **AM/RHS.** *Orchid Review* 98, 1155: 25.

1042. Rivière, M. (1926). **Relation d'un voyage à la Grande Ile de la Mer des Indes (Madagascar), L'** *Angraecum sesquipedale*. *Journal de La Société Nationale d'Horticulture de France* 27: 313 – 316.

1043. Rivois, Gaston. (1891). **Les orchidées populaires II,** *Angraecum sesquipedale*. *Journal des Orchidées* 2: 16 – 17.

1044. Robbins, Sarah. (1992). *Beclardia macrostachya* **a Madagascan orchid for connoisseurs.** *Orchid Review* 100, 1183: 167 – 178.

1045. Robbins, Sarah. (1990). *Graphorkis ecalcarata*. *Orchid Review,* 98, 1160: 176 – 179.
Graphorkis ecalcarata, recently awarded by the RHS is described in detail; its history is given together with a key to the genus. A new hybrid *Graphiella* Martialine (*G. scripta* x *Cymbidiella rhodochila*) is mentioned.

1046. Roberts, David. (1999). *Angraecum cadetii*. *Orchid Conservation News,* 1, 2: 13 – 14.
The conservation status of *Angraecum cadetii* in Mauritius is explained.

1047. Roberts, David. (2000). **Endangered orchid of Mauritius — Doomed like the Dodo.** *Scottish Orchid Society Bulletin*: 27 – 28.
The vegetation zones of Mauritius are described. Ecology and distribution of Orchids on the Island, collecting and cultivation, conservation and the future.

1048. Roberts, David. (2001). **Reproductive biology and conservation of the orchids of Mauritius.** PhD Thesis, University of Aberdeen: 290 pp.
The reproductive and conservation biology of the orchid flora of the Mascarene Islands, particularly Mauritius, were investigated with the aim of providing a large-scale comparative surveys of species-area relationship, fruiting success and demographic structure in the *Orchidaceae*.

1049. Roberts, David. (2003). **Old & new.** *Kew Scientist,* 24: 2.
A new species of *Aeranthes* from the Comoro Islands to be described by the author is illustrated (cfr. *A. virginales* ed).

1049b. Roberts, David L. (2005). ***Aeranthes virginalis* (*Orchidaceae*): a new species from the Comoro Islands.** *Kew Bulletin* 60: 139 – 141.
The new species is described and illustrated.

1051. Roberts, Jacqueline A, **Anuku**, Sharon, **Burdon**, Joanne, **Mathew**, Paul, **Mccough**, Noel H. & **Newman**, Andrew D. (2002). ***Cites orchid checklist 3.*** Royal Botanic Gardens, Kew: 233 pp.
A checklist of various genera including *Aerangis*, *Angraecum* and *Calanthe*. A listing by country is also provided.

1052. Robertson, S. A. (1989). ***Flowering plants of Seychelles.*** Royal Botanic Gardens, Kew: pp. 212.
An annotated checklist of the Angiosperms and Gymnosperms of the Seychelles.

1053. Rodigas, Émile. (1886). ***Angraecum citratum*** Thouars. *L'Illustration Horticole* 33: 59 – 60, t. 662.

1054. Rodigas, Émile. (1893). ***Eulophia pulchra.*** *L'Illustration Horticole* 181: 79.

1055. Rogers, A. E. T. (1944). **Notes on *Eulophiella.*** *American Orchid Society Bulletin* 12,10: 335 – 336.
Reprinted in *loc. cit.* 21: 77 – 78 (1952).

1056. Rolfe, Robert Allen. (1890). ***Angraecum citratum* Thouars.** *Lindenia* 5 (2): 95 (257). pl. 238.

1057. Rolfe, R. Allen. (1890). ***Phaius* × Cooksoni.** *Gardeners' Chronicle* ser. 3, 7: 388 – 389.

1058. Rolfe, R. Allen. (1890). ***Angraecum primulinum.*** *Gardeners' Chronicle* ser. 3, 7: 388.

1059. Rolfe, R. A. (1890). **Revue des orchidées nouvelles ou peu connues.** *Le Journal des Orchidées,* 1: 71.
A short note on *Angraecum* × *primulinum.*

1060. Rolfe, R. Allen. (1891). ***Eulophiella elisabethae* L. Lind. et Rolfe** *Lindenia* 7 (3): 77 (49), pl. 325.

1061. Rolfe, R. Allen. (1891). ***Phaius tuberculosus* Blume.** *Lindenia* 7 (3): 79 – 80 (52), pl. 326.

1062. Rolfe, R. Allen. (1891). **New and little known Madagascan plants, collected and enumerated by G. F. Scott Elliot. Orchidaceae.** *Journal of the Linnean Society* 29: 50 – 59 & 67, pl 11 – 12.

1063. Rolfe, R. A. (1891). **Revue des orchidées nouvelles ou peu connues.** *Le Journal des Orchidées,* 2: 165 – 166.
A short note on *Aeranthus brachycentron* and *Cirrhopetalum thouarsii.*

1064. Rolfe, R. Allen. (1892). ***Cymbidium humblotii* and *Cymbidium flabellatum.*** *Gardeners' Chronicle* ser. 3, 12: 8.

1065. Rolfe, R. Allen. (1892). ***Cymbidium humbloti* Rolfe n. sp. et *C. flabellatum* Lindl.** *L'Orchidophile* 12, 132: 161 – 162.

1066. Rolfe, R. Allen. (1892). ***Grammangis ellisi* Rchb.f.** *Lindenia* 8 (3): 7 – 888), pl. 338.

1067. Rolfe, R. Allen. (1892). ***Angraecum articulatum* Rchb.f.** *Lindenia* 8 (3): 91 (213), pl. 380.

1068. Rolfe, R. Allen, ed. (1893). ***Cynorchis grandiflora.*** *Orchid Review* 1, 2: 59 – 60.

1069. Rolfe, R. Allen, ed. (1893). **Notes.** *Orchid Review* 1, 7: 194.

1070. Rolfe, R. Allen. (1893). ***Eulophiella elisabethae.*** *Orchid Review* 1, 7: 207 – 208.

1071. Rolfe, R. Allen, ed. (1893). *Eulophiella elisabethae. Orchid Review* 1, 8: 234.

1072. Rolfe, R. Allen, ed. (1893). **Notes.** *Orchid Review* 1, 12: 354.

1073. Rolfe, R. Allen. (1893). **New orchids, decade 6.** *Bulletin of Miscellaneous Information Kew.* 169 – 173.

1074. Rolfe, R. Allen. (1894). **Notes.** *Orchid Review* 2, 16: 97 & 130.

1075. Rolfe, R. Allen. (1895). **New orchids, decade 14.** *Bulletin of Miscellaneous Information Kew.* 191 – 195.

1076. Rolfe, R. Allen, ed. (1895). *Eulophiella elisabethae. Orchid Review* 3, 28: 108.

1077. Rolfe, R. Allen. (1895). **List of garden orchids.** *Gardeners' Chronicle*, 18: 580 – 581.
The following affects the flora of Madagascar: p. 581, a description of *Cyrtopera biterculata* Rolfe. Habitat not known, flowered in Glasnevin in June 1890. Mr. Moore had purchased it at Stevens' Sale Rooms in 1881.

1078. Rolfe, R. Allen. (1896). **Revision of the genus** *Vanilla. Journal of the Linnean Society* 32: 439 – 476.

1079. Rolfe, R. Allen, ed. (1896). *Angraecum fournierae. Orchid Review* 4, 43: 196.

1080. Rolfe, R. Allen. (1897). *Angraecum eburneum* **and its allies.** *Orchid Review* 5, 49: 19 – 21.

1081. Rolfe, R. Allen. (1897). *Eulophiella peetersiana. Orchid Review* 5, 51: 67.

1082. Rolfe, R. Allen. (1897). *Eulophiella peetersiana. Orchid Review* 5, 52: 101 & 206.

1083. Rolfe, R. Allen. (1897). **Lecture notes, orchids abroad and at home.**
Orchid Review 5, 52: 105 – 109.

1084. Rolfe, R. Allen. (1897). *Aeranthes dentiens. Orchid Review* 5, 54: 164.

1085. Rolfe, R. Allen, ed. (1898). **Notes.** *Orchid Review* 6, 64: 99.

1086. Rolfe, R. Allen, ed. (1898). *Eulophiella peetersiana. Orchid Review* 6, 64: 104.

1087. Rolfe, R. Allen, ed. (1898). *Eulophiella peetersiana. Orchid Review* 6, 66: 177 – 178.

1088. Rolfe, R. Allen, ed. (1898). **Notes.** *Orchid Review* 6, 67: 194.

1089. Rolfe, R. Allen, ed. (1898). *Eulophiella peetersiana.Orchid Review* 6, 72: 357.

1090. Rolfe, R. Allen. (1898). *Orchidaceae.* in W. T. Thiselton-Dyer, ed., *Flora of Tropical Africa.*
Lovell Reeve & Co., London, 578 pp. Reprinted Bishen Singh Mahendra Pal Singh, India (1984).

1091. Rolfe, R. Allen. (1899). **Vanilla culture.** *Orchid Review.* 7, 7: 41 – 43.

1092. Rolfe, R. Allen. (1899). *Peristylis glaberrimus* **and** *P. madagascariensis. Orchid Review* 7, 73: 4.

1093. Rolfe, R. Allen, ed. (1899). **Orchids at the Royal Horticultural Society.** *Orchid Review* 7, 74: 62.

1094. Rolfe, R. Allen, ed. (1899). *Angraecum ellisii. Orchid Review* 7, 75: 81 – 82.

1095. Rolfe, R. Allen, ed. (1899). *Dies Orchidianae. Orchid Review* 7, 83: 328.

1096. Rolfe, R. Allen, ed.(1899). *Eulophiella peetersiana. Orchid Review* 7, 84: 358 – 359.

1097. **Rolfe**, R. Allen, ed. (1900). *Angraecum sanderianum. Orchid Review* 8, 89: 152.

1098. **Rolfe**, R. Allen. (1900). *Eulophiella hamelinii. Orchid Review* 8, 91: 197 – 198.

1099. **Rolfe**, R. Allen, ed. (1900). **The hybridist.** *Orchid Review* 8, 92: 227.

1100. **Rolfe**, R. Allen, ed. (1900). *Phaius tuberculosus. Orchid Review* 8, 93: 261.

1101. **Rolfe**, R. Allen in **Warpur**, M. G. (1901). **Habitats of Madagascar orchids.** *Orchid Review* 9, 97: 10 – 11. (Also see 148).

1102. **Rolfe**, R. Allen. (1901). *Cynorchis purpurascens. Orchid Review* 9, 97: 20.

1103. **Rolfe**, R. Allen, ed. (1901). **The hybridist.** *Orchid Review* 9, 98: 37.

1104. **Rolfe**, R. Allen. (1901). *Phaius tuberculosus. Orchid Review* 9, 98: 41 – 44.

1105. **Rolfe**, R. Allen, ed. (1901). *Angraecum sesquipedale. Orchid Review* 9, 98: 49 – 50.

1106. **Rolfe**, R. Allen, ed. (1901). *Phaius tuberculosus. Orchid Review* 9, 99: 72.

1107. **Rolfe**, R. Allen. (1901). *Aeranthes ramosa. Orchid Review* 9, 107: 352.

1108. **Rolfe**, R. Allen. (1901). **New Orchids, Decade 25.** *Bulletin of Miscellaneous Information Kew* 3: 149 – 150.

1109. **Rolfe**, Robert Allen. (1902). *Phaius fragrans. Orchid Review* 10, 109: 5.

1110. **Rolfe**, R. Allen, ed. (1902). *Aeranthes ramosa. Orchid Review* 10, 113: 143.

1111. **Rolfe**, R. Allen. (1902). *Cymbidium rhodochilum. Orchid Review* 10, 114: 184.

1112. **Rolfe**, R. Allen, ed. (1902). *Dies Orchidiani. Orchid Review* 10, 115: 193.

1113. **Rolfe**, R. Allen, ed. (1902). *Dies Orchidiani. Orchid Review* 10, 116: 227.

1114. **Rolfe**, Robert Allen. (1902). *Listrostachys bracteosa. Orchid Review* 10, 117: 6.

1115. **Rolfe**, R. Allen, ed. (1902). *Bulbophyllum hamelinii. Orchid Review* 10, 117: 284.

1116. **Rolfe**, R. Allen. (1903). **Events of 1902.** *Orchid Review* 11, 122: 36.

1117. **Rolfe**, R. Allen, ed. (1903). **A group of** *Phaius. Orchid Review* 11, 125: 136 – 138.

1118. **Rolfe**, R. Allen. (1903). *Bulbophyllum clavatum* **and its allies.** *Orchid Review* 11, 126: 190.

1119. **Rolfe**, R. Allen. (1903). **Notices of books.** *Orchid Review* 11, 127: 202 – 203.

1120. **Rolfe**, R. Allen, ed. (1903). **The hybridist.** *Orchid Review* 11, 127: 219.

1121. **Rolfe**, R. Allen. (1904). **The genus** *Mystacidium. Orchid Review* 12, 134: 46 – 47.

1122. **Rolfe**, R. Allen. (1904). *Vanilla humblotii. Orchid Review* 12, 139: 196 – 197.

1123. **Rolfe**, R. Allen, ed. (1904). **Societies.** *Orchid Review* 12, 141: 268 – 269.

1124. **Rolfe**, R. Allen, ed. (1904). *Angraecum sanderianum. Orchid Review* 12, 143: 337 – 338.

1125. **Rolfe**, R. Allen. (1905). *Phaius tetragonus* **and** *P. luridus. Orchid Review* 13, 149: 151 – 152.

1126. **Rolfe**, R. Allen. (1905). **Societies.** *Orchid Review.* 13, 152: 242 – 244.

1127. **Rolfe**, R. Allen. (1905). *Eulophia paniculata.* *Gardeners' Chronicle* ser. 3, 38: 197 – 198.

1128. **Rolfe**, Robert Allen. (1905). *Vanilla humblotii.* *Curtis's's Botanical Magazine* 131: t.7996.

1129. **Rolfe**, R. Allen. (1905). *Bulbophyllum crenulatum.* *Curtis's Botanical Magazine.* t.8000.

1130. **Rolfe**, R. Allen. (1906). *Polystachya bicolor.* *Kew Bulletin:* 114.

1131. **Rolfe**, R. Allen, ed. (1906). **Novelties of 1905.** *Orchid Review,* 14, 158: 40.

1132. **Rolfe**, R. Allen. (1906). *Cymbidium rhodochilum.* *Orchid Review,* 14,163: 209 – 210.

1133. **Rolfe**, R. Allen. (1906). **Novelties.** *Orchid Review,* 14, 163: 219 – 221.

1134. **Rolfe**, R. Allen. (1906). *Cirrhopetalum thouarsii.* *Orchid Review,* 14,164: 232 – 233.

1135. **Rolfe**, R. Allen, ed. (1906). **Orchids at Clare Lawn, East Sheen.** *Orchid Review,* 14, 166: 299 – 300.

1136. **Rolfe**, R. Allen, ed. (1906). *Cynorchis purpurascens.* *Orchid Review,* 14, 166: 305 – 306.

1137. **Rolfe**, R. Allen. (1906). **New orchids, decade 27.** *Kew Bulletin (Bulletin of Miscellaneous Information):* 88.

1138. **Rolfe**, R. Allen. (1906). **New orchids, decade 29.** *Kew Bulletin (Bulletin of Miscellaneous Information):* 378.

1139. **Rolfe**, R. Allen, ed. (1907). *Cynorchis lowiana.* *Orchid Review,* 15, 172: 121 – 122.

1140. **Rolfe**, R. Allen. (1907). *Cymbidium humblotii.* *Orchid Review,* 15, 176: 245 & 248.

1141. **Rolfe**, R. Allen. (1908). **New orchids, decade 31.** *Kew Bulletin, (Bulletin of Miscellaneous Information)* 1: 69.

1142. **Rolfe**, R. Allen. (1908). *Cynorchis.* *Orchid Review,* 16, 184: 112.

1143. **Rolfe**, R Allen, ed. (1908). *Angraecum hyaloides.* *Orchid Review,* 16, 185: 137.

1144. **Rolfe**, R. Allen, ed. (1908). **Societies.** *Orchid Review,* 16, 190: 306.

1145. **Rolfe**, R. Allen, ed. (1909). **Societies.** *Orchid Review,* 17, 202: 306.

1146. **Rolfe**, R. Allen. (1910). **XL.-New orchids: decade 36.** *Kew Bulletin (Bulletin of Miscellaneous Information).* 8: 280.

1147. **Rolfe** Robert Allen & **Hurst** Charles Chamberlain. (1909). *The orchid stud-book: An enumeration of hybrid orchids of artificial origin.* Frank Leslie & Co., Kew: 327 pp., 119 photos.

1148. **Rolfe**, R. Allen, ed. (1911). *Cynorchis purpurascens.* *Orchid Review,* 19, 225: 272 – 274.

1149. **Rolfe**, R. Allen. (1912). **The genus** *Eulophiella.* *Orchid Review,* 20: 136 – 140.

1150. **Rolfe**, R. Allen. (1912). **Novelties.** *Bulbophyllum trifarium.* *Orchid Review,* 20 238: 294.

1151. **Rolfe**, R. Allen. (1913). **Our note book.** *Orchid Review,* 21, 244: 105.

1152. **Rolfe**, R. Allen. (1913). **Orchids at Haywards Heath.** *Orchid Review,* 21, 245: 137 – 138.

1153. **Rolfe**, R. Allen. (1913). *Phaius tuberculosus.* *Orchid Review,* 21, 245: 144.

1154. Rolfe, R. Allen, ed. (1913). *Cymbidiella humblotii. Orchid Review*, 21, 245: 156.

1155. Rolfe, R. Allen, ed. (1913). *Angraecum recurvum. Orchid Review*, 21, 252: 369.

1156. Rolfe, R. Allen. (1913). Orchideae. In Sir William T. Thiselton-Dyer, ed., **Flora Capensis — being a systematic description of the plants of the Cape Colony, Caffria and Port Natal**, 5, 3: 262 – 263. Lovell Reeve & Co., London.

1157. Rolfe, R. Allen. (1914). **Culture of** *Phaius simulans. Orchid Review*, 22, 260: 254 – 255.

1158. Rolfe, R. Allen. (1914). *Grammangis ellisii. Orchid Review*, 22, 260: 273 – 274.

1159. Rolfe, R. Allen. (1914). **Angraecums in Madagascar.** *Orchid Review*, 22, 260: 283 – 285.

1160. Rolfe, R. Allen. (1915). *Eulophiella elisabethae. Orchid Review*, 26,303: 145 – 146.

1161. Rolfe, R. Allen, ed. (1916). *Cymbidium humblotii. Orchid Review*, 24, 284: 207.

1162. Rolfe, R. Allen. (1916). *Cymbidium humblotii. Orchid Review*, 26,285: 218.

1163. Rolfe, R. Allen, ed. (1917). *Angraecum bilobum. Orchid Review*, 25, 289: 9 – 12.

1164. Rolfe, R. Allen, ed. (1917). *Eulophiella* **Rolfei.** *Orchid Review*, 25, 291: 51 – 52.

1165. Rolfe, R. Allen. (1917). **Mistakes in orchidology.** *Orchid Review*, 25, 296: 182 – 184.

1166. Rolfe, R. Allen. (1918). *Cymbidiella rhodochila. Orchid Review*, 26,303: 57 – 59.

1167. Rolfe, R. Allen, ed. (1918). *Angraecum gracilipes. Orchid Review*, 26, 306: 129.

1168. Rolfe, R. Allen. (1918). *Angraecum gracilipes. Curtis's's Botanical Magazine* 144, 4 t.8758.

1169. Rolfe, R. Allen, ed. (1918). *Cymbidiella humblotii. Orchid Review*, 26, 306: 239 & frontispiece.

1170. Rolfe, R. Allen, ed. (1918). *Angraecum ellisii. Orchid Review*, 26, 306: 249.

1171. Rolfe, R. Allen. (1918). *Bulbophyllum hamelinii. Curtis's Botanical Magazine* 144: t.8785.

1172. Rolfe, R. Allen. (1918). **New orchids: decade XLVI.** *Kew Bulletin (Bulletin of Miscellaneous Information)*, 1: 234 – 238.
The description of *Bulbophyllum robustum* from Central Madagascar, collected by the Rev. R. Baron. This plant was first mentioned without description by Rolfe in the Botanical magazine t.8000 (see Rolfe, 1905).

1173. Rolfe, R. Allen. (1918). **New orchids.** *Orchid Review*, 26, 311 – 2: 232.
It is reported that *Bulbophyllum robustum* was described as a new species in Kew Bulletin (see Rolfe, 1907 & 1918).

1174. Rolfe, R. Allen. (1922). **New orchids: decas XLIX.** *Bulletin of Miscellaneous Information Kew* 1: 23 – 24

1175. Rose, James. (1996). **Flavor of the month:** *Aerangis. Orchids* (AOS.). 286 – 288.

1176. Röth, Jurgen. (2000). *Cymbidiella pardalina. Die Orchidee*, 51, 6: 695.
Four photographs illustrate the species, with a short history.

1177. Rotor, Gavaino II. (1983). *Aerangis citrata* **'Crestwood' CCM/AOS.** *American Orchid Society Bulletin*, 52, 2: 150.

1178. Rouillard, Guy & **Guého**, Joseph. (1990s). *Les plantes et leur histoire a L'Ile Maurice.* MSM Ltd.: 752 pp.

In this general work on the plants of Mauritius some of the more interesting native orchid species are discussed, habitats and short descriptions are given. The cultivation of a number of non-indigenous plants is also outlined.

1179. Roullet, Jacques. (1993). **Une corde bien emmelee.** *L' Orchidée*, 11, 43: 96 – 97.

1180. Royal Gardens, Kew. (1896). *Hand-list of orchids cultivated in the Royal Gardens.* HMSO, London: pp. 225.

1181. Royal Gardens, Kew. (1904). Preface to Second ed. by W.T.Thiselton-Dyer ?. **Hand-list of orchids cultivated in the Royal Gardens.** HMSO, London: pp. 229.

1182. Royal Gardens, Kew. (1962). *Hand-list of orchids cultivated in the Royal Gardens.* Third ed. HMSO, London: pp. 77.

Royal Horticultural Society, see Sander F. & Co. 1187.

1183. Salisbury, Edward James. (1957). **Henry Nicholas Ridley 1855 – 1956.** *Biographical Memoirs of Fellows of the Royal Society*, 3: 140 – 159.

1184. Sander & Sons. (1901?). *Sander's orchid guide.* Sander & Sons, St Albans: pp. 256.

1185. Sanders & Sons. (>1912). *A catalogue of orchids.* Sanders & Sons, St.Albans: pp. 110.

1186. Sanders. (1927). *Sander's orchid guide* **1927 edition.** Waterlow & Sons, London: pp. 451.

1187. Sander F. & Co. (1946 – 1995). **Sander's list of orchid hybrids.** F. Sander & Co., St. Albans (later published by the Royal Horticultural Society).

1188. Sander, David Fearnley, ed. (1970). *Eulophiella roempleriana.* *Orchid Review*, 78, 930: 350 & 351 – 353.

1189. Sander, David Fearnley. (1971). *Sobennikoffia fournerieriana.* *Orchid Review*, 79, 932: 52.

1190. Sander, David Fearnley, ed. (1971). *Eulophiella perrieri* **'Sherwood'.** *Orchid Review*, 79, 936: 161 – 162 &173.

1191. Sandhack, A. (1913). **Arbeitskalender für Juni.** *Orchis*, 7, 5: 63 – 64.

1192. Sanford, William W. (1968). **The genus** *Aerangis.* *American Orchid Society Bulletin*, 37, 6: 490 – 493.

1193. Sauvennet, G. (1897). **Note sur la culture du** *Phajus humbloti.*
La Semaine Horticole, 1, 33: 339 – 340.

1194. Schatz, George E. (1992). **The Race Between Deep Flowers And Long Tongues.** *Wings*, 16, 3: 19 – 21.

1195. Schiller, G. W. (1857). *Catalog der Orchideen-Sammumg von G. W. Schiller.* F.H. Nestler & Melle, Hamburg: 76 pp.

1196. Schlechter, Rudolf. (1898). *Orchidaceae africanae.* *Botanische Jahrbücher Syst.* 26: 330 – 344.

1197. Schlechter, Rudolf. (1898). *Orchidaceae africanae novae velminus cognitae.* *Englers Botanische Jahrbücher*, 24: 418 – 433.

1198. Schlechter, Rudolf. (1898). *Orchidaceae africanae.* *Englers Botanische Jahrbücher,* 26: 330 – 344. The description of *Angraecum anocentrum.*

1199. Schlechter, Rudolf. (1898). **Monographie der** *Disperideae.* *Bulletin de L'Herbier Bossier*, 6, 12: 929 – 955.

1200. Schlechter, Rudolf. (1899). **Revision der Gattung** *Holothrix*. *Österr. Bot. Zeitshrift*, 1: 1 – 16.

1201. Schlechter, Rudolf. (1901). **Monographie der** *Diseae*. *Englers Botanische Jahrbücher*, 31: 165 – 221.
A monograph of the Diseae, some species from Madagascar are included.

1202. Schlechter, Rudolf. (1906). *Orchidaceae africanae*, **IV.** *Englers Botanische Jahrbücher*, 38: 144 – 165.

1203. Schlechter, Rudolf. (1911). **Die** *Polychondreae* (*Neottiinae* Pfitz.) **und ihre systematische Einteilung.** *Englers Botanische Jahrbücher*, 45: 399 – 405.

1204. Schlechter, Rudolf. (1912). **Neue und seltene Garten-Orchideen.** *Orchis*, 6: 114 – 115, pl. 25.

1205. Schlechter, Rudolf. (1913). **Orchidacées de Madagascar.** *Orchidaceae Perrieranae Madagascarienses.* *Annals du Musée Colonial de Marseille*, (3rd series): 148 – 202, 24 pl. Also, *Extrait Annals du Musée Colonial de Marseille*. Augustin Challamel, Marseille & Paris.

1206. Schlechter, Rudolf. (1914). *Die Orchideen.* Verlag Paul Parey, Berlin. (Second editon, see 1308).

1207. Schlechter, Rudolf. (1915). *Orchidaceae Stolzianae.* *Botanische Jahrbücher*, 53: 483 – 597.

1208. Schlechter, Rudolf. (1915). **Kritische Aufzählung der bisher von Madagaskar, den Maskarenen, Komoren und Seychellen bekantgewordenen Orchidaceen.** *Beihefte zum Botanischen Centralblatt*, 33, 2: 390 – 440.

1209. Schlechter, Rudolf. (1915). **Die Gattungen** *Grammatophyllum* Bl. **und** *Grammangis* **Rchb.f.** *Orchis*, 9: 99 – 109, 115 – 122.

1210. Schlechter, Rudolf. (1916 – 1917). *Orchidaceae Perrierianae (Collectio secunda).* *Beihefte zum Botanischen Centralblatt*, 34, 2: 294 – 341 (pp. 1 – 341 were published in 1916, pp. 342 onwards in 1917).

1211. Schlechter, Rudolf. (1918). *Angraecum* **Wolterianum** Schltr. *Orchis*, 12, 3: 60 – 62.

1212. Schlechter, Rudolf. (1918). **XXXIII** *Orchidaceae novae et criticae.* **Decas. LI-LIII.** *Additamenta ad Orchideologiam Guatemalensem. Repertorium specierum novarum regni vegetabilis* 15, 427 – 433: 194 – 217.

1213. Schlechter, Rudolf. (1918). **Versuch einer natürlichen Neuordnung der afrikanischen angreakoiden Orchidaceen.** *Beihefte zum Botanischen Centrablatt*, 36 62 – 181.

1214. Schlechter, Rudolf. (1918). **Attempt at a natural new classification of the African angraecoid orchids.** Being a translation from: Versuch einer naturlichen neuordung der Afrikanischen Angraekoiden Orchidaceen, Beihefte Botanische Centralblatt XXXVI, Abt. 11: 62 – 181. Edited and translated by **Katz** H.J. & **Simmons** J. T. (1986). Australian Orchid Foundation: 105 pp.

1215. Schlechter, Rudolf. (1918 – 19). *Additamenta ad orchideologiam madagascarensem.* XLII *Orchidaceae novae et criticae. Decas* LV – LVII. *Repertorium specierum novarum regini vegetabilis* XV: 324 – 340.

1216. Schlechter, Rudolf. (1919). *Angraecum elephantinum.* *Notizblatt der Botanischer Gartens zu Berlin* 7, 67: 330.

1217. Schlechter, Rudolf. (1921) **Revision der gattungen** *Schizochilus* Sond. **und** *Brachycorythis* Ldl. *Beihefte zum Botanischen Centrablatt* 38, 2: 80 – 132.

1218. Schlechter, Rudolf. (1920). *Eulophiella peetersiana.* *Gartenflora* 14, 27.

1219. Schlechter, Rudolf. (1920). **Die Gattung** *Eulophiella* **Rolfe.** *Orchis* 14: 24 – 30.

1220. Schlechter, Rudolf. (1922). *Additamenta orchideologiam madagascarensem* **II.** *Repertorium Specierum Novarum Regni Vegetabilis* 18: 320 – 326.

1221. Schlechter, Rudolf. (1924 – 1925). *Orchidaceae Perrieranae. Ein Beitrag zur Orchideenkunde der Insel Madagasacar. Repertorium Specierum Novarum Regni Vegetabilis* 33: 391 pp. Reprint by Koeltz (1980).

1222. Schlechter, Rudolf. (1927). *Die Orchideen ihre Beschreibung, Kultur und Züchtung.* Berlin: 959 pp.

Schlechter, Rudolf. *Die Orchideen* ed.3. See Senghas 1259 & 1257.

1223. Schlechter, Rudolph, prepared by **Mansfeld** Rudolf. (1932). **Blütenanalysen neuer Orchideen III. Afrikanische und Madagassische Orchideen.** *Repertorium Specierum novarum regni vegetabilis* 68. Reprinted by O. Koeltz (1973).

1224. Schofield, J. (1981). **Orchids from Madagascar.** *Journal of the Orchid Society of Great Britain* 30, 4: 128 – 130.

1225. Schröter, C. (1938). **Noch einmal das "Madagassische Orchideenrätsel".** *Volks Hochschule* 7, 4: 126 – 127.

1226. Schumann, Karl Moritz. (1898). **Neue Arten der Siphonogamen 1898.** *Just's Botanischer Jahresbericht* 26, 1: 336.

1227. Schumann, Karl Moritz. (1899). **Neue Arten der Siphonogamen 1899.** *Just's Botanischer Jahresbericht* 27, 1: 67.

1228. Schuurman, Derek & **Ravelojaona** Nivo. (1997). *Madagascar.* New Holland. Cape Town, 128 pp. A tourist guide to the Island. Orchids are mentioned on p. 9.

1229. Scottie. (1891). *Angraecum eburneum. Gardeners' Chronicle* ser. 3, 10: 0.

1230. Scully, Robert M. (1963). **Angraecums I have grown.** *American Orchid Society Bulletin* 32, 2: 95 – 98.

1231. Scully, Robert M. (1964). *Angraecum sesquipedale. Orchid Review* 72, 852: 195.

1232. Seaton, Phillip. (1992). **Species portrait-***Angraecum magdalenae. Orchid Review* 100, 1187: 308 – 309.

1233. Seidenfaden, Gunnar. (1977) **Thalia Maravara and the rigid air-blossom.** *Botanical Museum Leaflets of Harvard University* 25, 2: 49 – 69.

1234. Seidenfaden, Gunnar. (1978). **Orchid genera in Thailand VI.** *Neottioideae* Ldl. *Dansk Botanisk Arkiv* 32, 2: 195.

1235. Senghas, Karlheinz. (1962). *Cymbidiella rhodochila* Rolfe. *Die Orchidee* 13, 1: 17 – 20.

1236. Senghas, Karlheinz. (1962). *Grammangis ellisii* Rchb.f. *Die Orchidee* 13, 1: 20 – 24.

1237. Senghas, Karlheinz. (1963). **Die angraekoiden Orchideen Afrikas und Madgaskars.** *Die Orchidee* 14, 2: 65 – 72.

1238. Senghas, Karlheinz. (1963). **Die Gattung** *Oeoniella.* **Die angraekoiden Orchideen Afrikas und Madagaskars II.** *Die Orchidee* 14, 5: 215 – 218.

1239. Senghas, Karlheinz. (1964). **Sur quelques orchidées nouvelles ou critiques de Madagascar.** *Adansonia* 4, 2: 301 – 314.

1240. Senghas, Karlheinz. (1964). **Die Gattung** *Graphorkis. Die Orchidee* 15: 61 – 66.

1241. Senghas, Karlheinz. (1964). **Uber die verbreitung von** *Acampe pachyglossa* Rchb.f. *Die Orchidee* 15: 2 – 165.

1242. Senghas, Karlheinz. (1964). *Nervilia bathiei*, eine neue Orchidee aus Madagascar. *Die Orchidee* 15: 17 – 222.

1243. Senghas, Karlheinz. (1964). *Nervilia bathiei. Die Orchidee* 16: 68 – 69.

1244. Senghas, Karlheinz. (1967). **Deux nouveau** *Eulophidium* **du Nord de Madagascar.** *Adansonia* ser. 2, 6, 4: 557 – 562.

1245. Senghas, Karlheinz. (1967). *Jumellea rossii*, **eine neue Orchidee von der Insel Réunion. (Die angraecoiden Orchideen Afrikas und Madagakars, V.,** *Jumellea*, **A.).** *Die Orchidee* 18, 5: 240 – 245.

1246. Senghas, Karlheinz. (1967). **Die Gattung** *Eulophidium* **Pfitzer, mit zwei neuen Arten aus Madagaskar.** *Die Orchidee* 18, 5: 245 – 250.

1247. Senghas, Karlheinz. (1969). *Aerangis buchlohii*, **eine neue Orchidee aus Madagaskar. (Die angraecoiden Orchideen Afrikas und Madagakars, VI.** *Die Orchidee* 20, 1: 12 – 16.

1248. Senghas, Karlheinz. (1970). *Cymbidiella flabellata* **(Th.) Rolfe.** *Die Orchidee* 21, 5: 286 – 289.

1249. Senghas, Karlheinz. (1972). **Die Gattung** *Plectrelminthus* **Raf. (Die angraecoiden Orchideen Afrikas und Madagakars, IX.** *Die Orchidee* 23: 221 – 227.

1250. Senghas, Karlheinz. (1973). *Eulophidium rauhii*, **ein pflanzengeographisch bedeutsamer Neufund aus Madagascar.** *Die Orchidee* 24, 2: 57 – 61.

1251. Senghas, Karlheinz. (1973). *Angraecum bosseri*, **ein neuer Stern von Madagaskar.** *Die Orchidee* 24: 191 – 193.

1252. Senghas, Karlheinz. (1975). **Standortvielfalt tropischer Orchideen, dargestellt am Meispiel Madagaskars.** *Wiss. Beitr. Universität Halle (DDR)*: 63.

1253. Senghas, Karlheinz. (1979). *Angraecum eburneum* **Bory 1804.** *Die Orchidee*, Orchideenkartei 30, 1: 120 – 121.

1254. Senghas, Karlheinz. (1980). *Angraecum longicalcar*: **Orchidee mit den längsten Nektarspornen.** *Mitteilungen aus dem Botanischen Garten St. Gallen* (Gärtnerische Botanischer Brief) 29,11: 39.

1255. Senghas, Karlheinz. (1982). **Sortiment und Kultur der Gattung Angraecum. 1. Teil: Die madagassischen Arten.** *Orchideen (DDR)* 16,1: 18 – 31.

1256. Senghas Karlheinz. (1983). *Aerangis seegeri*, **eine neue Orchidee aus Madagaskar.** *Die Orchidee* 34,1: 23 – 25.

1257. Senghas, Karlheinz, ed. (1984). in Schlechter, Rudolf. **Die Orchideen,** ed. 3, 1A, 15: 881 – 886.

1258. Senghas, Karlheinz. (1985). *Cribbia* **und** *Microterangis*, **zwei neue Orchideen Gattungen (Die angraekoiden Orchideen Afrikas und Madagaskars, XIV).** *Die Orchidee* 36, 1: 19 – 22.

1259. Senghas, Karlheinz. (1986). **Angraecinae. Aerangidinae.** in Rudolf Schlechter, *Die Orchideen*, 3rd. ed. Paul Parey, Hamburg.

1260. Senghas, Karlheinz. (1986). *Angraecum calceolus* **Thouars.** *Die Orchidee, Orchideenkartei*, 37: 51 – 52.

1261. Senghas, Karlheinz. (1987). **Die Gattung** *Aeranthes* **mit zwei neuer Arten aus Madagascar.** *Die Orchidee* 38, 1: 2 – 9.

1262. Senghas, Karlheinz. (1988). *Acampe pachyglossa* **Rchb.f 1881.** *Die Orchidee*, Orchideenkartei 39: 545 – 546.

1263. Senghas, Karlheinz. (1989). *Polystachya concreta* **(Jacq.) Garay & Sweet 1974.** *Die Orchidee*, Orchideenkartei 40, 3: 569 – 572.

1264. Senghas, Karlheinz. (1993). *Orchideen. Plants of Extremes, Contrasts, and Superlatives.* Paul Parey, Berlin & Hamburg: 182 pp.

1265. Senghas, Karlheinz. (1996). *Grammangis ellisii* **(Lindl.) Rchb.f. 1860.** *Die Orchidee*, Orchideenkartei 47, 1: 819 – 820.

1266. Senghas, Karlheinz. (1997). **Vielfalt tropischer und suptropischer Orchideenstandorte und ihre Übertragung in die Kultur, dargestellt am Beispiel der Gattund** *Angraecum* **in Madagaskar mit einer neuen Art,** *Angraecum dollii.* *Journal für den Orchideenfreund* 4, 1: 13 – 25.

1267. Senghas, Karlheinz. (1997). **Vielfalt tropischer und suptropischer Orchideenstandorte und ihre Übertragung in die Kultur, dargestellt am Beispiel der Gattund** *Angraecum* **in Madagaskar mit einer neuen Art,** *Angraecum dollii* **Teil II.** *Journal für den Orchideenfreund* 4, 2: 103 – 106.

1268. Senghas, Karlheinz. (1997). **Die Gattung** *Gastrorchis*, **mit einerneuen Art,** *Gastrorchis steinhardtiana.* *Journal für den Orchideenfreund* 4, 3: 130 – 134.

1269. Senghas, Karlheinz. (1998). **Vielfalt tropischer und suptropischer Orchideenstandorte und ihre Übertragung in die Kultur, dargestellt am Beispiel der Gattund** *Angraecum* **in Madagaskar Teil III.** *Journal für den Orchideenfreund*, 5, 2: 95 – 100.
The series continues with a further description of the Madagascan Highlands especially the higher altitude areas. The cultivation of orchids from the area is discussed.

1270. Senghas, Karlheinz. (1999). **Vielfalt tropischer und suptropischer Orchideenstandorte und ihre Übertragung in die Kultur, dargestellt am Beispiel der Gattund** *Angraecum* **in Madagaskar Teil IV.** *Journal für den Orchideenfreund*, 5, 1: 35 – 41.
The series on the orchids of Madagascar continues with the description of specialised habitats. To accompany the previous part *Nervilia bathiei* is depicted (p. 35 – 36). An extensive discussion of *Angraecum aloifolium* is given. It is explained that the species was found by Hilmar Doll and Rolf Hermann and was going to be described by the author as *A. rugosum.* Simultaneously the species was also found by Hermans & Cribb (see 466) who described it.

1271. Senghas, Karlheinz. (1999). **Vielfalt tropischer und suptropischer Orchideenstandorte und ihre Übertragung in die Kultur, dargestellt am Beispiel der Gattund** *Angraecum* **in Madagaskar Teil V.** *Journal für den Orchideenfreund*, 6, 2: 106 – 113.
The habitats of south-western Madagascar and a selection of orchids are described.

1272. Senghas, Karlheinz. (2000). **Zum Tode von Professor De. Werner Rauh.** *Journal für den Orchideenfreund*, 7, 4: 267 – 270.
Obituary of Werner Rauh, his career and major work, including that in Madagascar, are outlined.

1273. Senghas, Karlheinz. (2001). **Diversité des habitats tropicaux et subtropicaux des orchideés et conséquences sur la culture de celles-ci; exemple du genre** *Angraecum* **à Madagascar.** *Richardiana*, 1, 1: 26 – 37.
A translation into French of the series of articles in *Journal für den Orchideenfreund*.

1273a. Senghas, Karlheinz. (2003). **Orchideen der Trockengebiete.** *Die Sukkulentenwelt*, 8: 34 – 37.
A short analysis of the orchid found in the drier areas of Madagascar.

1274. Severin, Henry J. (1980) **Rare and exotic orchids.** In Sukshom M. R. Kashemsanta, ed., *Proceedings of the 9th World Orchid Conference.* 107 – 110. Amarin Press, Bangkok.

1275. Shaw, Julian. (2003). **The name** *Cynorkis* × *kewensis* **and the grex** *Cynorkis* **Kewensis.** *Orchid Review Supplement*, 111, 1252: 61.
The award of a Botanical Certificate by the RHS Orchid Committee to *Cynorkis lowiana* 'Madagascar' raised interesting questions of typography and nomenclature concerning its sole registered hybrid, *Cynorkis* × *kewensis*.

1276. **Sheehan**, Tom & Marion. (1970). **Orchid genera, Illustrated — XVIII.** *American Orchid Society Bulletin,* 39, 11: 1000.

1277. **Sheehan**, Tom & Marion. (1979). **Orchid genera; Illustrated. 71.** *Aeranthes.* *American Orchid Society Bulletin* 48, 9: 924.

1278. **Sheehan**, Tom & Marion. (1991). **Orchid genera illustrated 140.** *Oeoniella.* *American Orchid Society Bulletin* 60, 5: 459 – 460.

1279. **Sheehan**, Dr Thomas J. (1992). *Angraecum leonis.* *American Orchid Society Bulletin* 61, 7: 703.

1280. **Sheehan**, Tom & Marion. (1993). **Orchid genera illustrated.** *Cymbidiella.* *American Orchid Society Bulletin* 62,9: 918 – 919.

1281. **Sheehan**, Tom & Marion. (1994). *An illustrated survey of orchid genera.* Timber Press, Portland, Oregon: 415 pp.

1282. **Sheehan**, Thomas J. & **Huntington** Merritt. (2003) **Tasty orchids.** *Orchids (AOS),* 72, 12: 894.
A few edible orchids are discussed, including *Cynorkis.*

1283. **Sheehan**, Thomas J. & **Farace** Nancy. (2003). *Vanilla,* **the most versatile orchid.** *Orchids (AOS),* 72, 12: 936 – 939.
Vanilla cultivation and world markets are discussed. The medicinal properties of *V. madagascariensis* are also mentioned.

1284. **Shergold**, Murray. (1998). **Madagascar — today's land of yesteryear.** *Orchids Australia,* 10, 4: 45 – 48.
A broad account of a visit to Madagascar.

1285. **Shuttleworth**, Floyd S., **Zim**, Herbert S. & **Dillon**, Gordon W. (1970). *A Golden Guide. Orchids.* Golden Press, New York: 160 pp, numerous colour drawings.

1286. **Sibree**, James. (1890). *Madagascar.* Trübner & Co. Ludgate Hill, London: p. 372. 4 pl.

1287. **Sibree**, James. (1924). *Register of missionaries 1796 – 1923.* LMS. London: 708 pp.

1288. **Siegerist**, Emly S. (2001). *Bulbophyllums and their allies. A Grower's Guide.* Timber Press, Portland, Oregon: 251 pp.
A selective overview of the genus *Bulbophyllum,* the history of the genus is given, notes on cultivation and brief descriptions of the different sections. A few Madagascan species given a short description, some are illustrated.

1289. **Sigala**, P. **Forêt de Bébour, guide nature et flore des arbres et arbustres.** Office Nationale des Forets La Réunion. c. 1998.
A general nature guide.

1290. **Silbernagl**, Eugen. (1969). *Cryptopus elatus.* *Die Orchidee* 20, 6: 328 – 329.

1291. **Sitch**, John. (1997). **Orchids at Kew.** *Journal of the Orchid Society of Great Britain* 46, 4: 31 – 33.

1292. **Smith**, J. E. (1819). *Disperis.* In Abraham Rees, *Cyclopedia,* 11.

1293. **Smith**, J. J. (1905). *Die Orchideen von Java. Flora von Buitenzorg. 6.* Brill Leiden, 6: 672.

1294. **Smith**, Worthington G. (1875). *Angraecum ellisii.* *The Floral Magazine* 48 pl. 191.

1295. **Sonder**, G. (1846). *Enumeratio orchidearum, quas in Africa Australi Extratropica collegerunt C. F. Ecklon, Dr. et C. Zeyher.* *Linnaea,* 19: 71 – 112.

1296. **Speckmaier**, Manfred. (2000). **Elementos florales Africanos en Venezuela y el Neotropico. El genero** *Oeceoclades.* *Orquideophilo* 8, 2: 6 – 7.
Oeceoclades are described and illustrated including: *O. decaryana, O. spathulifera* and *O. roseovariegata.*

1297. Sprengel, Kurt P. J. (1826). *Systema vegetabilium* ed.10, vol. 3.

1298. Sprunger, Samuel, ed. (1986). ***Orchids from Curtis's Botanical Magazine.*** Cambridge University Press, Cambridge: 525 pp.
The plates from Curtis's's Botanical Magazine from 1787 to 1948 are reproduced. Their nomenclature is updated, but the text is not included. Introduction by Phillip Cribb, the history of the Magazine is outlined, the various artists are listed. A number of Madagascan orchids are included.

Sprunger, Samuel, ed. (1991). ***Orchids from the Botanical Register 1815 – 1847.*** Birkhauser Verlag, Basel: 326 pp. 2 vols.
An edited reprint of the orchids featured in the Botanical Register. A history of the Journal and its editor. See individual entries under Lindley 146, 770 – 775, 773, 777 – 779, 781 – 784, 787.

1299. Standard. ed. (1893). ***Eulophiella elisabethae.*** *London Standard* (letters) 25 Jul.

1300. Standard, ed. (1893). **Mr. Hamelin's adventures in Madagascar.** *London Standard*, Jul. 2: 6.

1301. Stein, B. (1888). ***Eulophia maculata* Rchb. fil.** *Gartenflora* 37: t.1285, 609 – 610.

1302. Stein, B. (1892). ***Stein's Orchideenbuch. Beschreibung, Abbildung und Kulturanweisung.*** Paul Parey Verlag, Berlin: 602 pp.

1303. Stern, William Louis & **Judd**, Walter S. (1999). **Comparative vegetative anatomy and systematics of *Vanilla* (*Orchidaceae*).** *Botanical Journal of the Linnean Society*, 131, 4: 353 – 382.
Vegetative anatomy of seventeen different species of *Vanilla* were studied; *V. madagascarienis* is included, various cross-sections are shown. An overview of literature on *Vanilla* is included. Cladistic relationships are discussed.

1304. Stern, William Louis & **Judd**, Walter S. (2002). **Systematic and comparative anatomy of *Cymbidieae* (*Orchidaceae*).** *Botanical Journal of the Linnean Society*, 139: 1 – 27.
The anatomy of Cymbideae is studied, velamen and pseudobulb vascular bundles are analysed. Cladistic analysis produces only a few tentative hypothesis of phylogenetic relationships among the 28 genera, showing the anatomical characters are of limited value in assessing affinities within the tribe.

1305. Steudel, Ernest. (1840 – 1). ***Nomenclator botanicus, seu Synonymia plantarum universaris.*** Part I, A-C. Part II, D-K. Stuttgart & Tubingen.

1306. Steudel, Ernest. (1821). ***Nomenclator botanicus.*** Part II, L-Z. Stuttgart & Tubingen.

1307. Stevens, Fred & Eileen. (1975). **Madagascar adventure.** *Orchid Review*, 83, 990: 386 – 389.

1308. Stevens, Fred & Eileen. (1976). **Madagascan adventure. 2.** *Orchid Review* 84, 994: 99 – 101.

1309. Stevenson, Steve. (1983). **Notes on exhibition table.** *Atlanta Orchid Society Newsletter* 11: 6 – 7.

1310. Stevenson, Steve. (1983). **Notes on exhibition table.** *Atlanta Orchid Society Newsletter* 12: 7 – 10.

1311. Stevenson, Steve. (1984). **Notes on exhibition table.** *Atlanta Orchid Society Newsletter* 12: 5.

1312. Stewart, Joyce. (1968). **Collecting orchids in the Comoro Islands.** *American Orchid Society Bulletin* 37, 10: 858 & 895 – 903.

1313. Stewart, Joyce. (1968) **Orchids of the Comoro Islands.** *Orchid Review*, 76, 899: 132 – 135.
A short geographical description of the Comoro Islands is followed by an enthralling account of the author's search for orchids. Mount Karthala is visited from where *Angraecum scottianum* is featured, a description and discussion of its habitat are given.

1314. Stewart, Joyce. (1968) **Orchids of the Comoro Island — 2.** *Orchid Review*, 76, 902: 227 – 229.
A further account of a collecting trip to Grande Comore, this contribution features *Angraecum leonis*.

1315. Stewart, Joyce. (1968) **Orchids of the Comoro Islands — 3.** *Orchid Review*, 76, 904: 288 – 289. The habitat, characteristics and cultivation of *Vanilla humblotii* are discussed in detail.

1316. Stewart, Joyce. (1969) **Orchids of the Comoro Islands — 4.** *Orchid Review*, 77, 912: 176 – 179. An in-depth discussion on the identity of *Angraecum germinyanum*. The differences from *A. ramosum* are explained together with its occurrence on the Comoro Islands. An in depth history of the species is given plus some cultural advice.

1317. Stewart, Joyce. (1969) **Orchids of the Comoro Islands — 5.** *Orchid Review*, 77, 914: 240 – 242. The series continues with a detailed discussion of the habitat, characteristics and cultivation of *Angraecum florulentum*.

1318. Stewart, Joyce. (1969) **Orchids of the Comoro Islands — 6.** *Orchid Review*, 77, 918: 382 – 383. *Jumellea comorensis* is described, together with its habitat on Grande Comore. Cultivation and other *Jumellea* species from the islands are also mentioned.

1319. Stewart, Joyce & **Campbell**, Bob. (1970). *Orchids of tropical Africa.* W.H. Allen, London: 117 pp., 45 colour photos.

1320. Stewart, Joyce. (1976). **The vandaceous group in Africa and Madagascar.** In K. Senghas, ed., *Proceedings of the 8th World Orchid Conference:* pp. 239 – 248. D.O.G., Frankfurt.

1321. Stewart, Joyce. (1975). **The genus *Grammangis* Rchb.f.** *Orchid Review* 83, 990: 403 – 405.

1322. Stewart, Joyce. (1976). *Angraecum* **Ol Tukai and some of its relatives.** *Orchid Review* 84, 998: 236 – 238.

1323. Stewart Joyce. (1977) **Orchids for Christmas.** *Orchid Review*, 85, 1014: 271. A flower arrangement of white African orchids is presented, it contains some from Madagascar.

1324. Stewart, Joyce. (1980). **The angraecoid orchids of the African region.** *American Orchid Society Bulletin* 49, 6: 621 – 629.

1325. Stewart, Joyce. (1981). *Eulophiella roempleriana. Flowering Plants of Africa* 46, 3 & 4: pl. 1832.

1326. Stewart, Joyce. (1981). *Cymbidiella falcigera. Flowering Plants of Africa* 46, 3 & 4: pl. 1833.

1327. Stewart, Joyce. (1981). *Angraecum sesquipedale* **var. angustifolium.** *Flowering Plants of Africa* 46, 3 & 4: pl. 1836.

1328. Stewart, Joyce & **Hennesy**, Esmé F. (1981). *Orchids of Africa.* Houghton Mifflin Co., Boston: 159 pp., 50 colour pl.

1329. Stewart, Joyce, **Linder**, H. P., **Schelpe**, E. A. & **Hall**, A. V. (1982). *Wild Orchids of Southern Africa.* Macmillan, Johannesburg, South Africa.: 397 pp., numerous colour photos.

1330. Stewart, Joyce. (1982). **Die *Aerangis*-Verwantschaft.** *Die Orchidee* 33,2: 48 – 57.

1331. Stewart, Joyce. (1983). **Portraits of pretty polystachyas.** *American Orchid Society Bulletin* 52, 11: 1138 – 1149.

1332. Stewart, Joyce. (1984). **Growing angraecoid orchids part – 1. Introduction.** *American Orchid Society Bulletin* 53, 7: 731 – 736.

1333. Stewart, Joyce. (1984). **Growing angraecoid orchids part – 2. *Angraecinae*.** *American Orchid Society Bulletin* 53, 8: 804 – 808.

1334. Stewart, Joyce (1985). In Tan Wie Kiat, ed., *Proceedings of the Eleventh World Orchid Conference* 1984. International Press Co., Singapore: 424 pp.

1335. **Stewart**, Joyce. (1986). **Stars of the Islands — A new look at the genus** *Aerangis* **in Madagascar and the Comoro Islands – 1.** *American Orchid Society Bulletin* 55, 8: 792 – 802.

1336. **Stewart**, Joyce. (1986). **Stars of the Islands — A new look at the genus** *Aerangis* **in Madagascar and the Comoro Islands – 2.** *American Orchid Society Bulletin* 55, 9: 902 – 909.

1337. **Stewart**, Joyce. (1986). **Stars of the Islands — A new look at the genus** *Aerangis* **in Madagascar and the Comoro Islands – 3.** *American Orchid Society Bulletin* 55, 10: 1008 – 1015.

1338. **Stewart**, Joyce. (1986). **Stars of the Islands-A new look at the genus** *Aerangis* **in Madagascar and the Comoro Islands – 4.** *American Orchid Society Bulletin* 55, 11: 1117 – 1125.

1339. **Stewart**, Joyce. (1986). *Eulocymbidiella:* **A new intergeneric hybrid.** *Orchid Review* 94, 1116: 337 – 339.

1340. **Stewart**, Joyce. (1987). *Aerangis spiculata.* *Kew Magazine* 4, 2: 69 – 74.

1341. **Stewart** Joyce. (1988) **Fine species at Westminster:** *Aerangis hyaloides.* *Orchid Review,* 96, 1131: 16 – 17.
A short history of *Aerangis hyaloides* on the occasion of it receiving an award from the RHS. Cultural advice is given.

1342. **Stewart**, Joyce. (1990). *Aerangis,* **stelle dele Isole.** *Orchis* 70

1343. **Stewart**, Joyce, ed. (1992). *Orchids at Kew.* HMSO, London: 155p., numerous colour photos.

1344. **Stewart**, Joyce & **Griffiths**, Mark, eds. (1995). *Manual of orchids.* Timber Press, Portland, Oregon: 388 pp., numerous drawings. Derived from A. Huxley, ed., *The New Royal Horticultural Society Dictionary.* The Royal Horticultural Society, 4 vols., 1992.

1345. **Stewart**, Joyce. (1996). *Orchids of Kenya.* St. Paul's Bibliographies, Winchester: 176 pp., numerous colour photos.

1346. **Stewart**, Joyce. (2002). **Stars of Africa and Madagascar: a new look at the genus** *Aerangis.* In Jean Clark *et al.*, eds, *Proceedings of the 16th World Orchid Conference.* 404 – 405. Vancouver Orchid Society.
A general overview of current thoughts on the genus, a few species from Madagascar are mentioned.

1346b. **Stewart**, Joyce, **Hermans**, Johan, **Campbell**, Bob. (2006). Angraecoid Orchids. Timber Press. Portland, 2004: pp. 431.
A well-illustrated book on the angraecoids of mainland Africa and Madagascar.

1347. **Stiles**, W. A. (1894). **Orchids.** *Scribner's Magazine* 15, 86: 190 – 203.

1348. **Strahn**, Wendy. (1989). *Plant Red Data Book for Rodrigues.* Koeltz Scientific Books, Konigstein.
The conservation status of the plants of Rodrigues is discussed. Each species has details on conservation status, distribution and habitat.

1349. **Strahn**, Wendy. (1996). **Mascarene Islands — an introduction.** *Curtis's Botanical Magazine,* 13, 4: 182 – 185.
An introduction to the Mascarene Islands is provided, including an overview of the geography, origin, topography, population densities, state of the flora and climate.

1350. **Strahn**, Wendy. (1996). **The vegetation of the Mascarene Islands.** *Curtis's Botanical Magazine,* 13, 4: 214 – 217.
The native vegetation types of Mauritius, Réunion and Rodrigues are described.

1351. **Strahn**, Wendy. (1996). **Botanical history of the Mascarene Islands.** *Curtis's Botanical Magazine,* 13, 4: 217 – 219.
The history of the study of the Mascarene Islands flora is presented and the effects of human occupation on the vegetation are discussed.

1352. Strahn, Wendy. (1996). **Conservation of the Flora of the Mascarene Islands.** *Curtis's Botanical Magazine,* 13, 4: 228 – 233.
The current threatened state of the Mascarene flora is described and some remedial measures are proposed.

1353. Strauss, Michael S. and **Koopowitz**, Harold. (1973). **Floral physiology of** *Angraecum* **1. Inheritance of post-pollination phenomena.** *American Orchid Society Bulletin* 42,6: 495 – 502.

1355. Studnicka, Miloslav. (2000). **Aus der Botanischen Garten Liberec:** *Angraecum elephantinum* **Schltr.** *Die Orchidee,* 51, 3: 293.
A short story on *Angraecum elephantinum,* its habitat is described.

1356. Summerhayes, Victor S. (1928). *Cymbidiella humblotii. Curtis's's Botanical Magazine,* 154: t.9216.

1357. Summerhayes Victor S. (1928). **New plants from the Seychelles.** *Bulletin of Miscellaneous Information Kew* 9: 388 – 395.

1358. Summerhayes, Victor S. (1931). **African orchids II.** *Bulletin of Miscellaneous Information, Kew* 7: 378 – 390.

1359. Summerhayes, Victor S. (1942). **African orchids XII.** *Botanical Museum Leaflets Harvard University* 10, 9: 257 – 299.

1360. Summerhayes, Victor S. (1943). **African orchids XIII-The leafless angraecoid orchids.** *Botanical Museum Leaflets Harvard University* 11, 5: 137 – 163.

1361. Summerhayes, Victor S. (1945). **African orchids XV.** *Botanical Museum Leaflets Harvard University* 11, 9: 249.

1362. Summerhayes, Victor S. (1945). **African orchids XVI.** *Botanical Museum Leaflets Harvard University* 12, 3: 89 – 116.

1363. Summerhayes, Victor S. (1947). **African orchids XVII.** *Kew Bulletin* 2, 2: 123 – 134.

1364. Summerhayes, Victor S. (1948). **African orchids XVIII.** *Kew Bulletin* 2: 277 – 302.

1365. Summerhayes, Victor S. (1949). **African orchids XIX.** *Kew Bulletin* 3, 3: 427 – 444.

1366. Summerhayes, Victor S. (1951). **A revision of the genus** *Angraecopsis.* *Botanical Museum Leaflets Harvard University* 14, 9: 240 – 261.

1367. Summerhayes, Victor S. (1951). **African orchids XX.** *Kew Bulletin* 6, 3: 461 – 475.

1368. Summerhayes, Victor S. (1953). **African orchids XXI.** *Kew Bulletin* 8, 1: 131, 161 – 162.

1369. Summerhayes, Victor S. (1953). **African orchids XXII.** *Kew Bulletin* 8, 4: 578.

1370. Summerhayes, Victor S. (1955). **A revision of the genus** *Brachycorythis.* *Kew Bulletin* 10, 2: 221 – 255.

1371. Summerhayes, Victor S. (1955). *Vanilla. Kew Bulletin* 10, 3: 494 – 496.

1372. Summerhayes, Victor S. (1956). **African orchids XXIII.** *Kew Bulletin* 11, 2: 217 – 236.

1373. Summerhayes, Victor S. (1957). **The genus** *Eulophidium* **Pfitzer.** *Bulletin du Jardin Botanique de l'Etat, Bruxelles* 27: 391 – 403.

1374. Summerhayes, Victor S. (1958). **African orchids XXV.** *Kew Bulletin* 13, 1: 57 – 87.

1375. Summerhayes, Victor S. & **Bullock**, A. A. (1960). **The application of the generic name** *Polystachya* Hook. *Taxon* 9, 5: 150 – 151.

1376. Summerhayes, Victor S. (1964). **African orchids XXIX.** *Kew Bulletin* 17, 1: 511 – 560.

1377. Summerhayes, Victor S. (1966). **African orchids XXX.** *Kew Bulletin* 20, 2: 165 – 199.

1378. Summerhayes, Victor S. (1968). **199.** *Orchidaceae.* In F. N. Hepper, ed., **Flora of West Tropical Africa.** Crown Agents, London: pp. 180 – 276. With b/w illustrations.

1379. Summerhayes, Victor S. (1968). *Orchidaceae.* **(Part 1).** In E. Milne-Redhead & R. M. Polhill, eds., *Flora of Tropical East Africa.* Crown Agents, London: 235 pp.

1379a. Supthut, Dieter. (2003). **Vom 17. bis zum 20 Jahrhundert.** *Die Sukkulentenwelt,* 8: 17 – 19.
A short history of botanical exploration in Madagascar. With a portrait of Etienne Flacourt, Alfred Grandidier and Werner Rauh.

1379b. Supthut, Dieter. (2003). **Ida Pfeiffer, Christian Rutenberg, Hans Bluntschli.** *Die Sukkulentenwelt,* 8: 21 – 25.
A detailed account of some key explorers involved in botanical exploration in Madagascar; Ida Pfeiffer and Christian Diedrich Rutenberg, with a portrait of both.

1380. Swan, W. (1898). *Angraecum leonis. Gardeners' Chronicle,* ser. 3, 23: 306.

1381. Swartz, Olaf. (1799). *Journal für die Botanic Herausgegeben von Medicinalrath Schrader.* 4.

1382. Swartz, Olaf. (1799). *Dianome Epidendri Generis* **Linn.** *Nova Acta regiae Societatis Scientiarum Upsliensis* 6: 72.

1383. Swartz, Olaf. (1800) *Koningl Svenska Vetenskaps Akademiens nya Handlinger* 21: 220, 230, 243 – 247.

1384. Swartz, Olaf. (1805). *Genera et species orchidearum, systematice coordinatarum.*

1385. Swartz, Olaf. (1806). *Flora Indiae Occidentalis* 3: 1231 – 2018. Erlang.

1386. Sylva, Alex K. J. (1936). **Notes on orchids cultivated in Ceylon.** *The Tropical Agriculturalist:* 215 – 216.

1387. Taylor, Mary Susan. (1991). *Flora and ethnobotany of Madagascar, contributions towards a taxonomic bibliography.* Missouri Botanical Garden, St Louis.

1388. Taylor, Peter. (1973). *Eulophiella elisabethae. Curtis's's Botanical Magazine* 179,4: t.656.

1389. Taylor, Peter. (1996). **Selected species.** *Aerangis fastuosa* Schltr. *Orchids Australia* 8, 3: 23.

1390. Teuscher, Henry. (1971). *Bulbophyllum maculosum* **(or** *B. guttulatum?***) and** *Cirrhopetalum umbellatum. American Orchid Society Bulletin* 40, 3: 218 – 222.

1391. Teuscher, Henry. (1974). *Phaius tankervilliae, Ph. woodfordii* **and** *Ph. pulchellus. American Orchid Society Bulletin* 43, 11: 981 – 988.

1392. Teuscher, Henry. (1976). **Collector's item.** *Angraecum, Neofinetia* **and** *Jumellea. American Orchid Society Bulletin* 45, 10: 896 – 909.

1393. Teuscher, Henry. (1977). *Eulophidium, Eulophia* **and the former genus** *Lissochilus. American Orchid Society Bulletin* 46, 2: 123 – 130.

1394. Thomas, Sarah. (1996). *Eulophia pulchra. Curtis's Botanical Magazine* 13, 2: 82 – 85, pl. 294.

1395. Thouars, Aubert Aubert du Petit-.(1804). *Plantes des Iles de l'Afrique Australe formant des genres nouveaux, ou perfectionnant les anciens; accompagnées de dissertations sur différens points de Botanique.* Paris.
The title was afterwards altered to: *Histoire des végétaux recueillis sur les Isles de France, La Réunion (Bourbon) et Madagascar.*

1396. Thouars, Aubert Aubert du Petit-.(1805). *Histoire des végétaux recueillis dans les Isles Australes d'Afrique.* Pp. xvi, 64; pls. iii – xx, xxv – xxx.
Pp. i – xvi and 1 – 24 are equivalent to pp. i – xvi and 15 – 38 of the 1804 edition. Plates xxv – xxx are shaded.

1397. Thouars, Aubert Aubert du Petit-. (1806). *Histoire des végétaux recueillis dans les Isles Australes d'Afrique.* Paris: 14 pp., 72; 24 pl.

1398. Thouars, Aubert Aubert du Petit-.(1804 – 1822). *Histoire particuliere des plantes orchidées recueillies sur les trois Iles Australes d'Afrique, de France, de Bourbon et de Madagascar.* 32 pp., 2 tabs. 110 b/w. drawings. L'Auteur, Arthus Bertrand, Treuttel & Wurtz, Paris. Reprinted in 1979 by Earl M. Coleman, New York.

1399. Thouars, Aubert Aubert du Petit-. (1809). **Extrait de trois Mémoires lus à la première classe de l'Institut, sur l'histoire des plantes orchidées des iles australes d'Afrique.** *Nouveau Bulletin des Sciences. Société Philomatique Paris* 1: 314 – 319.

1400. Thouars, Aubert Aubert du Petit-. (1819 – 1822). *Histoire particulière des plantes orchidées d'Afrique.* Paris.

1401. Thouars, Aubert Aubert du Petit-. (1819). *Revue Générale des Matériaux de Botanique et Autres.* Imprimerie de P. Gueffier, Paris.
A prospectus of Thouars' work by himself.

Thunberg, see Swartz 1381.

1402. Thurgood, F. W. (1931). **Eulophiellas and their cultivation.** *Orchid Review* 39, 457: 213 – 214.

1403. Toilliez-Genoud, J. (1958). **Sur une *Aeranthes* nouvelle de Madagascar.** *Naturaliste Malagache* 10: 19 – 20.

1404. Tolliez-Genoud, J., **Ursch**, E. & **Bosser**, J. (1960). **Contribution à l'étude des *Aeranthes* de Madagascar.** *Notulae Systematica* 16: 205 – 215.

1405. Tolliez-Genoud, J. & **Bosser**, J. (1960). **Contribution à l'étude des *Orchidaceae* de Madagascar I. Sur trois nouveaux *Angraecum*.** *Naturaliste Malagache*, 12: 9 – 16.

1406. Tolliez-Genoud, J. & **Bosser**, J. (1960). **Contribution à l'étude des *Orchidaceae* de Madagascar. *Bulbophyllum rauhi*, nouvelle Orchidée Malagache.** *Naturaliste Malagache*, 12: 17 – 19.

1407. Tolliez-Genoud, J. & **Bosser**, J. (1961). **Contribution à l'étude des *Orchidaceae* de Madagascar IV. Sur un *Angraecum* et un *Cynorchis* nouveau.** *Adansonia* 1, 100 – 105.

1408. Tolliez-Genoud, J. & **Bosser**, J. (1962). **Contribution à l'étude des *Orchidaceae* de Madagascar; III. Sur un *Neobathiea* et un *Cynorchis* nouveaux.** *Naturaliste Malagache* 13: 25 – 30.

1409. Trussell, Richard & Ann. (1996). **The mixed collection. Fragrant orchids.** *Orchid Review* 104, 1211: 288 – 292.

1410. Trussell, Richard & Ann. (1996). **The mixed collection. The weird and wonderful.** *Orchid Review* 104, 1212: 343 & 345.

1411. Turkel, Marni & **Christenson**, Eric A. (2001). *Microterangis hariotiana.* *Orchids (AOS)*, 70, 8:.737.
A history, short description and cultural advice on *Microterangis hariotiana*.

1412. **Tyson Northern** Rebbeca. (1980). *Miniature orchids.* Van Nostrand Reinhold, New York: pp. 189.

1413. **Ursch**, E. & **Toilliez-Genoud**, J. (1950). **Les** *Gastrorchis* (Orchidées) **du Jardin Botanique de Tsimbazaza.** *Naturaliste Malagache* 2, 2: 147 – 161.

1414. **Ursch**, E. & **Toilliez-Genoud**, J. (1951). **Les calanthes** (Orchidées) **du Jardin Botanique de Tsimbazaza.** *Naturaliste Malagache* 3, 2: 99 – 111.

1415. **Ursch**, E. & **Toilliez-Genoud**, J. (1953). **Une nouvelle** *Eulophiella* **de Madagascar.** *Naturaliste Malagache* 5, 2, 149 – 150.

1416. **Van Der Cingel**, N. A. (1987). **Orchideeën en evolutie. 1. Over het ontstaan van soorten.** *Orchideeën* 49, 2: 25 – 32.

1417. **Vanderwalt**, B. C. (1975). **The elusive butterfly of the Comores.** *Suid-afrikaanse Orgideejoernaal* 6: 4

1418. **Vanderwalt**, B. C. (1976). *Oeniella polystachys.* *South African Orchid Journal* 7, 2: 52.

1419. **Van Ede**, Gerrit. (1980). **A for Africa;** *Angraecum*; *Aerangis*; *Aeranthes.* In M. R., Kashemsanta Sushom, ed., *Proceedings of the 9th World Orchid Conference, Bangkok*: 249 – 252.

1420. **Van Houtte**, Louis, ed. (1861). *Angraecum sesquipedale.* *Flore des Serres* 11: t.1413 – 1414: 49.

1421. **Van Houtte**, Louis, ed. (1861) *Grammatophyllum ellisii.* *Flore des Serres* 11: 257 t.1488 – 1489.

1422. **Van Houtte**, Louis. (1867) *Vanilla phalaenopsis* **Rchb. f.** *Flore des Serres* 17: 97 – 98, t.1769 – 1770.

1423. **Van Zijll De Jong**, J. (1961). **Bij de Voorpagina:** *Angraecum sesquipedale* Thou. *Orchideeën* 23, 2: 23 – 24.

1424. **Van Zuylen**, Bert. (1996). *Aerangis.* *Orchideeën* 1: 8 – 9.

1425. **Vaschalde**, Claude. (1981). **Hybrides intergeneriques chez les angraecoides.** *L'Orchidophile*, 12, 46: 1802 – 1805.
Intergeneric hybridisation of the angraecoids is discussed; chromosome numbers and generic differences are investigated. Some existing hybrids plus crosses made by Fred Hillerman, are listed.

1426. **Vaughan**, R. E. (1958). **Wencelaus Bojer.** *Proceedings of the Royal Society of Arts & Sciences of Mauritius* 2, 1: 73 – 98.

1427. **Veitch**, James. (1887 – 94). *A Manual of Orchidaceous Plants.* James Veitch & Sons, Chelsea: 2 vols.

1428. **Veitch**, James Herbert. (1896). *A Traveller's Notes.* James Veitch & Sons, Chelsea, 219 pp.
Notes by the nurseryman of a tour through India, Malaysia, Japan, Korea, Australia and New Zealand during 1891 – 1893. *Angraecum sesquipedale* is recorded at Georgetown Public Gardens and Adelaide Botanic Garden.

1429. **Verdoorn**, I. C. (1954). *Acampe pachyglossa.* *Flowering Plants of Africa*, 30: t.1175.
A detailed illustration, history and description of the species.

1430. **Verdoorn**, I. C. (19??). *Nervilia grandiflora.* *Flowering Plants of Africa*, t.1312.
A detailed history and description of *Nervilia grandiflora* (= *Nervilia bicarinata*) to accompany the plate.

1431. **Vermeulen**, J. J. (1987). **A taxonomic revision of the continental African** *Bulbophyllinae.* *Orchid Monographs*, 2, 1 – 300.

Vermeulen, also see 384.

1432. Veyret, Yvonne. (1967). **L'apomixie chez le** *Cynorchis lilacina* Ridley (Orchidacées). *Compte Rendu Academie, Sciences, Paris* D, 265, 22: 1713 – 1716.

1433. Veyret, Yvonne.(1971). **Études embryologiques dans le genre** *Cynorkis* (*Orchidaceae*). *Adansonia* sér.2, 12, 3: 389 – 402.

1434. Viguier, René & **Humbert**, Henri. (1924 – 1925). **Plantes recoltèes à Madagascar en 1912.** *Bulletin de la Société Linnéenne de Normandie* ser. 7, 7: 193 – 208.

1435. Viguier, René. (1927). **Orchidées de Madagascar.** *Archives Botanique Bulletin Mensuel* 1,1.

1436. Voeltzkow, A. (1917). *Flora und Fauna der* **Comoren**: 429 – 432 & 443 – 444. Berlin.

1437. Vöth, Walter (1965). **Zu** *Nervilia bathiei*. *Die Orchidee* 16, 2: 68 – 69.

1438. Waki, Melvin Z. (1994). **FCC's of the Honolulu Orchid Society.** *Hawaii Orchid Journal* 23, 3: 11 – 12.

1439. Walford, James F. & **Hunt**, Peter F. (1972). *A Book of Orchid Paintings.* The Medici Society, London: 112 pp.

1440. Walters, Ian. (1972). **Culture of the angraecums and their allies.***Australian Orchid Review* 37, 1: 34 – 41.

1441. Warner, H. H. (1934). *Angraecum* **and** *Xanthopan.* *American Orchid Society Bulletin* 3, 3: 42 – 43.

1442. Warner, Robert, **William**, Benjamin Samuel & **Moore**, Thomas. (1882 – 1897). *The Orchid Album.* Victoria & Paradise Nursery, London.

1443. Warpur, M. G. (1901). **Habitats of Madagascar orchids.** *Orchid Review* 9, 97: 10 – 11, also see 1101.

1444. Warren, Richard. (1994). **Brazilian orchids**. *Equatorial Plants Newsletter* 11, 2: 15 – 16.

1445. Watson, James, ed. (1993). *Vanilla planifolia.* *American Orchid Society Bulletin* 62, 4: 415.

1446. Watson, W. (1890). *Cymbidium loise-chauvieri.* *Garden & Forest* 3: 153.

1447. Watson, W. (1890). *Angraecum citratum.* *Garden & Forest* 3: 216.

1448. Watson, W. & **Bean**, W. (1890) *Orchids. Their culture and management.* Upcott Gill, London: 554 pp., 3 colour pl.

1449. Watson, W. (1893). **Orchid Notes.** *Garden & Forest* 6: 336.

1450. Watson, W. (1896). **Foreign correspondence. London letter.** *Garden & Forest* 9: 514.

1451. Watson, W. (1900). **A new** *Cynorchis.* *The Garden*: 375 – 376.

1452. Watson, W. (1900). **Orchid notes and gleanings.** *Gardeners' Chronicle* ser. 3, 28: 335.

1453. Watson, W. & **Chapman**, H. J. (1903). *Orchids. Their culture and management.* Upcott Gill, London: 560 pp., 20 colour pl.

1454. Weathers, John. (1901). *Cynorchis purpurascens.* *The Gardeners' Chronicle*, ser. 3, 29: 86, fig. 37.

1455. Weathers, John. (1901). *Phaius tuberculosus,* **& co.** *Gardener's Chronicle*, ser. 3, 29: 82.

1456. Weathers, P. (1893). **The new orchid,** "*Eulophiella elisabethae*", **of Madagascar.** *London Standard* Oct. 27: 3.

Wendland, see **Kraenzlin** 648, 651.

1457. White, Judy. (1992). **Flights of fancy.** *American Orchid Society Bulletin* 61, 8: 774 – 775.

1458. White, W. H. (1904). **Cultural notes on the eulophiellas.** *Flora & Sylva* 2, 18: 261 – 263. Also, *Orchid Digest* 36, 4: 123 – 124.

1459. White, W. H. (1925). *Angraecum distichum* and *A. pectinatum.* *Orchid Review,* 33, 388: 303.

1459a. Whitman M., **Randrimanidry**, J.J., **Medler**, M. & **Rabakoandrianina**, E. (2005). **Fire impact on terrestrial and epilithic orchids, case study from the Ambatolava Inselberg granite outcrop Madagascar.** World Orchid Conference Dijon, Poster.
A study of the fire response between *Cynorkis* and *Angraecinae* inhabiting the same Inselberg habitat. *Angraecum sororium, Jumellea rigida, Cynorkis uniflora, Cynorkis fastigiata* and *Cynorkis angustipetala* are investigated.

1460. Willdenow, Carl Ludwig von. (1805). *Species plantarum.* Ed. 4, 4. G. C. Nauk, Berlin.

1461. Willemet, Pierre Remi (1796). *Herbarium Mauritianum.* Wolff, Leipzig.

1462. Williams, Benjamin Samuel. (1852). *The orchid-grower's manual.* 104 pp. Chapman & Hall, London.

1463. Williams, Benjamin Samuel. (1862). *The orchid grower's manual.* 2nd ed.: 160 pp. Chapman & Hall, London.

1464. Williams, Benjamin Samuel. (1868). *The orchid grower's manual.* 3rd ed.: 249 pp. Victoria and Paradise Nursery, London.

1465. Williams, Benjamin Samuel. (1871). *The orchid grower's manual.* 4th ed.: 300 pp., 17 drawings. Victoria & Paradise Nursery, London.

1466. Williams, Benjamin Samuel. (1877). *The orchid grower's manual.* 5th ed.: 336 pp., 52 drawings. Victoria & Paradise Nurseries, London.

1467. Williams, Benjamin Samuel. (1885). *The orchid grower's manual.* 6th. ed.: 659 pp., numerous pl. Victoria and Paradise Nursery, London.

1468. Williams, Henry. (1894). *The orchid grower's manual.* 7th. ed.: 796 pp. Victoria & Paradise Nursery, London.

1469. Williams, Brian. (1976). **Stairway to the "stars". A cultural guide for angraecoids.** *Orchid Review* 84, 991: 20 – 23.

1470. Williams, Brian, ed. (1977). *Aerangis cryptodon* 'Remi' **AM/RHS.** *Orchid Review* 85, 1003: 3 & 81.

1471. Williams, Brian, ed. (1977). *Aerangis pumilio* 'Bucklebury' **AM/RHS.** *Orchid Review* 85, 1006: 99.

1472. Williamson, Graham. (1977). *The Orchids of South Central Africa.* J. M. Dent & Sons, London: 237 pp., drawings and photos.

1473. Wilson, Gurney, ed. (1910). *Angraecum sesquipedale.* *Orchid World* 1, 1: 10 – 11.

1474. Wilson, Gurney, ed. (1912). *Angraecum citratum.* *Orchid World* 2, 8: 170.

1475. Wilson, Gurney, ed. (1912). **Some notes on Madagascar.** *Orchid World* 2, 11: 257.

1476. Wilson, Gurney, ed. (1913). *Cymbidium humblotii.* *Orchid World* 3: 172.

1477. Wilson, Gurney, ed. (1913). *Angraecum sesquipedale.* *Orchid World* 3, 9: 195.

1478. **Wilson**, Gurney, ed. (1914). *Angraecum rectum.* Orchid World 4, 4: 80 – 81.

1479. **Wilson**, Gurney, ed. (1914). *Angraecum citratum.* Orchid World 4, 7: 152 – 153.

1480. **Wilson**, Gurney, ed. (1916). *Orchid World* 6, 6.

1481. **Wilson**, Gurney, ed. (1916). *Eulophiella.* Orchid World 6, 7: 150.

1482. **Wilson**, Gurney, ed. (1916). *Cymbidium humblotii.* Orchid World 6, 8: 176.

1483. **Wilson**, Gurney, ed. (1916). *Phaius humblotii.* Orchid World 6, 8: 182 – 183.

1484. **Wilson**, Gurney, ed. (1916). *Cymbidium humblotii.* Orchid World 6: 222 – 223.

1485. **Wilson**, Gurney, ed. (1921). *Phaius* **Clive.** Orchid Review 29, 341: 139 – 140.

1486. **Wilson**, Gurney, ed. (1922). *Angraecum eburneum* **and** *A. superbum.* Orchid Review 30, 344: 43.

1487. **Wilson**, Gurney, ed. (1922). **Royal Horticultural Society.**
Orchid Review 30, 348: 161 &190.

1488. **Wilson**, Gurney, ed. (1922). **New orchids.** *Orchid Review,* 30, 353: 343-.
New orchids are mentioned, including *Agrostophyllum seychellarum* and *Microstylis thomassettii* from the Seychelles.

1489. **Wilson**, Gurney, ed. (1925). *Phaius tuberculosus.* Orchid Review, 33, 382: 98.

1490. **Wilson**, Gurney, ed. (1925). **Auction sale catalogues.** *Orchid Review* 33, 385: 202 – 4.

1491. **Wilson**, Gurney, ed. (1925). *Phaius humblotii.* Orchid Review 33, 386: 246 – 7.
An earlier issue (382: 98) has more extensive cultural advice.

1492. **Wilson**, Gurney, ed. (1925) *Angraecum scottianum.* Orchid Review, 33, 388: 290.

1493. **Wilson**, Gurney, ed.(1926). *Angraecum sesquipedale.* Orchid Review 34, 396: 177 – 178.

1494. **Wilson**, Gurney, ed. (1926). *Angraecum eburneum.* Orchid Review 34, 397: 207.

1495. **Wilson**, Gurney, ed.(1926). *Angraecum citratum.* Orchid Review 34, 400: 318 – 319.

1496. **Wilson**, Gurney, ed. (1927). *Eulophiella elizabethae.* Orchid Review 35, 411: 278.

1497. **Wilson**, Gurney, ed. (1929). *Eulophiella roempleriana (Peetersiana).* Orchid Review 37, 436: 298 – 300.

1498. **Wilson**, Gurney, ed. (1931). *Cymbidiella (Cymbidium) humblotii.* Orchid Review 39, 453: 77.

1499. **Wilson**, Gurney, ed. (1931). *Eulophiella* **Rolfei.** Orchid Review 39, 455: 141 – 142.

1500. **Wilson**, Gurney, ed. (1931). *Cirrhopetalum thouarsii.* Orchid Review 39, 458: 247 – 248.

1501. **Wilson**, Gurney, ed. (1931). *Angraecum gracilipes.* Orchid Review 39, 459: 263 – 264.

1502. **Witt**, Otto N. (1910). *Angraecum sanderianum* Rchb.f. Orchis 4, 8: 120 – 121.

1503. **Wittmack**, L. ed. (1892). **Neue und empfehlenswerte Pflanzen etc.** Gartenflora 41, 15: 412 – 415.

1504. **Wittmack**, L. (1893). **Die internationale Gartenbau-Ausstellung in Gent.** Gartenflora 41, 9: 304 – 310.

1505. Wodrich, Karsten H. K. (1997). *Growing South African indigenous orchids.* A. A. Balkema. Rotterdam: 253 pp.
An extensive account on the cultivation of South African orchids. Both terrestrials and epiphytes are discussed. Climatic conditions, habitat, housing of plants in cultivation, composts, watering, fertilisers, pests are explained in detail. A large number of species are covered.

1506. Woodward, B. B. (1900). **Bibliographical notes XXIII:-Du Petit-Thouars.** *Journal of Botany* 38: 392 – 400.

1507. Yearsley, Graham. (2001). **Reverend William Ellis.** *Orchid Review,* 109, 1241: 288 – 294.
A tangled biography of the missionary.

1508. Young, W. H. (1897). **Calender operations for February.** *Orchid Review* 5: 59.

1509. Zelenko, H. (2001). **Orchids in art.** *Cymbidiella pardalina. Orchids (AOS),* 70, 2: 110 – 111.
A short note with cultural details to accompany Angela Mirro's painting of *Cymbidiella pardalina.*

1510. Zelenko, H. (2003). **Orchids in art.** *Angraecum sesquipedale. Orchids (AOS),* 72, 9: 682 – 683.
A short note on the species to accompany a painting by Patricia Kessler.

1510a. Zelenko, Harry. (2005). **Orchids in Art.** *Angraecum magdalenae* **Schltr. And H. Perrier.** *Orchids,* 74, 1: 32 – 33.
A short note with cultural details to accompany Linda Walsh Petchnik's painting of *Angraecum magdalenae.*

Zollinger, Heinrich, see Moritzi, Alexandre 830.

INDEX TO THE BIBLIOGRAPHY

Only references to subjects not covered by the checklist are included; i.e. people, cultivation of genera, ecology, hybrids and history not directly connected with a particular species. Index entries refer to the content of the publication, not necessarily to key words in the Bibliography. Only main revisions of genera have been included.

A

Acampe, 1281, 1344
Acampe, cultivation, 50, 941
Addie, Mme., 198
Aerangis, 390, 893, 1281, 1344
Aerangis Callikot, 529
Aerangis, cultivation, 50, 147, 184, 191, 258, 444, 529, 530, 680, 941, 944, 1175
Aerangis, revision of, 889, 1335, 1336, 1337, 1338
Aerangis Rhodostrata, 860
Aerangis Spicusticta, 460, 534
Aeranthes, 806, 807, 893, 1277, 1281
Aeranthes Grandianna, 529
Aeranthes, cultivation, 50, 51, 444, 529, 941, 944, 1259, 1261, 1277
Africa, Central, orchids of, 384
Africa, East, orchids of, 219, 221, 914, 918, 922, 1379
Africa, orchids of, 282, 683, 1090, 1214, 1319
Africa, Southern, orchids of, 768, 786, 1156, 1329, 1472, 1505
Africa, West, orchids of, 1378
Africa, Zambesi, orchids of, 682, 685
Afzelia, orchids on, 308
Afzelius, K. R., 193, 1215
Agauria salicifolia, 877
Albizzia fastigiata, orchids on, 412, 1319, 1443
Alleizette, d' Ch., 193, 529
Ambrella, 893
Anatomy of flower, 385
Andriamaharo, Rémy, 427, 712, 724
Andringitra, orchids of, 740, 874
Angola, orchids of, 975, 976
Angraecoids, cultivation, 508, 513, 522, 527, 529, 531, 532, 536, 581, 932, 1224, 1332, 1469
Angraecoids, hybrids, 295, 514
Angraecoids, morphology, 1320, 1334
Angraecoids, revision of, 292, 1213, 1214
Angraecopsis, 893
Angraecopsis, cultivation, 941
Angraecopsis, revision of, 955, 1366
Angraecum, 510, 808, 893, 916, 924, 1276, 1281, 1386, 1392
Angraecum Alabaster, 1344
Angraecum, cultivation, 38, 50, 107, 178, 251, 334, 444, 519, 523, 529, 533, 591, 698, 711, 941, 944, 1255, 1440
Angraecum, distribution, 208, 924
Angraecum Eburlena, 529
Angraecum Eburscott, 529
Angraecum Lady Lisa, 1344
Angraecum Lemforde White Beauty, 678, 1409
Angraecum Longiscott, 529
Angraecum Malagasy, 529
Angraecum Ol Tukai, 295, 1322

Angraecum Orchidglade, 508, 529
Angraecum, revision of, 314, 891, 1121
Angraecum Scotticom, 529
Angraecum Sesquibert, 529
Angraecum Stephanie, 529
Angraecum Superlena, 529, 679
Angraecum Supertans, 529
Angraecum Veitchii, 265, 442, 599, 935, 1025, 1035, 1093, 1103, 1147, 1211, 1231, 1344, 1486
Angraecum Vigulena, 295, 534
Angraecum Wolterianum, 1211, 1486
Angranthes Christina, 529
Angranthes Compactolena, 295
Angranthes Grandalena, 295, 529, 534
Angranthes Grandivag, 529
Angranthes Primera, 529
Ankafina, area, 193
Ankarana, Tsingy, orchid of, 193, 459, 1250
Ankaratra, orchids of, 874
Antananarivo Bot. Garden — see Tsimbazaza
Arkle, J.W., 1105
Ashburton, Dowager Lady, 544, 1084
Asplenium nidus, orchids on, 556, 928, 1148, 1443
Auxopus, 893
Awarded plants, Belgium, 694
Awarded plants, Germany, 848
Awarded plants, Hawaii, 402, 1438
Awarded plants, Switzerland, 849
Awarded plants, United Kingdom, 19, 25, 32, 59, 64, 67, 70, 71, 72, 73, 74, 75, 76, 77, 78, 79, 80, 81, 82, 83, 84, 85, 88, 192, 198, 199, 328, 331, 341, 358, 367, 370, 373, 461, 472, 475, 476, 481, 487, 488, 490, 492, 573, 704, 752, 851, 933, 937, 1037, 1038, 1039, 1040, 1041, 1056, 1061, 1099, 1123, 1132, 1140, 1144, 1145, 1152, 1156, 1161, 1162, 1165, 1167, 1169, 1188, 1190, 1388, 1442, 1470, 1471, 1478, 1484, 1487, 1499
Awarded plants, United States, 3, 4, 508, 515, 709, 1021, 1179, 1231
Aye-Aye, 151

B

Baillon, Prof. H., 529, 568, 1098
Baker, John Gilbert, 41, 568
Balfour, Dr. I. B., 825, 1114
Baobabs, *Angraecum* on, 1427
Baron, Rev. Richard, 193, 271, 467, 484, 529, 617, 568, 758, 1017, 1072, 1128, 1479, 1501
Bateman, James, 241, 302
Bauer, Franz, 46
Beclardia, 893
Benthamia, cultivation, 444
Benthamia, revision of, 879
Benzene analysis, 613
Berlin, Botanical Garden, 1019

Berlin, Herbarium, 164
Bernier, Charles, 529
Betsileo, area, 193
Beaumois, Chateau, 634
Billiard, Mr., 35, 349
Biodiversity, 42, 426, 883
Biographies of nurserymen, 1023
Biology, of orchids, 893
Bleu, Alfred, 1127, 1131
Boiteau, Pierre, 894
Boivin, Louis Hyacinthe, 529
Bojer, bibliography, 1426
Bojer, Wenceslas, 529, 568, 1426
Bory, Jean Baptiste de St. Vincent, 50, 199, 529, 675
Bosser, Jean, 484, 529, 1327
Bot. Garden, Antananarivo — see Tsimbazaza
Botanic Garden, Kew — see Kew
Botanical research, 568
Bouriquet, Gilbert, 568
Brachycorythis, 1370
Braun, Johannes, 553, 651, 652, 1129, 1204
Breon, Mr., 529
Breton, Le, 25
Brodbeck, Edmon, 724
Brooke, James & Co., 415
Broughton Hall — see Clowes, John
Brown, Johannes, 1082
Brownlee, Rev. J., 786
Brownleea, 893
Brownleea, cultivation, 444
Brownleea, revision of, 764
Bruce Scott, Mr., 947
Buchanan-Hamilton drawing, 1233
Buchloh, Prof., 50, 1244
Bulbophyllum, 893, 1288
Bulbophyllums, clinging to bark, 888
Bulbophyllum, cultivation, 50, 444, 941
Bulbophyllums, of Manongarivo, 888
Bulbophyllums, of Tsaratanana, 888
Bulbophyllum, revision of, 108, 610, 888
Bull, Mr., 174
Bultel, Mr., 704
Burrage, Albert, 161

C

Calanthe, 893
Calanthe, cultivation, 50, 444, 1344
Calanthe, revision of, 1414
Camboue, Mr., 529
Campenon, Mr., 529
Capuron, René, 529
Cataloguing of Malagasy orchids, 220
Catat, L., 529
Cèdres, Les, 257
Centre National de Recherche Appliqué, 170
Ceylon, connection, 1125
Chamaeangis — also see Microterangis, 893
Chamaeangis, cultivation, 50, 444
Chapelier, Mr., 529
Charlesworth, nurserymen, 189, 188, 192, 289, 374, 1106, 1151, 1152, 1153, 1402, 1478, 1479
Cheapside Auction — see Protheroe & Morris
Cheirostylis, 893
Cheirostylis, revision of, 884
Chelsea, Show, 289
Chiswick Gardens, 327

Chromosone numbers, 15, 21, 514, 595, 1259, 1339
Cirrhopetalum, 893
Cirrhopetalum, taxonomy, 319
CITES, 1051
Clare Lawn, orchids at, 602, 1135, 1139
Climate of Madagascar, 529, 700, 816, 957
Clowes, John, 45, 432, 561
Coates, T., 1442
Cogniaux, Alfred Céléstin, 1110
Collecting — see exploration, orchid hunting
Commerson, Phillibert, 529, 961
Comoro Islands, orchids of, 272, 619, 1006, 1007, 1008, 1012, 1312, 1313, 1314, 1315, 1316, 1317, 1318, 1436
Comparison with non-Malagasy species, 893
Conservation, 14, 42, 155, 220, 229, 277, 279, 280, 410, 424, 426, 456, 465, 470, 484, 507, 815, 847, 862, 951, 1046, 1051, 1291
Cookson, Norman C., 379, 752, 1057, 1099, 1100, 1117
Cordemoy, Eugène Jacob de, 568
Corning collection, 1474
Corymborkis, revision of, 312, 954
Corymbus, 893
Cost, of orchids — see price of orchids
Cours, G, 529, 624, 900
Cowan & Co., 210
Crassulacean Acid Metabolism (CAM), 634, 635
Crawshay, D. B., 1442
Cryptoprocta ferox, 22, 1300
Cryptopus, 126, 185, 893
Cryptopus, cultivation, 126, 529
Cryptopus, revision of, 114
Cryptopus Nouchka, 126
Cultivation — see under individual genera
Curnow, Richard, 928, 361
Curtis Botanical Magazine, 45, 1298
Curtis, Charles, 50, 233, 628, 1023, 1094
Cymbidieae, 1304
Cymbidiella, 612, 1277, 1281, 1344
Cymbidiella, cultivation, 444, 452, 499, 529, 587, 711, 846, 941, 944, 1344
Cymbidiella Kori Dingeman, 452, 499
Cymbidiella, revision of, 317, 1212
Cymphiella Hiroshima Peace, 15
Cynorkis, 390, 893, 921, 1281
Cynorkis, cultivation, 366, 444, 944
Cynorkis, embryology, 1432, 1433
Cynorkis Kewensis, 366, 1120, 1139, 1142, 1147, 1275
Cynorkis, medicinal use, 693
Cynorkis, revision of, 878
Cytotaxonomy, 21

D

Dahlemer Botanic Garden, 1216
D'Alleizette, Ch. — see Alleizette
Danguy, Paul, 567
Darwin, Charles — also see pollination, 50, 150, 302, 713, 714, 952, 1251, 1493
Day, John, 50, 226, 244, 343, 344, 988, 1094, 1294
Deans Cowan, Rev. William, 49, 52, 193, 246, 484, 568, 1030, 1031, 1033, 1287, 1478
Deans Cowan, drawings, 246
Decary, Raymond, 193, 484, 529, 568, 894
Delephila nerii, 725

Disa, 893
Disa, cultivation, 444
Disa, revision of, 765, 766
Disperis, revision of, 686, 886, 1199
Distribution of orchids, 280, 538, 883, 893, 1208
Dormen, Mr., 174
Douliot, H., 529
Drake del Castillo, Emmanuel, 529
Drill Hall, orchids at the, 25, 186, 1454
Durchud, Madeleine, 50
du Petit Thouars — see Thouars

E

Ecological niches, 510
Economic use of orchids, 310
Ellis, Rev. William, 45, 50, 183, 226, 275, 286, 287, 302, 337, 484, 486, 529, 562, 564, 734, 736, 746, 748, 749, 792, 978, 1067, 1094, 1159, 1301, 1420, 1421, 1442, 1465, 1507
Embryology, 1432, 1433
Endemism, 893, 1208, 1221
Epiphora pubescens, 1286
Ericaceous scrub, orchids in — also see *Philippia*, 124, 877, 888, 894
Exploration, 272, 1270
Eulocymbidiella, 452
Eulocymbidiella Susan Orenstein, 499, 1339
Eulophia, 683, 1393
Eulophia, cultivation, 50, 179, 444, 1344
Eulophia, revision of, 141, 882, 1373
Eulophidium — also see *Oeceoclades*, 1244, 1246, 1393
Eulophidium, revision of, 1373
Eulophiella, 893, 1055
Eulophiella, cultivation, 50, 444, 467, 498, 941, 1083, 1344, 1402
Eulophiella, history, 301, 438, 467, 1219
Eulophiella, revision of, 112, 1219
Eulophiella Rolfei, 15, 189, 231, 300, 405, 434, 441, 498, 529, 593, 813, 1022, 1164, 1402, 1499
Eurangis Grass Valley, 534
Exploration — also see orchid hunting, 42, 272, 705, 902
Export tax on orchids, 706
Exposition Internationale, Paris, 8
Eyman, Mr., 571

F

Faham, 101, 310, 690, 701, 897, 928, 974
Fairfield Orchids, 415
Franham Castle, 338
Ferko, Dr. P., 1216
Fertilisation — see Pollination
Finet, Achille Eugène, 36, 400
Flacourt, Etienne de, 484
Floral regions, 42
Floral structure, 639
Flower arrangement, 1323
Flowering times in cultivation, 431
Forbes, John, 45, 50, 529, 561, 771, 778
Forsyth Major, Dr. Charles Immanuel, 193, 568
Fournier, Louis, 11, 12, 178, 803, 805
Fox, Dr., 1032
Fragrance, of orchids, 135, 204, 415, 613, 700, 779, 837, 940, 952, 1035, 1409
François, Edmond, 894

G

Galeola, 893
Gastrodia, 893
Gastrorchis — also see *Phaius*
Gastrorchis Elizabeth Castle, 490
Geay, Francois, 193, 208, 529, 927
Gent — see Ghent
Geographical origin, 883
Geography, of Madagascar, 308, 612, 700, 816
Geological areas of Madagascar, 308, 570
Germiny, Compte de, 1442
Ghent exhibition, 10, 377, 694, 756, 759, 1504
Gilpin, Miss Helen, 193, 986
Glasnevin, Royal Botanic Garden, 199, 1107, 1115, 1116, 1127, 1131, 1133, 1150, 1171
Glen Ridge, Boston, orchids at, 953
Godefroy-Lebeuf, Alexandre, 704
Goodyera, 893
Goodyera, revision of, 884
Goodyerinae, 853
Grammangis, 893, 502
Grammangis, cultivation, 50, 444, 502, 941, 944, 1344
Grammangis, revision of, 112, 436, 443, 1209
Grandidier, Alfred, 529, 568
Graphiella, 452
Graphiella Martialine, 499, 1045
Graphorkis, cultivation, 50, 444, 1344
Graphorkis, revision of, 1240, 1368
Greve, Mr., 529
Gussonea — also see *Microcoelia*, 893
Gymnochilus, revision of, 884

H

Habenaria, cultivation, 50
Habenaria, revision of, 644
Habitat of orchids, 1101
Hadrangis, 126
Halictid bees, pollinators, 909, 911
Hamelin, L, 23, 43, 300, 377, 429, 462, 467, 523, 630, 633, 753, 756, 758, 759, 1055, 1072, 1085, 1098, 1116, 1122, 1128, 1171, 1299, 1300, 1347, 1456
Hamilton, J., 1076
Hardy, David, 1327
Hautegente, Chateau de (Hamelin), 1098
Hawk-moth — also see *Xanthophan*, 30, 838, 839, 1194
Heckel, Mr., 568
Hederorkis, revision of, 123
Heidelberg, Botanic Garden, 666, 1239
Herrenhausen, Botanic Garden, 647, 655, 1204
Hildebrandt Johann Maria, 52, 193, 529, 548, 568, 667, 982, 991, 997, 1001, 1019, 1024
Hilsenberg, Karl Theodor, 529
Hintzy — see *Afzelia*
History, 226, 272, 484
Hochreutiner, Bénédict Pierre Georges, 568
Holland House, show, 1140
Holothrix, revision of, 1200
Hooker, Sir. W., 286, 735
Horne, Mr., 565
Horticulture Internationale — see Linden
Humbert, Henri, 193, 484, 529, 624, 894, 902
Humbert, orchids collected by, 901
Humbert, specimen numbers, 624
Humblot, Henry, 1490

N

Neale, J. J., 243
Neobathiea, 893
Neobathiea, cultivation, 50, 444, 529, 1259, 1344
Neobathiea, revision of, 114
Nephela oenopion, 725
Nervilia, 158, 893, 907
Nervilia, cultivation, 444, 1344
Nervilia, ecology, 909
Nervilia, habitat, 909
Nervilia, pollination, 909
Nervilia, revision of, 610, 908, 909, 910, 1203
Nomenclature, 408
Nomina conservanda, 408
Nonioides sp., 907
North, Marianne, 54
Nurserymen. Biographies, 1023

O

Oberonia, 893
Oberonia, cultivation, 1344
O'Brien, Mr., 752
Oeceoclades, cultivation, 50, 382, 1244, 1246, 1344
Oeceoclades, revision of, 141, 318, 882, 1373
Oeonia, 893
Oeonia, cultivation, 50, 382, 444, 455, 529, 1259, 1344
Oeonia, revision of, 129
Oeoniella, 893, 1277, 1281
Oeoniella, cultivation, 50, 444, 529, 1344
Oeoniella, revision of, 1238
Ombrarium at Tsimbazaza, 99
Orchid hunting — also see exploration, 42, 151, 429, 526, 529, 571, 618, 619, 811, 816, 902, 1042, 1083, 1113, 1300, 1307, 1308
Orchidoideae, 945
Orchid trade — see trade & price of orchids
Orchidvale collection — see Burrage, Albert
Ornamental species, 897

P

Pandanus, orchids on, 308, 556, 706, 1102, 1148, 1152, 1454
Panogena lingens, polliantor, 58, 835, 943
Parc Bot., Antanarivo — see Tsimbazaza
Paris, Jardin des Plantes herbarium, 409
Paris, Museum Nat. d'Histoire Naturelle, 404, 704, 896, 929, 1432
Parker, Dr. G. W., 39, 529
Partington, B. J., 1442
Pauwels, Th. & Co., 159, 1149
Peeters, A., 199, 200, 371, 650, 704, 845, 949, 1088, 1107, 1110, 1118
Perieriella, 893
Perieriella, revision of, 125
Perrier de la Bathie, H., 50, 153, 193, 305, 484, 529, 568, 1205, 1210, 1221
Pescatore J. - P., 571, 743, 789, 790, 979
Pesomeria tetragona, 781
Peters, Dr., 964
Peyrot, Dr. J. P., 724
Petit Thouars — see Thouars
Pfitzer, Ernst — manuscripts by, 661
Phaiocalanthe Berryana, 1147

Phaiocalanthe Colmanii, 1147
Phaius, 134, 158, 587, 893, 1117, 1257, 1281, 1344
Phaius amabilis, 330, 1147
Phaius Chapmanii, 1147
Phaius Clive, 110, 189, 1147, 1485
Phaius Cooksoni(i), 1, 252, 330, 381, 925, 949, 1057, 1117, 1147, 1185, 1302, 1427, 1442, 1468
Phaius Cooksoni(i) var. *splendidus*, 1185
Phaius Crawshawianus, 1147
Phaius, cultivation, 33, 50, 252, 444, 529, 711
Phaius Doris, 1147
Phaius Harold, 1117, 1147
Phaius, hybrids, 26, 330, 1491
Phaius Marthae, 330, 539, 751, 1147
Phaius Norman(i), 189, 199, 328, 330, 362, 378, 381, 1117, 1147, 1485
Phaius Norman var. *aureus*, 199
Phaius Normani var. *rosea*, 199, 328, 763
Phaius Oakwoodensis, 372, 1099, 1147, 1185
Phaius Owenianus, 1147
Phaius Phoebe, 1185, 1147
Phaius, revision of, 120, 1376, 1413
Phaius Ruby, 1147
Phaius Wigianus, 1147
Philippia as host, 877, 880, 957, 1248, 1259
Photography, 485
Pickergill, W. C., 180
Platylepis, 893
Platylepis, revision of, 884
Plectrelgraecum Mannerhill, 529
Pogonia — also see *Nervilia*, 571
Poisson, Dr. H, 193, 529
Pollination, 18, 20, 58, 241, 403, 508, 531, 584, 639, 725, 824, 835, 836, 837, 838, 839, 840, 842, 880, 881, 907, 909, 911, 943, 952, 1194, 1416, 1441
Polystachya, 893, 1281, 1331, 1344, 1349, 1350
Polystachya, revision of, 50, 665, 884, 941
Popow, Neboijscha, 153
Post-pollination phenomena, 17, 18, 20, 1353
Price of orchids, 183, 189, 210, 371, 415, 694, 704, 928, 1022, 1088, 1184, 1185
Propagation of orchids, 179, 428, 536, 1402
Protheroe & Morris, Auctioneers, 203, 371, 377, 1069, 1088, 1450
Protocryptoferox — also see *Cryptoprocta*, 429, 1347
Putnam, Charles S. Letters by, 265

Q

Queen's Palace, Antananarivo, 250
Quesnel, Mr., 743

R

Raphia farinifera as host, 132, 308, 447, 696, 720, 877
Rauh, Prof. Werner, 50, 1239, 1269, 1272
Rayband, Mr., 571
Reichenbach, Prof. H. G., 571, 667, 814
Research, state of orchid, 1435
Réunion, orchids of, 126, 149b, 167, 168, 169, 182, 206, 207, 310, 689, 690, 692, 705, 789, 824, 825, 1028, 1289, 1398
Richard, Achille, 529
Ridley, Henry Nicholas, 246, 540, 1183, 1478
Ridley, list of publications, 1183
Riviere, Auguste, 1042